STUDENT'S SOLUTIONS MANUAL

NANCY S. BOUDREAU
Bowling Green State University

STATISTICS
TWELFTH EDITION

James T. McClave
Info Tech, Inc.
University of Florida

Terry Sincich
University of South Florida

PEARSON

Boston Columbus Indianapolis New York San Francisco Upper Saddle River
Amsterdam Cape Town Dubai London Madrid Milan Munich Paris Montreal Toronto
Delhi Mexico City Sao Paulo Sydney Hong Kong Seoul Singapore Taipei Tokyo

Reproduced by Pearson from electronic files supplied by the author.

Copyright © 2013, 2009, 2006 Pearson Education, Inc.
Publishing as Pearson, 75 Arlington Street, Boston, MA 02116.

ISBN-13: 978-0-321-75597-1
ISBN-10: 0-321-75597-9

2 3 4 5 6 V0SV 18 17 16 15

www.pearsonhighered.com

PEARSON

Table of Contents

Preface

This solutions manual is designed to accompany the text, *Statistics*, Twelfth Edition, by James T. McClave and Terry Sincich. It provides solutions to the odd-numbered exercises for each chapter in the text.

Other methods of solution may also be appropriate; however, the author has presented one that she believes to be the most instructive to the beginning statistics student. The student should first attempt to solve the assigned exercises without help from this manual. Then, if unsuccessful, the solution in the manual will clarify points necessary to the solution. The student who successfully solves an exercise should still refer to the manual's solution. Many points are clarified and expanded upon to provide maximum insight into and benefit from each exercise.

Instructors will also benefit from the use of this manual. It will save time in preparing presentations of the solutions and possibly provide another point of view regarding their meaning.

Some of the exercises are subjective in nature and thus omitted from the Answer Key at the end of *Statistics*. The subjective decisions regarding these exercises have been made and are explained by the author. Solutions based on these decisions are presented; the solution to this type of exercise is often most instructive. When an alternative interpretation of an exercise may occur, the author has often addressed it and given justification for the approach taken.

Nancy S. Boudreau
Bowling Green State University
Bowling Green, Ohio

Statistics, Data, and Statistical Thinking

1.1   Statistics is a science that deals with the collection, classification, analysis, and interpretation of information or data. It is a meaningful, useful science with a broad, almost limitless scope of applications to business, government, and the physical and social sciences.

1.3   The first element of inferential statistics is the population of interest. The population is a set of existing units. The second element is one or more variables that are to be investigated. A variable is a characteristic or property of an individual population unit. The third element is the sample. A sample is a subset of the units of a population. The fourth element is the inference about the population based on information contained in the sample. A statistical inference is an estimate, prediction, or generalization about a population based on information contained in a sample. The fifth and final element of inferential statistics is the measure of reliability for the inference. The reliability of an inference is how confident one is that the inference is correct.

1.5   Quantitative data are measurements that are recorded on a meaningful numerical scale. Qualitative data are measurements that are not numerical in nature; they can only be classified into one of a group of categories.

1.7   A population is a set of existing units such as people, objects, transactions, or events. A sample is a subset of the units of a population.

1.9   An inference without a measure of reliability is nothing more than a guess. A measure of reliability separates statistical inference from fortune telling or guessing. Reliability gives a measure of how confident one is that the inference is correct.

1.11   The data consisting of the classifications A, B, C, and D are qualitative. These data are nominal and thus are qualitative. After the data are input as 1, 2, 3, and 4, they are still nominal and thus qualitative. The only differences between the two data sets are the names of the categories. The numbers associated with the four groups are meaningless.

1.13   a.   The experimental units for this study are the 15 earthquake sites.

   b.   From 1940 to 1995, there were many more than 15 earthquakes. Thus, these 15 earthquakes represent a sample.

   c.   There are 3 variables in this problem. The variable 'type of ground motion' has 3 levels (short, long, or forward directive). Thus, this variable is qualitative. The variable 'earthquake magnitude' is measured on a Richter scale which results in a meaningful number. Thus this variable is quantitative. The variable peak ground acceleration is measured in feet per second and is quantitative.

1.15   a.   For question 1, the type of data collected is qualitative. The answer to the question will be either 'yes' or 'no'. For question 2, the type of data collected is quantitative. The answer to the question will be a number. For question 3, the type of data collected is qualitative. The answer to the question will be either 'yes' or 'no'.

b.   The data collected from the 1,010 adults represents a sample. The population of interest is all adults in the U.S. Only 1,010 adults in the U.S. were actually questioned.

1.17.   This study represents a descriptive statistical study. Data were collected and used to portray or describe the condition of the U.S. Treasury in 1861. No inferences were made.

1.19   a.   The town where the sample was collected is a name and thus, is a qualitative variable.

b.   The type of water supply is a category. This is a qualitative variable.

c.   The acidic level is a meaningful number. Thus, this is a quantitative variable.

d.   The turbidity level is measured on a numerical scale and is a quantitative variable.

e.   The temperature is measured on a numerical scale and is a quantitative variable.

f.   The number of fecal coliforms per 100 milliliters is measured on a numerical scale and is thus, a quantitative variable.

g.   The free chlorine-residual (milligrams per liter) is measured on a numerical scale and is a quantitative variable.

h.   The presence of hydrogen sulphide is measured on a categorical scale (yes or no). This is a qualitative variable.

1.21   a.   Length of maximum span can take on values such as 15 feet, 50 feet, 75 feet, etc. Therefore, it is quantitative.

b.   The number of vehicle lanes can take on values such as 2, 4, etc. Therefore, it is quantitative.

c.   The answer to this item is "yes" or "no," which are not numeric. Therefore, it is qualitative.

d.   Average daily traffic could take on values such as 150 vehicles, 3,579 vehicles, 53,295 vehicles, etc. Therefore, it is quantitative.

e.   Condition can take on values "good," "fair," or "poor," which are not numeric. Therefore, it is qualitative.

f.   The length of the bypass or detour could take on values such as 1 mile, 4 miles, etc. Therefore, it is quantitative.

g.   Route type can take on values "interstate," U.S.," "state," "county," or "city," which are not numeric. Therefore, it is qualitative.

1.23   a.   The data collection method used by the cancer researchers is a designed experiment. Those participating in the study were randomly assigned to one of the two screening methods.

b. The experimental units in this study are the 50,000 smokers.

c. The variable measured in this study is the age at which the scanning method first detects a tumor. Since age is represented by a meaningful number, it is quantitative.

d. The population of interest is all smokers in the U.S. The sample of interest is the set of 50,000 smokers participating in the study.

e. The inference of interest is the difference in mean age at which each of the scanning methods first detects a tumor.

1.25   a. The data collection method used by the researchers was a designed experiment. Half of the boxers received the massage and half did not.

b. The experimental units are the amateur boxers.

c. There are two variables measured on the boxers – heart rate and blood lactate level. Both of these variables are measured on a numeric scale and thus, are quantitative.

d. The inferences drawn from the analysis are: there is no difference in the mean heart rates between the two groups of boxers (those receiving massage and those not receiving massage) and there is no difference in the mean blood lactate levels between the two groups of boxers. Thus, massage did not affect the recovery rate of the boxers.

e. No. Only amateur boxers were used in the experiment. Thus, all inferences relate only to boxers.

1.27.   a. Since only students were sampled, the population of interest would be all students. The sample would be the 155 students who participated in the experiment.

b. The data were collected using a designed experiment. Each of the participants was randomly assigned to a treatment (emotional state).

c. The researchers inferred that a higher proportion of students in the guilty-state group chose not to repair the car than those in the neutral-state and anger-state groups.

d. One factor that could affect the reliability of the inference from this study is how representative the sample is. The sample was 155 volunteer students. Students in psychology classes typically have to volunteer for a certain number of experiments. These students may not take the study seriously or may not be representative of all students.

1.29.   a. Data for age would be quantitative. Age is measured on a numerical scale. Data for gender would be qualitative. Gender is either 'male' or 'female'. Data for dating experience is quantitative. The number of dates is measured on a numerical scale. Data for the extent to which a student was willing to tell parents about a dating issue would be qualitative. The responses could be 'never tell', 'rarely tell', 'sometimes tell', 'almost always tell', and 'always tell'.

    b. If students were recruited from a primarily European American middle-class school district, then the responses would probably not be typical of all high school students. Upper-class students and lower-class students' behaviors might be quite different from middle-class students' behaviors. In addition, the behaviors of minority students might be quite different from that of European American students. Thus, the inferences from this study may not be reflective of the entire population of high school students.

1.31. a. The data collection method for this study is observational. Specifically, a survey was conducted to collect the data.

    b. The data collected for this study are qualitative. The responses were either 'Obama' or 'Palin'.

    c. The results of this study are very suspect. The sample used was self-selected. Only those readers who elected to respond to the question were included. People who respond to these types of surveys typically have very strong views and may not be representative of the entire population. In addition, readers who might want to respond but who do not have access to an online computer would not be able to participate in the survey. Thus, the results would very likely not be representative of the population in general.

1.33 a. Although eating oat bran can reduce cholesterol, oat bran must be the only thing eaten. Reporting that eating oat bran is an easy and cheap way to lower your cholesterol implies that if you add oat bran to your diet, you can reduce your cholesterol, which may not be the case at all.

    b. One would need to collect data on a sample of women who gave birth to babies with birth defects. Then, each woman in the study would also be asked whether she was a victim of domestic violence while pregnant or not. The data collection method for this study would be an observational study. Specifically, one would use a survey to collect the data.

    c. In this study, only the results of the most positive response were reported. Not that many high school girls would 'Always' be happy with the way they were. To be more representative of those who are happy with the way they were, one should combine the results of those who responded 'Always true', 'Sort of true', and 'Sometimes true'.

    d. In this study, the parents were asked questions about whether they ever cut the size of meals, whether they ever ate less than they felt they should, and whether they ever relied on a limited number of foods to feed their children because they were running out of money. From the responses to these questions, the researchers concluded that one in four American children under age 12 is hungry or at risk of hunger. The children were never asked if they were hungry or not. Just because one eats less than one feels he/she should eat does not necessarily mean the children are hungry. To collect the appropriate data, the researchers should survey children under age 12 and ask them if they ever start their day hungry and if they ever go to bed hungry.

Methods for Describing Set of Data

2.1   The class frequency is the number of observations in the data set falling in a particular class. The class relative frequency is the class frequency divided by the total number of observations in the data set. The class percentage is the class relative frequency multiplied by 100.

2.3   In a bar graph, a bar or rectangle is drawn above each class of the qualitative variable corresponding to the class frequency or class relative frequency. In a Pareto diagram, the bars of the bar graph are arranged in order from the largest to the smallest from left to right.

2.5   a.   To find the frequency for each class, count the number of times each letter occurs. The frequencies for the three classes are:

Class	Frequency
X	8
Y	9
Z	3
Total	20

b.   The relative frequency for each class is found by dividing the frequency by the total sample size. The relative frequency for the class X is 8/20 = .40. The relative frequency for the class Y is 9/20 = .45. The relative frequency for the class Z is 3/20 = .15.

Class	Frequency	Relative Frequency
X	8	.40
Y	9	.45
Z	3	.15
Total	20	1.00

c.   The frequency bar chart is:

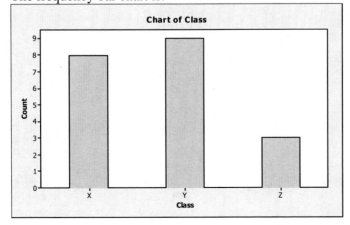

d.   The pie chart for the relative frequency distribution is:

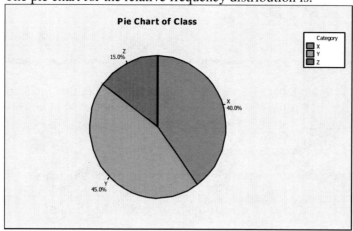

2.7   a.   The variable of interest in this study is degree of tooth wear. The variable is qualitative. Possible responses include 'Unknown', 'Heavy', Moderate-heavy', etc.

b.   The number of cheek teeth in each wear category is:

Degree of Wear	Frequency	Relative Frequency
Unknown	5	5/18 = .278
Unworn	2	2/18 = .111
Slight	4	4/18 = .222
Light-moderate	2	2/18 = .111
Moderate	3	3/18 = .167
Moderate-heavy	1	1/18 = .056
Heavy	1	1/18 = .056
Total	**18**	**18/18 = 1**

c.   To compute the relative frequency, you divide the frequency by the total number of observations. For this example, the total number of observations is 18. The relative frequency for 'Unknown' is 5/18 = .278. The rest of the calculations are done in a similar manner and appear in the table in part b.

d.   Using MINITAB, a relative frequency bar graph is:

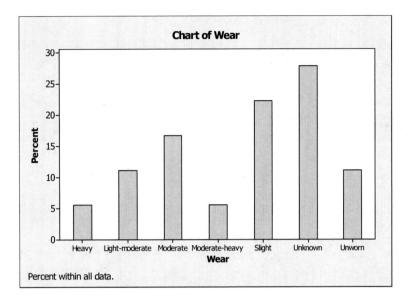

e.   A Pareto diagram is a bar graph with the categories arranged in order from the largest to the smallest from left to right. The Pareto diagram for this data is:

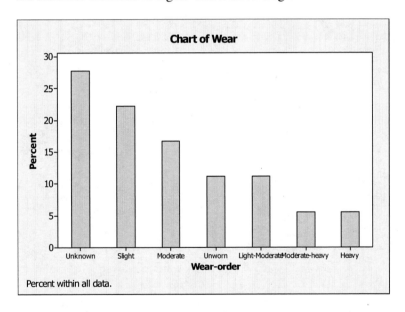

f.   The degree of wear category that occurred most often is 'Unknown'.

2.9   a.   The proportion of books at reading level 1 is found by dividing the number of books at level 1 by the total number of books, or 39/266 = .147.

b.   For reading level 2, the proportion of books is 76/266 = .286. For reading level 3, the proportion of books is 50/266 = .188. For reading level 4, the proportion of books is 87/266 = .327. For reading level 5, the proportion of books is 11/266 = .041. For reading level 6, the proportion of books is 3/266 = .011.

c.   The sum of the proportions is $.147 + .286 + .188 + .327 + .041 + .011 = 1.000$.

d.   Using MINITAB, a bar graph of the data is:

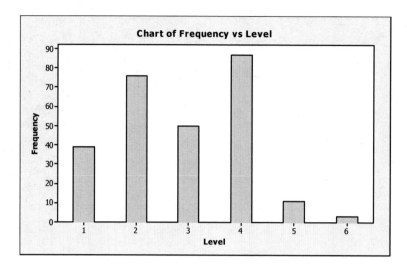

e.   Using MINITAB, the Pareto diagram is:

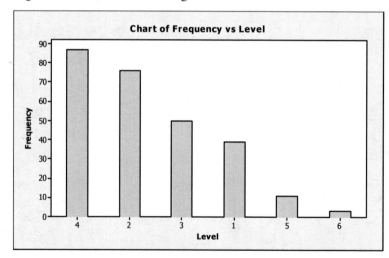

From this diagram, the reading level that occurs the most frequently is level 4.

2.11   a.   To construct a relative frequency table for data, we must find the relative frequency for each species of rhinos. To find the relative frequency, divide the frequency by the total population size, 17,800. The relative frequency for African Black rhinos is $3,610/17,800 = .203$. The rest of the relative frequencies are found in a similar manner and are reported in the table.

Rhino species	Population Estimate	Relative Frequency
African Black	3,610	3610 / 17800 = .203
African White	11,330	11330 / 17800 = .637
(Asian) Sumatran	300	300 / 17800 = .017
(Asian) Javan	60	60 / 17800 = .003
(Asian) Indian	2,500	2500 / 17800 = .140
Total	**17,800**	**1.000**

b.   The relative frequency bar chart is:

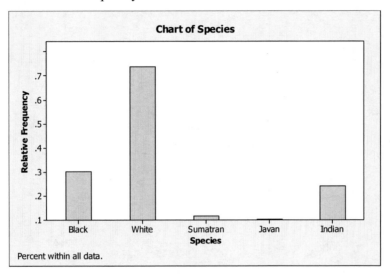

c.   The proportion of rhinos that are African is (3,610 + 11,330)/17,800 = .839.

The proportion of rhinos that are Asian is (300 + 60 + 2,500)/17,800 = .161.

2.13   a.   From the summary table, the proportion of melt ponds that had landfast ice is 38.89 / 100 = .3889.

b.   Yes. From the summary table 17.46% of the melt ponds had first-year ice.

c.   Using MINITAB, the Pareto diagram is:

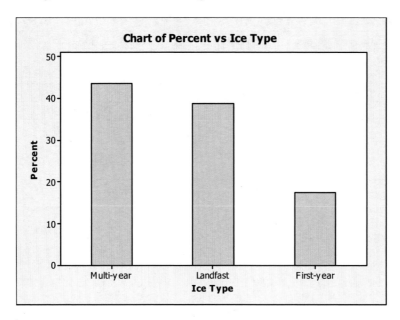

The most commonly found type of ice is multi-year ice, followed by landfast ice. The least frequently found type of ice is first-year ice.

2.15   Using MINITAB, pie charts for the two sets of data are:

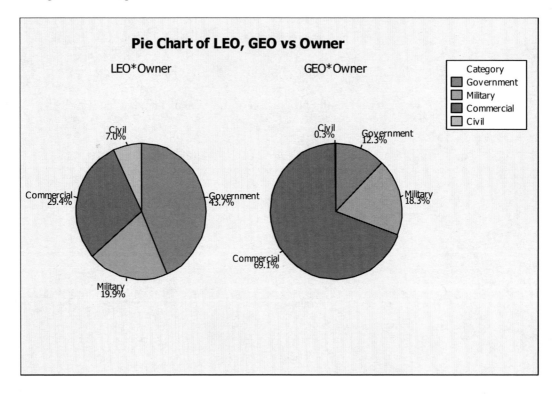

In the LEO, most of the satellites are owned by the government (43.7%), while in GEO, most of the satellites are commercially owned (69.1%). About the same percentage of both GEO and LEO satellites are owned by the military (18.3% and 19.9% respectively).

2.17   First, construct a frequency table. There are 5 categories of highest degree obtained. The relative frequencies are found by dividing the frequencies by the total number of CEO's in the sample, which is 40. The table is:

Highest Degree Obtained	Frequency	Relative Frequency
None	6	6 / 40 = .150
Bachelors	14	14 / 40 = .350
MBA	11	11 / 40 = .275
Masters	3	3 / 40 = .075
PhD	3	3 / 40 = .075
Law	3	3 / 40 = .075
Total	**40**	**1.00**

Using MINITAB, a pie chart of the data is:

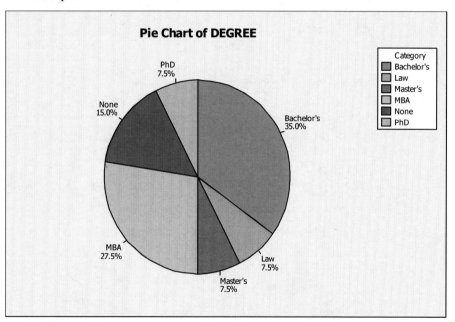

Half of the CEO's have some type of advanced degree (.275 + .075 + .075 + .075 = .500), while half do not (.350 + .150 = .500).

2.19   Using MINITAB, a bar graph of the data is:

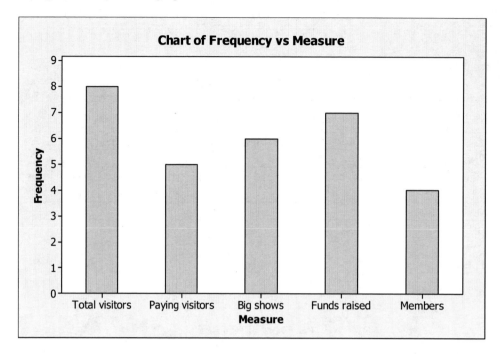

The researcher concluded that "there is a large amount of variation within the museum community with regards to . . . performance measurement and evaluation". From the data, there are only 5 different performance measures. I would not say that this is a large amount. Within these 5 categories, the number of times each is used does not vary that much. I would disagree with the researcher. There is not much variation.

2.21   a.   Using MINITAB, pie charts for Well Class, Aquifer, and Detect MTBE are:

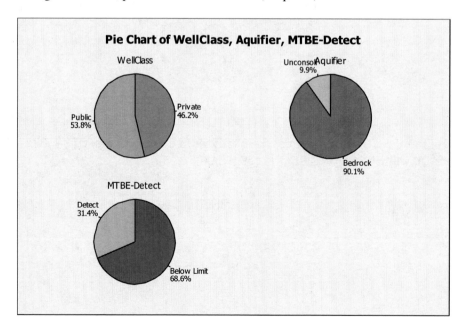

b. Using MINITAB, the side-by-side bar charts to compare the proportions of contaminated wells for private and public wells classes are:

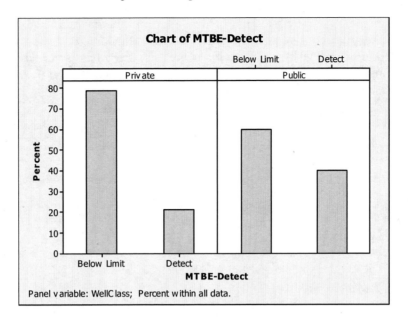

c. Using MINITAB, the side-by-side bar charts to compare the proportions of contaminated wells for bedrock and unconsolidated aquifers are:

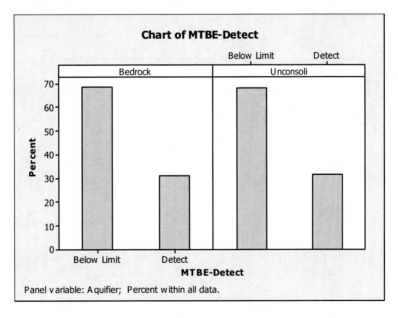

d. For the pie charts in part **a**, a little more than half of the wells are public (53.8%) and a little less than half are private (46.2%). Most of the aquifers are bedrock (90.1%) and very few are unconsolidated (9.9%). About two-thirds of the wells are not contaminated (below limit – 68.6%) and about one-third are contaminated (detect – 31.4%).

For part **b**, a larger proportion of public wells are contaminated than private wells.

For part **c**, about the same proportion of bedrock and unconsolidated aquifers are contaminated.

2.23   In a dot plot, the values are placed on the horizontal axis. The numerical value of each measurement is located on the horizontal axis by a dot above the respective value. When values repeat, the dots are placed one above the other. In a stem-and-leaf display, the stems are the left-most digits of the measurements in the data set. They are arranged vertically from the smallest to the largest. The leaves are the right-most digit of the measurements in the data set. Each leaf is placed beside the appropriate stem. For each stem, the leaves are arranged in order from the smallest to the largest..

2.25   In a histogram, a class interval is a range of numbers above which the frequency of the measurements or relative frequency of the measurements is plotted.

2.27   a.   The original data set has $1 + 3 + 5 + 7 + 4 + 3 = 23$ observations.

        b.   For the bottom row of the stem-and-leaf display:

           The stem is 0.
           The leaves are 0, 1, 2.
           The numbers in the original data set are 0, 1, and 2.

        c.   The dot plot corresponding to all the data points is:

2.29   To find the number of measurements for each measurement class, multiply the relative frequency by the total number of observations, $n = 500$. The frequency table is:

Measurement Class	Relative Frequency	Frequency
.5 – 2.5	.10	500(.10) = 50
2.5 – 4.5	.15	500(.15) = 75
4.5 – 6.5	.25	500(.25) = 125
6.5 – 8.5	.20	500(.20) = 100
8.5 – 10.5	.05	500(.05) = 25
10.5 – 12.5	.10	500(.10) = 50
12.5 – 14.5	.10	500(.10) = 50
14.5 – 16.5	.05	500(.05) = 25
		500

The frequency histogram is:

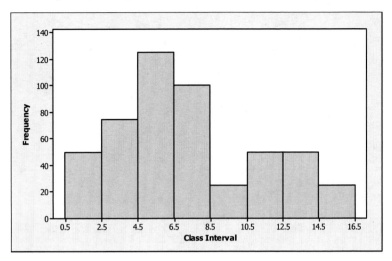

2.31   a.   This graph is a frequency histogram.

b.   There are $16 + 22 = 38$ armadillos with body lengths between 87 and 91 centimeters.

c.   The proportion of 38 armadillos with body lengths between 87 and 91 centimeters is the number observed divided by the total number of captured turtles or $38/80 = .475$.

d.   The number of illegal armadillos is $2 + 4 + 3 + 1 + 1 = 11$. The proportion is $11/80 = .1375$.

2.33   a.   Using MINITAB, the stem-and-leaf display is as follows.

Character Stem-and-Leaf Display

```
Stem-and-leaf of No. Book   N  = 14
Leaf Unit = 1.0

      1      1  6
      5      2  0124
      6      2  8
    (3)      3  044
      5      3  9
      4      4  002
      1      4
      1      5  3
```

b.   The leaves that correspond to students who earned an "A" grade are underlined in the graph above. Those students who earned A's tended to read the most books.

2.35   a.   Using MINITAB, the stem-and-leaf display of the data is:

Stem-and-Leaf Display: Score

```
Stem-and-leaf of Score   N  = 186
Leaf Unit = 1.0


    1     6    9
    1     7
    2     7    3
    3     7    4
    3     7
    4     7    8
    4     8
    5     8    3
    7     8    44
   11     8    6667
   17     8    888999
   24     9    0001111
   41     9    2222222222333333
   62     9    44444455555555555555
  (37)    9    6666666666666667777777777777777777777
   87     9    88888888888888888888888888899999999999999999999999999999999
   27    10    000000000000000000000000000
```

The stems are 60, 70, 80, … 100. The leaves are units from 0 to 9.

   b.   From the stem-and-leaf display, we see that there are only 7 observations with sanitation scores less than the acceptable score of 86. The proportion of ships that have an accepted sanitation standard would be (186 – 7) / 186 = .962.

   c.   The sanitation score of 69 is highlighted in the stem-and-leaf display in part **a**.

2.37   Using MINITAB, a histogram of the data is:

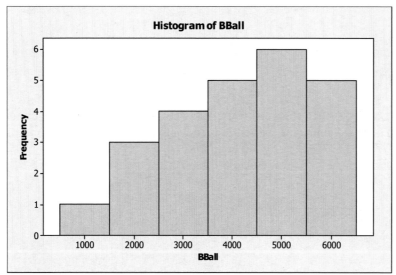

2.39   The dot plot for these data is:

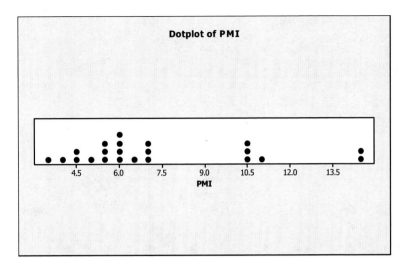

From the dot plot, most of the PMI's range from 3 to 7.5 (16 of the 22). The rest of the PMI's range from 10 to 15.

2.41   Using MINITAB, the stem-and-leaf display for the duration is:

Character Stem-and-Leaf Display

Stem-and-leaf of Duration N = 18
Leaf Unit = 0.10

```
 7    1 0000000
 8    2 0
(2)   3 00
 8    4 000
 5    5
 5    6
 5    7
 5    8
 5    9 0
 4   10 0
 3   11 0
 2   12 00
```

The leaves that correspond to sit-ins where at least one arrest was made are highlighted in the graph above. The pattern revealed does not support the theory that sit-ins of longer duration are more likely to lead to arrests.

2.43   Using MINITAB, a histogram of the data is:

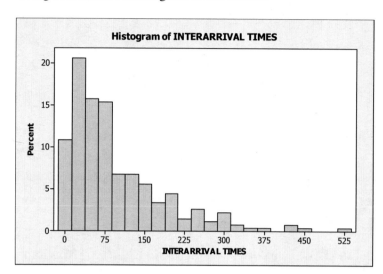

This histogram looks very similar to the one shown in the problem. Thus, there appears that there was minimal or no collaboration or collusion from within the company. We could conclude that the phishing attack against the organization was probably not an "inside job".

2.45   a. $\sum x = 3 + 8 + 4 + 5 + 3 + 4 + 6 = 33$

     b. $\sum x^2 = 3^2 + 8^2 + 4^2 + 5^2 + 3^2 + 4^2 + 6^2 = 175$

     c. $\sum (x-5)^2 = (3-5)^2 + (8-5)^2 + (4-5)^2 + (5-5)^2 + (3-5)^2 + (4-5)^2 + (6-5)^2 = 20$

     d. $\sum (x-2)^2 = (3-2)^2 + (8-2)^2 + (4-2)^2 + (5-2)^2 + (3-2)^2 + (4-2)^2 + (6-2)^2 = 71$

     e. $\left(\sum x \right)^2 = (3 + 8 + 4 + 5 + 3 + 4 + 6)^2 = 33^2 = 1089$

2.47   a. $\sum x = 6 + 0 + (-2) + (-1) + 3 = 6$

     b. $\sum x^2 = 6^2 + 0^2 + (-2)^2 + (-1)^2 + 3^2 = 50$

     c. $\sum x^2 - \dfrac{\left(\sum x \right)^2}{5} = 50 - \dfrac{6^2}{5} = 50 - 7.2 = 42.8$

2.49   Three measures of central tendency are the mean, the median, and the mode.

2.51   The two factors that impact the accuracy of the sample mean as an estimate of the population mean are sample size and variability or spread of the data. For the sample size, the larger the sample size, the more accurate the estimate. For variability, all other factors remaining constant, the more variable the data the less accurate the estimate of the mean.

2.53   a.   For a distribution that is skewed to the left, the mean is less than the median.

    b.   For a distribution that is skewed to the right, the mean is greater than the median.

    c.   For a symmetric distribution, the mean and median are equal.

2.55   Assume the data are a sample. The mode is the observation that occurs most frequently. For this sample, the mode is 15, which occurs 3 times.

The sample mean is:

$$\bar{x} = \frac{\sum x}{n} = \frac{18+10+15+13+17+15+12+15+18+16+11}{11} = \frac{160}{11} = 14.545$$

The median is the middle number when the data are arranged in order. The data arranged in order are: 10, 11, 12, 13, 15, 15, 15, 16, 17, 18, 18. The middle number is the 6th number, which is 15.

2.57   a. $\bar{x} = \dfrac{\sum x}{n} = \dfrac{85}{10} = 8.5$

    b. $\bar{x} = \dfrac{400}{16} = 25$

    c. $\bar{x} = \dfrac{35}{45} = .78$

    d. $\bar{x} = \dfrac{242}{18} = 13.44$

2.59   The sample mean is:

$$\bar{x} = \frac{\sum_{i=1}^{13} x_i}{n} = \frac{10.94+13.71+11.38+...+6.77}{13} = \frac{126.32}{13} = 9.72$$

The average rebound length of these 13 observations is 9.72.

The sample median is found by finding the middle observation once the data are arranged in order. The middle number is 10.94. Thus, the sample median is 10.94. Half of all rebound lengths are longer than 10.94 and half are shorter.

2.61   a.   The mean is $\bar{x} = \dfrac{\sum_{i=1}^{n} x_i}{n} = \dfrac{53+42+40+...+16}{14} = \dfrac{443}{14} = 31.64$. This is the average number of books read per student.

To find the median, the data must be arranged in order. In this problem, the data are already arranged in order. There are a total of 14 observations, which is an even number. The median is the average of the middle 2 numbers which are 30 and 34. The median is $\dfrac{34+30}{2} = \dfrac{64}{2} = 32$. Half of the students read more than 32 books and half read fewer.

The mode is the observation appearing the most. In this data set, the mode is 34 and 40 because each appears 2 times in the data set. The most frequent number of books read is either 34 or 40.

b.   Since the mean and the median are almost the same, the distribution of the data set is approximately symmetric. This can be verified by the stem-and-leaf display of Exercise 2.33.

2.63   a.   The sample mean radioactivity level is:

$$\bar{x} = \frac{\sum x}{n} = \frac{-43.75}{9} = -4.861$$

The median is the middle observation once they have been ordered. The 5th observation is -4.85. Thus the median is -4.85.

The mode is -5.00.

b.   The average radioactivity level is -4.861. Half of the radioactivity levels are less than -4.85 and half are greater. The modal observation occurred 2 times.

2.65   a.   The mean response was 5.87. The average rating (across the 15 respondents) of the statement "The Training Game is a great way for students to understand the animal's perspective during training" was 5.87. Since the highest rating the statement could receive was 7, this indicated that on the average, the students rather strongly agreed with this statement.

The mode response was 6. More students rated the statement with a 6 than any other rating. A rating of 6 indicated a rather strong agreement to the statement.

b.   Since the mode is only slightly larger than the mean, there probably is no skewness present. If any skewness exists, the data would probably be skewed to the left since the mode is larger than the mean and because the mode is very close to the largest value possible.

2.67   a.   The sample mean is $\bar{x} = \dfrac{\sum\limits_{i=1}^{n} x_i}{n} = \dfrac{18.12 + 19.48 + 19.36 + \ldots + 16.20}{18} = \dfrac{296.99}{18} = 16.499$.

The average depth of all the teeth collected is 16.499 mm. If the largest depth measurement were doubled, the mean would increase.

b. There are 18 measurements in the data set so the median is the average of the middle 2 numbers once the data have been arranged in order. The 9th and 10th observations when arranged in order are 16.12 and 16.20. The average is $\dfrac{16.12+16.20}{2}=\dfrac{32.32}{2}=16.16$.

Thus, the median is 16.16. Half of the depths are larger than 16.16 and half are smaller. If the largest depth measurement were doubled, the median would not change. It would stay the same.

c. Since no observation occurs more than once, then we say there is either no mode or we say that all of the observations are modes.

2.69 a. The mean cylinder power measurements for student #11 is:

$$\bar{x}=\frac{\sum x}{n}=\frac{-.08+(-.06)+...+(-.16)}{25}=\frac{-3.86}{25}=-.1544$$

The median is the middle number once the data have been arranged in order:
$-1.07, -.21, -.20, -.17, -.17, -.17, -.16, -.16, -.16, -.15, -.12, -.12, -.11, -.10,$
$-.09, -.09, -.09, -.08, -.08, -.07, -.07, -.06, -.06, -.06, -.04$

The median is $-.11$.

The mode is the value with the highest frequency. Since $-.17, -.16, -.09, -.06$ each occur 3 times, all are modes.

From the printout, the mean is -0.1544 and the median is -0.11. (All the numbers in the table are negative numbers.) Since these two values are very close together, both are good representatives of the middle of the data set.

b. The outlier is -1.07.

c. After deleting -1.07, the mean is:

$$\bar{x}=\frac{\sum x}{n}=\frac{(-.08)+(-.06)+(-.15)+\cdots+(-.16)}{24}=\frac{-2.79}{24}=-.11625$$

The median is the average of the middle two numbers once the data are arranged in order.

The middle two numbers are 10 and 11. The median is $\dfrac{(-.10)+(-.11)}{2}=-.105$

The modes remain the same.

The mean changes from $-.1544$ to $-.11625$ while the median changes from $-.11$ to $-.105$. The mean changes much more than the median when the outlier is removed.

2.71 The range of a data set is the difference between the largest and smallest measurements.

2.73   The sample variance is the sum of the squared deviations from the sample mean divided by the sample size minus 1. The population variance is the sum of the squared deviations from the population mean divided by the population size.

2.75   If the standard deviation increases, this implies that the data are more variable.

2.77   a.   Range $= 42 - 37 = 5$

$$s^2 = \frac{\sum x^2 - \frac{\left(\sum x\right)^2}{n}}{n-1} = \frac{7935 - \frac{199^2}{5}}{5-1} = 3.7 \qquad s = \sqrt{3.7} = 1.92$$

b.   Range $= 100 - 1 = 99$

$$s^2 = \frac{\sum x^2 - \frac{\left(\sum x\right)^2}{n}}{n-1} = \frac{25{,}795 - \frac{303^2}{9}}{9-1} = 1{,}949.25 \qquad s = \sqrt{1{,}949.25} = 44.15$$

c.   Range $= 100 - 2 = 98$

$$s^2 = \frac{\sum x^2 - \frac{\left(\sum x\right)^2}{n}}{n-1} = \frac{20{,}033 - \frac{295^2}{8}}{8-1} = 1{,}307.84 \qquad s = \sqrt{1{,}307.84} = 36.16$$

2.79   This is one possibility for the two data sets.

Data Set 1: 1, 1, 2, 2, 3, 3, 4, 4, 5, 5
Data Set 2: 1, 1, 1, 1, 1, 5, 5, 5, 5, 5

$$\bar{x}_1 = \frac{\sum x}{n} = \frac{1+1+2+2+3+3+4+4+5+5+}{10} = \frac{30}{10} = 3$$
$$\bar{x}_2 = \frac{\sum x}{n} = \frac{1+1+1+1+1+5+5+5+5+5}{10} = \frac{30}{10} = 3$$

Therefore, the two data sets have the same mean. The variances for the two data sets are:

$$s_1^2 = \frac{\sum x^2 - \frac{\left(\sum x\right)^2}{n}}{n-1} = \frac{110 - \frac{30^2}{10}}{9} = \frac{20}{9} = 2.2222$$

$$s_2^2 = \frac{\sum x^2 - \frac{\left(\sum x\right)^2}{n}}{n-1} = \frac{130 - \frac{30^2}{10}}{9} = \frac{40}{9} = 4.4444$$

The dot diagram for the two data sets are shown.

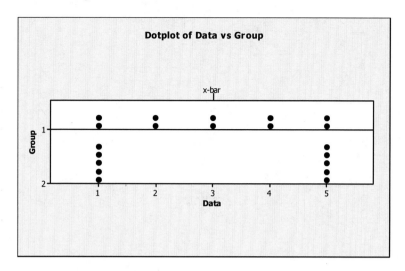

2.81   a.   Range = 3 − 0 = 3

$$s^2 = \frac{\sum x^2 - \frac{(\sum x)^2}{n}}{n-1} = \frac{15 - \frac{7^2}{5}}{5-1} = 1.3 \qquad s = \sqrt{1.3} = 1.1402$$

b.   After adding 3 to each of the data points,

Range = 6 − 3 = 3

$$s^2 = \frac{\sum x^2 - \frac{(\sum x)^2}{n}}{n-1} = \frac{102 - \frac{22^2}{5}}{5-1} = 1.3 \qquad s = \sqrt{1.3} = 1.1402$$

c.   After subtracting 4 from each of the data points,

Range = −1 − (−4) = 3

$$s^2 = \frac{\sum x^2 - \frac{(\sum x)^2}{n}}{n-1} = \frac{39 - \frac{(-13)^2}{5}}{5-1} = 1.3 \qquad s = \sqrt{1.3} = 1.1402$$

d.   The range, variance, and standard deviation remain the same when any number is added to or subtracted from each measurement in the data set.

2.83   a.   From the printout, the range is 51.26.

b.   From the printout, the variance is 128.57.

c.   From the printout, the standard deviation is 11.34.

d.   If we are interested in just these 76 captured turtles, then the data would represent a population. The variance would be represented by σ^2 and the standard deviation would be represented by σ.

2.85   a.   The sample variance is:

$$s^2 = \frac{\sum x^2 - \frac{\left(\sum x\right)^2}{n}}{n-1} = \frac{4295 - \frac{375^2}{35}}{35-1} = 8.151$$

The standard deviation is $s = \sqrt{8.151} = 2.855$

   b.   The sample variance is:

$$s^2 = \frac{\sum x^2 - \frac{\left(\sum x\right)^2}{n}}{n-1} = \frac{2631 - \frac{275^2}{33}}{33-1} = 10.604$$

The standard deviation is $s = \sqrt{10.604} = 3.256$

   c.   The sample variance is:

$$s^2 = \frac{\sum x^2 - \frac{\left(\sum x\right)^2}{n}}{n-1} = \frac{1881 - \frac{241^2}{37}}{37-1} = 8.646$$

The standard deviation is $s = \sqrt{8.646} = 2.940$

   d.   The DM dosage group appears to have the most variability since the variance of the DM dosage group is larger than the variances of the other 2 groups. The honey dosage group appears to have the least variability since the variance of the honey dosage group is smaller than the variances of the other 2 groups. However, the variance of the honey dosage group and the control group are very similar.

2.87   a.   The range is the difference between the largest and smallest observations and is 19.70 – 13.25 = 6.45 mm. If the largest depth measurement were doubled, the range would increase. The new range would be 39.40 – 13.25 = 26.15.

b.   The variance is:

$$s^2 = \frac{\sum x^2 - \frac{\left(\sum x\right)^2}{n}}{n-1} = \frac{4966.173 - \frac{296.99^2}{18}}{18-1} = 3.883$$

If the largest depth measurement were doubled, the variance would increase. The new variance would be:

$$s^2 = \frac{\sum x^2 - \frac{\left(\sum x\right)^2}{n}}{n-1} = \frac{6130.443 - \frac{316.69^2}{18}}{18-1} = 32.861$$

c.   The standard deviation is $s = \sqrt{3.883} = 1.970$. If the largest depth measurement were doubled, the standard deviation would increase. The new standard deviation would be $s = \sqrt{32.861} = 5.732$

2.89   a.   Range = 11 – 1 = 10

$$s^2 = \frac{\sum x^2 - \frac{\left(\sum x\right)^2}{n}}{n-1} = \frac{450 - \frac{78^2}{20}}{20-1} = 7.674$$

$$s = \sqrt{s^2} = \sqrt{7.674} = 2.770$$

b.   Dropping the largest measurement:

Range = 9 – 1 = 8

$$s^2 = \frac{\sum x^2 - \frac{\left(\sum x\right)^2}{n}}{n-1} = \frac{329 - \frac{67^2}{19}}{19-1} = 5.152$$

$$s = \sqrt{s^2} = \sqrt{5.152} = 2.270$$

By dropping the largest observation from the data set, the range decreased from 10 to 8, the variance decreased from 7.674 to 5.1520 and the standard deviation decreased from 2.770 to 2.270.

c.   Dropping the largest and smallest measurements:

Range = 9 – 1 = 8

$$s^2 = \frac{\sum x^2 - \frac{\left(\sum x\right)^2}{n}}{n-1} = \frac{328 - \frac{66^2}{18}}{18-1} = 5.059$$

$$s = \sqrt{s^2} = \sqrt{5.059} = 2.249$$

By dropping the largest and smallest observations from the data set, the range decreased from 10 to 8, the variance decreased from 7.674 to 5.059 and the standard deviation decreased from 2.770 to 2.249.

2.91   a.   The unit of measurement of the variable of interest is dollars (the same as the mean and standard deviation). Based on this, the data are quantitative.

b.   Since no information is given about the shape of the data set, we can only use Chebyshev's rule.

$900 is 2 standard deviations below the mean, and $2100 is 2 standard deviations above the mean. Using Chebyshev's rule, at least 3/4 of the measurements (or $3/4 \times 200 = 150$ measurements) will fall between $900 and $2100.

$600 is 3 standard deviations below the mean and $2400 is 3 standard deviations above the mean. Using Chebyshev's rule, at least 8/9 of the measurements (or $8/9 \times 200 = 178$ measurements) will fall between $600 and $2400.

$1200 is 1 standard deviation below the mean and $1800 is 1 standard deviation above the mean. Using Chebyshev's rule, nothing can be said about the number of measurements that will fall between $1200 and $1800.

$1500 is equal to the mean and $2100 is 2 standard deviations above the mean. Using Chebyshev's rule, at least 3/4 of the measurements (or $3/4 \times 200 = 150$ measurements) will fall between $900 and $2100. It is possible that all of the 150 measurements will be between $900 and $1500. Thus, nothing can be said about the number of measurements between $1500 and $2100.

2.93   According to the Empirical Rule:

a.   Approximately 68% of the measurements will be contained in the interval $\bar{x} - s$ to $\bar{x} + s$.

b.   Approximately 95% of the measurements will be contained in the interval $\bar{x} - 2s$ to $\bar{x} + 2s$.

c.   Essentially all the measurements will be contained in the interval $\bar{x} - 3s$ to $\bar{x} + 3s$.

2.95   Using Chebyshev's rule, at least 8/9 of the measurements will fall within 3 standard deviations of the mean. Thus, the range of the data would be around 6 standard deviations. Using the Empirical Rule, approximately 95% of the observations are within 2 standard deviations of the mean. Thus, the range of the data would be around 4 standard deviations. We would expect the standard deviation to be somewhere between Range/6 and Range/4.

For our data, the range = $760 - 135 = 625$.

The Range/6 = $625/6 = 104.17$ and Range/4 = $625/4 = 156.25$.

Therefore, I would estimate that the standard deviation of the data set is between 104.17 and 156.25.

It would not be feasible to have a standard deviation of 25. If the standard deviation were 25, the data would span $625/25 = 25$ standard deviations. This would be extremely unlikely.

2.97   a. $$\bar{x} = \frac{\sum x}{n} = \frac{17800}{186} = 95.699$$

$$s^2 = \frac{\sum x^2 - \frac{\left(\sum x\right)^2}{n}}{n-1} = \frac{1707998 - \frac{17800^2}{186}}{186-1} = 24.633$$

$$s = \sqrt{s^2} = \sqrt{24.633} = 4.963$$

b. $\bar{x} \pm s \Rightarrow 95.699 \pm 4.963 \Rightarrow (90.736, 100.662)$

$\bar{x} \pm 2s \Rightarrow 95.699 \pm 2(4.963) \Rightarrow 95.699 \pm 9.926 \Rightarrow (85.773, 105.625)$

$\bar{x} \pm 3s \Rightarrow 95.699 \pm 3(4.963) \Rightarrow 95.699 \pm 14.889 \Rightarrow (80.810, 110.588)$

c.   There are 166 out of 186 observations in the first interval. This is (166/186)*100% = 89.2%. There are 179 out of 186 observations in the second interval. This is (179/186)*100% = 96.2%. There are 182 out of 186 observations in the second interval. This is (182/186)*100% = 97.8%.

The percentages for the first 2 intervals are somewhat larger than what we would expect using the Empirical Rule. The Empirical Rule indicates that approximately 68% of the observations will fall within 1 standard deviation of the mean. It also indicates that approximately 95% of the observations will fall within 2 standard deviations of the mean. Chebyshev's Theorem says that at least ¾ or 75% of the observations will fall within 2 standard deviations of the mean and at least 8/9 or 88.8% of the observations will fall within 3 standard deviations of the mean. It appears that our observed percentages agree with Chebyshev's Theorem better that the Empirical Rule.

2.99   a.   The mean is 79 and the standard deviation is 23. Thus, a measurement of 102 is 1 standard deviation above the mean. If nothing is known about the shape of the distribution, we must use Chebyshev's Rule to describe the distribution. Using this, we know nothing about the number of observations within 1 standard deviation of the mean, and thus, know nothing about the percentage of observations less than 102.

b.   If we assume that the distribution is mound-shaped, we can use the Empirical Rule to describe the distribution. If the distribution is mound-shaped, then it is symmetric. Thus, we know that half of the observations will be below 79. We know that about 68% of the observations are within 1 standard deviation of the mean. Therefore, about 34%

of the observations will be between the mean and 1 standard deviation above the mean, or between 79 and 102. Thus, the percentage of observations less than 102 is about 50% + 34% = 84%.

2.101   For parts a and b, we must use Chebyshev's Rule to describe the data sets. For both the handrubbers and the handwashers, the standard deviation is greater than the mean. Since no observations can be less than 0, we know that the smallest observation in the data sets is less than 1 standard deviation from the mean. Thus, the distributions cannot be symmetric, but rather skewed to the right. We cannot use the Empirical Rule.

    a.   Using Chebyshev's Rule, we know that at least $1-\dfrac{1}{2^2}=1-\dfrac{1}{4}=\dfrac{3}{4}=.75$ of the observations fall within 2 standard deviations of the mean.
Thus, the interval $\bar{x}\pm 2s$ will contain at least .75 of the observations.

        For handrubbers: $\bar{x}\pm 2s \Rightarrow 35\pm 2(59)\Rightarrow 35\pm 118\Rightarrow (0,\ 153)$

    b.   For handwashers: $\bar{x}\pm 2s \Rightarrow 69\pm 2(106)\Rightarrow 69\pm 212\Rightarrow (0,\ 281)$

    c.   Since the interval for handwashers is much larger that that for handrubbers and the sample mean for handrubbers is less than that for handwashers, we can conclude that handrubbing is more effective than handwashing for reducing bacterial counts.

2.103   a.   Since no information is given about the distribution of the velocities of the Winchester bullets, we can only use Chebyshev's rule to describe the data. We know that at least 3/4 of the velocities will fall within the interval:

        $\bar{x}\pm 2s \Rightarrow 936\pm 2(10)\Rightarrow 936\pm 20\Rightarrow (916,\ 956)$

        Also, at least 8/9 of the velocities will fall within the interval:

        $\bar{x}\pm 3s \Rightarrow 936\pm 3(10)\Rightarrow 936\pm 30\Rightarrow (906,\ 966)$

    b.   Since a velocity of 1,000 is much larger than the largest value in the second interval in part **a**, it is very unlikely that the bullet was manufactured by Winchester.

2.105   a.   Regardless of the shape of the distribution, most of the observations will fall within 3 standard deviations of the mean. Thus, for the SAT-Math scores, an interval likely to contain a student's change in score is:

        $\bar{x}\pm 3s \Rightarrow 19\pm 3(65)\Rightarrow 19\pm 195\Rightarrow (-176,\ 214)$

    b.   Regardless of the shape of the distribution, most of the observations will fall within 3 standard deviations of the mean. Thus, for the SAT-Verbal scores, an interval likely to contain a student's change in score is:

        $\bar{x}\pm 3s \Rightarrow 7\pm 3(49)\Rightarrow 7\pm 147\Rightarrow (-140,\ 154)$

c. For the SAT-Verbal, the maximum increase in scores is about 154 points. For the SAT-Math, the maximum increase in scores is approximately 214. Thus, a student is more likely to get a 140-point increase on the SAT-Math test.

2.107 Since we do not know if the distribution of the heights of the trees is mound-shaped, we need to apply Chebyshev's rule. We know $\mu = 30$ and $\sigma = 3$. Therefore,

$$\mu \pm 3\sigma \Rightarrow 30 \pm 3(3) \Rightarrow 30 \pm 9 \Rightarrow (21, 39)$$

According to Chebyshev's rule, at least 8/9 or .89 of the tree heights on this piece of land fall within this interval and at most $\frac{1}{9}$ or .11 of the tree heights will fall above the interval.

However, the buyer will only purchase the land if at least $\frac{1000}{5000}$ or .20 of the tree heights are at least 40 feet tall. Therefore, the buyer should not buy the piece of land.

2.109 Using the definition of a percentile:

	Percentile	Percentage Above	Percentage Below
a.	75th	25%	75%
b.	50th	50%	50%
c.	20th	80%	20%
d.	84th	16%	84%

2.111 a. $z = \dfrac{x - \bar{x}}{s} = \dfrac{40 - 30}{5} = 2$ (sample)   2 standard deviations above the mean.

   b. $z = \dfrac{x - \mu}{\sigma} = \dfrac{90 - 89}{2} = .5$ (population)   .5 standard deviations above the mean.

   c. $z = \dfrac{x - \mu}{\sigma} = \dfrac{50 - 50}{5} = 0$ (population)   0 standard deviations above the mean.

   d. $z = \dfrac{x - \bar{x}}{s} = \dfrac{20 - 30}{4} = -2.5$ (sample)   2.5 standard deviations below the mean.

2.113   Since the element 40 has a z-score of -2 and 90 has a z-score of 3,

$$-2 = \frac{40 - \mu}{\sigma} \text{ and } 3 = \frac{90 - \mu}{\sigma}$$

$$\Rightarrow -2\sigma = 40 - \mu \qquad \Rightarrow 3\sigma = 90 - \mu$$
$$\Rightarrow \mu - 2\sigma = 40 \qquad \Rightarrow \mu + 3\sigma = 90$$
$$\Rightarrow \mu = 40 + 2\sigma$$

By substitution,
$$40 + 2\sigma + 3\sigma = 90$$
$$\Rightarrow 5\sigma = 50$$
$$\Rightarrow \sigma = 10$$

By substitution, $\mu = 40 + 2(10) = 60$

Therefore, the population mean is 60 and the standard deviation is 10.

2.115   The percentile ranking for the age of 25 years in the distribution of all ages of licensed drivers stopped by police is $100 - 74 = 26$ percentile.

2.117   a.   The z-score associated with a score of 30 is $z = \dfrac{x - \bar{x}}{s} = \dfrac{30 - 39}{6} = -1.50$. This means that a score of 30 is 1.5 standard deviations below the mean.

b.   The z-score associated with a score of 39 is $z = \dfrac{x - \bar{x}}{s} = \dfrac{39 - 39}{6} = 0$. Half or .5 of the observations are below a score of 39.

2.119 a.   Using SAS, the output is:

```
                          EGG LENGTH

                    The UNIVARIATE Procedure
                       Variable:  length

                          Moments

N                       130     Sum Weights               130
Mean              60.6538462     Sum Observations         7885
Std Deviation     43.9861168     Variance            1934.77847
Skewness           2.1091362     Kurtosis            4.70026314
Uncorrected SS        727842     Corrected SS        249586.423
Coeff Variation   72.5199136     Std Error Mean      3.85783765

                  Basic Statistical Measures

           Location                    Variability

        Mean     60.65385     Std Deviation          43.98612
        Median   49.50000     Variance                   1935
        Mode     35.00000     Range              220.00000
                              Interquartile Range    32.00000


            Tests for Location: Mu0=0
```

```
Test                -Statistic-     -----p Value------

Student's t     t  15.72224     Pr > |t|    <.0001
Sign            M         65     Pr >= |M|   <.0001
Signed Rank     S     4257.5     Pr >= |S|   <.0001


              Quantiles (Definition 5)

              Quantile       Estimate

              100% Max         236.0
              99%              218.0
              95%              160.0
              90%              122.0
              75% Q3            67.0
              50% Median        49.5
              25% Q1            35.0
              10%              23.0
              5%               19.0
              1%               16.0
              0% Min           16.0


                 EGG LENGTH

            The UNIVARIATE Procedure
              Variable:  length

             Extreme Observations

     ----Lowest----         ----Highest---

     Value      Obs         Value      Obs

     16.0       102          195       128
     16.0        90          205       123
     17.0       101          216       122
     18.0       103          218       129
     18.5       100          236       130


                 Missing Values

                        -----Percent Of-----
      Missing                         Missing
      Value      Count    All Obs       Obs

        .          2       1.52       100.00
```

The 10th percentile egg length is 23. This means that 10% of all egg lengths are less than 23 and 90% are greater than 23.

b.   From the printout, $\bar{x} = 60.65$ and $s = 43.99$. The z-score corresponding to the moas (*P. australis*) bird species' egg length is:

$$z = \frac{x - \bar{x}}{s} = \frac{205 - 60.65}{43.99} = 3.28$$

The z-score for moas is 3.28 standard deviations above the mean. This is a fairly large value for a z-score. This indicates that the egg length for moas could be very unusual.

2.121   a.   The *z*-score for Harvard is 5.08. This means that the productivity score for Harvard was 5.08 standard deviations above the mean. This is an unusually high *z*-score.

b.   The *z*-score for Howard is -0.81. This means that the productivity score for Howard was 0.81 standard deviations below the mean.

c.   Yes. If only 44 out of 129 schools had positive z-scores, this indicates that the data are skewed to the right. Other indications that the data are skewed to the right are the *z*-scores associated with the smallest and largest scores. The largest z-score is 5.08 and the smallest is -0.81. The largest observation is much further from the mean than the smallest observation.

Using MINITAB, a histogram of the data is:

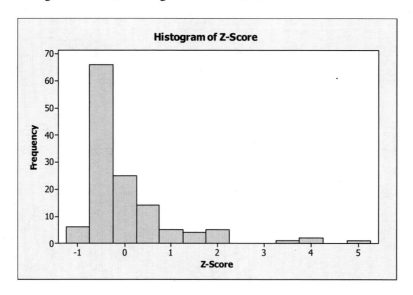

The data are, in fact, skewed to the right.

2.123   a.   From the problem, $\mu = 2.7$ and $\sigma = .5$

$$z = \frac{x - \mu}{\sigma} \Rightarrow z\sigma = x - \mu \Rightarrow x = \mu + z\sigma$$

For $z = 2.0$, $x = 2.7 + 2.0(.5) = 3.7$

For $z = -1.0$, $x = 2.7 - 1.0(.5) = 2.2$

For $z = .5$, $x = 2.7 + .5(.5) = 2.95$

For $z = -2.5$, $x = 2.7 - 2.5(.5) = 1.45$

b.   For $z = -1.6$, $x = 2.7 - 1.6(.5) = 1.9$

c. If we assume the distribution of GPAs is approximately mound-shaped, we can use the Empirical Rule.

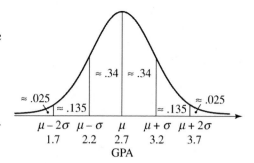

From the Empirical Rule, we know that $\approx .025$ or $\approx 2.5\%$ of the students will have GPAs above 3.7 (with $z = 2$). Thus, the GPA corresponding to summa cum laude (top 2.5%) will be greater than 3.7 ($z > 2$).

We know that $\approx .16$ or $\approx 16\%$ of the students will have GPAs above 3.2 ($z = 1$). Thus, the limit on GPAs for cum laude (top 16%) will be greater than 3.2 ($z > 1$).

We must assume the distribution is mound-shaped.

2.125   An observation that is unusually large or small relative to the data values we want to describe is an outlier.

2.127   The hinges of a box plot are the upper quartile and the lower quartile (top and bottom of the rectangle).

2.129   To determine if the measurements are outliers, compute the z-score.

a. $z = \dfrac{x - \bar{x}}{s} = \dfrac{65 - 57}{11} = .727$ Since this z-score is less than 3 in magnitude, 65 is not an outlier.

b. $z = \dfrac{x - \bar{x}}{s} = \dfrac{21 - 57}{11} = -3.273$ Since this z-score is more than 3 in magnitude, 21 is an outlier.

c. $z = \dfrac{x - \bar{x}}{s} = \dfrac{72 - 57}{11} = 1.364$ Since this z-score is less than 3 in magnitude, 72 is not an outlier.

d. $z = \dfrac{x - \bar{x}}{s} = \dfrac{98 - 57}{11} = 3.727$

Since this z-score is more than 3 in magnitude, 98 is an outlier.

2.131   a. The median is approximately 4.

b. Q_L (Lower quartile) is approximately 3 and Q_U (Upper quartile) is approximately 6.

c. The interquartile range is IQR $= Q_U - Q_L \approx 6 - 3 = 3$.

d. The data set is skewed to the right since the right whisker is longer than the left and there are outlying observations to the right.

e. 50% of the observations lie to the right of the median. 75% of the observations lie to the left of the upper quartile.

f. The outliers in the data set are 12, 13, and 16.

2.133   a.   The approximate 25th percentile PASI score before treatment is 10. The approximate median before treatment is 15. The approximate 75th percentile PASI score before treatment is 27.5.

b.   The approximate 25th percentile PASI score after treatment is 3.5. The approximate median after treatment is 5. The approximate 75th percentile PASI score after treatment is 7.5.

c.   Since the 75th percentile after treatment is lower than the 25th percentile before treatment, it appears that the ichthyotherapy is effective in treating psoriasis.

2.135   a.   The z-score is $z = \dfrac{x - \bar{x}}{s} = \dfrac{3.3 - 7.3}{3.18} = -1.26$.

b.   A PMI score of 3.3 would not be considered an outlier. The z-score is -1.26. A z-score this small is not considered unusual.

2.137   a.   Using MINITAB, the boxplot is:

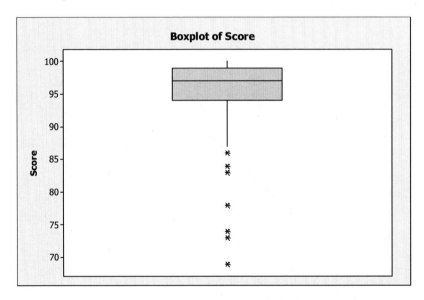

From the boxplot, there appears to be 7 outliers: 69, 73, 74, 78, 83, 84, and 86. However, there are 2 observations with a value of 84 and 3 observations with a value of 86. Thus, there appear to be 10 outliers.

b.   From Exercise 2.97, $\bar{x} = 95.699$ and $s = 4.963$. Since the data are skewed to the left, we will consider observations more than 3 standard deviations from the mean to be outliers. An observation with a z-score of 3 would have the value:

$$z = \frac{x - \bar{x}}{s} \Rightarrow 3 = \frac{x - 95.699}{4.963} \Rightarrow 3(4.963) = x - 95.699 \Rightarrow 14.889 = x - 95.699 \Rightarrow x = 110.588$$

An observation with a z-score of -3 would have the value:

$$z = \frac{x - \bar{x}}{s} \Rightarrow -3 = \frac{x - 95.699}{4.963} \Rightarrow -3(4.963) = x - 95.699 \Rightarrow -14.889 = x - 95.699 \Rightarrow x = 80.810$$

Observations greater than 10.588 or less than 80.81 would be considered outliers. Using this criterion, the following observations would be outliers: 69, 73, 74, and 78.

c. No, these methods do not agree. Using the boxplot, 10 observations were identified as outliers. Using the *z*-score method, only 4 observations were identified as outliers. Since the data are very highly skewed to the left, the *z*-score method may not be appropriate.

2.139   Using MINITAB, the boxplots of the 3 groups are:

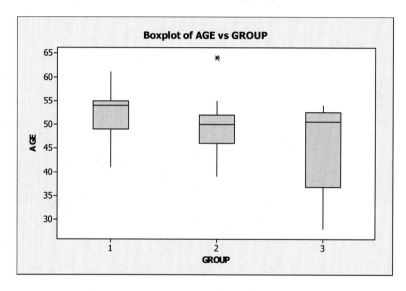

From the boxplots, there appears to be only 1 outlier. That outlier is in the second group.

2.141   a.   Using MINITAB, the side-by-side box plots are:

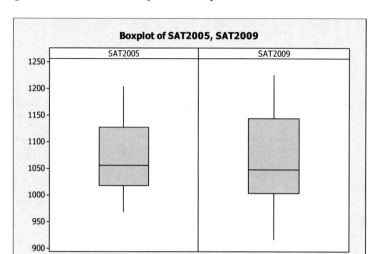

b.   Using MINITAB, the descriptive statistics are:

Descriptive Statistics: SAT2005, SAT2009

Variable	N	Mean	StDev	Minimum	Q1	Median	Q3	Maximum
SAT2005	51	1077.0	67.9	968.0	1018.0	1056.0	1127.0	1204.0
SAT2009	51	1073.5	81.4	917.0	1003.0	1048.0	1144.0	1225.0

The standard deviation for 2005 is 67.9 and the standard deviation for 2009 is 81.4. Also, the IQR for 2005 is $Q_U - Q_L = 1127 - 1018 = 109$ while the IQR for 2009 is $Q_U - Q_L = 1144 - 1003 = 141$. Thus, the variability for 2009 is slightly greater than that for 2005.

c.   Since there are no observations outside the inner fences for either year, there are no outliers.

2.143   A bivariate relationship is a relationship between 2 quantitative variables.

2.145   A positive association between two variables means that as one variable increases, the other variable tends to also increase. A negative association between two variables means that as one variable increases, the other variable tends to decrease.

2.147   Using MINITAB, the scatterplot is:

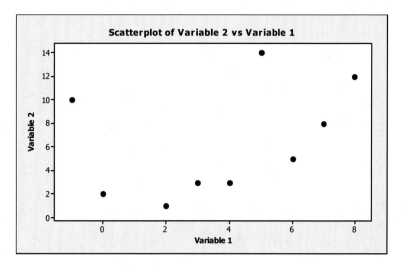

From the scatterplot, there does not appear to be much of a trend between variable 1 and variable 2. There is a slight positive linear trend - as variable 1 increases, variable 2 tends to increase. However, this relationship appears to be very weak.

2.149   If we constructed a scatterplot of the 2009 SAT scores versus the 2005 SAT scores, we would expect there to be a positive linear trend to the data. If a state had a high mean SAT score in 2005, we would expect that the state would also have a high mean SAT score in 2009. Using MINITAB, the scatterplot of the data is:

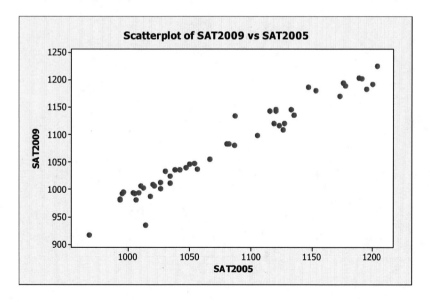

2.151   Using MINITAB, the scatterplot of the data is:

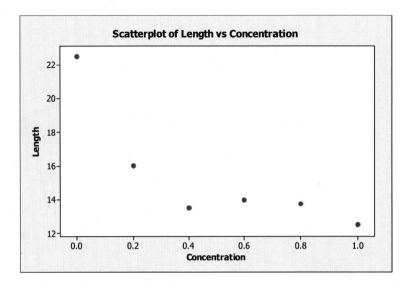

There appears to be a negative trend to the data. As the concentration increases, the wicking length tends to decrease. It appears that the relationship may not be linear, but rather curvilinear.

2.153   a.   Using MINITAB, a scatterplot of the data is:

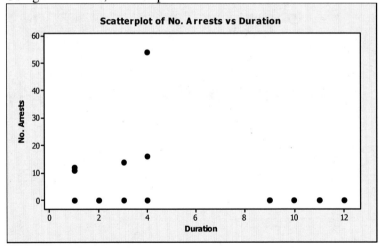

There does not appear to be much of a trend between the number of arrests and the duration of the sit-in.

b.   Using MINITAB, a scatterplot of the data using only those observations where there was at least one arrest is:

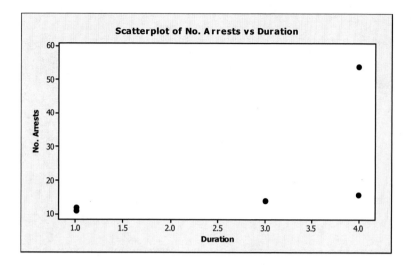

There appears to be a positive relationship between the number of arrests and the duration of the sit-in. As the duration increases, the number of arrests tends to increase.

c.   Since there are only 5 observations used in the scatterplot in part **b**, the reliability is suspect.

2.155   a.   Using MINITAB, a scatterplot of the data is:

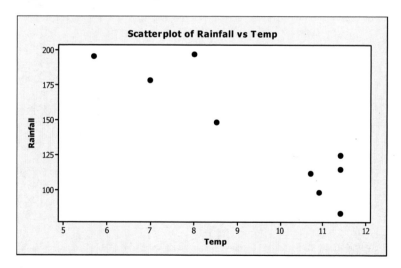

There appears to be a negative relationship between the annual rainfall and the maximum daily temperature. As the maximum daily temperature increases, the amount of annual rainfall tends to decrease.

b.   Using MINITAB, a scatterplot of the relationship between annual rainfall and total plant cover is:

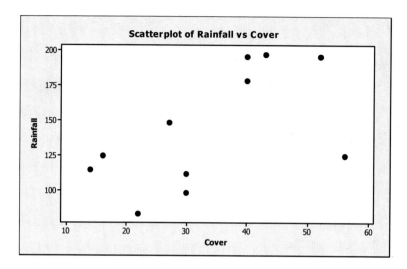

There appears to be a positive relationship between annual rainfall and total plant cover. As the total plant cover increases, the annual rainfall tends to increase.

Using MINITAB, a scatterplot of the relationship between annual rainfall and number of ant species is:

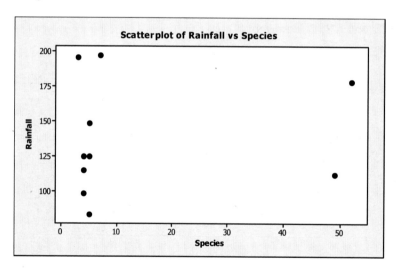

There does not appear to be a relationship between annual rainfall and number of ant species.

Using MINITAB, a scatterplot of the relationship between annual rainfall and species diversity index is:

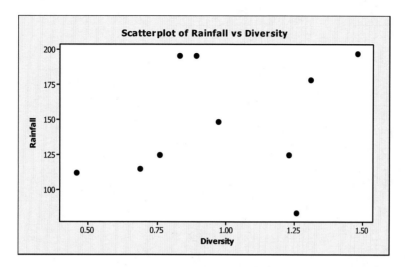

There appears to be a positive relationship between annual rainfall and species diversity index.   As the species diversity index increases, the annual rainfall tends to increase. However, this relationship appears to be very weak.

2.157   Using MINITAB, a graph of the accuracy scores versus the driving distance is as follows:

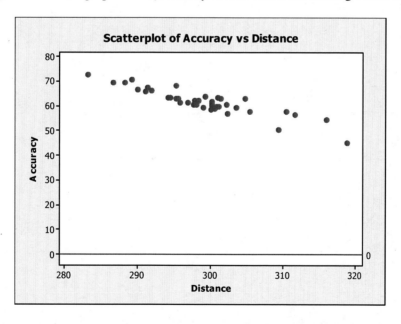

Yes, the golfer's concern is valid. From the graph, as the driving distance increases, the accuracy decreases.

2.159   The median is preferred to the mean as a measure of central tendency when the data set is skewed either right or left. If the data set has some extreme observations, the median better measure of central tendency than the mean.

2.161   A stem-and-leaf display is generally preferred over a histogram when the data set is relatively small.

2.163   One technique for distorting information on a graph is by stretching the vertical axis by starting the vertical axis somewhere above 0.

2.165   a. $z = \dfrac{x - \mu}{\sigma} = \dfrac{50 - 60}{10} = -1$

$z = \dfrac{70 - 60}{10} = 1$

$z = \dfrac{80 - 60}{10} = 2$

b. $z = \dfrac{x - \mu}{\sigma} = \dfrac{50 - 60}{5} = -2$

$z = \dfrac{70 - 60}{5} = 2$

$z = \dfrac{80 - 60}{5} = 4$

c. $z = \dfrac{x - \mu}{\sigma} = \dfrac{50 - 40}{10} = 1$

$z = \dfrac{70 - 40}{10} = 3$

$z = \dfrac{80 - 40}{10} = 4$

d. $z = \dfrac{x - \mu}{\sigma} = \dfrac{50 - 40}{100} = .1$

$z = \dfrac{70 - 40}{100} = .3$

$z = \dfrac{80 - 40}{100} = .4$

2.167   a. $s^2 = \dfrac{\sum x^2 - \dfrac{\left(\sum x\right)^2}{n}}{n - 1} = \dfrac{246 - \dfrac{63^2}{22}}{22 - 1} = 3.1234$

b. $s^2 = \dfrac{\sum x^2 - \dfrac{\left(\sum x\right)^2}{n}}{n - 1} = \dfrac{666 - \dfrac{106^2}{25}}{25 - 1} = 9.0233$

c. $s^2 = \dfrac{\sum x^2 - \dfrac{\left(\sum x\right)^2}{n}}{n - 1} = \dfrac{76 - \dfrac{11^2}{7}}{7 - 1} = 9.7857$

2.169   a. $\sum x = 4 + 6 + 6 + 5 + 6 + 7 = 34$

$\sum x^2 = 4^2 + 6^2 + 6^2 + 5^2 + 6^2 + 7^2 = 198$

$\bar{x} = \dfrac{\sum x}{n} = \dfrac{34}{6} = 5.67$

$s^2 = \dfrac{\sum x^2 - \dfrac{\left(\sum x\right)^2}{n}}{n-1} = \dfrac{198 - \dfrac{34^2}{6}}{6-1} = \dfrac{5.3333}{5} = 1.0667$

$s = \sqrt{1.0667} = 1.03$

    b. $\sum x = -1 + 4 + (-3) + 0 + (-3) + (-6) = -9$

$\sum x^2 = (-1)^2 + 4^2 + (-3)^2 + 0^2 + (-3)^2 + (-6)^2 = 71$

$\bar{x} = \dfrac{\sum x}{n} = \dfrac{-9}{6} = -\$1.5$

$s^2 = \dfrac{\sum x^2 - \dfrac{\left(\sum x\right)^2}{n}}{n-1} = \dfrac{71 - \dfrac{(-9)^2}{6}}{6-1} = \dfrac{57.5}{5} = 11.5$

$s = \sqrt{11.5} = \$3.39$

    c. $\sum x = \dfrac{3}{5} + \dfrac{4}{5} + \dfrac{2}{5} + \dfrac{1}{5} + \dfrac{1}{16} = 2.0625$

$\sum x^2 = \left(\dfrac{3}{5}\right)^2 + \left(\dfrac{4}{5}\right)^2 + \left(\dfrac{2}{5}\right)^2 + \left(\dfrac{1}{5}\right)^2 + \left(\dfrac{1}{16}\right)^2 = 1.2039$

$\bar{x} = \dfrac{\sum x}{n} = \dfrac{2.0625}{5} = .4125\%$

$s^2 = \dfrac{\sum x^2 - \dfrac{\left(\sum x\right)^2}{n}}{n-1} = \dfrac{1.2039 - \dfrac{2.0625^2}{5}}{5-1} = \dfrac{.3531}{4} = .0883\%\text{ squared}$

$s = \sqrt{.0883} = .297\%$

    d.   (a)   Range = $7 - 4 = 3$

        (b)   Range = $\$4 - (\$-6) = \$10$

        (c)   Range = $\dfrac{4}{5}\% - \dfrac{1}{16}\% = \dfrac{64}{80}\% - \dfrac{5}{80}\% = \dfrac{59}{80}\% = .7375\%$

2.171   Using MINITAB, the scatterplot is:

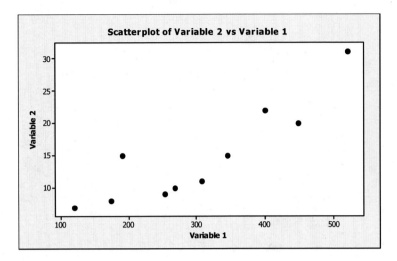

From the scatterplot, it appears that there is a positive trend. As variable 1 increases, variable 2 also tends to increase.

2.173   Using MINITAB, the dot plot of the data is:

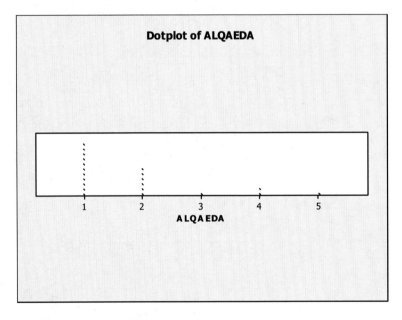

The most frequent number of attacks per incident is 1. There is only 1 incident with 3 attacks and only 1 incident with 5 attacks.

2.175   a.   The sample mean is 603.7. This is the average of the 98 observations. The standard deviation is StDev = 185.4. We would expect most of the observations to fall within 2 standard deviations of the mean. The minimum value is 216. This is the smallest observation in the data set. The Q1 value is 475. This is the 25th percentile. Twenty-five percent of all observations in the data set are less than or equal to 475. The median is 605.0. Half of the observations in the data set are above 605 and half are below. The Q3 value is 724.3. This is the 75th percentile. Seventy-five percent of all observations in the data set are less than or equal to 724.3. The maximum value is 1240.0. This is the value of the largest observation in the data set.

b. The z-score is: $z = \dfrac{x - \bar{x}}{s} = \dfrac{408 - 603.7}{185.4} = -1.06$

A head-injury rating of 408 is less than the mean head-injury rating. It is a little more than one standard deviation below the mean.

2.177   a.   Using MINITAB, a bar chart is:

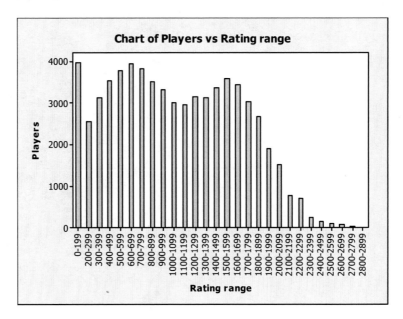

b. The total number of players with ratings of 2,000 or higher is 1 + 13 + 66 + 87 + 133 + 231 + 691 + 783 + 1516 = 3521. The percentage of players is found by dividing the number of players with ratings of 2,000 or higher by the total number of players which is 65,455 and then multiplying by 100%. The percentage is (3,521 / 65,455)*100% = 5.38%.

c. The average rating is 1068. This is a measure of central tendency or a rating of an average player.

2.179   a.   The data collection method was a survey.

b. Since the data were numbers (percentage of US labor and materials), the variable is quantitative. Once the data were collected, they were grouped into 4 categories.

c.   Using MINITAB, a pie chart of the data is:

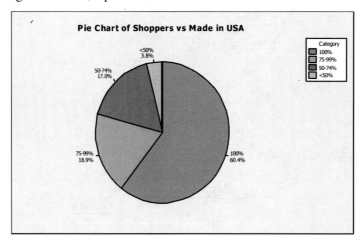

About 60% of those surveyed believe that "Made in USA" means 100% US labor and materials.

2.181   a.   Using MINITAB, the Pareto diagram is:

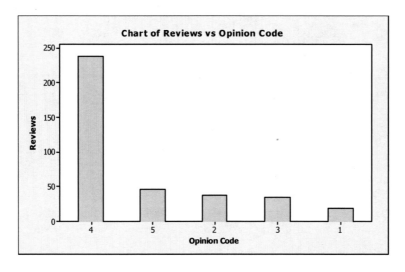

The opinion that occurred most often was "favorable/recommended" with 238 responses. The total number of responses was 19 + 37 + 35 + 238 + 46 = 375. The proportion of books receiving a "favorable/recommended" opinion is 238/375 = .635.

b.   Books receiving either a 4 (favorable/recommended or a 5 (outstanding/significant) were reviewed as favorable and recommended for purchase. The total number of books receiving a rating of 4 or 5 is 238 + 46 = 284. The proportion of books receiving these ratings is 284/375 = .757. This proportion is more than .75 or 75%. Thus, the statement made is correct.

2.183 a.   Using MINITAB, a scatterplot of the data is:

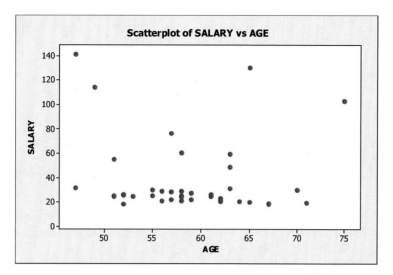

There is not much evidence of a trend between a CEO's salary and age.

b.   According to Chebyshev's Rule, at least $1-\dfrac{1}{k^2}$ of the measurements will fall within k standard deviations of the mean (for k > 1). Thus, we can say nothing about the percentage of measurements that will fall within .75 standard deviations of the mean.

At least $1-\dfrac{1}{2.5^2}=1-.16=.84$ or 84% of the measurements will fall within 2.5 standard deviations of the mean. At least $1-\dfrac{1}{4^2}=1-.0625=.9375$ or 93.75% of the measurements will fall within 4 standard deviations of the mean.

c.   Using MINITAB, the descriptive statistics are:

Descriptive Statistics: AGE, SALARY

Variable	N	Mean	StDev	Minimum	Q1	Median	Q3	Maximum
AGE	40	58.68	6.51	47.00	53.50	58.00	63.00	75.00
SALARY	40	38.58	31.34	18.39	21.45	25.36	31.28	141.36

Next, we compute the intervals for age:

$\bar{x}\pm.75s \Rightarrow 58.68\pm.75(6.51) \Rightarrow 58.68\pm4.88 \Rightarrow (53.80,\ 63.56)$

Twenty-two of the 40 or (22/40)100% = 55% of the ages fall within .75 standard deviations of the mean.

$\bar{x}\pm2.5s \Rightarrow 58.68\pm2.5(6.51) \Rightarrow 58.68\pm16.28 \Rightarrow (42.41,\ 74.96)$

Thirty-nine of the 40 or (39/40)100% = 97.5% of the ages fall within 2.5 standard deviations of the mean. This is at least 84% that we found in part b.

$$\bar{x} \pm 4s \Rightarrow 58.68 \pm 4(6.51) \Rightarrow 58.68 \pm 26.04 \Rightarrow (32.64, 84.72)$$

Forty of the 40 or (40/40)100% = 100% of the ages fall within 4 standard deviations of the mean. This is at least 93.7% that we found in part b.

d.   Now, we compute the intervals for salary:

$$\bar{x} \pm .75s \Rightarrow 38.58 \pm .75(31.34) \Rightarrow 38.58 \pm 23.51 \Rightarrow (15.08, 62.09)$$

Thirty-five of the 40 or (35/40)100% = 87.5% of the ages fall within .75 standard deviations of the mean.

$$\bar{x} \pm 2.5s \Rightarrow 38.58 \pm 2.5(31.34) \Rightarrow 38.58 \pm 78.35 \Rightarrow (-39.77, 116.93)$$

Thirty-eight of the 40 or (38/40)100% = 95% of the salaries fall within 2.5 standard deviations of the mean. This is at least 84% that we found in part b.

$$\bar{x} \pm 4s \Rightarrow 38.58 \pm 4(31.34) \Rightarrow 38.58 \pm 125.36 \Rightarrow (-86.78, 163.94)$$

Forty of the 40 or (40/40)100% = 100% of the salaries fall within 4 standard deviations of the mean. This is at least 93.7% that we found in part b.

2.185   A relative frequency bar graph is used to depict the data:

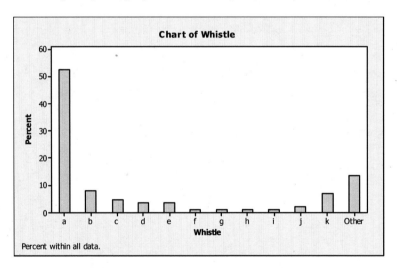

From the bar graph, over half of the whistle types were "Type a." The next most frequent category was the "Other types" with a relative frequency of about .14. Whistle types b and k were the next most frequent. None of the other whistle types had relative frequencies higher than .05.

2.187   a.   Using MINITAB, the boxplot is:

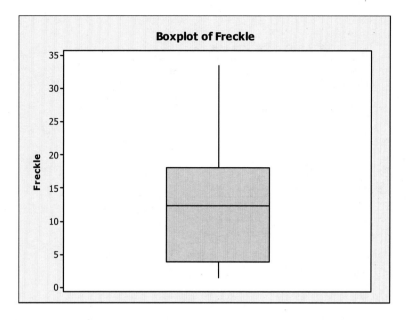

From the boxplot, there are no outliers detected.

b.   Since there are no outliers, we cannot find the z-score for the points identified as outliers.

2.189   a.   For each transect, three variables were measured. The number of seabirds found is quantitative.   The length of the transect is also quantitative. Whether or not the transect was in an oiled area is qualitative.

b.   The experimental unit is the transect.

c.   A pie chart of the oiled and unoiled areas is:

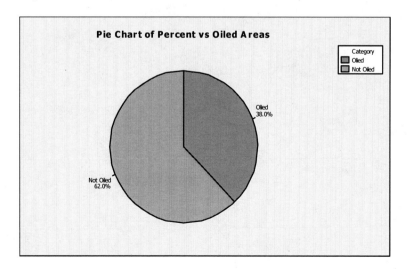

d. Using MINITAB, a scattergram of the data is:

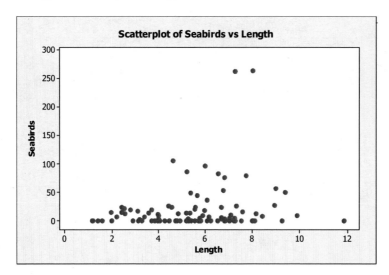

e. The mean density for the unoiled area is 3.27, while the mean for the oiled area is 3.495. The median for the unoiled area is .89 and is .70 for the oiled area. These are both fairly similar.

f. Using Chebyshev's Theorem, at least 75% of the observations will fall within 2 standard deviations of the mean. This interval for unoiled areas would be:

$$\bar{x} \pm 2s \Rightarrow 3.27 \pm 2(6.7) \Rightarrow 3.27 \pm 13.4 \Rightarrow (-10.13, 16.67)$$

g. Using Chebyshev's Theorem, at least 75% of the observations will fall within 2 standard deviations of the mean. This interval for oiled areas would be:

$$\bar{x} \pm 2s \Rightarrow 3.495 \pm 2(5.968) \Rightarrow 3.495 \pm 11.939 \Rightarrow (-8.441, 15.431)$$

h. From the above two intervals, we know that at least 75% of the observations for the unoiled area will fall between −10.31 and 16.67 and at most 25% of the observations will fall above 15.431 for the oiled areas. Thus, the unoiled areas would be more likely to have a seabird density of 16.

2.191   a.   Using Minitab, a stem-and-leaf display of the data is:

```
Stem-and-leaf of VELOCITY      N = 51
Leaf Unit = 100

  1    18 4
  3    18 79
 12    19 001112444
 18    19 566788
 20    20 12
 21    20 7
 21    21
 23    21 99
 (5)   22 11344
 23    22 5666777777889
 10    23 001222344
  1    23
  1    24
  1    24 9
```

   b.   From this stem-and-leaf display, it is fairly obvious that there are two different distributions since there are two groups of data.

   c.   Since there appears to be two distributions, we will compute two sets of numerical descriptive measures. We will call the group with the smaller velocities A1775A and the group with the larger velocities A1775B.

   For A1775A:

   $$\bar{x} = \frac{\sum x}{n} = \frac{408,707}{21} = 19,462.2$$

   $$s^2 = \frac{\sum x^2 - \frac{\left(\sum x\right)^2}{n}}{n-1} = \frac{7,960,019,531 - \frac{408,707^2}{21}}{21-1} = 283,329.3$$

   $$s = \sqrt{283,329.3} = 532.29$$

   For A1775B:

   $$\bar{x} = \frac{\sum x}{n} = \frac{685,154}{30} = 22,838.5$$

   $$s^2 = \frac{\sum x^2 - \frac{\left(\sum x\right)^2}{n}}{n-1} = \frac{15,656,992,942 - \frac{685,154^2}{30}}{30-1} = 314,694.88$$

   $$s = \sqrt{314,694.88} = 560.98$$

    d.   To determine which of the two clusters this observation probably belongs to, we will compute z-scores for this observation for each of the two clusters.

For A1775A:
$$z = \frac{20,000 - 19,462.2}{532.29} = 1.01$$

Since this z-score is so small, it would not be unlikely that this observation came from this cluster.

For A1775B:

$$z = \frac{20,000 - 22,838.5}{560.98} = -5.06$$

Since this z-score is so large (in magnitude), it would be very unlikely that this observation came from this cluster.

Thus, this observation probably came from the cluster A1775A.

2.193   First, compute the z-score for 1.80: $z = \frac{x - \mu}{\sigma} = \frac{1.80 - 2.00}{.08} = -2.5$

If the data are actually mound-shaped, it would be very unusual (less than 2.5%) to observe a batch with 1.80% zinc phosphide if the true mean is 2.0%. Thus, if we did observe 1.8%, we would conclude that the mean percent of zinc phosphide in today's production is probably less than 2.0%.

2.195   a.   Due to the "elite" superstars, the salary distribution is skewed to the right. Since this implies that the median is less than the mean, the players' association would want to use the median.

    b.   The owners, by the logic of part **a**, would want to use the mean.

2.197   The answers to this will vary. Some things that should be included in the discussion are:

Even though 4,500 completed questionnaires were returned, the response rate was extremely small, only 4.5%. Also, it is very likely that this is not a representative sample of the population of American women. First, the questionnaires were not distributed to a random sample of women from the U.S. The questionnaires were sent specifically to certain groups and the directors of those groups were asked to distribute the surveys to members of their organizations. Thus, if a woman did not belong to one of the targeted groups, she had a very small chance of participating in the survey.

In these types of mail-in surveys, usually only those with very strong opinions respond to the surveys. Thus, those who actually responded to the survey were probably not the same as the population in general. The results are very likely not very reliable.

Probability

3.1   An experiment is an act or process of observation that leads to a single outcome that cannot be predicted with certainty.

3.3   A sample space of an experiment is a collection of all its sample points.

3.5   Two probability rules for sample points are:

    1.   All sample point probabilities **must** lie between 0 and 1, i.e. $0 \leq p_i \leq 1$.

    2.   The probabilities of all sample points within a sample space **must** sum to 1, i.e., $\sum p_i = 1$.

3.7   The probability of an event A is calculated by summing the probabilities of the sample points in the sample space for A.

3.9   a.   Since the probabilities must sum to 1,

$$P(E_3) = 1 - P(E_1) - P(E_2) - P(E_4) - P(E_5) = 1 - .1 - .2 - .1 - .1 = .5$$

    b. $P(E_3) = 1 - P(E_1) - P(E_2) - P(E_4) - P(E_5) = 1 - P(E_3) - P(E_2) - P(E_4) - P(E_5)$

        $\Rightarrow 2P(E_3) = 1 - P(E_2) - P(E_4) - P(E_5)$

        $\Rightarrow 2P(E_3) = 1 - .1 - .2 - .1 \Rightarrow 2P(E_3) = .6 \Rightarrow P(E_3) = .3$

    c. $P(E_3) = 1 - P(E_1) - P(E_2) - P(E_4) - P(E_5) = 1 - .1 - .1 - .1 - .1 = .6$

3.11 $P(A) = P(1) + P(2) + P(3) = .05 + .20 + .30 = .55$

       $P(B) = P(1) + P(3) + P(5) = .05 + .30 + .15 = .50$

       $P(C) = P(1) + P(2) + P(3) + P(5) = .05 + .20 + .30 + .15 = .70$

3.13   a. $\dbinom{5}{2} = \dfrac{5!}{2!(5-2)!} = \dfrac{5!}{2!3!} = \dfrac{5 \cdot 4 \cdot 3 \cdot 2 \cdot 1}{2 \cdot 1 \cdot 3 \cdot 2 \cdot 1} = 10$

    b. $\dbinom{6}{3} = \dfrac{6!}{3!(6-3)!} = \dfrac{6!}{3!3!} = \dfrac{6 \cdot 5 \cdot 4 \cdot 3 \cdot 2 \cdot 1}{3 \cdot 2 \cdot 1 \cdot 3 \cdot 2 \cdot 1} = 20$

    c. $\dbinom{20}{5} = \dfrac{20!}{5!(20-5)!} = \dfrac{20!}{5!15!} = \dfrac{20 \cdot 19 \cdot 18 \cdots 2 \cdot 1}{5 \cdot 4 \cdot 3 \cdot 2 \cdot 1 \cdot 15 \cdot 14 \cdot 13 \cdots 2 \cdot 1} = 15{,}504$

3.15 a. If we denote the marbles as B_1, B_2, R_1, R_2, R_3, then the ten equally likely sample points in the sample space would be:

$$S: \begin{bmatrix} (B_1,B_2),(B_1,R_1),(B_1,R_2),(B_1,R_3),(B_2,R_1) \\ (B_2,R_2),(B_2,R_3),(R_1,R_2),(R_1,R_3),(R_2,R_3) \end{bmatrix}$$

Notice that order is ignored, as the only concern is whether or not a marble is selected.

b. Each of these ten would be equally likely, implying that each occurs with a probability 1/10.

c. $P(A) = \dfrac{1}{10}$ $P(B) = 6\left(\dfrac{1}{10}\right) = \dfrac{6}{10} = \dfrac{3}{5}$ $P(C) = 3\left(\dfrac{1}{10}\right) = \dfrac{3}{10}$

3.17 a. There are 6 sample points for this experiment: Blue, Orange, Green, Yellow, Brown, and Red.

b. According to the Mars Corporation, reasonable probabilities would be:

P(Blue) = .24, P(Orange) = .20, P(Green) = .16, P(Yellow) = .14, P(Brown) = .13, and P(Red) = .13.

c. P(Brown) = .13.

d. P(Red, Green, or Yellow) = P(Red) + P(Green) + P(Yellow) = .13 + .16 + .14 = .43.

e. P(Not Blue) = 1 – P(Blue) = 1 - .24 = .76.

3.19 a. Let A = {chicken passes inspection with fecal contamination}. $P(A) = 1 / 100 = .01$.

b. Yes. The relative frequency of passing inspection with fecal contamination is 306/32,075 = .0095 ≈ .01.

3.21 a. There are 5 sample points for this experiment: None, 1 or 2, 3-5, 6-9, and 10 or more.

b. We can use the percentages given as reasonable estimates of the probabilities: P(None) = .25, P(1 or 2) = .31, P(3-5) = .25, P(6-9) = .05, and P(10 or more) = .14.

c. P(more than 2) = P(3-5) + P(6-9) + P(10 or more) = .25 + .05 + .14 = .44.

3.23 P(Interview or Grounded Theory) = (5,079 + 537) / 7,506 = 5,616 / 7,506 = .748

3.25 P(tooth shows slight or moderate amount of wear)

= P(tooth shows slight amount of wear) + P(tooth shows moderate amount of wear)

= 4/18 + 3/18 = 7/18 = .389.

3.27 a. There are a total of $\dbinom{5}{3} = \dfrac{5!}{3!(5-3)!} = \dfrac{5!}{3!2!} = \dfrac{5 \cdot 4 \cdot 3 \cdot 2 \cdot 1}{3 \cdot 2 \cdot 1 \cdot 2 \cdot 1} = 10$ ways to get 3-grill displays.

These 10 display combinations are:

1, 2, 3 1, 3, 4 2, 3, 4 3, 4, 5
1, 2, 4 1, 3, 5 2, 3, 5
1, 2, 5 1, 4, 5 2, 4, 5

However, since grill #2 must be selected, there are only 6 possibilities:

1, 2, 3 2, 3, 4
1, 2, 4 2, 3, 5
1, 2, 5 2, 4, 5

b. We can estimate the probabilities by using the relative frequency for each sample point. The relative frequency is found by dividing the frequency by the total sample size of 124. These estimates are contained in the following table:

Grill Display Combination	Number of Students	Probability
1-2-3	35	.282
1-2-4	8	.065
1-2-5	42	.339
2-3-4	4	.032
2-3-5	1	.008
2-4-5	34	.274
TOTAL	124	1.0000

c. Of the 6 sample points, only 3 of them contain Grill #1.

P(Grill #1 chosen) = P(1-2-3 or 1-2-4 or 1-2-5) = .282 + .065 + .339 = .686.

3.29 a. There are a total of $\binom{8}{2} = \frac{8!}{2!(8-2)!} = \frac{8!}{2!6!} = \frac{8\cdot7\cdot6\cdot5\cdot4\cdot3\cdot2\cdot1}{2\cdot1\cdot6\cdot5\cdot4\cdot3\cdot2\cdot1} = 28$ possible Quinella bets.

b. If all the players are of equal ability, then the probability of getting any combination is 1/28.

3.31 The total number of pairs of bullets is a combination of 1,837 bullets taken 2 at a time or:

$$\binom{1,837}{2} = \frac{1,837!}{2!1,835!} = \frac{1,837(1,836)}{2} = 1,686,366$$

The probability of finding a match or a false positive is 693 / 1,686,366 = .000411.

This probability is very small. The confidence in the FBI's forensic evidence should be very high.

3.33 a. $\binom{6}{2} = \frac{6!}{2!(6-2)!} = \frac{6!}{2!4!} = \frac{6\cdot5\cdot4\cdot3\cdot2\cdot1}{2\cdot1\cdot4\cdot3\cdot2\cdot1} = 15$

b. $\binom{6}{3} = \dfrac{6!}{3!(6-3)!} = \dfrac{6!}{3!3!} = \dfrac{6 \cdot 5 \cdot 4 \cdot 3 \cdot 2 \cdot 1}{3 \cdot 2 \cdot 1 \cdot 3 \cdot 2 \cdot 1} = 20$

c. $\binom{6}{4} = \dfrac{6!}{4!(6-4)!} = \dfrac{6!}{4!2!} = \dfrac{6 \cdot 5 \cdot 4 \cdot 3 \cdot 2 \cdot 1}{4 \cdot 3 \cdot 2 \cdot 1 \cdot 2 \cdot 1} = 15$

d. $\binom{6}{5} = \dfrac{6!}{5!(6-5)!} = \dfrac{6!}{5!1!} = \dfrac{6 \cdot 5 \cdot 4 \cdot 3 \cdot 2 \cdot 1}{5 \cdot 4 \cdot 3 \cdot 2 \cdot 1 \cdot 1} = 6$

e.   In addition to the combinations computed in parts a – d, we need to find the number of 0-drug combinations, 1-drug combination, and 6-drug combinations.

$\binom{6}{0} = \dfrac{6!}{0!(6-0)!} = \dfrac{6!}{0!6!} = \dfrac{6 \cdot 5 \cdot 4 \cdot 3 \cdot 2 \cdot 1}{1 \cdot 6 \cdot 5 \cdot 4 \cdot 3 \cdot 2 \cdot 1} = 1$

$\binom{6}{1} = \dfrac{6!}{1!(6-1)!} = \dfrac{6!}{1!5!!} = \dfrac{6 \cdot 5 \cdot 4 \cdot 3 \cdot 2 \cdot 1}{1 \cdot 5 \cdot 4 \cdot 3 \cdot 2 \cdot 1} = 6$

$\binom{6}{6} = \dfrac{6!}{6!(6-6)!} = \dfrac{6!}{6!0!} = \dfrac{6 \cdot 5 \cdot 4 \cdot 3 \cdot 2 \cdot 1}{6 \cdot 5 \cdot 4 \cdot 3 \cdot 2 \cdot 1 \cdot 1} = 1$

The total number of ways the 6 drugs can be combined is $15 + 20 + 15 + 6 + 1 + 6 + 1 = 64$.

3.35   The union of 2 events A and B is the event that occurs if either A or B or both occur on a single performance of the experiment.

3.37   The complement of an event A is the event that A does not occur – that is, the event consisting of all sample points that are not in event A.

3.39   The Additive Rule of Probability is: The probability of the union of events A and B is the sum of the probability of events A and B minus the probability of the intersection of events A and B, that is $P(A \cup B) = P(A) + P(B) - P(A \cap B)$.

3.41   The Additive Rule of Probability for mutually exclusive events is: If 2 events A and B are mutually exclusive, the probability of the union of A and B equals the sum of the probabilities of A and B; that is, $P(A \cup B) = P(A) + P(B)$.

3.43 a. A: {*HHH, HHT, HTH, THH, TTH, THT, HTT*}
              B: {*HHH, TTH, THT, HTT*}
        $A \cup B$: {*HHH, HHT, HTH, THH, TTH, THT, HTT*}
            A^c: {*TTT*}
      $A \cap B$: {*HHH, TTH, THT, HTT*}

b. If the coin is fair, then each of the 8 possible outcomes are equally likely, with probability 1/8.

$$P(A) = \frac{7}{8} \qquad P(B) = \frac{4}{8} = \frac{1}{2} \qquad P(A \cup B) = \frac{7}{8}$$

$$P(A^c) = \frac{1}{8} \qquad P(A \cap B) = \frac{4}{8} = \frac{1}{2}$$

c. $P(A \cup B) = P(A) + P(B) - P(A \cap B) = \frac{7}{8} + \frac{1}{2} - \frac{1}{2} = \frac{7}{8}$

d. No. $P(A \cap B) = \frac{1}{2}$ which is not 0.

3.45 a. $P(A) = P(E_1) + P(E_2) + P(E_3) + P(E_5) + P(E_6) = \frac{1}{5} + \frac{1}{5} + \frac{1}{5} + \frac{1}{20} + \frac{1}{10} = \frac{15}{20} = \frac{3}{4}$

b. $P(B) = P(E_2) + P(E_3) + P(E_4) + P(E_7) = \frac{1}{5} + \frac{1}{5} + \frac{1}{20} + \frac{1}{5} = \frac{13}{20}$

c. $P(A \cup B) = P(E_1) + P(E_2) + P(E_3) + P(E_4) + P(E_5) + P(E_6) + P(E_7)$

$$= \frac{1}{5} + \frac{1}{5} + \frac{1}{5} + \frac{1}{20} + \frac{1}{20} + \frac{1}{10} + \frac{1}{5} = 1$$

d. $P(A \cap B) = P(E_2) + P(E_3) = \frac{1}{5} + \frac{1}{5} = \frac{2}{5}$

e. $P(A^c) = 1 - P(A) = 1 - \frac{3}{4} = \frac{1}{4}$

f. $P(B^c) = 1 - P(B) = 1 - \frac{13}{20} = \frac{7}{20}$

g. $P(A \cup A^c) = P(E_1) + P(E_2) + P(E_3) + P(E_4) + P(E_5) + P(E_6) + P(E_7)$

$$= \frac{1}{5} + \frac{1}{5} + \frac{1}{5} + \frac{1}{20} + \frac{1}{20} + \frac{1}{10} + \frac{1}{5} = 1$$

h. $P(A^c \cap B) = P(E_4) + P(E_7) = \frac{1}{20} + \frac{1}{5} = \frac{5}{20} = \frac{1}{4}$

3.47   a. $P(A) = .50 + .10 + .05 = .65$

     b. $P(B) = .10 + .07 + .50 + .05 = .72$

     c. $P(C) = .25$

     d. $P(D) = .05 + .03 = .08$

     e. $P(A^c) = .25 + .07 + .03 = .35$ (Note: $P(A^c) = 1 - P(A) = 1 - .65 = .35$)

     f. $P(A \cup B) = P(B) = .10 + .07 + .50 + .05 = .72$

     g. $P(A \cap B) = P(A) = .50 + .10 + .05 = .65\,5$

     h.   Two events are mutually exclusive if they have no sample points in common or if the probability of their intersection is 0.

        $P(A \cap B) = .50 + .10 + .05 = .65$. Since this is not 0, A and B are not mutually exclusive.

        $P(A \cap C) = 0$. Since this is 0, A and C are mutually exclusive.

        $P(A \cap D) = .05$. Since this is not 0, A and D are not mutually exclusive.

        $P(B \cap C) = 0$. Since this is 0, B and C are mutually exclusive.

        $P(B \cap D) = .05$. Since this is not 0, B and D are not mutually exclusive.

        $P(C \cap D) = 0$. Since this is 0, C and D are mutually exclusive.

3.49   a.   A Venn diagram of the data is:

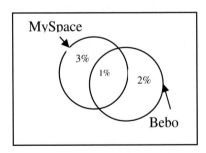

     b.   Define the following events:
        A: {UK citizen visits MySpace}
        B: {UK citizen visits Bebo}

        $P(A \cup B) = P(A) + P(B) - P(A \cap B) = .04 + .03 - .01 = .06$

c. $P\left[A^c \cap B^c\right] = 1 - P(A \cup B) = 1 - .06 = .94$

3.51   a.   The sample points of this experiment are the locations where toxic chemical incidents occurred. They are:

School laboratory, In Transit, Chemical plant, Non-chemical plant, and Other.

b.   Reasonable probabilities would be the percents of the incidents changed to proportions.

P(School laboratory) = .06, P(In Transit) = .26, P(Chemical plant) = .21,
P(Non-chemical plant) = .35, P(Other) = .12

c. P(School laboratory) = .06

d. P(Chemical plant or Non-chemical plant) = .21 + .35 = .56

e. P(Not occur In Transit) = 1 – P(In Transit) = 1 – .26 = .74

3.53   a.   A Venn Diagram that illustrates the results of the gene profiling analysis is:

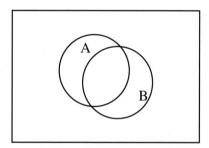

b.   From the problem, we know $P(A)$ = .41, $P(B)$ = .42, and $P(A \cap B)$ = .40
$P(A \cup B) = P(A) + P(B) - P(A \cap B) = .41 + .42 - .40 = .43$.

c. P(Neither) = $P(A \cup B)^c = 1 - P(A \cup B) = 1 - .43 = .57$

3.55   a.   Define the following events:
A: {Student choose stated option}
B: {Student did not choose stated option}
C: {Emotion state is Guilt}
D: {Emotion state is Anger}
E: {Emotion state is Neutral}

$$P(C) = \frac{45 + 12}{171} = \frac{57}{171} = .333$$

b. $$P(A) = \frac{45 + 8 + 7}{171} = \frac{60}{171} = .351$$

c. $P(C \cap A) = \dfrac{45}{171} = .263$

d. $P(C \cup A) = P(C) + P(A) - P(C \cap A) = .333 + .351 - .263 = .421$

3.57   a.   Define the following events:

F: {Fight}
N: {No fight}
W: {Initiator Wins}
T: {No Clear Winner}
L: {Initiator Loses}

$P(F \cap W) = 26/167 = .156$

b. $P(N) = 103/167 = .617$

c. $P(T) = 35/167 = .210$

d. $P(F \cup L) = (26 + 23 + 15 + 11)/167 = 75/167 = .449$

e.   Yes. $P(T \cap L) = 0$.

3.59   a. $P(A) = \dfrac{1,465}{2,143} = .684$

b. $P(B) = \dfrac{265}{2,143} = .124$

c.   No. $P(A \cap B) = \dfrac{194}{2,143} = .091$. Since this is not 0, events A and B are not mutually exclusive.

d. $P(A^c) = 1 - P(A) = 1 - .684 = .316$.

e. $P(A \cup B) = P(A) + P(B) - P(A \cap B) = .684 + .124 - .091 = .717$

f. $P(A \cap B) = \dfrac{194}{2,143} = .091$

3.61   a. $P(\text{responds to all 3 source odors}) = .19$

b. $P(\text{responds to kairomone}) = .19 + .025 + .025 + .165 = .405$

c. $P(\text{responds to Mups A and Mups B but not kairomone}) = .21 + .275 + .11 = .595$

3.63   A conditional probability is the probability of an event based on another event, while an unconditional probability is not based on another event.

3.65 $P(A|B) = \dfrac{P(A \cap B)}{P(B)}$

3.67   a. $P(A|B) = \dfrac{P(A \cap B)}{P(B)} = \dfrac{.1}{.2} = .5$

   b. $P(B|A) = \dfrac{P(A \cap B)}{P(A)} = \dfrac{.1}{.4} = .25$

   c.   No, events A and B are not independent. If A and B were independent, then $P(A|B) = P(A)$. For this problem, $P(A|B) = .5$ and $P(A) = .4$.

3.69   a. $P(A \cap B) = P(A)P(B) = .4(.2) = .08$ (Since A and B are independent)

   b. $P(A|B) = \dfrac{P(A \cap B)}{P(B)} = \dfrac{.08}{.2} = .4$

   c. $P(A \cup B) = P(A) + P(B) - P(A \cap B) = .4 + .2 - .08 = .52$

3.71   a. $P(A) = P(E_1) + P(E_2) + P(E_3) = .1 + .1 + .2 = .4$

   $P(B) = P(E_2) + P(E_3) + P(E_5) = .1 + .2 + .1 = .4$

   $P(A \cap B) = P(E_2) + P(E_3) = .1 + .2 = .3$

   b. $P(E_1|A) = \dfrac{P(E_1 \cap A)}{P(A)} = \dfrac{.1}{.4} = .25$

   $P(E_2|A) = \dfrac{P(E_2 \cap A)}{P(A)} = \dfrac{.1}{.4} = .25$

   $P(E_3|A) = \dfrac{P(E_3 \cap A)}{P(A)} = \dfrac{.2}{.4} = .5$

   We are given that $P(E_1) = .1$, $P(E_2) = .1$, and $P(E_3) = .2$
   Thus, $P(E_1) = P(E_2)$ and $P(E_3) = 2P(E_1)$
   From above, $P(E_1|A) = P(E_2|A)$ and $P(E_3|A) = 2P(E_1|A)$

   $P(E_1|A) + P(E_2|A) + P(E_3|A) = .25 + .25 + .50 = 1.00$

   c.   Using the sum of the conditional probabilities,

   $P(B|A) = P(E_2|A) + P(E_3|A) = .25 + .50 = .75$

Using the formula,

$$P(B \mid A) = \frac{P(A \cap B)}{P(A)} = \frac{.3}{.4} = .75$$

3.73   The 36 possible outcomes obtained when tossing two dice are listed below:

(1, 1) (1, 2) (1, 3) (1, 4) (1, 5) (1, 6)
(2, 1) (2, 2) (2, 3) (2, 4) (2, 5) (2, 6)
(3, 1) (3, 2) (3, 3) (3, 4) (3, 5) (3, 6)
(4, 1) (4, 2) (4, 3) (4, 4) (4, 5) (4, 6)
(5, 1) (5, 2) (5, 3) (5, 4) (5, 5) (5, 6)
(6, 1) (6, 2) (6, 3) (6, 4) (6, 5) (6, 6)

A: {(1, 2), (1, 4), (1, 6), (2, 1), (2, 3), (2, 5), (3, 2), (3, 4), (3, 6), (4, 1), (4, 3), (4, 5), (5, 2), (5, 4), (5, 6), (6, 1), (6, 3), (6, 5)}

B: {(3, 6), (4, 5), (5, 4), (5, 6), (6, 3), (6, 5), (6, 6)}

$A \cap B$: {(3, 6), (4, 5), (5, 4), (5, 6), (6, 3), (6, 5)}

If A and B are independent, then $P(A)P(B) = P(A \cap B)$.

$$P(A) = \frac{18}{36} = \frac{1}{2} \qquad P(B) = \frac{7}{36} \qquad P(A \cap B) = \frac{6}{36} = \frac{1}{6}$$

$P(A)P(B) = \frac{1}{2} \cdot \frac{7}{36} = \frac{7}{72} \neq \frac{1}{6} = P(A \cap B)$. Thus, A and B are not independent.

3.75   Let W_1 and W_2 represent the two white chips, R_1 and R_2 represent the two red chips, and B_1 and B_2 represent the two blue chips. The sample space is:

W_1W_2	W_2R_1	R_1B_1
W_1R_1	W_2R_2	R_1B_2
W_1R_2	W_2B_1	R_2B_1
W_1B_1	W_2B_2	R_2B_2
W_1B_2	R_1R_2	B_1B_2

Assuming each event is equally likely, each event will have a probability of 1/15.

Then, $P(A) = P(W_1W_2) + P(R_1R_2) + P(B_1B_2) = 3\left(\frac{1}{15}\right) = \frac{3}{15} = \frac{1}{5}$

$$P(B) = P(R_1R_2) = \frac{1}{15}$$

$$P(C) = P(W_1W_2) + P(W_1R_1) + P(W_1R_2) + P(W_1B_1) + P(W_1B_2) + P(W_2R_1)$$
$$+ P(W_2R_2) + P(W_2B_1) + P(W_2B_2) + P(R_1R_2) + P(R_1B_1) + P(R_1B_2)$$
$$+ P(R_2B_1) + P(R_2B_2)$$
$$= 14\left(\frac{1}{15}\right) = \frac{14}{15}$$

$$P(A \cap B) = P(R_1R_2) = \frac{1}{15}$$

$$P(A^c) = 1 - P(A) = 1 - \frac{1}{5} = \frac{4}{5}$$

$$P(A^c \cap B) = 0$$

$$P(B \cap C) = P(R_1R_2) = \frac{1}{15}$$

$$P(A \cap C) = P(W_1W_2) + P(R_1R_2) = 2\left(\frac{1}{15}\right) = \frac{2}{15}$$

$$P(A^c \cap C) = P(W_1R_1) + P(W_1R_2) + P(W_1B_1) + P(W_1B_2) + P(W_2R_1)$$
$$+ P(W_2R_2) + P(W_2B_1) + P(W_2B_2) + P(R_1B_1) + P(R_1B_2)$$
$$+ P(R_2B_1) + P(R_2B_2)$$
$$= 12\left(\frac{1}{15}\right) = \frac{12}{15} = \frac{4}{5}$$

$$P(B\,|\,A) = \frac{P(A \cap B)}{P(A)} = \frac{\frac{1}{15}}{\frac{1}{5}} = \frac{1}{3} \qquad P(B\,|\,A^c) = \frac{P(A^c \cap B)}{P(A^c)} = \frac{0}{\frac{4}{5}} = 0$$

$$P(B\,|\,C) = \frac{P(B \cap C)}{P(C)} = \frac{\frac{1}{15}}{\frac{14}{15}} = \frac{1}{14}$$

$$P(A\,|\,C) = \frac{P(A \cap C)}{P(C)} = \frac{\frac{2}{15}}{\frac{14}{15}} = \frac{1}{7}$$

$$P(C\,|\,A^c) = \frac{P(A^c \cap C)}{P(A^c)} = \frac{\frac{4}{5}}{\frac{4}{5}} = 1$$

3.77 a. Define the following events:
A: {adult owns at least one gun}
B: {adult owns a hand gun}

From the exercise, we know that $P(A) = .26$ and $P(B\,|\,A) = .05$. Thus, $P(A) = .26$.

b. $P(A \cap B) = P(B \mid A)P(A) = .26(.05) = .013$

3.79   Define the following events:

C: {Speeding is cause of fatal crash}
M: {Missing curve is cause of fatal crash}

From the problem, $P(C) = .3$ and $P(C \cap M) = .12$

$$P(M \mid C) = \frac{P(C \cap M)}{P(C)} = \frac{.12}{.30} = .40$$

3.81   a.   Define the following events:
A: {response indicates a perceived unfairness}
B: {angry emotion}

$$P(A) = \frac{594}{10,797} = .055$$

b. $P(B \mid A) = \dfrac{127}{594} = .214$

3.83   The probability that at least one acquires gastroenteritis is equal to 1 minus the probability that neither acquires gastroenteritis. We will assume that the events that you acquire gastroenteritis and the event that your friend acquires gastroenteritis are independent.

$P(\text{at least one acquires gastroenteritis}) = 1 - P(\text{neither acquires gastroenteritis})$
$= 1 - P(\text{you do not acquire gastroenteritis})P(\text{friend does not acquire gastroenteritis})$
$= 1 - \left(\dfrac{994}{1000}\right)\left(\dfrac{993}{1000}\right) = 1 - .987 = .013$

3.85   Define the following event:

R: {Fish is red snapper}

From the problem, $P(R^c) = .77$

a. $P(R) = 1 - P(R^c) = 1 - .77 = .23$

b. $P(\text{At least one customer is served red snapper}) = 1 - P(\text{No customers are served red snapper}) = 1 - P(R^c \cap R^c \cap R^c \cap R^c \cap R^c) = 1 - .77(.77)(.77)(.77)(.77) = 1 - .271 = .729$

3.87   a. $P(\text{Species extinct}) = 38 / 132 = .2879$

b.   If 9 species are chosen and all are extinct, then for the 10th pick, there are only
$132 - 9 = 123$ species to pick from, of which $38 - 9 = 29$ are extinct. P(Species extinct
on 10th selection | first 9 species are extinct) $= 29 / 123 = .236$.

3.89   a.   Define the following events:
A: {ambulance travels to location A under 8 minutes}
B: {ambulance travels to location B under 8 minutes}
C: {ambulance is busy}

From the exercise, we know that $P(A\,|\,C^c) = .58$, $P(B\,|\,C^c) = .42$, and $P(C) = .3$.

$$P(A) = P(A\,|\,C^c)P(C^c) = .58(1 - .3) = .58(.7) = .406$$

b. $P(B) = P(B\,|\,C^c)P(C^c) = .42(1 - .3) = .42(.7) = .294$

3.91 $P(A\,|\,B) = 0$

3.93   a.   The 8 outcomes are:

A and C win first round matches and A wins final (A defeats B, C defeats D, A defeats C)
A and C win first round matches and C wins final (A defeats B, C defeats D, C defeats A)
A and D win first round matches and A wins final (A defeats B, D defeats C, A defeats D)
A and D win first round matches and D wins final (A defeats B, D defeats C, D defeats A)
B and C win first round matches and B wins final (B defeats A, C defeats D, B defeats C)
B and C win first round matches and C wins final (B defeats A, C defeats D, C defeats B)
B and D win first round matches and B wins final (B defeats A, D defeats C, B defeats D)
B and D win first round matches and D wins final (B defeats A, D defeats C, D defeats B)

b.   If all the players are of equal ability, then each of the above outcomes are equally likely.
Each outcome has a probability of 1/8 of occurring. $P(A$ wins$) = 2/8 = .25$.

c.   There are 2 outcomes where A wins the tournament:

A and C win first round matches and A wins final (A defeats B, C defeats D, A defeats C)
A and D win first round matches and A wins final (A defeats B, D defeats C, A defeats D)

$P(A$ wins$) = P(A$ defeats $B \cap C$ defeats $D \cap A$ defeats $C)$
$\qquad\qquad\quad + P(A$ defeats $B \cap D$ defeats $C \cap A$ defeats $D)$

$\qquad = P(A$ defeats $B)P(C$ defeats $D)P(A$ defeats $C)$
$\qquad\qquad + P(A$ defeats $B)P(D$ defeats $C)P(A$ defeats $D)$

$\qquad = .9(.4)(.7) + .9(1 - .4)(.6) = .252 + .324 = .576$

3.95   a.   Define the following event:

   A: {Psychic picks box with crystal}
   If psychic is just guessing, $P(A) = 1/10 = .1$

   b.   P(Psychic guesses correct at least once in 7 trials)
         $= 1 - P$(Psychic does not guess correct in 7 trials)

   $$= 1 - P(A^c \cap A^c \cap A^c \cap A^c \cap A^c \cap A^c \cap A^c) = 1 - .9(.9)(.9)(.9)(.9)(.9)$$
   $$= 1 - .478 = .522$$
   (This assumes that the trials are independent)

   c.   If the psychic is just guessing, then P(Psychic does not guess correct in 7 trials) $= .478$.

         Thus, if the person was not a psychic (and was merely guessing) we would expect the person to guess wrong in all seven trials about half the time. This would not be a rare event. Thus, this outcome would support the notion that the person is guessing.

3.97   a.   The total number of outcomes is a combination of 14 numbers taken 4 at a time.

   $$\binom{14}{4} = \frac{14!}{4!(14-4)!} = \frac{14!}{4!10!} = \frac{14 \cdot 13 \cdot 12 \cdots 2 \cdot 1}{4 \cdot 3 \cdot 2 \cdot 1 \cdot 10 \cdot 9 \cdots 2 \cdot 1} = 1,001$$

   b.   The probability that any team obtains the first pick in the draft is that team's number of combinations divided by 1000. The probabilities are:

NBA Lottery Team	Number of Combinations	Probability
Worst Record	250	.250
2nd worst record	200	.200
3rd worst record	157	.157
4th worst record	120	.120
5th worst record	89	.089
6th worst record	64	.064
7th worst record	44	.044
8th worst record	29	.029
9th worst record	18	.018
10th worst record	11	.011
11th worst record	7	.007
12th worst record	6	.006
13th worst record	5	.005
Total	1000	1.000

   c.   Define the following events:

   W_{i1}: {ith worst team gets 1st pick}
   W_{12}: {Worst team gets 2nd pick}

   If the 2nd worst team obtains the number 1 pick, then that team is removed from the group that could obtain the number 2 pick. Thus, the 200 combinations assigned to the team with the 2nd worst record are removed from the available combinations. Now, there are only 800 possible combinations. The probability that the team with the worst record will get the 2nd pick in the draft given that the team with the 2nd worst record gets the first pick would be $P(W_{12} \mid W_{21}) = 250 / 800 = .313$.

d. If the 3rd worst team obtains the number 1 pick, then that team is removed from the group that could obtain the number 2 pick. Thus, the 157 combinations assigned to the team with the 3rd worst record are removed from the available combinations. Now, there are only $1000 - 157 = 843$ possible combinations. The probability that the team with the worst record will get the 2nd pick in the draft given that the team with the 3rd worst record gets the first pick would be $P(W_{12} \mid W_{31}) = 250 / 843 = .297$.

e. We need to find $P(W_{12} \mid W_{11}^c) = \dfrac{P(W_{12} \cap W_{11}^c)}{P(W_{11}^c)} = \dfrac{P(W_{12})}{1 - P(W_{11})}$

Now, $P(W_{11}^c) = 1 - P(W_{11}) = 1 - .250 = .750$

$$P(W_{12}) = P(W_{12} \mid W_{21})P(W_{21}) + P(W_{12} \mid W_{31})P(W_{31})$$
$$+ P(W_{12} \mid W_{41})P(W_{41}) + \cdots + + P(W_{12} \mid W_{13\,1})P(W_{13\,1})$$

To aid in the computation, we will use the following table:

NBA Lottery Team	Number of Combinations	Probability $P(W_{i1})$	$P(W_{12} \mid W_{i1})$	$P(W_{12} \mid W_{i1})P(W_{i1})$
Worst Record	250	.250		
2nd worst record	200	.200	.313	.0626
3rd worst record	157	.157	.297	.0466
4th worst record	120	.120	.284	.0341
5th worst record	89	.089	.274	.0244
6th worst record	64	.064	.267	.0171
7th worst record	44	.044	.262	.0115
8th worst record	29	.029	.257	.0075
9th worst record	18	.018	.255	.0046
10th worst record	11	.011	.253	.0028
11th worst record	7	.007	.252	.0018
12th worst record	6	.006	.252	.0015
13th worst record	5	.005	.251	.0013
Total	1000	1.000		.2158

Thus,
$$P(W_{12}) = P(W_{12} \mid W_{21})P(W_{21}) + P(W_{12} \mid W_{31})P(W_{31})$$
$$+ P(W_{12} \mid W_{41})P(W_{41}) + \cdots + + P(W_{12} \mid W_{13\,1})P(W_{13\,1}) = .2158$$

Finally, $P(W_{12} \mid W_{11}^c) = \dfrac{P(W_{12} \cap W_{11}^c)}{P(W_{11}^c)} = \dfrac{P(W_{12})}{1 - P(W_{11})} = \dfrac{.2158}{.750} = .288$

3.99   A representative sample exhibits characteristics typical of those possessed by the population of interest. The most common way to obtain a representative sample is to use a random sample.

3.101   A blind study is one where the participants in the study do not know what treatment they are receiving.

3.103   a. $\binom{600}{3} = \frac{600!}{(3!)(597!)} = \frac{(600)(599)(598)}{(1)(2)(3)} = 35,820,200$

b.   All sets are equally likely in random sampling, so that the probability of any particular sample is:

$$\frac{1}{35,820,200}$$

c.   Answers will vary. Suppose we use Table I and start in row 18, column 5, the first 3 digits and go down. We have to ignore any numbers that are greater than 600. We will list all the numbers and highlight the numbers that are ignored for each sample of size 3:

949	496	144	763	470
585	781	982	078	133
099	812	789	061	587
143	642	826	277	197
741	827	533	988	248
242	464	278	188	469
873	672	742	174	846
073	073	101	530	444
964	299	954	709	267
264	319	142	496	422
664	253	417	889	863
264	707	967	482	189
943	383	897	772	679
773	533	337	774	308
561	919	512	893	040
552	876	819	312	200
886	493	816	232	023
129		309	426	846
301			091	396
491				016
				344
				232

Notice that no sample of size 3 is repeated. Since the probability in part b was so small, we would not expect any duplicate samples.

d.   Using MINITAB, the generated sample of size 3 is 36, 362 and 76.

3.105 a. Decide on a starting point on the random number table. Then take the first *n* numbers reading down, and this would be the sample. Group the digits on the random number table into groups of 7 (for part **b**) or groups of 4 (for part **c**). Eliminate any duplicates and numbers that begin with zero since they are not valid telephone numbers.

       b. Starting in Row 6, column 5, take the first 10 seven-digit numbers reading down. The telephone numbers are:

           277-5653
           988-7231
           188-7620
           174-5318
           530-6059
           709-9779
           496-2669
           889-7433
           482-3752
           772-3313

       c. Starting in Row 10, column 7, take the first 5 four-digit numbers reading down. The 5 telephone numbers are:

           373-3886
           373-5686
           373-1866
           373-3632
           373-6768

3.107 Using MINITAB, we will generate 50 random numbers between 1 and 100. These 50 women will receive the kiwifruit breakfast. The others will receive the banana breakfast.

       The answers can vary. One possible solution follows. Seventy random numbers between 1 and 100 were generated using MINITAB to account for duplicate numbers. After deleting all the duplicate numbers, there were 56 random numbers remaining. The first fifty numbers were kept. They were:

        1   4   7   9 11 12 17 18 19 20 23 26 28 29 32 33 34 36 38 39 40 43 46 51 56
       57 58 61 64 67 69 70 71 72 73 74 75 80 81 82 84 85 86 89 90 91 94 95 96 97

       Thus, the women numbered above will receive the kiwifruit breakfast. The others will receive the banana breakfast.

3.109 Answers can vary. One possible solution follows. First, we will number the students from 1 to 120. Then we will generate 90 random numbers. The first 30 generated numbers will be assigned to group 1. The next 30 generated numbers will be assigned to group 2. The next 30 generated numbers will be assigned to group 3. The remaining 30 students will be assigned to group 4. Using MINITAB, we will generate 150 random numbers between 1 and 120. After deleting all the duplicate numbers, there were 93 random numbers remaining. The first 90 numbers were kept. The first 30 numbers sorted were:

1 2 4 6 18 20 25 33 36 41 46 52 61 66 67 70 78 84 86 90 92 94 103 104 106 107 110 112 113 120

Students assigned to these numbers were assigned to group 1.

The next 30 numbers sorted were:

5 7 8 11 12 22 23 29 40 43 47 48 49 51 54 58 59 60 63 65 75 79 80 82 85 89 93 100 101 109

Students assigned to these numbers were assigned to group 2.

The next 30 numbers sorted were:

3 10 13 15 17 26 27 31 35 37 39 42 44 55 56 62 68 74 77 87 95 96 97 99 102 105 114 117 118 119

Students assigned to these numbers were assigned to group 3. The remaining 30 students were assigned to group 4.

3.111   First, number the intersections from 1 to 5000. Using the random number table, select a starting point that contains 4 digits. Following either the row or column, select successive 4 digit numbers until 50 different 4 digit numbers between 0001 and 5000 are selected. The intersections corresponding to these 50 four digit numbers are selected for sampling.

The second method requires the rows to be numbered from 00 to 99 and the columns to be numbered from 1 to 50. Using the random number table, select a starting point that contains 2 digits. Following either the row or column, select successive 2 digit numbers in sets of 2. The first 2 digit number will correspond to the row number. The second 2 digit number must be between 1 and 50 and corresponds to the column number. This procedure is followed until 50 pairs of 2 digit numbers are selected that correspond to 50 different intersections.

3.113   Suppose a basketball team has a total of 10 players. There are 5 distinct positions on the team – center, forward 1, forward 2, shooting guard, and point guard. How many different ways could a coach arrange the starting lineup, assuming that all players could play all positions?

We can apply the Permutations Rule to find the number of combinations.

$$P_n^N = N(N-1)(N-2)\cdots(N-n+1) = \frac{N!}{(N-n)!} = \frac{10!}{(10-5)!}$$

$$= \frac{10!}{5!} = \frac{10 \cdot 9 \cdots 1}{5 \cdot 4 \cdot 3 \cdot 2 \cdot 1} = 10 \cdot 9 \cdot 8 \cdot 7 \cdot 6 = 30,240.$$

3.115   The Combinations Rule does not take order into account, while the Permutations Rule does take order into account.

3.117   Since the number of possible outcomes from tossing a coin is always two on each flip, the multiplicative rule would yield a product of a series of 2's.

a. $(2)(2) = 2^2 = 4$

b. $(2)(2)(2) = 2^3 = 8$

c. $(2)(2)(2)(2)(2) = 2^5 = 32$

d. 2^n

3.119 a. $\dbinom{7}{3} = \dfrac{7!}{(3!)(4!)} = \dfrac{7 \cdot 6 \cdot 5}{3 \cdot 2 \cdot 1} = 35$

b. $\dbinom{6}{2} = \dfrac{6!}{(2!)(4!)} = \dfrac{6 \cdot 5}{2 \cdot 1} = 15$

c. $\dbinom{30}{2} = \dfrac{30!}{(2!)(28!)} = \dfrac{30 \cdot 29}{2 \cdot 1} = 435$

d. $\dbinom{10}{8} = \dfrac{10!}{(8!)(2!)} = \dfrac{10 \cdot 9}{2 \cdot 1} = 45$

e. $\dbinom{q}{r} = \dfrac{q!}{r!(q-r!)}$

3.121 a. We want to select 3 grills from the 5 possible grills. The total number of possible 3-grill-displays is a combination of 5 things taken 3 at a time or

$$\binom{5}{3} = \frac{5!}{3!(5-3)!} = \frac{5(4)(3)(2)(1)}{3(2)(1)(2)(1)} = 10.$$

b. If Grill #2 must be in the display, then we have to choose 2 grills from the remaining 4. Thus, the number of different displays with Grill #2 included is

$$\binom{4}{2} = \frac{4!}{2!(4-2)!} = \frac{(4)(3)(2)(1)}{(2)(1)(2)(1)} = 6$$

c. Now order matters and Grill #2 must be in the display. If Grill #2 is in the first position, then there are $1(4)(3) = 12$ ways to order the three grills. If Grill #2 is in the second position, then there are $4(1)(3) = 12$ ways to order the 3 grills. If Grill #2 is in the third position, then there are $4(3)(1) = 12$ ways to order the 3 grills. Thus, there are a total of $12 + 12 + 12 = 36$ ways to view the grills.

3.123 a. Since order is not important, the number of choices is:

$$\binom{8}{5} = \frac{8!}{5!3!} = 56$$

b. Order is important because it is necessary to distinguish between the positions (guard, forward, and center). However, the two players chosen as guards and the two forwards are indistinguishable once they are chosen. That is, choosing player #1 as the first guard and player #3 as the second guard results in the same team as choosing player #3 for the first guard and player #1 for the second guard. Denote the five positions to be filled as

G_1, G_2, F_1, F_2, and C. The number of distinct teams, assuming each position was distinct is:

$$P_5^8 = \frac{8!}{3!} = 8(7)(6)(5)(4) = 6720$$

However, since G_1 and G_2 are indistinguishable, this number must be divided by 2, or $6720/2 = 3360$ possible teams. Similarly, F_1 and F_2 are indistinguishable, so that there are in fact only $3360/2 = 1680$ possible teams.

c. In this situation, all five positions are distinct. Thus, there are $P_5^8 = 6720$ possible teams.

3.125 a. To visit all 4 cities and then return home, the order of the visits is important. The number of different orders is a permutation of 4 things taken 4 at a time or

$$P_4^4 = \frac{4!}{(4-4)!} = \frac{4 \cdot 3 \cdot 2 \cdot 1}{1} = 24$$

b. If the salesperson visits B first, then he only has 2 choices for the next city (D or E), since B and C are not connected. Once in D or E, he has 2 choices for the next city (if in D, can choose C or E; if in E, can choose D or C). Once in the third city, there is only one choice for the fourth city. The possible trips with B first are $2 \times 2 \times 1 = 4$: BDEC, BDCE, BEDC, BECD.

Similarly, if city C is the first city, there are 2 choices for the second city (D or E), 2 choices for the third, and one for the fourth. Thus, there are $2 \times 2 \times 1 = 4$ choices: CDEB, CDBE, CEDB, CEBD.

If city D is the first chosen, then to visit all four cities exactly once, city B or C must be chosen next. The third city must be E, and the fourth city the only one left. Thus, there are $2 \times 1 \times 1 = 2$ choices: DBEC, DCEB.

Similarly, if city E is the first chosen, then B or C must be the second city, D must be the third, and the fourth city must be the one remaining. Thus, there are $2 \times 1 \times 1 = 2$ choices: EBDC, ECDB.

The total number of routes are $4 + 4 + 2 + 2 = 12$.

3.127 We will select 2 women from the 50 who were assigned to the kiwifruit breakfast. There will be a combination of 50 things taken 2 at a time or

$$\binom{50}{2} = \frac{50!}{2!(50-2)!} = \frac{50(49)(48)\cdots(1)}{(2)(1)(48)(47)(46)\cdots(1)} = 1225 \text{ ways to select 2 women from this group.}$$

There will also be 1225 ways to select 2 women from the 50 who were assigned to the banana breakfast. The total number of ways to select 2 women from each group is $1225(1225) = 1,500,625$.

3.129 a. The total number of distributors that make up Elaine's group is $6(5) = 30$.

b. The total number of distributors that make up Elaine's group now is $6(5)(7)(5) = 1,050$.

3.131 a. From the problem, we have three students and three hospitals. Suppose we consider the three hospitals as three positions. Using the permutations rule, there are

$$P = P_3^3 = \frac{3!}{(3-3)!} = \frac{3 \cdot 2 \cdot 1}{1} = 6 \text{ different assignments possible.}$$

These assignments are:

$$[S_1H_A \quad S_2H_B \quad S_3H_C] \quad [S_1H_A \quad S_3H_B \quad S_2H_C]$$
$$[S_2H_A \quad S_1H_B \quad S_3H_C] \quad [S_2H_A \quad S_3H_B \quad S_1H_C]$$
$$[S_3H_A \quad S_1H_B \quad S_2H_C] \quad [S_3H_A \quad S_2H_B \quad S_1H_C]$$

b. Assuming that all of the above assignments are equally likely, the probability of any one of them is 1/6. The probability that student #1 is assigned to hospital B is 2/6 = 1/3.

3.133 a. There are a total of $2 \times 3 \times 3 = 18$ maintenance organization alternatives.

b. If all alternatives are equally likely, then each has a probability of 1/18 of being selected. If 4 of these alternatives are feasible, then P(feasible alternative) = 4/18 = .222.

3.135 The professor will choose 5 of 10 questions. Of these 10 questions, the student has prepared answers for 7, and has not prepared the other three. The 5 questions on the exam can be chosen in

$$\binom{10}{5} = \frac{10!}{5!5!} = 252 \text{ ways}$$

a. If the student has prepared all 5 exam questions, the professor has chosen 5 from the 7 prepared questions. This can be done in

$$\binom{7}{5} = 21 \text{ ways}$$

Hence, the desired probability is 21/252.

b. Define the following events.

 A: {The student is prepared for no questions}
 B: {The student is prepared for 1 question}
 C: {The student is prepared for 2 questions}

A will occur if the professor picks 0 questions from the 7 prepared and 5 from the 3 unprepared. This is impossible, as is B. C will occur if the professor picks 2 questions from the 7 prepared and 3 questions from the 3 unprepared. This can happen in

$$\binom{7}{2}\binom{3}{3} = 21(1) = 21 \text{ ways}$$

Hence, P(prepared for less than 3) = $P(A) + P(B) + P(C) = 0 + 0 + 21/252 = 21/252$

c.   The professor must pick 4 from the 7 prepared questions and 1 from the remaining 3 questions. This can happen in

$$\binom{7}{4}\binom{3}{1} = 35(3) = 105 \ \text{ways}$$

Hence, P(prepared for exactly 4) = 105/252

3.137   a.   The number of hands possible is:

$$\binom{52}{5} = \frac{52!}{5!47!} = \frac{52 \cdot 51 \cdot 50 \cdot 49 \cdot 48}{5 \cdot 4 \cdot 3 \cdot 2 \cdot 1} = 2,598,690$$

b.   Every set of five cards would be equally likely. Therefore, the probability of any one type of hand can be calculated by finding the number of ways of obtaining that particular type of hand and dividing by the total number of hands possible. For example,

$$P(A) = \frac{\text{Number of ways of obtaining a flush}}{\text{Number of hands possible}}$$

To determine the number of flushes possible, it is necessary to recognize that there would be the same number of flushes in spades as hearts, diamond, or clubs. If we can determine the number of ways of obtaining a flush in any particular suit, then four times that amount would be the number of flushes possible. However, since there are 13 cards of each suit, it follows that the number of flushes for any particular suit is:

$$\binom{13}{5} = \frac{13!}{5!8!} = 1,287$$

Then the total number of flushes possible would be:

4(1,287)= 5,148

Then $P(A) = \dfrac{5,148}{2,598,960} \approx .002$

c.   To compute $P(B)$, it is necessary to recognize that there are 10 different sets of cards' values which result in straights:

{A, 2, 3, 4, 5}
{2, 3, 4, 5, 6}
{3, 4, 5, 6, 7}
{4, 5, 6, 7, 8}
{5, 6, 7, 8, 9}
{6, 7, 8, 9, 10}
{7, 8, 9, 10, J}
{8, 9, 10, J, Q}
{9, 10, J, Q, K}
{10, J, Q, K, A}

We can determine the total number of possible straights by determining the number of ways that any one particular type of straight could occur and then multiply this amount by 10. But the number of ways of obtaining any one straight, say {10, J, Q, K, A}, is done easily by the multiplicative rule. Since there would be one 10 from four to be selected, and one jack from four to be selected, etc., the number of ways of obtaining a straight of the form {10, J, Q, K, A} is:

$$4(4)(4)(4)(4) = 1024$$

There would then be $10(1024) = 10{,}240$ possible straights, so that

$$P(B) = \frac{10{,}240}{2{,}598{,}960} \approx .00394$$

d.   There are only four straight flushes of each type. For example, a straight flush of the type {10, J, Q, K, A} can occur in spades, hearts, diamonds, and clubs. Since there are 10 types of straights, there are $4(10) = 40$ possible straight flushes.

Thus, $P(A \cap B) = \dfrac{40}{2{,}598{,}960} = .0000154$

3.139   If events A and B are independent, then the probability that event B occurs does not depend on whether event A has occurred or not. If events A and B are independent, then $P(B \mid A) = P(B)$.

3.141   a. $P(B_1 \cap A) = P(A \mid B_1)P(B_1) = .3(.75) = .225$

b. $P(B_2 \cap A) = P(A \mid B_2)P(B_2) = .5(.25) = .125$

c. $P(A) = P(B_1 \cap A) + P(B_2 \cap A) = .225 + .125 = .35$

d. $P(B_1 \mid A) = \dfrac{P(B_1 \cap A)}{P(A)} = \dfrac{.225}{.35} = .643$

e. $P(B_2 \mid A) = \dfrac{P(B_2 \cap A)}{P(A)} = \dfrac{.125}{.35} = .357$

3.143   If A is independent of B_1, B_2, and B_3, then $P(A \mid B_1) = P(A) = .4$.

Then $P(B_1 \mid A) = \dfrac{P(A \mid B_1)P(B_1)}{P(A)} = \dfrac{.4(.2)}{.4} = .2$

3.145   Define the following events:

T: {Athlete illegally uses testosterone}
P: {Test for testosterone is positive}
N: {Athlete does not illegally use testosterone}

a. $P(P|T) = \dfrac{50}{100} = .50$

b. $P(P^c|N) = 1 - P(P|N) = 1 - \dfrac{9}{900} = 1 - .01 = .99$

c. $P(T|P) = \dfrac{P(P|T)P(T)}{P(P)}$.

Now, $P(P) = P(P|N)P(N) + P(P|T)P(T) = .01(.9) + .5(.1) = .009 + .05 = .059$

$P(T|P) = \dfrac{P(P|T)P(T)}{P(P)} = \dfrac{.5(.1)}{.059} = \dfrac{.05}{.059} = .847$

3.147   Using Bayes' Rule, the probability is:

$$P(D|N) = \dfrac{P(D \cap N)}{P(N)} = \dfrac{P(D)P(N|D)}{P(D)P(N|D) + P(D^c)P(N|D^c)}$$

$$= \dfrac{\dfrac{1}{80}\left(\dfrac{1}{2}\right)}{\dfrac{1}{80}\left(\dfrac{1}{2}\right) + \dfrac{79}{80}(1)} = \dfrac{\dfrac{1}{160}}{\dfrac{1}{160} + \dfrac{79}{80}} = \dfrac{\dfrac{1}{160}}{\dfrac{1}{160} + \dfrac{158}{160}} = \dfrac{\dfrac{1}{160}}{\dfrac{1}{160} + \dfrac{158}{160}} = \dfrac{\dfrac{1}{160}}{\dfrac{159}{160}} = \dfrac{1}{159}$$

3.149   Define the following events:
       D: {sample is dolomite}
       S: {sample is shale}
       G: {Gamma ray greater than 60}

From the exercise
$P(D) = \dfrac{476}{771} = .61738$, $P(S) = \dfrac{295}{771} = .38262$, $P(G|D) = \dfrac{34}{476} = .07143$, and $P(G|S) = \dfrac{280}{295} = .94915$.

$$P(D|G) = \dfrac{P(D \cap G)}{P(G)} = \dfrac{P(D)P(G|D)}{P(D)P(G|D) + P(D^c)P(G|D^c)}$$

$$= \dfrac{.6174(.0714)}{.6174(.0714) + .3826(.9492)} = \dfrac{.04408}{.04408 + .36316} = \dfrac{.04408}{.40724} = .1082$$

Thus, the probability that the area should be mined is .1082.

3.151   From Exercise 3.90, we defined the following events:

I: {Intruder}
N: {No intruder}
A: {System A sounds alarm}
B: {System B sounds alarm}

Also, from Exercise 3.90, $P(A \cap B \mid I) = .855$ and $P(A \cap B \mid N) = .02$.

Thus,
$$P(A \cap B) = P(A \cap B \mid I)P(I) + P(A \cap B \mid N)P(N) = .855(.4) + .02(.6) = .342 + .012 = .354$$

$$P(I \cap A \cap B) = P(A \cap B \mid I)P(I) = .855(.4) = .342$$

$$P(I \mid A \cap B) = \frac{P(I \cap A \cap B)}{P(A \cap B)} = \frac{.342}{.354} = .966$$

3.153   a.   If $\dfrac{P(T \mid E)}{P(T^c \mid E)} < 1$, then $P(T^c \mid E) > P(T \mid E)$. Thus, T^c is more likely to occur than T.

This supports the theory that more than 2 bullets were used.

b.   Using Bayes' Rule, we know:

$$P(T \mid E) = \frac{P(T)P(E \mid T)}{P(T)P(E \mid T) + P(T^c)P(E \mid T^c)} \text{ and } P(T^c \mid E) = \frac{P(T^c)P(E \mid T^c)}{P(T^c)P(E \mid T^c) + P(T)P(E \mid T)}$$

If we take the ratio of these 2 equations, we get:

$$\frac{P(T \mid E)}{P(T^c \mid E)} = \frac{\dfrac{P(T)P(E \mid T)}{P(T)P(E \mid T) + P(T^c)P(E \mid T^c)}}{\dfrac{P(T^c)P(E \mid T^c)}{P(T^c)P(E \mid T^c) + P(T)P(E \mid T)}} = \frac{P(T)P(E \mid T)}{P(T^c)P(E \mid T^c)}$$

3.155   a. $A \cup B$

b. B^c

c. $A \cap B$

d. $A^c \mid B$

3.157   a.   Since events *A* and *B* are mutually exclusive, $P(A \cap B) = 0$.

$$P(A \mid B) = \frac{P(A \cap B)}{P(B)} = \frac{0}{.3} = 0$$

b. If Events A and B are independent, then $P(A|B) = P(A)$. From part **a**, we know $P(A|B) = 0$. However, we also know $P(A) = .2$. Thus, events A and B are not independent.

3.159   We know $P(A|B) = \dfrac{P(A \cap B)}{P(B)}$. Thus, $P(B) = \dfrac{P(A \cap B)}{P(A|B)} = \dfrac{.4}{.8} = .5$.

3.161   a.

(1, H)	(2, 1)	(3, H)	(4, 1)	(5, H)	(6, 1)
(1, T)	(2, 2)	(3, T)	(4, 2)	(5, T)	(6, 2)
	(2, 3)		(4, 3)		(6, 3)
	(2, 4)		(4, 4)		(6, 4)
	(2, 5)		(4, 5)		(6, 5)
	(2, 6)		(4, 6)		(6, 6)

b.   Each simple event is an intersection of two independent events. Each simple event whose first element is 1, 3, or 5 has probability

$$\left(\frac{1}{6}\right)\left(\frac{1}{2}\right) = \frac{1}{12}$$

while each simple event whose first element is 2, 4, or 6 has probability

$$\left(\frac{1}{6}\right)\left(\frac{1}{6}\right) = \frac{1}{36}$$

c. $P(A) = P\{(1,H),(3,H),(5,H)\} = \dfrac{1}{12} + \dfrac{1}{12} + \dfrac{1}{12} = \dfrac{3}{12} = \dfrac{1}{4}$

$P(B) = P\{(1,H),(1,T),(3,H),(3,T),(5,H),(5,T)\} = \dfrac{6}{12} = \dfrac{1}{2}$

d. A^c: {all except (1, H), (3, H), and (5, H)}
B^c: {(2, 1), (2, 2), (2, 3), (2, 4), (2, 5), (2, 6), (4, 1), (4, 2), (4, 3), (4, 4), (4, 5), (4, 6), (6, 1), (6, 2), (6, 3), (6, 4), (6, 5), (6, 6)}
$A \cap B$: {(1, H), (3, H), (5, H)}
$A \cup B$: {(1, H), (1, T), (3, H), (3, T), (5, H), (5, T)}

e. $P(A^c) = 1 - P(A) = 1 - \dfrac{1}{4} = \dfrac{3}{4}$

$P(B^c) = 1 - P(B) = 1 - \dfrac{1}{2} = \dfrac{1}{2}$

$P(A \cap B) = P(A) = \dfrac{1}{4}$

$P(A \cup B) = P(A) + P(B) - P(A \cap B) = \dfrac{1}{4} + \dfrac{1}{2} - \dfrac{1}{4} = \dfrac{1}{2}$

$P(A \mid B) = \dfrac{P(A \cap B)}{P(B)} = \dfrac{1/4}{1/2} = \dfrac{1}{2}$

$P(B \mid A) = \dfrac{P(A \cap B)}{P(A)} = \dfrac{1/4}{1/4} = 1$

f. $P(A \cap B) \neq 0$, so that A and B are not mutually exclusive.

$P(A \mid B) \neq P(A)$, so that A and B are not independent.

3.163 a. Because events A and B are independent, we have:

$P(A \cap B) = P(A)P(B) = .3(.1) = .03$

Thus, $P(A \cap B) \neq 0$, and the two events cannot be mutually exclusive.

b. $P(A \mid B) = \dfrac{P(A \cap B)}{P(B)} = \dfrac{.03}{.1} = .3$ $P(B \mid A) = \dfrac{P(A \cap B)}{P(A)} = \dfrac{.03}{.3} = .1$

c. $P(A \cup B) = P(A) + P(B) - P(A \cap B) = .3 + .1 - .03 = .37$

3.165 Use the relative frequency as an estimate for the probability. Thus, P(physically assaulted) = 600/12,000 = .05.

3.167 From the problem, to find probabilities, we must convert the percents to proportions by dividing by 100. Thus, P(1 star) = 0, P(2 stars) = .0408, P(3 stars) = .1735, P(4 stars) = .6020, and P(5 stars) = .1837.

a. False. The probability of an event cannot be 4. The probability of an event must be between 0 and 1.

b. True. P(4 or 5 stars) = P(4 stars) + P(5 stars) = .6020 + .1837 = .7857.

c. True. No cars have a rating of 1 star, thus P(1 star) = 0.

d. False. P(2 stars) = .0408 and P(5 stars) = .1837. Since .0408 us smaller than .1837, the car has a better chance of having a 5-star rating than a 2-star rating.

3.169   Suppose we define the following event:

A = {eighth-grader scores above 655 on mathematics assessment test}

Then the probability that a randomly selected eighth-grader has a score of 655 or below on the mathematics assessment test is:

$P(A^c) = 1 - P(A) = 1 - .05 = .95$.

3.171   a.   Define the following event:

A: {Beech tree is damaged by fungi}

$P(A) = \dfrac{49}{188} = .261$

b.   There would be 3 sample points for this experiment: trunk, leaves, and branch. We can convert the percentages to proportions and use these to estimate the probabilities.

Sample Points	Probabilities
Trunk	.85
Leaves	.10
Branch	.05
Total	**1.00**

3.173   a.   The sample space is the listing of all possible nearshore bar conditions. The sample space is {Single, shore parallel; Other; Planar}.

b.   Define the following events:

SP: {Single, shore parallel}
P: {Planar}
O: {Other}

Assuming that all the sample points are equally likely, $P(SP) = 2/6 = .333$.
$P(O) = 2/6 = .333$. $P(P) = 2/6 = .333$

c. $P(P \cup SP) = 4/6 = .667$

d.   The sample space is the listing of all possible beach conditions. The sample space is {No dunes/flat; Bluff/scarp; Single dune; Not observed}.

e.   Define the following events:

N: {No dunes/flat}
B: {Bluff/scarp}
D: {Single dune}
NO: {Not observed}

Assuming that all the sample points are equally likely, $P(N) = 2/6 = .333$.
$P(B) = 1/6 = .167$. $P(D) = 2/6 = .333$. $P(NO) = 1/6 = .167$.

f. $P(N^C) = 1 - P(NC) = 1 - .333 = .667$.

3.175 a. The event $A \cap B$ is the event the outcome is black and odd. The event is $A \cap B$: {11, 13, 15, 17, 29, 31, 33, 35}

b. The event $A \cup B$ is the event the outcome is black or odd or both. The event $A \cup B$ is {2, 4, 6, 8, 10, 11, 13, 15, 17, 20, 22, 24, 26, 28, 29, 31, 33, 35, 1, 3, 5, 7, 9, 19, 21, 23, 25, 27}

c. Assuming all events are equally likely, each has a probability of 1/38.

$$P(A) = 18\left(\frac{1}{38}\right) = \frac{18}{38} = \frac{9}{19}$$

$$P(B) = 18\left(\frac{1}{38}\right) = \frac{18}{38} = \frac{9}{19}$$

$$P(A \cap B) = 8\left(\frac{1}{38}\right) = \frac{8}{38} = \frac{4}{19}$$

$$P(A \cup B) = 28\left(\frac{1}{38}\right) = \frac{28}{38} = \frac{14}{19}$$

$$P(C) = 18\left(\frac{1}{38}\right) = \frac{18}{38} = \frac{9}{19}$$

d. The event $A \cap B \cap C$ is the event the outcome is odd and black and low. The event $A \cap B \cap C$ is {11, 13, 15, 17}.

e. $P(A \cup B) = P(A) + P(B) - P(A \cap B) = \frac{9}{19} + \frac{9}{19} - \frac{4}{19} = \frac{14}{19}$. Events A and B are not mutually exclusive because $P(A \cap B) \neq 0$.

f. $P(A \cap B \cap C) = 4\left(\frac{1}{38}\right) = \frac{4}{38} = \frac{2}{19}$

g. The event $A \cup B \cup C$ is the event the outcome is odd or black or low. The event $A \cup B \cup C$ is:

{1, 2, 3, ... , 29, 31, 33, 35}
or
{All simple events except 00, 0, 30, 32, 34, 36}

h. $P(A \cup B \cup C) = 32\left(\frac{1}{38}\right) = \frac{32}{38} = \frac{16}{19}$

3.177   a. $P(E_1 | error) = \dfrac{P(E_1 \cap error)}{P(error)}$

$$= \dfrac{P(error | E_1)P(E_1)}{P(error | E_1)P(E_1) + P(error | E_2)P(E_2) + P(error | E_3)P(E_3)}$$

$$= \dfrac{.01(.30)}{.01(.30) + .03(.20) + .02(.50)} = \dfrac{.003}{.003 + .006 + .01} = \dfrac{.003}{.019} = .158$$

b. $P(E_2 | error) = \dfrac{P(E_2 \cap error)}{P(error)}$

$$= \dfrac{P(error | E_2)P(E_2)}{P(error | E_1)P(E_1) + P(error | E_2)P(E_2) + P(error | E_3)P(E_3)}$$

$$= \dfrac{.03(.20)}{.01(.30) + .03(.20) + .02(.50)} = \dfrac{.006}{.003 + .006 + .01} = \dfrac{.006}{.019} = .316$$

c. $P(E_3 | error) = \dfrac{P(E_3 \cap error)}{P(error)}$

$$= \dfrac{P(error | E_3)P(E_3)}{P(error | E_1)P(E_1) + P(error | E_2)P(E_2) + P(error | E_3)P(E_3)}$$

$$= \dfrac{.02(.50)}{.01(.30) + .03(.20) + .02(.50)} = \dfrac{.01}{.003 + .006 + .01} = \dfrac{.01}{.019} = .526$$

d.   If there was a serious error, the probability that the error was made by engineer 3 is .526. This probability is higher than for any of the other engineers. Thus engineer #3 is most likely responsible for the error.

3.179   a.   Define the following event:

F_i: {Player makes a foul shot on i^{th} attempt} $P(F_i) = .8$

$$P(F_i^c) = 1 - P(F_i) = 1 - .8 = .2$$

The event "the player scores on both shots" is $F_1 \cap F_2$. If the throws are independent, then:

$$P(F_1 \cap F_2) = P(F_1)P(F_2) = .8(.8) = .64$$

The event "the player scores on exactly one shot" is $(F_1 \cap F_2^c) \cup (F_1^c \cap F_2)$

$$P\left[(F_1 \cap F_2^c) \cup (F_1^c \cap F_2)\right] = P(F_1 \cap F_2^c) + P(F_1^c \cap F_2)$$

$$= P(F_1)P(F_2^c) + P(F_1^c)P(F_2)$$

$$= .8(.2) + .2(.8) = .16 + .16 = .32$$

The event "the player scores on neither shot" is $(F_1^c \cap F_2^c)$.

$$P(F_1^c \cap F_2^c) = P(F_1^c)P(F_2^c) = .2(.2) = .04$$

b.   We know $P(F_1)=.8$, $P(F_2\,|\,F_1)=.9$, and $P(F_2\,|\,F_1^c)=.7$

The probability the player scores on both shots is:

$$P(F_1\cap F_2)=P(F_2\,|\,F_1)P(F_1)=.9(.8)=.72$$

The probability the player scores on exactly one shot is:

$$P(F_1\cap F_2^c)+P(F_1^c\cap F_2)=P(F_2^c\,|\,F_1)P(F_1)+P(F_2\,|\,F_1^c)P(F_1^c)$$
$$=\left[1-P(F_2\,|\,F_1)\right]P(F_1)+P(F_2\,|\,F_1^c)P(F_1^c)$$
$$=(1-.9)(.8)+.7(.2)=.08+.14=.22$$

The probability the player scores on neither shot is:

$$P(F_1^c\cap F_2^c)=P(F_2^c\,|\,F_1^c)P(F_1^c)=\left[1-P(F_2\,|\,F_1^c)\right]P(F_1^c)=(1-.7)(.2)=.06$$

c.   Two consecutive foul shots are probably dependent. The outcome of the second shot probably depends on the outcome of the first.

3.181   Define the following events:

A: {Seed carries single spikelets}
B: {Seed carries paired spikelets}
C: {Seed produces ears with single spikelets}
D: {Seed produces ears with paired spikelets}

From the problem, $P(A)=.4$, $P(B)=.6$, $P(C\,|\,A)=.29$, $P(D\,|\,A)=.71$, $P(C\,|\,B)=.26$, and $P(D\,|\,B)=.74$.

a. $P(A\cap C)=P(C\,|\,A)P(A)=.29(.4)=.116$

b. $P(D)=P(A\cap D)+P(B\cap D)=P(D\,|\,A)P(A)+P(D\,|\,B)P(B)=.71(.4)+.74(.6)$
$$=.284+.444=.728$$

3.183   Define the following events:

A_1: {Component 1 in System A works properly}
A_2: {Component 2 in System A works properly}
A_3: {Component 3 in System A works properly}
B_1: {Component 1 in System B works properly}
B_2: {Component 2 in System B works properly}
B_3: {Component 3 in System B works properly}
B_4: {Component 4 in System B works properly}
A: {System A works properly}
B: {System B works properly}
C: {Subsystem C works properly}
D: {Subsystem D works properly}

$P(A_1)=1-P(A_1^c)=1-.12=.88$ $P(B_1)=1-P(B_1^c)=1-.1=.9$
$P(A_2)=1-P(A_2^c)=1-.09=.91$ $P(B_2)=1-P(B_2^c)=1-.1=.9$

$$P(A_3)=1-P(A_3^c)=1-.11=.89 \qquad\qquad P(B_3)=1-P(B_3^c)=1-.1=.9$$
$$P(B_4)=1-P(B_4^c)=1-.1=.9$$

a. $P(A)=P(A_1\cap A_2\cap A_3)=P(A_1)P(A_2)P(A_3)=.88(.91)(.89)=.7127$
  (since the three components operate independently)

b. $P(A^c)=1-P(A)=1-.7127-.2873$

c. Subsystem C works properly if both components B_1 and B_2 work properly.

$$P(C)=P(B_1\cap B_2)=P(B_1)P(B_2)=.9(.9)=.81$$
  (since the 2 components operate independently)

Similarly, subsystem D works properly if both components B_3 and B_4 work properly.

$$P(D)=P(B_3\cap B_4)=P(B_3)P(B_4)=.9(.9)=.81$$

System B operates properly if either subsystem C or subsystem D operates properly.

$$P(B)=P(C\cup D)=P(C)+P(D)-P(C\cap D)=P(C)+P(D)-P(C)P(D)$$
$$=.81+.81-.81(.81)=.9639$$

d. The probability that exactly one subsystem in System B fails is:

$$P(C\cap D^c)+P(C^c\cap D)=P(C)P(D^c)+P(C^c)P(D)$$
$$=.81(1-.81)+(1-.81)(.81)=.1539+.1539=.3078$$

e. System B fails to operate only if both subsystems C and D fail.

$$P(B^c)=P(C^c\cap D^c)=P(C^c)P(D^c)=(1-.81)(1-.81)=.0361$$

f. If System B operates correctly 99% of the time, then it fails 1% of the time. The probability that one of the subsystems fails is 1 - .81 = .19. The probability that n subsystems fail is $.19^n$. Thus, we must find n such that

$$.19^n\le.01 \Rightarrow n\ge3$$

3.185   Define the following event:

D: {Chip is defective}

From the Exercise, $P(S_1)=.15$, $P(S_2)=.05$, $P(S_3)=.10$, $P(S_4)=.20$, $P(S_5)=.12$, $P(S_6)=.20$, and $P(S_7)=.18$. Also, $P(D|S_1)=.001$, $P(D|S_2)=.0003$, $P(D|S_3)=.0007$, $P(D|S_4)=.006$, $P(D|S_5)=.0002$, $P(D|S_6)=.0002$, and $P(D|S_7)=.001$.

a. We must find the probability of each supplier given a defective chip.

$$P(S_1 | D) = \frac{P(S_1 \cap D)}{P(D)} =$$

$$\frac{P(D | S_1)P(S_1)}{P(D | S_1)P(S_1) + P(D | S_2)P(S_2) + P(D | S_3)P(S_3) + P(D | S_4)P(S_4) + P(D | S_5)P(S_5) + P(D | S_6)P(S_6) + P(D | S_7)P(S_7)}$$

$$= \frac{.001(.15)}{.001(.15) + .0003(.05) + .0007(.10) + .006(.20) + .0002(.12) + .0002(.02) + .001(.18)}$$

$$= \frac{.00015}{.00015 + .000015 + .00007 + .0012 + .000024 + .00004 + .00018} = \frac{.00015}{.001679} = .0893$$

$$P(S_2 | D) = \frac{P(S_2 \cap D)}{P(D)} = \frac{P(D | S_2)P(S_2)}{P(D)} = \frac{.0003(.05)}{.001679} = \frac{.000015}{.001679} = .0089$$

$$P(S_3 | D) = \frac{P(S_3 \cap D)}{P(D)} = \frac{P(D | S_3)P(S_3)}{P(D)} = \frac{.0007(.10)}{.001679} = \frac{.00007}{.001679} = .0417$$

$$P(S_4 | D) = \frac{P(S_4 \cap D)}{P(D)} = \frac{P(D | S_4)P(S_4)}{P(D)} = \frac{.006(.20)}{.001679} = \frac{.0012}{.001679} = .7147$$

$$P(S_5 | D) = \frac{P(S_5 \cap D)}{P(D)} = \frac{P(D | S_5)P(S_5)}{P(D)} = \frac{.0002(.12)}{.001679} = \frac{.000024}{.001679} = .0143$$

$$P(S_6 | D) = \frac{P(S_6 \cap D)}{P(D)} = \frac{P(D | S_6)P(S_6)}{P(D)} = \frac{.0002(.20)}{.001679} = \frac{.00004}{.001679} = .0238$$

$$P(S_7 | D) = \frac{P(S_7 \cap D)}{P(D)} = \frac{P(D | S_7)P(S_7)}{P(D)} = \frac{.001(.18)}{.001679} = \frac{.00018}{.001679} = .1072$$

Of these probabilities, .7147 is the largest. This implies that if a failure is observed, supplier number 4 was most likely responsible.

b. If the seven suppliers all produce defective chips at the same rate of .0005, then $P(D|S_i)$ =.0005 for all i = 1, 2, 3, ... 7 and $P(D)$ = .0005.

For any supplier i, $P(S_i \cap D) = P(D | S_i)P(S_i) = .0005P(S_i)$ and

$$P(S_i | D) = \frac{P(S_i \cap D)}{P(D)} = \frac{P(D | S_i)P(S_i)}{.0005} = \frac{.0005P(S_i)}{.0005} = P(S_i)$$

Thus, if a defective is observed, then it most likely came from the supplier with the largest proportion of sales (probability). In this case, the most likely supplier would be either supplier 4 or supplier 6. Both of these have probabilities of .20.

3.187   Define the following events:

A: {Man never smoked cigars}

B: {Man formerly smoked cigars}

C: {Man currently smokes cigars}

D: {Man died from Cancer}

a. $P(A \cap D) = \dfrac{782}{137,243} = .006$

b. $P(B \cap D) = \dfrac{91}{137,243} = .0007$

c. $P(C \cap D) = \dfrac{141}{137,243} = .001$

d. $P(D|C) = \dfrac{P(D \cap C)}{P(C)} = \dfrac{141/137,243}{7,866/137,243} = \dfrac{141}{7,866} = .018$

e. $P(D|A) = \dfrac{P(D \cap A)}{P(A)} = \dfrac{782/137,243}{121,529/137,243} = \dfrac{782}{121,529} = .006$

3.189   a.   The odds in favor of Smarty Jones are $\dfrac{1}{3}$ to $\left(1 - \dfrac{1}{3}\right)$ or $\dfrac{1}{3}$ to $\dfrac{2}{3}$ or 1 to 2

b.   If the odds are 1 to 1, $P(\text{Smarty Jones will win}) = \dfrac{1}{1+1} = \dfrac{1}{2} = .5$

c.   If the odds against Smarty Jones winning are 3 to 2, the odds for Smarty Jones winning are 2 to 3. The probability Smarty Jones will win is $\dfrac{2}{2+3} = \dfrac{2}{5} = .4$

3.191   Define the following event:

A: {Antigens match}

From the problem, $P(A) = .25$ and $P(A^c) = 1 - P(A) = 1 - .25 = .75$

a.   The probability that one sibling has a match is $P(A) = .25$.

b. The probability that all three will match is:

$$P(AAA) = P(A \cap A \cap A) = P(A)P(A)P(A) = .25^3 = .0156$$

c. The probability that none of the three match is:

$$P(A^c A^c A^c) = P(A^c \cap A^c \cap A^c) = P(A^c)P(A^c)P(A^c) = .75^3 = .4219$$

d. For this part, $P(A) = .001$ and $P(A^c) = 1 - P(A) = 1 - .001 = .999$

$$P(A) = .001$$

$$P(AAA) = P(A \cap A \cap A) = P(A)P(A)P(A) = .001^3 = .000000001$$

$$P(A^c A^c A^c) = P(A^c \cap A^c \cap A^c) = P(A^c)P(A^c)P(A^c) = .999^3 = .9970$$

3.193 We will work with the following events:

    A: {Woman is pregnant}
    B: {Pregnancy test is positive}

The given probabilities can be written as:

$$P(A) = .75, \quad P(B \mid A) = .99, \quad P(B \mid A^c) = .02$$

$$P(A \mid B) = \frac{P(A \cap B)}{P(B)} = \frac{P(B \mid A)P(A)}{P(A \cap B) + P(A^c \cap B)}$$

Then,

$$= \frac{P(B \mid A)P(A)}{P(B \mid A)P(A) + P(B \mid A^c)P(A^c)}$$

$$= \frac{(.99)(.75)}{(.99)(.75) + (.02)(.25)} = \frac{.7425}{.7475} \approx .993$$

3.195 We first determine the number of simple events for this experiment. According to the partitions rule, the number of ways of dealing 52 cards so that each of the four players receives 13 cards is:

$$\frac{52!}{13!13!13!13!} = 5.3644738 \times 10^{28}$$

If the cards are well shuffled, then the probability of each simple event is:

$$\frac{1}{5.3644738 \times 10^{-28}}$$

We now need to count the number of simple events that result in one player receiving all the diamonds, another all the hearts, another all the spades, and another all the clubs. To do this, first consider a particular ***ordering*** that yields the desired result:

    Player #1 receives 13 diamonds
    Player #2 receives 13 hearts

Player #3 receives 13 spades
Player #4 receives 13 clubs

This particular ***ordered*** event can happen in

$$\binom{13}{13}\binom{13}{13}\binom{13}{13}\binom{13}{13}=1$$

way. However, the ***ordering*** of the players is not relevant to us.

Since there are $P_4^4 = \dfrac{4!}{(4-4)!} = 24$ possible orderings of the players, we conclude that the

probability of the rare event observed in Dubuque is:

$$\frac{24}{5.3644738\times10^{28}} \approx 4.4739\text{x}10^{-28}$$

3.197   First, we will list all possible sample points for placing a car (C) and 2 goats (G) behind doors
#1, #2, and #3. If the first position corresponds to door #1, the second position corresponds
to door #2, and the third position corresponds to door #3, the sample space is:

(C G G)   (G C G)   (G G C)

Now, suppose you pick door #1. Initially, the probability that you will win the car is 1/3 –
only one of the sample points has a car behind door #1.

The host will now open a door behind which is a goat. If you pick door #1 in the first sample
point (C G G), the host will open either door #2 or door #3. Suppose he opens door #3 (it
really does not matter). If you pick door #1 in the second sample point (G C G), the host will
open door #3. If you pick door #1 in the third sample point (G G C), the host will open door
#2. Now, the new sample space will be:

(C G)   (G C)   (G C)

where the first position corresponds to door #1 (the one you chose) and the second position
corresponds to the door that was not opened by the host.

Now, if you keep door #1, the probability that you win the car is 1/3. However, if you switch
to the remaining door, the probability that you win the car is now 2/3. Based on these
probabilities, it is to your advantage to switch doors.

The above could be repeated by selecting door #2 initially or door #3 initially. In either of
these cases, again, the probability of winning the car is 1/3 if you do not switch and 2/3 if you
switch. Thus, Marilyn was correct.

3.199   If a coin is balanced, then $P(H) = P(T) = .5$. To find the probability of any sequence, you find
the probability of the intersection of the simple events. Thus,

$$P(HHHHHHHHHH) = P(H \cap H \cap H \cap H \cap H \cap H \cap H \cap H \cap H \cap H)$$
$$= P(H)P(H)P(H)P(H)P(H)P(H)P(H)P(H)P(H)P(H)$$
$$= .5(.5)(.5)(.5)(.5)(.5)(.5)(.5)(.5) = .00098$$

Similarly,
$$P(HHTTHTTHHH) = P(H \cap H \cap T \cap T \cap H \cap T \cap T \cap H \cap H \cap H)$$
$$= P(H)P(H)P(T)P(T)P(H)P(T)P(T)P(H)P(H)P(H)$$
$$= .5(.5)(.5)(.5)(.5)(.5)(.5)(.5)(.5) = .00098$$

Similarly,
$$P(TTTTTTTTTT) = P(T \cap T \cap T \cap T \cap T \cap T \cap T \cap T \cap T \cap T)$$
$$= P(T)P(T)P(T)P(T)P(T)P(T)P(T)P(T)P(T)$$
$$= .5(.5)(.5)(.5)(.5)(.5)(.5)(.5)(.5) = .00098$$

Thus, any of these 3 sequences are equally likely. In fact, any specific sequence has the same probability.

Now, suppose we compute the probability of getting all heads or all tails:

P(All heads or all tails) = P(All heads $\cup$ All tails)

$= P$(All heads) $+ P$(All tails) $- P$(All heads $\cup$ All tails) $= .00098 + .00098 - 0 = .00196$

Consequently, the probability of getting a mix of heads and tails is:

P(10 coin tosses result in a mix of heads and tails) = 1 – P(All heads or all tails)

$= 1 - .00196 = .99804$

So, even though any particular sequence has a probability of .00098, we just found out the probability of getting all heads or all tails is only .00196 and the probability of getting a mix of heads and tails is .99804. If we know that one of these sequences actually occurred, then we would conclude that it was probably the one with the mix of heads and tails because the probability of a mix is close to 1.

Discrete Random Variables

4.1   A random variable is a rule that assigns one and only one value to each simple event of an experiment.

4.3   a.   Since we can count the number of words spelled correctly, the random variable is discrete.

        b.   Since we can assume values in an interval, the amount of water flowing through the Hoover Dam in a day is continuous.

        c.   Since time is measured on an interval, the random variable is continuous.

        d.   Since we can count the number of bacteria per cubic centimeter in drinking water, this random variable is discrete.

        e.   Since we cannot count the amount of carbon monoxide produced per gallon of unleaded gas, this random variable is continuous.

        f.   Since weight is measured on an interval, weight is a continuous random variable.

4.5   a.   The reaction time difference is continuous because it lies within an interval.

        b.   Since we can count the number of violent crimes, this random variable is discrete.

        c.   Since we can count the number of near misses in a month, this variable is discrete.

        d.   Since we can count the number of winners each week, this variable is discrete.

        e.   Since we can count the number of free throws made per game by a basketball team, this variable is discrete.

        f.   Since distance traveled by a school bus lies in some interval, this is a continuous random variable.

4.7   The values x can assume are 0, 1, 2, 3, 4, …. Thus, x is a discrete random variable.

4.9   Annual rainfall can take on any value in an interval. Thus, annual rainfall is a continuous random variable. The number of ant species can assume only a countable number of outcomes. Thus, number of ant species is a discrete random variable.

4.11   Answers will vary. An example of a discrete random variable of interest to a psychologist might be the number of times a person has violent dreams in a week.

4.13   Answers will vary. An example of a discrete random variable of interest to a hospital nurse might be the number of medications a patient receives in 24 hours.

4.15   The probability distribution of a discrete random variable can be represented by a table, graph, or formula that specifies the probability associated with each possible value the random variable can assume.

4.17   a. $p(22) = .25$

b. $p(20) + p(24) = .15 + .20 = .35$

c. $P(x \le 23) = p(20) + p(21) + p(22) + p(23) = .15 + .10 + .25 + .30 = .80$

4.19   a. $P(x \le 12) = P(x = 10) + P(x = 11) + P(x = 12) = .2 + .3 + .2 = .7$

b. $P(x > 12) = 1 - P(x \le 12) = 1 - .7 = .3$

c. $P(x \le 14) = P(x = 10) + P(x = 11) + P(x = 12) + P(x = 13) + P(x = 14)$
$= .2 + .3 + .2 + .1 + .2 = 1$

d. $P(x = 14) = .2$

e. $P(x \le 11 \text{ or } x > 12) = P(x \le 11) + P(x > 12)$
$= P(x = 10) + P(x = 11) + P(x = 13) + P(x = 14)$
$= .2 + .3 + .1 + .2 = .8$

4.21   a.   The simple events are (where H = head, T = tail):

HHH HHT HTH THH HTT THT TTH TTT
x = # heads   3   2   2   2   1   1   1   0

b.   If each event is equally likely, then P(simple event) $= \dfrac{1}{n} = \dfrac{1}{8}$

$p(3) = \dfrac{1}{8}$, $p(2) = \dfrac{1}{8} + \dfrac{1}{8} + \dfrac{1}{8} = \dfrac{3}{8}$, $p(1) = \dfrac{1}{8} + \dfrac{1}{8} + \dfrac{1}{8} = \dfrac{3}{8}$, and $p(0) = \dfrac{1}{8}$

c.

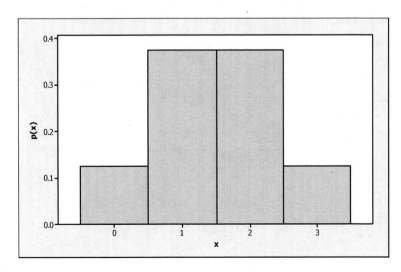

d. $P(x = 2 \text{ or } x = 3) = p(2) + p(3) = \dfrac{3}{8} + \dfrac{1}{8} = \dfrac{4}{8} = \dfrac{1}{2}$

4.23   a. $p(0) + p(1) + p(2) + p(3) + p(4) = .09 + .30 + .37 + .20 + .04 = 1.00$

b. $P(x = 3 \text{ or } 4) = p(3) + p(4) = .20 + .04 = .24$

c. $P(x < 2) = p(0) + p(1) = .09 + .30 = .39$

4.25   a.   There are 4 sample points: Boy-Boy, Boy-Girl, Girl-Boy, and Girl-Girl.

b.   If all the sample points are equally likely, then each would have a probability of ¼.

c. $P(x = 0) = p(0) = \dfrac{1}{4},\ P(x = 1) = p(1) = \dfrac{1}{4} + \dfrac{1}{4} = \dfrac{2}{4} = \dfrac{1}{2},\ P(x = 2) = p(2) = \dfrac{1}{4}$

d.   Using the proportions in the table as approximate probabilities, the distribution of x is:
$P(x = 0) = p(0) = .222,\ P(x = 1) = p(1) = .259 + .254 = .513,\ P(x = 2) = p(2) = .265$

4.27   a. $p(1) = (.23)(.77)^{1-1} = .23(.77)^0 = .23$.   The probability that a contaminated cartridge is selected on the first sample is .23.

b. $p(5) = (.23)(.77)^{5-1} = .23(.77)^4 = .081$. The probability that the first contaminated cartridge will be selected on the 5th sample is .081.

c. $P(x \geq 2) = 1 - P(x \leq 1) = 1 - p(1) = 1 - .23 = .77$. The probability that the first contaminated cartridge will be selected on the second trial or later is .77.

4.29   a. x is the sum of the grill numbers. To find the probability distribution for x, we divide the number of students choosing the particular grill display combination divided by the total number of students or 124. The table below shows the probability distribution for x.

Grill Display Combination	Number of Students	x	$p(x)$
1-2-3	35	6	35/124 = .282
1-2-4	8	7	8/124 = .065
1-2-5	42	8	42/124 = .339
2-3-4	4	9	4/124 = .032
2-3-5	1	10	1/124 = .008
2-4-5	34	11	34/124 = .274
Total	**124**		**1.00**

b. $P(x > 10) = P(x = 11) = .274$.

4.31   The sample space of the experiment would be:

$$S: \{BBB, BBG, BGB, GBB, GGB, GBG, BGG, GGG\}$$

where B and G represent a boy and girl respectively. Since the female parent always donates an X chromosome, the gender of any child is determined by the chromosome donated by the father; an X chromosome donated by the father will produce a girl, a Y, a boy. Each child therefore has a .5 probability of being either gender (as might be expected). From this, it follows that each of the above simple events has a probability of 1/8. If we let z represent the number of male offspring, the probability distribution of z is:

z	0	1	2	3
$p(z)$	1/8	3/8	3/8	1/8

Then, $P(\text{At least one boy}) = P(z \geq 1) = 1 - P(z = 0) = 1 - 1/8 = 7/8$

4.33   a.   Each point in the system can have one of 2 status levels, "free" or "obstacle". Define the following events:

A_F: {Point A is free} A_O: {Point A is obstacle}
B_F: {Point B is free} B_O: {Point B is obstacle}
C_F: {Point C is free} C_O: {Point C is obstacle}

Thus, the sample points for the space are:

$A_F B_F C_F$, $A_F B_F C_O$, $A_F B_O C_F$, $A_F B_O C_O$, $A_O B_F C_F$, $A_O B_F C_O$, $A_O B_O C_F$, $A_O B_O C_O$

The variable x is the total number of links that are "free". Thus, x can have values 0, 1, or 2.

b.   Since it is stated that the probability of any point in the system having a "free" status is .5, the probability of any point having an "obstacle" status is also .5, Thus, the probability of each of the sample points above is $P(A_i B_i C_i) = .5(.5)(.5) = .125$.

The values of x, the number of free links in the system, for each sample point are listed below. A link is free if both the points are free. Thus, a link from A to B is free if A is free and B is free. A link from B to C is free if B is free and C is free.

Sample point	x	Probability
$A_F B_F C_F$	2	.125
$A_F B_F C_O$	1	.125
$A_F B_O C_F$	0	.125
$A_F B_O C_O$	0	.125
$A_O B_F C_F$	1	.125
$A_O B_F C_O$	0	.125
$A_O B_O C_F$	0	.125
$A_O B_O C_O$	0	.125

The probability distribution for x is:

x	Probability
0	.625
1	.250
2	.125

4.35   The expected value of a random variable, $E(x)$, does not always equal a specific value of the random variable, x. The $E(x)$ is the mean of the probability distribution and does not have to equal a specific value of x.

4.37   a. $\mu = E(x) = \sum xp(x) = 1(.2) + 2(.4) + 4(.2) + 10(.2) = .2 + .8 + .8 + 2 = 3.8$

      b. $\sigma^2 = E[(x-\mu)^2] = \sum (x-\mu)^2 p(x)$
$$= (1-3.8)^2(.2) + (2-3.8)^2(.4) + (4-3.8)^2(.2) + (10-3.8)^2(.2)$$
$$= 1.568 + 1.296 + .008 + 7.688 = 10.56$$

      c. $\sigma = \sqrt{10.59} = 3.2496$

      d.   The average value of x over many trials is 3.8.

      e.   No. The random variable can only take on values 1, 2, 4, or 10.

      f.   Yes. It is possible that μ can be equal to an actual value of x.

4.39   a.   It would seem that the mean of both would be 1 since they both are symmetric distributions centered at 1.

      b.   The distribution of x seems more variable since there appears to be greater probability for the two extreme values of 0 and 2 than there is in the distribution of y.

      c.   For x: $\mu = E(x) = \sum xp(x) = 0(.3) + 1(.4) + 2(.3) = 0 + .4 + .6 = 1$
$$\sigma^2 = E[(x-\mu)^2] = \sum (x-\mu)^2 p(x)$$
$$= (0-1)^2(.3) + (1-1)^2(.4) + (2-1)^2(.3) = .3 + 0 + .3 = .6$$

            For y: $\mu = E(y) = \sum yp(y) = 0(.1) + 1(.8) + 2(.1) = 0 + .8 + .2 = 1$
$$\sigma^2 = E[(y-\mu)^2] = \sum (y-\mu)^2 p(y)$$
$$= (0-1)^2(.1) + (1-1)^2(.8) + (2-1)^2(.1) = .1 + 0 + .1 = .2$$

The variance for x is larger than that for y.

4.41   From Exercise 4.25, we found: $p(0) = .222, \ p(1) = .513, \ p(2) = .265$.

$$E(x) = \sum xp(x) = 0(.222) + 1(.513) + 2(.265) = 0 + .513 + .530 = 1.043$$

The average number of boys in two-children families is 1.043.

4.43   a. $E(x) = \sum xp(x) = 0(.09) + 1(.30) + 2(.37) + 3(.20) + 4(.04) = 0 + .30 + .74 + .60 + .16 = 1.8$

In a random sample of 4 homes, the average number of homes with high dust mite levels is 1.8.

b. $\sigma^2 = E[(x - \mu)^2] = \sum (x - \mu)^2 p(x)$

$$= (0 - 1.8)^2 (.09) + (1 - 1.8)^2 (.30) + (2 - 1.8)^2 (.37) + (3 - 1.8)^2 (.20) + (4 - 1.8)^2 (.04)$$

$$= .2916 + .1920 + .0148 + .2880 + .1936 = .98$$

$$\sigma = \sqrt{.98} = .9899$$

c. $\mu \pm 2\sigma \Rightarrow 1.8 \pm 2(.9899) \Rightarrow 1.8 \pm 1.9798 \Rightarrow (-.1798, \ 3.7798)$

$$P(-.1798 < x < 3.7798) = p(0) + p(1) + p(2) + p(3) = .09 + .03 + .37 + .20 = .96$$

Chebyshev's Theorem says that the interval $\mu \pm 2\sigma$ will contain at least .75 of the data. The Empirical Rule says that approximately .95 of the data will be contained in the interval. Both Chebyshev's Theorem and the Empirical Rule fit the distribution.

4.45   a.   Define the following events:

M: {Miami Beach, FL}
C: {Coney Island, NY}
S: {Surfside, CA}
B: {Monmouth Beach, NJ}
O: {Ocean City, NJ}
L: {Spring Lake, NJ}

The possible pairs of beach hotspots that can be selected are:

MC, MS, MB, MO, ML, CS, CB, CO, CL, SB, SO, SL, BO, BL, OL

b.   If we are just selecting 2 hotspots from 6, each of the 15 combinations are equally likely. Thus, each would have a probability of 1/15.

c. The value of x for each of the pairs is:

Pair	x
MC	0
MS	0
MB	1
MO	0
ML	1
CS	0
CB	1
CO	0

Pair	x
CL	1
SB	1
SO	0
SL	1
BO	1
BL	2
OL	1

d. The probability distribution for x is:

x	$p(x)$
0	6/15
1	8/15
2	1/15

e. $\mu = E(x) = \sum xp(x) = 0(6/15) + 1(8/15) + 2(1/15) = 10/15 = .667$

The average total number of hotspots with a planar nearshore bar condition in each pair is .667.

4.47   For a \$5 bet, you will either win \$5 or lose \$5 (−\$5). The probability distribution for the net winnings is:

x	$p(x)$
−5	20/38
5	18/38

$\mu = E(x) = \sum xp(x) = -5\left(\dfrac{20}{38}\right) + 5\left(\dfrac{18}{38}\right) = -.263$

Over a large number of trials, the average winning for a \$5 bet on red is -\$0.263.

4.49   Let x = bookie's earnings per dollar wagered. Then x can take on values \$1 (you lose) and \$−5 (you win). The only way you win is if you pick 3 winners in 3 games. If the probability of picking 1 winner in 1 game is .5, then $P(www) = p(w)p(w)p(w) = .5(.5)(.5) = .125$ (assuming games are independent).

Thus, the probability distribution for x is:

x	$p(x)$
\$1	.875
\$−5	.125

$$E(x) = E(x) = \sum xp(x) = 1(.875) - 5(.125) = .875 - .625 = \$.25$$

On the average, the bookie will earn \$.25 for each bet made.

4.51   The formula for $p(x)$ for a binomial random variable with $n = 7$ and $p = .2$ is:

$$p(x) = \binom{n}{x} p^x q^{n-x} = \binom{7}{x} .2^x .8^{7-x} \qquad (x = 0,1,2,...,7)$$

4.53   a. $\dfrac{6!}{2!(6-2)!} = \dfrac{6!}{2!4!} = \dfrac{6 \cdot 5 \cdot 4 \cdot 3 \cdot 2 \cdot 1}{(2 \cdot 1)(4 \cdot 3 \cdot 2 \cdot 1)} = 15$

b. $\binom{5}{2} = \dfrac{5!}{2!(5-2)!} = \dfrac{5!}{2!3!} = \dfrac{5 \cdot 4 \cdot 3 \cdot 2 \cdot 1}{(2 \cdot 1)(3 \cdot 2 \cdot 1)} = 10$

c. $\binom{7}{0} = \dfrac{7!}{0!(7-0)!} = \dfrac{7!}{0!7!} = \dfrac{7 \cdot 6 \cdot 5 \cdot 4 \cdot 3 \cdot 2 \cdot 1}{(1)(7 \cdot 6 \cdot 5 \cdot 4 \cdot 3 \cdot 2 \cdot 1)} = 1$

(Note: 0! = 1)

d. $\binom{6}{6} = \dfrac{6!}{6!(6-6)!} = \dfrac{6!}{6!0!} = \dfrac{6 \cdot 5 \cdot 4 \cdot 3 \cdot 2 \cdot 1}{(6 \cdot 5 \cdot 4 \cdot 3 \cdot 2 \cdot 1)(1)} = 1$

e. $\binom{4}{3} = \dfrac{4!}{3!(4-3)!} = \dfrac{4!}{3!1!} = \dfrac{4 \cdot 3 \cdot 2 \cdot 1}{(3 \cdot 2 \cdot 1)(1)} = 4$

4.55   a. $P(x=1) = \dfrac{5!}{1!4!}(.2)^1(.8)^4 = \dfrac{5 \cdot 4 \cdot 3 \cdot 2 \cdot 1}{(1)(4 \cdot 3 \cdot 2 \cdot 1)}(.2)^1(.8)^4 = 5(.2)^1(.8)^4 = .4096$

b. $P(x=2) = \dfrac{4!}{2!2!}(.6)^2(.4)^2 = \dfrac{4 \cdot 3 \cdot 2 \cdot 1}{(2 \cdot 1)(2 \cdot 1)}(.6)^2(.4)^2 = 6(.6)^2(.4)^2 = .3456$

c. $P(x=0) = \dfrac{3!}{0!3!}(.7)^0(.3)^3 = \dfrac{3 \cdot 2 \cdot 1}{(1)(3 \cdot 2 \cdot 1)}(.7)^0(.3)^3 = 1(.7)^0(.3)^3 = .027$

d. $P(x=3) = \dfrac{5!}{3!2!}(.1)^3(.9)^2 = \dfrac{5 \cdot 4 \cdot 3 \cdot 2 \cdot 1}{(3 \cdot 2 \cdot 1)(2 \cdot 1)}(.1)^3(.9)^2 = 10(.1)^3(.9)^2 = .0081$

e. $P(x=2) = \dfrac{4!}{2!2!}(.4)^2(.6)^2 = \dfrac{4 \cdot 3 \cdot 2 \cdot 1}{(2 \cdot 1)(2 \cdot 1)}(.4)^2(.6)^2 = 6(.4)^2(.6)^2 = .3456$

f. $P(x=1) = \dfrac{3!}{1!2!}(.9)^1(.1)^2 = \dfrac{3 \cdot 2 \cdot 1}{(1)(2 \cdot 1)}(.9)^1(.1)^2 = 3(.9)^1(.1)^2 = .027$

4.57   a. $\mu = np = 25(.5) = 12.5$

$\sigma^2 = np(1-p) = 25(.5)(.5) = 6.25$
$\sigma = \sqrt{\sigma^2} = \sqrt{6.25} = 2.5$

b. $\mu = np = 80(.2) = 16$

$\sigma^2 = np(1-p) = 80(.2)(.8) = 12.8$
$\sigma = \sqrt{\sigma^2} = \sqrt{12.8} = 3.578$

c. $\mu = np = 100(.6) = 60$

$\sigma^2 = np(1-p) = 100(.6)(.4) = 24$
$\sigma = \sqrt{\sigma^2} = \sqrt{24} = 4.899$

d. $\mu = np = 70(.9) = 63$

$\sigma^2 = np(1-p) = 70(.9)(.1) = 6.3$
$\sigma = \sqrt{\sigma^2} = \sqrt{6.3} = 2.510$

e. $\mu = np = 60(.8) = 48$

$\sigma^2 = np(1-p) = 60(.8)(.2) = 9.6$
$\sigma = \sqrt{\sigma^2} = \sqrt{9.6} = 3.098$

f. $\mu = np = 1,000(.04) = 40$

$\sigma^2 = np(1-p) = 1,000(.04)(.96) = 38.4$
$\sigma = \sqrt{\sigma^2} = \sqrt{38.4} = 6.197$

4.59   a. $P(x < 10) = P(x \le 9) = 0$

b. $P(x \ge 10) = 1 - P(x \le 9) = 1 - .002 = .998$

c. $P(x = 2) = P(x \le 2) - P(x \le 1) = .206 - .069 = .137$

4.61 x is a binomial random variable with $n = 4$.

a.   If the probability distribution of x is symmetric, $p(0) = p(4)$ and $p(1) = p(3)$.

Since $p(x) = \binom{n}{x} p^x q^{n-x}$ $x = 0, 1, \ldots, n,$

When $n = 4$,

$$\binom{4}{0}p^0 q^4 = \binom{4}{4}p^4 q^0 \implies \frac{4!}{0!4!}p^0 q^4 = \frac{4!}{4!0!}p^4 q^0 \implies q^4 = p^4 \implies p = q$$

Since $p + q = 1$, $p = .5$

Therefore, the probability distribution of x is symmetric when $p = .5$.

b. If the probability distribution of x is skewed to the right, then the mean is greater than the median. Therefore, there are more small values in the distribution (0, 1) than large values (3, 4). Therefore, p must be smaller than .5. Let $p = .2$ and the probability distribution of x will be skewed to the right.

c. If the probability distribution of x is skewed to the left, then the mean is smaller than the median. Therefore, there are more large values in the distribution (3, 4) than small values (0, 1). Therefore, p must be larger than .5. Let $p = .8$ and the probability distribution of x will be skewed to the left.

d. In part **a**, x is a binomial random variable with $n = 4$ and $p = .5$.

$$p(x) = \binom{4}{x}.5^x .5^{4-x} \qquad x = 0, 1, 2, 3, 4$$

$$p(0) = \binom{4}{0}.5^0 .5^4 = \frac{4!}{0!4!}.5^4 = 1(.5)^4 = .0625$$

$$p(1) = \binom{4}{1}.5^1 .5^3 = \frac{4!}{1!3!}.5^4 = 4(.5)^4 = .25$$

$$p(2) = \binom{4}{2}.5^2 .5^2 = \frac{4!}{2!2!}.5^4 = 6(.5)^4 = .375$$

$p(3) = p(1) = .25$ (since the distribution is symmetric)

$p(4) = p(0) = .0625$

The probability distribution of x in tabular form is:

x	0	1	2	3	4
$p(x)$	.0625	.25	.375	.25	.0625

$\mu = np = 4(.5) = 2$

The graph of the probability distribution of x when $n = 4$ and $p = .5$ is as follows.

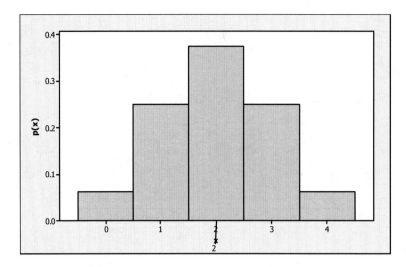

In part **b**, x is a binomial random variable with $n = 4$ and $p = .2$.

$$p(x) = \binom{4}{x}.2^x.8^{4-x} \quad x = 0, 1, 2, 3, 4$$

$$p(0) = \binom{4}{0}.2^0.8^4 = 1(1).8^4 = .4096$$

$$p(1) = \binom{4}{1}.2^1.8^3 = 4(.2)(.8)^3 = .4096$$

$$p(2) = \binom{4}{2}.2^2.8^2 = 6(.2)^2(.8)^2 = .1536$$

$$p(3) = \binom{4}{3}.2^3.8^1 = 4(.2)^3(.8) = .0256$$

$$p(4) = \binom{4}{4}.2^4.8^0 = 1(.2)^4(1) = .0016$$

The probability distribution of x in tabular form is:

x	0	1	2	3	4
$p(x)$	.4096	.4096	.1536	.0256	.0016

$$\mu = np = 4(.2) = .8$$

The graph of the probability distribution of x when $n = 4$ and $p = .2$ is as follows:

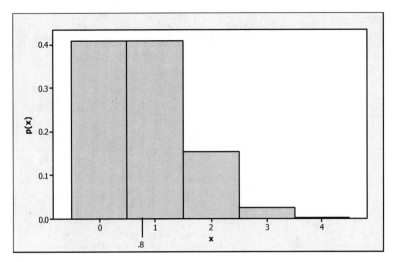

In part **c**, x is a binomial random variable with $n = 4$ and $p = .8$.

$$p(x) = \binom{4}{x} .8^x .2^{4-x} \quad x = 0, 1, 2, 3, 4$$

$$p(0) = \binom{4}{0} .8^0 .2^4 = 1(1).2^4 = .0016$$

$$p(1) = \binom{4}{1} .8^1 .2^3 = 4(.8)(.2)^3 = .0256$$

$$p(2) = \binom{4}{2} .8^2 .2^2 = 6(.8)^2 (.2)^2 = .1536$$

$$p(3) = \binom{4}{3} .8^3 .2^1 = 4(.8)^3 (.2) = .4096$$

$$p(4) = \binom{4}{4} .8^4 .2^0 = 1(.8)^4 (1) = .4096$$

The probability distribution of x in tabular form is:

x	0	1	2	3	4
$p(x)$	.0016	.0256	.1536	.4096	.4096

Note: The distribution of x when $n = 4$ and $p = .2$ is the reverse of the distribution of x when $n = 4$ and $p = .8$.

$$\mu = np = 4(.8) = 3.2$$

The graph of the probability distribution of x when $n = 4$ and $p = .8$ is as follows:

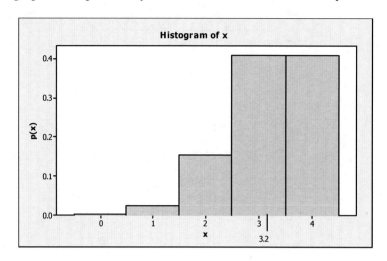

e.   In general, when $p = .5$, a binomial distribution will be symmetric regardless of the value of n. When p is less than .5, the binomial distribution will be skewed to the right; and when p is greater than .5, it will be skewed to the left. (Refer to parts **a**, **b**, and **c**.)

4.63   a.   For this experiment, there are $n = 100$ identical trials. Each trial can result in 2 possible outcomes (mouse responds positively or mouse does not respond positively). Let $S =$ mouse responds positively and $F =$ mouse does not respond positively. Then, $P(S) = p = .4$ and $P(F) = q = 1 - .4 = .6$. We can assume the trials are independent. Thus, the conditions of a binomial random variable hold.

b. $E(x) = \mu = np = 100(.4) = 40$

c. $V(x) = \sigma^2 = npq = 100(.4)(.6) = 24$

d.   Since $p = .4$ is close to .5, the distribution should be fairly symmetric. Thus, we would expect most of the observations to fall within 2 standard deviation of the mean. The standard deviation is $\sigma = \sqrt{24} = 4.899$. The interval is

$$\mu \pm 2\sigma \Rightarrow 40 \pm 2(4.899) \Rightarrow 40 \pm 9.798 \Rightarrow (30.202, 49.798)$$

4.65   a.   We will check the characteristics of a binomial random variable:

1.   This experiment consists of $n = 20$ identical trials.

2.   There are only 2 possible outcomes for each trial. SBIRS detects intruding object (S) or not (F).

3.   The probability of S remains the same from trial to trial. In this case, $p = P(S) = .8$ for each trial.

4.   The trials are independent. Whether SBIRS detects one intruding object should not effect whether it detects another.

5. x = number of intruding objects detected by SBIRS in 20 trials.

b. $n = 20$ and $p = P(S) = .8$

c. Using the Table II, Appendix A with $n = 20$ and $p = .8$,

$$P(x=15)=1-P(x\leq15)-P(x\leq14)=.370-.196=.174.$$

d. $P(x\geq15)=1-P(x\leq14)=1-.196=.804$

e. $E(x)=\mu=E(x)=np=20(.8)=16$. The average number of intruding objects detected in 20 trials is 16.

4.67 a. We will check the characteristics of a binomial random variable:

1. This experiment consists of $n = 15$ identical trials.

2. There are only 2 possible outcomes for each trial. Hotel guests are aware and participate in the hotel's conservation efforts (S) or not (F).

3. The probability of S remains the same from trial to trial. In this case, $p = P(S) = .66(.72) = .4752$ for each trial.

4. The trials are independent. Whether one hotel guest is aware and participates in the hotel's conservation efforts should not effect whether another guest is aware and participates.

5. x = number of hotel guests who are aware and participate in the hotel's conservation efforts in 15 trials.

The characteristics of the binomial random variable are met.

b. Define the following events:
G: {hotel guest is aware of the hotel's "green" conservation program}
P: {hotel guest participates in the conservation program}

We are given $P(G)=.66$ and $P(P|G)=.72$.

$$P(S)=P(P\cap G)=P(P|G)P(G)=.66(.72)=.4752.$$

c. $P(x \geq 10) = p(10) + p(11) + p(12) + p(13) + p(14) + p(15)$

$$= \binom{15}{10}.45^{10}(.55)^{15-10} + \binom{15}{11}.45^{11}(.55)^{15-11} + \binom{15}{12}.45^{12}(.55)^{15-12}$$

$$+ \binom{15}{13}.45^{13}(.55)^{15-13} + \binom{15}{14}.45^{14}(.55)^{15-14} + \binom{15}{15}.45^{15}(.55)^{15-10}$$

$$= \frac{15!}{10!5!}.45^{10}(.55)^5 + \frac{15!}{11!4!}.45^{11}(.55)^4 + \frac{15!}{12!3!}.45^{12}(.55)^3$$

$$+ \frac{15!}{13!2!}.45^{13}(.55)^2 + \frac{15!}{14!1!}.45^{14}(.55)^1 + \frac{15!}{15!0!}.45^{15}(.55)^0$$

$$= .0515 + .0191 + .0052 + .0010 + .0001 + .0000 = .0769$$

4.69   a.   Let x = number of commissioners favoring an issue in 4 trials. Then x is a binomial random variable with $n = 4$ and $p = .5$.

$$P(x = 2) = \binom{4}{2}.5^2.5^{4-2} = \frac{4!}{2!(4-2)!}.5^2.5^2 = \frac{4(3)(2)(1)}{2(1)(2)(1)}.25(.25) = .375$$

b.   Let x = number of commissioners favoring an issue in 2 trials. Then x is a binomial random variable with $n = 2$ and $p = .5$.

$$P(x = 1) = \binom{2}{1}.5^1.5^{2-1} = \frac{2!}{1!(2-1)!}.5^1.5^1 = \frac{2(1)}{1(1)}.5(.5) = .5$$

4.71   a.   Let x = number of women out of 15 that have been abused. Then x is a binomial random variable with $n = 15$, S = woman has been abused, F = woman has not been abused, $P(S) = p = \frac{1}{3}$ and $q = 1 - p = 1 - \frac{1}{3} = \frac{2}{3}$.

$$P(x \geq 4) = 1 - P(x < 4) = 1 - P(x = 0) - P(x = 1) - P(x = 2) - P(x = 3)$$

$$= 1 - \frac{15!}{0!15!}\left(\frac{1}{3}\right)^0\left(\frac{2}{3}\right)^{15} - \frac{15!}{1!14!}\left(\frac{1}{3}\right)^1\left(\frac{2}{3}\right)^{14} - \frac{15!}{2!13!}\left(\frac{1}{3}\right)^2\left(\frac{2}{3}\right)^{13} - \frac{15!}{3!12!}\left(\frac{1}{3}\right)^3\left(\frac{2}{3}\right)^{12}$$

$$= 1 - .0023 - .0171 - .0599 - .1299 = .7908$$

b. Using Table II, Appendix A, with $n = 15$ and $p = .1$,

$$P(x \geq 4) = 1 - P(x \leq 3) = 1 - .944 = .056$$

c. We sampled 15 women and actually found that 4 had been abused. If $p = 1/3$, the probability of observing 4 or more abused women is .7908. If $p = .1$, the probability of observing 4 or more abused women is only .056. If $p = .1$, we would have seen a very unusual event because the probability is so small (.056). If $p = 1/3$, we would have seen an event that was very common because the probability was very large (.7908). Since we normally do not see rare events, the true probability of abuse is probably close to 1/3.

4.73   a. If the psychic is just guessing, she has 1 chance out of 10 of guessing the correct decision. Thus, $p = 1/10 = .1$.

b. Let x = number of correct decisions in 7 trials. Then x is a binomial random variable with $n = 7$ and $p = .1$. The expected number of correct decisions in 7 trials is

$$E(x) = np = 7(.1) = .7 .$$

Thus, if the psychic is just guessing, we would expect her to make less than 1 correct decision in 7 trials.

c. $P(x = 0) = \binom{7}{0} .1^0 .9^7 = .4783$

d. Now, let $p = .5$. $P(x = 0) = \binom{7}{0} .5^0 .5^7 = .0078$

e. Yes. If the psychic is just guessing, the probability that she makes no correct decisions is .4783. Thus, if the psychic, in fact, made no correct decisions, this would not be an unusual event if she is guessing. However, on the other hand, if she does have ESP ($p = .5$) and made no correct decisions, this would be a very unusual event ($P(x = 0) = .0078$). The evidence indicates the psychic probably is just guessing and does not have ESP.

4.75   a. We must assume that the probability that a specific type of ball meets the requirements is always the same from trial to trial and the trials are independent. To use the binomial probability distribution, we need to know the probability that a specific type of golf ball meets the requirements.

b. For a binomial distribution,

$$\mu = np$$
$$\sigma = \sqrt{npq}$$

In this example, n = two dozen = $2 \cdot 12 = 24$.

$p = .10$ (Success here means the golf ball *does not* meet standards.)
$q = .90$
$\mu = \mu = np = 24(.10) = 2.4$
$\sigma = \sqrt{npq} = \sqrt{24(.10)(.90)} = 1.47$

c.   In this situation,

p = Probability of success
  = Probability golf ball *does* meet standards
  = .90
$q = 1 - .90 = .10$
$n = 24$
$E(y) = \mu = np = 24(.90) = 21.6$
$\sigma = \sqrt{npq} = \sqrt{24(.10)(.90)} = 1.47$ (Note that this is the same as in part **b.**)

4.77   The four characteristics of a Poisson random variable are:

a.   The experiment consists of counting the number of times a certain event occurs during a given unit of time or in a given area or volume (or weight, distance, or any other unit of measurement).
b.   The probability that an event occurs in a given unit of time, area, or volume is the same for all the units.
c.   The number of events that occur in one unit of time, area, or volume is independent of the number that occur in other units.
d.   The mean (or expected) number of events in each unit is denoted by the Greek letter lambda, λ.

4.79   For the Poisson probability distribution

$$p(x) = \frac{3^x e^{-3}}{x!} \qquad (x = 0, 1, 2, \ldots)$$

the value of λ is 3.

4.81   a.   In order to graph the probability distribution, we need to know the probabilities for the possible values of x. Using Table III of Appendix A with $\lambda = 3$:

$p(0) = .050$
$p(1) = P(x \leq 1) - P(x = 0) = .199 - .050 = .149$
$p(2) = P(x \leq 2) - P(x \leq 1) = .423 - .199 = .224$
$p(3) = P(x \leq 3) - P(x \leq 2) = .647 - .423 = .224$
$p(4) = P(x \leq 4) - P(x \leq 3) = .815 - .647 = .168$
$p(5) = P(x \leq 5) - P(x \leq 4) = .916 - .815 = .101$
$p(6) = P(x \leq 6) - P(x \leq 5) = .966 - .916 = .050$
$p(7) = P(x \leq 7) - P(x \leq 6) = .988 - .966 = .022$
$p(8) = P(x \leq 8) - P(x \leq 7) = .996 - .988 = .008$
$p(9) = P(x \leq 9) - P(x \leq 8) = .999 - .996 = .003$
$p(10) = P(x \leq 9) - P(x \leq 8) = 1 - .999 = .001$

The probability distribution of x in graphical form is:

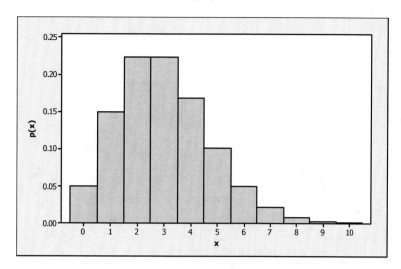

b. $\mu = \lambda = 3$

$\sigma^2 = \lambda = 3$

$\sigma = \sqrt{3} = 1.7321$

4.83 $\mu = \lambda = 1.5$

Using Table III of Appendix A:

a. $P(x \leq 3) = .934$

b. $P(x \geq 3) = 1 - P(x \leq 2) = 1 - .809 = .191$

c. $P(x = 3) = P(x \leq 3) - P(x \leq 2) = .934 - .809 = .125$

d. $P(x = 0) = .223$

e. $P(x > 0) = 1 - P(x = 0) = 1 - .223 = .777$

f. $P(x > 6) = 1 - P(x \leq 6) = 1 - .999 = .001$

4.85   a.   To graph the Poisson probability distribution with $\lambda = 3$, we need to calculate $p(x)$ for $x = 0$ to 10. Using Table III, Appendix A,

$p(0) = .050$
$p(1) = P(x \leq 1) - P(x = 0) = .199 - .050 = .149$
$p(2) = P(x \leq 2) - P(x \leq 1) = .423 - .199 = .224$
$p(3) = P(x \leq 3) - P(x \leq 2) = .647 - .423 = .224$
$p(4) = P(x \leq 4) - P(x \leq 3) = .815 - .647 = .168$
$p(5) = P(x \leq 5) - P(x \leq 4) = .916 - .815 = .101$
$p(6) = P(x \leq 6) - P(x \leq 5) = .966 - .916 = .050$
$p(7) = P(x \leq 7) - P(x \leq 6) = .988 - .966 = .022$
$p(8) = P(x \leq 8) - P(x \leq 7) = .996 - .988 = .008$

$$p(9) = P(x \le 9) - P(x \le 8) = .999 - .996 = .003$$
$$p(10) = P(x \le 10) - P(x \le 9) \approx 1 - .999 = .001$$

The graph is shown here.

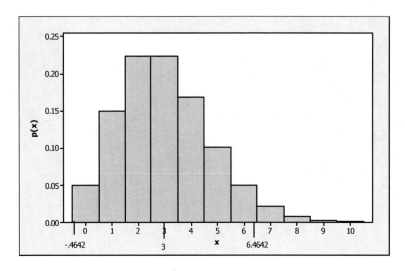

b. $\mu = \lambda = 3$

$\sigma = \sqrt{\lambda} = \sqrt{3} = 1.7321$

$\mu \pm 2\sigma \Rightarrow 3 \pm 2(1.7321) \Rightarrow 3 \pm 3.4642 \Rightarrow (-.4642, \ 6.4642)$

c. $P(\mu - 2\sigma < x < \mu + 2\sigma) = P(-.4642 < x < 6.4642) = P(0 \le x \le 6) = P(x \le 6) = .966$

4.87   a.   Using Table III, Appendix A, with $\lambda = 1$, $P(x = 0) = .368$.

b.   Using Table III, Appendix A, with $\lambda = 1$, $P(x > 1) = 1 - P(x \le 1) = 1 - .736 = .264$.

c.   Using Table III, Appendix A, with $\lambda = 1$, $P(x \le 2) = .920$.

4.89   a.   Using Table III, Appendix A, with $\lambda = 1.2$, $P(x = 0) = .301$

b.   Using Table III, Appendix A, with $\lambda = 1.2$, $P(x \ge 2) = 1 - P(x \le 1) = 1 - .663 = .337$

4.91   a. $P(x = 0) = \dfrac{\lambda^0 e^{-\lambda}}{0!} = \dfrac{1.6^0 e^{-1.6}}{0!} = .202$

b. $P(x = 1) = \dfrac{\lambda^1 e^{-\lambda}}{1!} = \dfrac{1.6^1 e^{-1.6}}{1!} = .323$

c. $E(x) = \mu = \lambda = 1.6$

$\sigma^2 = \lambda = 1.6 \qquad \sigma = \sqrt{1.6} = 1.26$

4.93   a. $P(x < 3) = P(x \le 2) = .125$ from Table III, Appendix A, with $\lambda = 5$.

b. $E(x) = \mu = \lambda = 5$. The average number of blocked calls during the peak hour is 5.

c. $P(x=0) = .007$ from Table III, Appendix A, with $\lambda = 5$. The probability of no blocked calls in the peak hour is very small. Thus, it is not likely that there will be no blocked calls during the peak hour.

d.   Since the probability that $x = 0$ is so small, we would infer that we have either seen a very rare event or that the value of λ is not 5 but a value smaller than 5.

4.95   a.   Using Table III and $\lambda = 6.2$, $P(x=2) = P(x \le 2) - P(x \le 1) = .054 - .015 = .039$
$P(x=6) = P(x \le 6) - P(x \le 5) = .574 - .414 = .160$
$P(x=10) = P(x \le 10) - P(x \le 9) = .949 - .902 = .047$

b.   The plot of the distribution is:

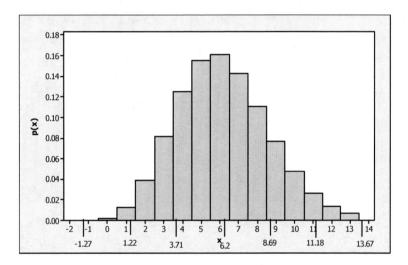

c. $\mu = \lambda = 6.2$, $\sigma = \sqrt{\lambda} = \sqrt{6.2} = 2.490$
$\mu \pm \sigma \Rightarrow 6.2 \pm 2.49 \Rightarrow (3.71,\ 8.69)$
$\mu \pm 2\sigma \Rightarrow 6.2 \pm 2(2.49) \Rightarrow 6.2 \pm 4.98 \Rightarrow (1.22,\ 11.18)$
$\mu \pm 3\sigma \Rightarrow 6.2 \pm 3(2.49) \Rightarrow 6.2 \pm 7.47 \Rightarrow (-1.27,\ 13.67)$

See the plot in part **b**.

d.   First, we need to find the mean number of customers per hour. If the mean number of customers per 10 minutes is 6.2, then the mean number of customers per hour is $6.2(6) = 37.2 = \lambda$.

$\mu = \lambda = 37.2$ and $\sigma = \sqrt{\lambda} = \sqrt{37.2} = 6.099$
$\mu \pm 3\sigma \Rightarrow 37.2 \pm 3(6.099) \Rightarrow 37.2 \pm 18.297 \Rightarrow (18.903,\ 55.498)$

Using Chebyshev's Rule, we know at least 8/9 or 88.9% of the observations will fall within 3 standard deviations of the mean. The number 75 is way beyond the 3 standard deviation limit. Thus, it would be very unlikely that more than 75 customers entered the store per hour on Saturdays.

4.97   Let x = number of cars that will arrive within the next 30 minutes. If the average number of
         cars that arrive in 60 minutes is 10, the average number of cars that arrive in 30 minutes is
         $\lambda = 5$.

         In order for no cars to be in line at closing time, no more than 1 car can arrive in the next 30
         minutes.

         $P(x \le 1) = .040$ from Table III, Appendix A, with $\lambda = 5$.

         Since this probability is so small, it is very likely that at least one car will be in line at closing
         time. The probability that at least one car will be in line at closing is $1 - .040 = .960$.

4.99   A binomial random variable is characterized by n trials where the trials are identical. In order
         for the trials to be identical, we have to sample from an infinite population or sample from a
         finite population with replacement. The probability of success on any trial remains constant
         from trial to trial and the trials are independent. A hypergeometric random variable is
         characterized by n trials where the elements are selected from a finite population without
         replacement. The probability of success on any trial depends on what happened on previous
         trials, and thus the trials are not independent.

         Both a binomial random variable and a hypergeometric random variable are characterized by
         n trials, 2 possible outcomes on each trial (S or F), and the random variable represents the
         number of successes in n trials.

4.101   a. $P(x=1) = \dfrac{\binom{r}{x}\binom{N-r}{n-x}}{\binom{N}{n}} = \dfrac{\binom{3}{1}\binom{5-3}{3-1}}{\binom{5}{3}} = \dfrac{\frac{3!}{1!2!}\frac{2!}{2!0!}}{\frac{5!}{3!2!}} = \dfrac{3(1)}{10} = .3$

         b. $P(x=3) = \dfrac{\binom{r}{x}\binom{N-r}{n-x}}{\binom{N}{n}} = \dfrac{\binom{3}{3}\binom{9-3}{5-3}}{\binom{9}{5}} = \dfrac{\frac{3!}{3!0!}\frac{6!}{2!4!}}{\frac{9!}{5!4!}} = \dfrac{1(15)}{126} = .119$

         c. $P(x=2) = \dfrac{\binom{r}{x}\binom{N-r}{n-x}}{\binom{N}{n}} = \dfrac{\binom{2}{2}\binom{4-2}{2-2}}{\binom{4}{2}} = \dfrac{\frac{2!}{2!0!}\frac{2!}{0!2!}}{\frac{4!}{2!2!}} = \dfrac{1(1)}{6} = .167$

         d. $P(x=0) = \dfrac{\binom{r}{x}\binom{N-r}{n-x}}{\binom{N}{n}} = \dfrac{\binom{2}{0}\binom{4-2}{2-0}}{\binom{4}{2}} = \dfrac{\frac{2!}{0!2!}\frac{2!}{2!0!}}{\frac{4!}{2!2!}} = \dfrac{1(1)}{6} = .167$

4.103   With $N = 12$, $n = 8$ and $r = 6$, x can take on values 2, 3, 4, 5, or 6.

a. $$P(x = 2) = \frac{\binom{r}{x}\binom{N-r}{n-x}}{\binom{N}{n}} = \frac{\binom{6}{2}\binom{12-6}{8-2}}{\binom{12}{8}} = \frac{\frac{6!}{2!4!}\frac{6!}{6!0!}}{\frac{12!}{8!4!}} = \frac{15(1)}{495} = .030$$

$$P(x = 3) = \frac{\binom{r}{x}\binom{N-r}{n-x}}{\binom{N}{n}} = \frac{\binom{6}{3}\binom{12-6}{8-3}}{\binom{12}{8}} = \frac{\frac{6!}{3!3!}\frac{6!}{5!1!}}{\frac{12!}{8!4!}} = \frac{20(6)}{495} = .242$$

$$P(x = 4) = \frac{\binom{r}{x}\binom{N-r}{n-x}}{\binom{N}{n}} = \frac{\binom{6}{4}\binom{12-6}{8-4}}{\binom{12}{8}} = \frac{\frac{6!}{4!2!}\frac{6!}{4!2!}}{\frac{12!}{8!4!}} = \frac{15(15)}{495} = .455$$

$$P(x = 5) = \frac{\binom{r}{x}\binom{N-r}{n-x}}{\binom{N}{n}} = \frac{\binom{6}{5}\binom{12-6}{8-5}}{\binom{12}{8}} = \frac{\frac{6!}{5!1!}\frac{6!}{3!3!}}{\frac{12!}{8!4!}} = \frac{6(20)}{495} = .242$$

$$P(x = 6) = \frac{\binom{r}{x}\binom{N-r}{n-x}}{\binom{N}{n}} = \frac{\binom{6}{6}\binom{12-6}{8-6}}{\binom{12}{8}} = \frac{\frac{6!}{6!0!}\frac{6!}{2!4!}}{\frac{12!}{8!4!}} = \frac{1(15)}{495} = .030$$

The probability distribution of x in tabular form is:

x	$p(x)$
2	.030
3	.242
4	.455
5	.242
6	.030

b. $$\mu = \frac{nr}{N} = \frac{8(6)}{12} = 4$$

$$\sigma^2 = \frac{r(N-r)n(N-n)}{N^2(N-1)} = \frac{6(12-6)8(12-8)}{12^2(12-1)} = \frac{1,152}{1,584} = .7273$$

$$\sigma = \sqrt{.7273} = .853$$

c. $\mu \pm 2\sigma \Rightarrow 4 \pm 2(.853) \Rightarrow 4 \pm 1.706 \Rightarrow (2.294, \; 5.706)$

The graph of the distribution is:

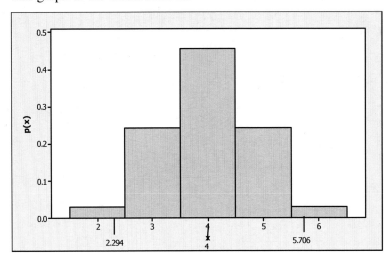

d. $P(2.294 < x < 5.706) = P(3 \le x \le 5) = .242 + .455 + .242 = .939$

4.105   For this problem, $N = 100$, $n = 10$, and $x = 4$.

a.   If the sample is drawn without replacement, the hypergeometric distribution should be used. The hypergeometric distribution requires that sampling be done without replacement.

b.   If the sample is drawn with replacement, the binomial distribution should be used. The binomial distribution requires that sampling be done with replacement.

4.107   a.   This is a hypergeometric distribution.

b.   For this experiment, $r = 20$, $N = 100$, and $n = 3$.

c. $p(0) = \dfrac{\dbinom{20}{0}\dbinom{80}{3-0}}{\dbinom{100}{3}} = \dfrac{\dfrac{20!}{0!20!}\dfrac{80!}{3!77!}}{\dfrac{100!}{3!97!}} = \dfrac{1 \cdot \dfrac{80 \cdot 79 \cdot 78 \cdots 1}{3 \cdot 2 \cdot 1 \cdot 77 \cdot 76 \cdot 75 \cdots 1}}{\dfrac{100 \cdot 99 \cdot 98 \cdots 1}{3 \cdot 2 \cdot 1 \cdot 97 \cdot 96 \cdot 95 \cdots 1}} = \dfrac{82,160}{161,700} = .508$

d. $p(1) = \dfrac{\dbinom{20}{1}\dbinom{80}{3-1}}{\dbinom{100}{3}} = \dfrac{\dfrac{20!}{1!19!}\dfrac{80!}{2!78!}}{\dfrac{100!}{3!97!}}$

$= \dfrac{\dfrac{20 \cdot 19 \cdot 18 \cdots 1}{1 \cdot 19 \cdot 18 \cdot 17 \cdots 1} \cdot \dfrac{80 \cdot 79 \cdot 78 \cdots 1}{2 \cdot 1 \cdot 78 \cdot 77 \cdot 76 \cdots 1}}{\dfrac{100 \cdot 99 \cdot 98 \cdots 1}{3 \cdot 2 \cdot 1 \cdot 97 \cdot 96 \cdot 95 \cdots 1}} = \dfrac{63,200}{161,700} = .391$

e. $p(2) = \dfrac{\binom{20}{2}\binom{80}{3-2}}{\binom{100}{3}} = \dfrac{\frac{20!}{2!18!}\frac{80!}{1!79!}}{\frac{100!}{3!97!}}$

$= \dfrac{\frac{20\cdot19\cdot18\cdots1}{2\cdot1\cdot18\cdot17\cdot16\cdots1}\cdot\frac{80\cdot79\cdot78\cdots1}{1\cdot79\cdot78\cdot77\cdots1}}{\frac{100\cdot99\cdot98\cdots1}{3\cdot2\cdot1\cdot97\cdot96\cdot95\cdots1}} = \dfrac{15,200}{161,700} = .094$

f. $p(3) = \dfrac{\binom{20}{3}\binom{80}{3-3}}{\binom{100}{3}} = \dfrac{\frac{20!}{3!17!}\frac{80!}{0!80!}}{\frac{100!}{3!97!}}$

$= \dfrac{\frac{20\cdot19\cdot18\cdots1}{3\cdot2\cdot1\cdot17\cdot16\cdot15\cdots1}\cdot\frac{80\cdot79\cdot78\cdots1}{1\cdot80\cdot79\cdot78\cdots1}}{\frac{100\cdot99\cdot98\cdots1}{3\cdot2\cdot1\cdot97\cdot96\cdot95\cdots1}} = \dfrac{1,140}{161,700} = .007$

4.109   a.   Let x = number of facilities chosen in a sample of 10 that treat hazardous waste on-site. For this problem, $N = 209$, $r = 8$, and $n = 10$.

$\mu = E(x) = \dfrac{rn}{N} = \dfrac{8(10)}{209} = .383$

If 10 facilities are repeatedly selected, the average number of facilities in the sample of 10 that treat hazardous waste on-site is .383.

b. $P(x=4) = \dfrac{\binom{r}{x}\binom{N-r}{n-x}}{\binom{N}{n}} = \dfrac{\binom{8}{4}\binom{209-8}{10-4}}{\binom{209}{10}} = \dfrac{\frac{8!}{4!4!}\cdot\frac{201!}{6!195!}}{\frac{209!}{10!199!}} = .0002$

4.111   Let x = number of "clean" cartridges selected in 5 trials. For this problem, $N = 158$, $r = 122$, and $n = 5$.

$P(x=5) = \dfrac{\binom{r}{x}\binom{N-r}{n-x}}{\binom{N}{n}} = \dfrac{\binom{122}{5}\binom{36}{0}}{\binom{158}{5}} = \dfrac{\frac{122!}{5!17!}\frac{36!}{0!36!}}{\frac{158!}{5!53!}} = .2693$

4.113   Let x = number of extinct bird species in $n = 10$ trials. Then x is a hypergeometric random variable with $N = 132$, $r = 38$, and $n = 10$.

a. $P(x = 5) = \dfrac{\binom{38}{5}\binom{132-38}{10-5}}{\binom{132}{10}} = \dfrac{\dfrac{38!}{5!(33)!} \cdot \dfrac{94!}{5!(89)!}}{\dfrac{132!}{10!(122)!}} = \dfrac{501,942(54,891,018)}{3.12058705x10^{14}} = .0883$

b. $P(x \le 1) = P(x = 0) + P(x = 1) = \dfrac{\binom{38}{0}\binom{132-38}{10-0}}{\binom{132}{10}} + \dfrac{\binom{38}{1}\binom{132-38}{10-0}}{\binom{132}{10}}$

$= \dfrac{\dfrac{38!}{0!(38)!} \cdot \dfrac{94!}{10!(84)!}}{\dfrac{132!}{10!(122)!}} + \dfrac{\dfrac{38!}{1!(37)!} \cdot \dfrac{94!}{9!(85)!}}{\dfrac{132!}{10!(122)!}} = \dfrac{9.041256842x10^{12}}{3.12058705x10^{14}} + \dfrac{4.041973645x10^{13}}{3.12058705x10^{14}}$

$= .0290 + .1295 = .1585$

4.115   Let x = number of spoiled bottles in the sample of 3. Since the sampling will be done without replacement, x is a hypergeometric random variable with $N = 12$, $n = 3$, and $r = 1$.

$$P(x = 1) = \dfrac{\binom{r}{x}\binom{N-r}{n-x}}{\binom{N}{n}} = \dfrac{\binom{1}{1}\binom{12-1}{3-1}}{\binom{12}{3}} = \dfrac{\dfrac{1!}{1!0!} \cdot \dfrac{11!}{2!9!}}{\dfrac{12!}{3!9!}} = \dfrac{55}{220} = .25$$

4.117   Let x = number of grants awarded to the north side in 140 trials. The random variable x has a hypergeometric distribution with $N = 743$, $n = 140$, and $r = 601$.

a. $\mu = E(x) = \dfrac{nr}{N} = \dfrac{140(601)}{743} = 113.24$

$\sigma^2 = \dfrac{r(N-r)n(N-n)}{N^2(N-1)} = \dfrac{601(743-601)140(743-140)}{743^2(743-1)} = 17.5884$

$\sigma = \sqrt{17.5884} = 4.194$

b.   If the grants were awarded at random, we would expect approximately 113 to be awarded to the north side. We observed 140. The z-score associated with 140 is:

$z = \dfrac{x - \mu}{\sigma} = \dfrac{140 - 113.24}{4.194} = 6.38$

Because this *z*-score is so large, it would be extremely unlikely to observe all 140 grants to the north side if they are randomly selected. Thus, we would conclude that the grants were not randomly selected.

4.119   a.   Poisson

     b.   Binomial

     c.   Binomial

4.121   From Table II, Appendix A:

     a. $P(x=14)=P(x\leq14)-P(x\leq13)=.584-.392=.192$

     b. $P(x\leq12)=.228$

     c. $P(x>12)=1-P(x\leq12)=1-.228=.772$

     d. $P(9\leq x\leq18)=P(x\leq18)-P(x\leq8)=.992-.005=.987$

     e. $P(8<x<18)=P(x\leq17)-P(x\leq8)=.965-.005=.960$

     f. $\mu=np=20(.7)=1.4$

       $\sigma^2=npq=20(.7)(.3)=4.2$, $\sigma=\sqrt{4.2}=2.049$

     g. $\mu\pm2\sigma\Rightarrow14\pm2(2.049)\Rightarrow14\pm4.098\Rightarrow(9.902,\ 18.098)$

$$P(9.902<x<18.098)=P(10\leq x\leq18)=P(x\leq18)-P(x\leq9)$$
$$=.992-.017=.975$$

4.123   a. $P(x=1)=\binom{3}{1}(.1)^1(.9)^{3-1}=\frac{3!}{1!2!}(.1)^1(.9)^2=3(.1)(.81)=.243$

     b. $P(x=4)=P(x\leq4)-P(x\leq3)=.238-.107=.131$ from Table II, Appendix A.

     c. $P(x=0)=\binom{2}{0}(.4)^0(.6)^{2-0}=\frac{2!}{0!2!}(1)(.6)^2=.36$

     d. $P(x=4)=P(x\leq4)-P(x\leq3)=.969-.812=.157$ from Table II, Appendix A.

     e. $P(x=12)=P(x\leq12)-P(x\leq11)=.184-.056=.128$ from Table II, Appendix A.

     f. $P(x=8)=P(x\leq8)-P(x\leq7)=.954-.833=.121$ from Table II, Appendix A.

4.125   Using Table III, Appendix A,

     a.   When $\lambda=2$, $p(3)=P(x\leq3)-P(x\leq2)=.857-.677=.180$

b.   When $\lambda = 1$, $p(4) = P(x \le 4) - P(x \le 3) = .996 - .981 = .015$

c.   When $\lambda = 5$, $p(2) = P(x \le 2) - P(x \le 1) = .986 - .910 = .076$

4.127   a.   To find the probabilities associated with each value of x, we divide the frequency associated with each value by the total sample size, 743. The probabilities appear in the table:

First Digit	Frequency of Occurrence	Probability
1	109	109 / 743 = .1467
2	75	75 / 743 = .1009
3	77	77 / 743 = .1036
4	99	99 / 743 = .1332
5	72	72 / 743 = .0969
6	117	117 / 743 = .1575
7	89	89 / 743 = .1198
8	62	62 / 743 = .0834
9	43	43 / 743 = 0579
Total	743	1.0000

b. $\mu = E(x) = \sum xp(x) = 1(.1467) + 2(.1009) + 3(.1036) + 4(.1332) + 5(.0969)$
$+ 6(.1575) + 7(.1198) + 8(.0834) + 9(.0579) = 4.6485$

c.   Over a very large number of trials, the average first significant digit is 4.6485.

4.129   a.   For this experiment, there are $n = 200$ smokers (n identical trials). For each smoker, there are 2 possible outcomes: S = smoker enters treatment program and F = smoker does not enter treatment program. The probability of S (smoker enters treatment program) is the same from trial to trial. This probability is $P(S) = p = .05$. $P(F) = 1 - P(S) = 1 - .05 = .95$. The trials are independent and x = number of smokers entering treatment program in 200 trials. Thus, x is a binomial random variable.

b. $p = P(S) = .05$. Of all the smokers, only .05 or 5% enter treatment programs.

c. $E(x) = np = 200(.05) = 10$. For all samples of 200 smokers, the average number of smokers who enter a treatment program will be 10.

4.131   The random variable x would have a binomial distribution:

1. n identical trials. A sample of married women was taken and would be considered small compared to the entire population of married women, so the trials would be very close to identical.

2.   There are only two possible outcomes for each trial. In this experiment, there are only two possible outcomes: the woman would marry the same man (S) or she would not (F).

3. The probability of Success stays the same from trial to trial. Here, $P(S) = .50$ and $P(F) = .50$. In reality, these probabilities would not be exactly the same from trial to trial, but rounded off to 4 decimal places, they would be the same.

4. The trials are independent. In this experiment, the trials would not be exactly independent because we would be sampling without replacement from a finite population. However, if the sample size is fairly small compared to the population size, the trials will be essentially independent.

5. The binomial random variable x would be the number of successes in n trials. For this experiment, x = the number of women out of those sampled who would marry the same man again.

Thus, x would possess (approximately) a binomial distribution.

4.133   a.   In order for this to be a valid probability distribution, $\sum p(x) = 1$ and $0 \le p(x) \le 1$ for all values of x.

In this distribution, $0 \le p(x) \le 1$ for all values of x and

$\sum p(x) = .051 + .099 + .093 + .635 + .122 = 1.000$. Thus, this is a valid probability distribution.

b. $P(x = 1) = .051$

c. $P(x \ge 4) = P(x = 4) + P(x = 5) = .635 + .122 = .757$

d. $P(x = 2 \text{ or } x = 3) = P(x = 2) + P(x = 3) = .099 + .093 = .192$

e. $E(x) = \mu = \sum xp(x) = 1(.051) + 2(.099) + 3(.093) + 4(.635) + 5(.122) = 3.678$

f.   The average rating of all books rated is 3.678.

4.135   a.   From the problem, x is a binomial random variable with $n = 10$ and $p = .8$.

$P(x = 3) = P(x \le 3) - P(x \le 2) = .001 - .000 = .001$ (from Table II, Appendix A)

b. $P(x \le 7) = .322$

c. $P(x > 4) = 1 - P(x \le 4) = 1 - .006 = .994$

4.137   a. $\mu = E(x) = np = 800(.65) = 520$
$\sigma = \sqrt{npq} = \sqrt{800(.65)(.35)} = 13.491$

b.   Half of the 800 food items is 400. A value of $x = 400$ would have a z-score of:

$$z = \frac{x - \mu}{\sigma} = \frac{400 - 520}{13.491} = -8.895$$

Since the *z*-score associated with 400 items is so small (−8.895), it would be virtually impossible to observe less than half without any traces of pesticides if the 65% value was correct.

4.139   Let *x* = number of parents who yell at their child before, during, or after a meet in 20 trials. Then *x* has a binomial distribution with *n* = 20 and *p* = .05.

$$P(x \geq 1) = 1 - P(x = 0) = 1 - .358 = .642 \quad \text{(Using Table II, Appendix A)}$$

4.141   a. $\sigma = \sqrt{\lambda} = \sqrt{4} = 2$

b. $P(x > 10) = 1 - P(x \leq 10) = 1 - .997 = .003$ from Table III, Appendix A, with $\lambda = 4$. Since the probability is so small (.003), it would be very unlikely that the plant would yield a value that would exceed the EPA limit.

4.143   Using Table II with *n* = 25 and *p* = .8:

a. $P(x < 15) = P(x \leq 14) = .006$

b.   Since the probability of such an event is so small when *p* = .8, if less than 15 insects die we would conclude that the insecticide is not as effective as claimed.

4.145   We know that $\mu = np = n(.4) = .4n$ and $\sigma = \sqrt{npq} = \sqrt{n(.4)(.6)} = \sqrt{.24n}$. Since *p* = .4 is close to .5, the distribution of *x* will be close to symmetrical. Therefore, we can use the Empirical Rule. We know that approximately 95% of all observations will fall within 2 standard deviation of the mean. Thus, approximately 97.5% of the observations will fall above two standard deviations below the mean. We will set 100 equal to the mean minus 2 standard deviations and solve for n:

$$.4n - 2\sqrt{.24n} = 100$$

$$\Rightarrow n - 5\sqrt{.24n} = 250$$

$$\Rightarrow n - 250 = 5\sqrt{.24n}$$

$$\Rightarrow n^2 - 500n + 62500 = 6n$$

$$\Rightarrow n^2 - 506n + 62500 = 0$$

Now, using Pythagorean Theorem to solve for n, we get:

$$n = \frac{506 \pm \sqrt{506^2 - 4(62500)}}{2} = \frac{506 \pm 77.69}{2} \Rightarrow n = 214 \text{ or } n = 292$$

Since the mean is *np,* and for this exercise the mean must be greater than 100, we know that *n* cannot be 214. Thus, *n* has to be 292.

4.147   There will not be enough beds if x = the number of newly admitted patients exceeds 4. For $\lambda = 2.5$,

$$P(x > 4) = 1 - P(x \le 4) = 1 - \left\{ P(x = 0) + P(x = 1) + P(x = 2) + P(x = 3) + P(x = 4) \right\}$$

$$= 1 - \left(\frac{2.5^0 \cdot e^{-2.5}}{0!} + \frac{2.5^1 \cdot e^{-2.5}}{1!} + \frac{2.5^2 \cdot e^{-2.5}}{2!} + \frac{2.5^3 \cdot e^{-2.5}}{3!} + \frac{2.5^4 \cdot e^{-2.5}}{4!} \right)$$

$$= 1 - (.0821 + .2052 + .2565 + .2138 + .1336) = 1 - .8912 = .1088$$

Continuous Random Variables

5.1   A uniform random variable is a continuous random variable that appears to have equally likely outcomes over its range of possible values.

5.3   a.   For a uniform random variable,

$$f(x) = \begin{cases} \dfrac{1}{d-c} = \dfrac{1}{30-10} = \dfrac{1}{20} & 10 \le x \le 30 \\ 0 & \text{otherwise} \end{cases}$$

b. $\mu = \dfrac{c+d}{2} = \dfrac{10+30}{2} = 20$

$\sigma = \dfrac{d-c}{\sqrt{12}} = \dfrac{30-10}{\sqrt{12}} = 5.774$

c.

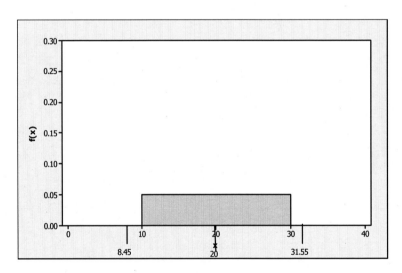

$\mu \pm 2\sigma \Rightarrow 20 \pm 2(5.774) \Rightarrow 20 \pm 11.584 \Rightarrow (8.452, 31.548)$

5.5   a.   For a uniform random variable,

$$f(x) = \begin{cases} \dfrac{1}{d-c} = \dfrac{1}{4-2} = \dfrac{1}{2} & 2 \le x \le 4 \\ 0 & \text{otherwise} \end{cases}$$

b. $\mu = \dfrac{c+d}{2} = \dfrac{2+4}{2} = 3$ $\sigma = \dfrac{d-c}{\sqrt{12}} = \dfrac{4-2}{\sqrt{12}} = .577$

c. $\mu \pm \sigma \Rightarrow 3 \pm .577 \Rightarrow (2.423, 3.577)$

$$P(2.423 \le x \le 3.577) = (3.577 - 2.423)\frac{1}{2} = .577$$

d. $P(x > 2.78) = (4 - 2.78)\frac{1}{2} = .61$

e. $P(2.4 \le x \le 3.7) = (3.7 - 2.4)\frac{1}{2} = .65$ ✓

f. $P(x < 2) = 0$

5.7　For the uniform random variable,

$$f(x) = \begin{cases} \dfrac{1}{d-c} = \dfrac{1}{200-100} = \dfrac{1}{100} & 100 \le x \le 200 \\ 0 & \text{otherwise} \end{cases}$$

a. $\mu = \dfrac{c+d}{2} = \dfrac{100+200}{2} = 150$　　　$\sigma = \dfrac{d-c}{\sqrt{12}} = \dfrac{200-100}{\sqrt{12}} = 28.87$

$\mu \pm 2\sigma \Rightarrow 150 \pm 2(28.87) \Rightarrow 150 \pm 57.74 \Rightarrow (92.26, 207.74)$

$P(x < 92.26) + P(x > 207.74) = 0 + 0 = 0$

b. $\mu \pm 3\sigma \Rightarrow 150 \pm 3(28.87) \Rightarrow 150 \pm 86.61 \Rightarrow (63.39, 236.61)$

$P(63.9 < x < 236.61) = P(100 \le x \le 200) = 1$

c. $P(92.26 < x < 207.74) = P(100 \le x \le 200) = 1$

5.9　a.　Since x is a uniform random variable on the interval 1 to 3, $c = 1$ and $d = 3$.

$E(x) = \dfrac{c+d}{2} = \dfrac{1+3}{2} = \dfrac{4}{2} = 2$. The average amount of uranium in all reservoirs is 2 parts per million.

b.　$P(2 < x < 2.5) = \dfrac{2.5-2}{3-1} = \dfrac{.5}{2} = .25$

c.　$P(x \le 1.75) = \dfrac{1.75-1}{3-1} = \dfrac{.75}{2} = .375$

5.11   a.   Let x be the number of anthrax spores detected. Then x has a uniform distribution on the interval from 0 to 10. Thus, $c = 0$ and $d = 10$.

$$f(x) = \begin{cases} \dfrac{1}{d-c} = \dfrac{1}{10-0} = \dfrac{1}{10} & \text{for } 0 \leq x \leq 10 \\ 0 & \text{otherwise} \end{cases}$$

$$P(x \leq 8) = (8-0)\left(\dfrac{1}{10}\right) = \dfrac{8}{10} = .8$$

  b. $P(2 \leq x \leq 5) = (5-2)\left(\dfrac{1}{10}\right) = \dfrac{3}{10} = .3$

5.13   a.   Suppose no bolt-on trace elements are used. Let x be a uniform random variable on the interval 260 to 290. From this, $c = 260$ and $d = 290$.

$$f(x) = \begin{cases} \dfrac{1}{d-c} = \dfrac{1}{290-260} = \dfrac{1}{30} & 260 \leq x \leq 290 \\ 0 & \text{otherwise} \end{cases}$$

$$P(280 \leq x \leq 284) = (284-280)\left(\dfrac{1}{30}\right) = 4\left(\dfrac{1}{30}\right) = .1333$$

Suppose bolt-on trace elements are used. Let y be a uniform random variable on the interval 278 to 285. From this, $c = 278$ and $d = 285$.

$$f(y) = \begin{cases} \dfrac{1}{d-c} = \dfrac{1}{285-278} = \dfrac{1}{7} & 278 \leq y \leq 285 \\ 0 & \text{otherwise} \end{cases}$$

$$P(280 \leq y \leq 284) = (284-280)\left(\dfrac{1}{7}\right) = 4\left(\dfrac{1}{7}\right) = .5714$$

  b.   Suppose no bolt-on trace elements are used.

$$P(x \leq 268) = (268-260)\left(\dfrac{1}{30}\right) = 8\left(\dfrac{1}{30}\right) = .2667$$

Suppose bolt-on trace elements are used.

$$P(x \leq 268) = (0)\left(\dfrac{1}{7}\right) = 0$$

5.15   Let x be cycle availability. Then x has a uniform distribution on the interval from 0 to 1. Thus, $c = 0$ and $d = 1$.

$$f(x) = \begin{cases} \dfrac{1}{d-c} = \dfrac{1}{1-0} = 1 & \text{for } 0 \le x \le 1 \\ 0 & \text{otherwise} \end{cases}$$

mean: $E(x) = \mu = \dfrac{c+d}{2} = \dfrac{0+1}{2} = .5$ The average cycle availability is .5.

standard deviation: $\sigma = \dfrac{d-c}{\sqrt{12}} = \dfrac{1-0}{\sqrt{12}} = .2887$

10th percentile: $P(x \le k) = .10 \Rightarrow (k-0)(1) = .10 \Rightarrow k = .1$. 10% of all values of cycle availability are less than or equal to .10.

Lower quartile: $P(x \le k) = .25 \Rightarrow (k-0)(1) = .25 \Rightarrow k = .25$. 25% of all values of cycle availability are less than or equal to .25.

Upper quartile: $P(x \le k) = .75 \Rightarrow (k-0)(1) = .75 \Rightarrow k = .75$. 75% of all values of cycle availability are less than or equal to .75.

5.17   a.   The amount dispensed by the beverage machine is a continuous random variable since it can take on any value between 6.5 and 7.5 ounces.

b.   Since the amount dispensed is random between 6.5 and 7.5 ounces, x is a uniform random variable.

$$f(x) = \dfrac{1}{d-c} \quad (c \le x \le d)$$

$$\dfrac{1}{d-c} = \dfrac{1}{7.5-6.5} = \dfrac{1}{1} = 1$$

Therefore, $f(x) = \begin{cases} 1 & (6.5 \le x \le 7.5) \\ 0 & \text{otherwise} \end{cases}$

The graph is as follows:

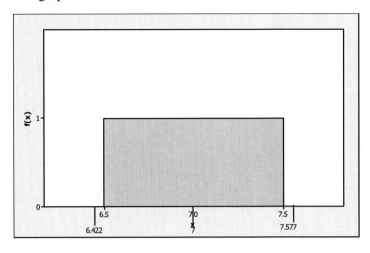

c. $\mu = \dfrac{c+d}{2} = \dfrac{6.5+7.5}{2} = \dfrac{14}{2} = 7$

$\sigma = \dfrac{d-c}{\sqrt{12}} = \dfrac{7.5-6.5}{\sqrt{12}} = .2887$

$\mu \pm 2\sigma \Rightarrow 7 \pm 2(.2887) \Rightarrow 7 \pm .577 \Rightarrow (6.422,\ 7.577)$

d. $P(x \geq 7) = (7.5-7)(1) = .5$

e. $P(x < 6) = 0$

f. $P(6.5 \leq x \leq 7.25) = (7.25-6.5)(1) = .75$

g.   The probability that the next bottle filled will contain more than 7.25 ounces is:

$P(x > 7.25) = (7.5-7.25)(1) = .25$

The probability that the next 6 bottles filled will contain more than 7.25 ounces is:

$P[(x > 7.25) \cap (x > 7.25) \cap (x > 7.25) \cap (x > 7.25) \cap (x > 7.25) \cap (x > 7.25)]$
$= [P(x > 7.25)]^6 = .25^6 = .0002$

5.19   Let x = number of inches a gouge is from one end of the spindle. Then x has a uniform distribution with $f(x)$ as follows:

$$f(x) = \begin{cases} \dfrac{1}{d-c} = \dfrac{1}{18-0} = \dfrac{1}{18} & 0 \leq x \leq 18 \\ 0 & \text{otherwise} \end{cases}$$

In order to get at least 14 consecutive inches without a gouge, the gouge must be within 4 inches of either end. Thus, we must find:

$P(x < 4) + P(x > 14) = (4-0)(1/18) + (18-14)(1/18) = 4/18 + 4/18 = 8/14 = .4444$

5.21   A normal distribution with $\mu = 0$ and $\sigma = 1$ is a standard normal distribution.

5.23   a. $P(0 < z < 2.00) = .4772$
            (from Table IV, Appendix A)

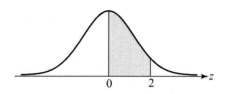

        b. $P(0 < z < 1.00) = .3413$
            (from Table IV, Appendix A)

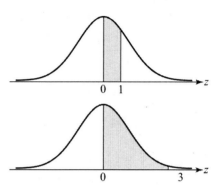

        c. $P(0 < z < 3) = .4987$
            (from Table IV, Appendix A)

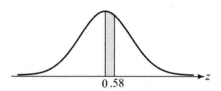

        d. $P(0 < z < .58) = .2190$
            (from Table IV, Appendix A)

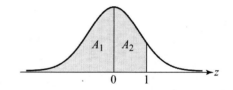

5.25   a. $P(z = 1) = 0$, since a single point does not have an area.

        b. $P(z \leq 1) = P(z \leq 0) + P(0 < z \leq 1)$
                    $= A_1 + A_2$
                    $= .5 + .3413 = .8413$
                    (Table IV, Appendix A)

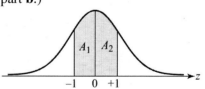

        c. $P(z < 1) = P(z \leq 1) = .8413$ (Refer to part **b**.)

        d. $P(z > 1) = 1 - P(z \leq 1) = 1 - .8413 = .1587$ (Refer to part **b**.)

        e. $P(-1 \leq z \leq 1) = P(-1 \leq z \leq 0) + P(0 \leq z \leq 1)$
                    $= A_1 + A_2$
                    $= .3413 + .3413 = .6826$

        f. $P(-2 \leq z \leq 2) = P(-2 \leq z \leq 0) + P(0 \leq z \leq 2)$
                    $= A_1 + A_2$
                    $= .4772 + .4772 = .9544$

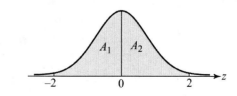

g. $P(-2.16 \leq z \leq 0.55) = P(-2.16 \leq z \leq 0) + P(0 \leq z \leq 0.55)$

$= A_1 + A_2$

$= .4846 + .2088 = .6934$

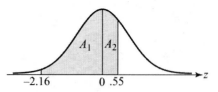

h. $P(-.42 < z < 1.96) = P(-.42 \leq z \leq 0) + P(0 \leq z \leq 1.96)$

$= A_1 + A_2$

$= .1628 + .4750 = .6378$

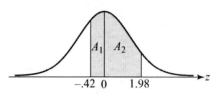

5.27　a.　$z = 1$

b.　$z = -1$

c.　$z = 0$

d.　$z = -2.5$

e.　$z = 3$

5.29　Using Table IV, Appendix A:

a.　$P(z \geq z_0) = .05$
$A_1 = .5 - .05 = .4500$
Looking up the area .4500 in Table IV gives
$z_0 = 1.645$.

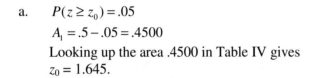

b.　$P(z \geq z_0) = .025$
$A_1 = .5 - .025 = .4750$
Looking up the area .4750 in Table IV
gives $z_0 = 1.96$.

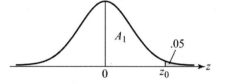

c.　$P(z \leq z_0) = .025$
$A_1 = .5 - .025 = .4750$
Looking up the area .4750 in Table IV gives
$z = 1.96$. Since z_0 is to the left of 0, $z_0 = -1.96$.

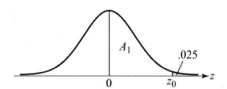

d.　$P(z \geq z_0) = .10$
$A_1 = .5 - .10 = .4000$
Looking up the area .4000 in Table IV
gives $z_0 = 1.28$.

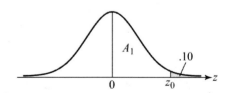

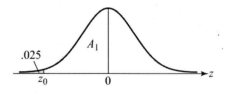

e. $P(z > z_0) = .10$
$A_1 = .5 - .10 = .4000$
$z_0 = 1.28$ (same as in **d**)

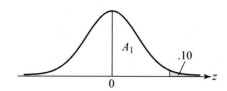

5.31 a. $P(10 \le x \le 12) = P\left(\dfrac{10-11}{2} \le z \le \dfrac{12-11}{2}\right)$

$= P(-0.50 \le z \le 0.50)$
$= A_1 + A_2$
$= .1915 + .1915 = .3830$

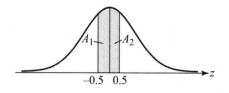

b. $P(6 \le x \le 10) = P\left(\dfrac{6-11}{2} \le z \le \dfrac{10-11}{2}\right)$

$= P(-2.50 \le z \le -0.50)$
$= P(-2.50 \le z \le 0) - P(-0.50 \le z \le 0)$
$= .4938 - .1915 = .3023$

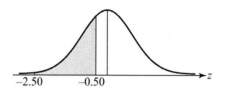

c. $P(13 \le x \le 16) = P\left(\dfrac{13-11}{2} \le z \le \dfrac{16-11}{2}\right)$

$= P(1.00 \le z \le 2.50)$
$= P(0 \le z \le 2.50) - P(0 \le z \le 1.00)$
$= .4938 - .3413 = .1525$

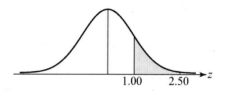

d. $P(7.8 \le x \le 12.6) = P\left(\dfrac{7.8-11}{2} \le z \le \dfrac{12.6-11}{2}\right)$

$= P(-1.60 \le z \le 0.80)$
$= A_1 + A_2$
$= .4452 + .2881 = .7333$

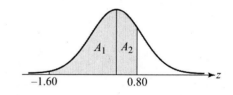

e. $P(x \ge 13.24) = P\left(z \ge \dfrac{13.24-11}{2}\right)$

$= P(z \ge 1.12)$
$= A_2 = .5 - A_1$
$= .5000 - .3868 = .1314$

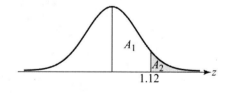

f. $P(x \geq 7.62) = P\left(z \geq \dfrac{7.62-11}{2}\right)$

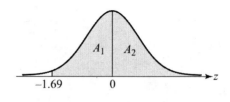

$= P(z \geq -1.69)$

$= A_1 + A_2$

$= .4545 + .5000 = .9545$

5.33 a. $P(x < x_0) = .10 \Rightarrow P\left(z < \dfrac{x_0 - 30}{8}\right)$

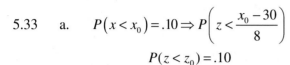

$P(z < z_0) = .10$

$A_1 = .5 - .10 = .4000$

Looking up the area .4000 in Table IV gives $z_0 = 1.28$. Since z_0 is to the left of 0, $z_0 = -1.28$.

$z_0 = -1.28 = \dfrac{x_0 - 30}{8} \Rightarrow x_0 = 8(-1.28) + 30 = 19.76$

b. $P(x < x_0) = .80 \Rightarrow P\left(z < \dfrac{x_0 - 30}{8}\right)$

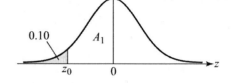

$P(z < z_0) = .80$

$A_1 = .80 - .5 = .3000$

Looking up the area .3000 in Table IV gives $z_0 = 0.84$.

$z_0 = 0.84 = \dfrac{x_0 - 30}{8} \Rightarrow x_0 = 8(.84) + 30 = 36.72$

c. $P(x > x_0) = .01 \Rightarrow P\left(z > \dfrac{x_0 - 30}{8}\right)$

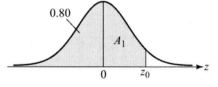

$P(z > z_0) = .01$

$A_1 = .5 - .01 = .4900$

Looking up the area .4900 in Table IV gives $z_0 = 2.33$.

$z_0 = 2.33 = \dfrac{x_0 - 30}{8} \Rightarrow x_0 = 8(2.33) + 30 = 48.64$

5.35 The random variable x has a normal distribution with $\sigma = 25$.

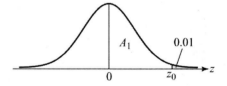

We know $P(x > 150) = .90$. So, $A_1 + A_2 = .90$.

Since $A_2 = .50$, $A_1 = .90 - .50 = .40$. Look up the area .40 in the body of Table IV; (take the closest value) $z_0 = -1.28$.

To find μ, substitute all the values into the z-score formula:

$$z = \frac{x-\mu}{\sigma} \Rightarrow -1.28 = \frac{150-\mu}{25} \Rightarrow \mu = 150 + 25(1.28) = 182$$

5.37 a. $z = \dfrac{x-\mu}{\sigma} = \dfrac{10-11}{4} = \dfrac{-1}{4} = -.25$

b. $P(10 < x < 15) = P\left(\dfrac{10-11}{4} < z < \dfrac{15-11}{4}\right) = P(-.25 < z < 1)$
$$= P(-.25 < z < 0) + P(0 < z < 1)$$
$$= .0987 + .3413 = .4400$$
(Using Table IV, Appendix A)

c. $P(x > 20) = P\left(z > \dfrac{20-11}{4}\right) = P(z > 2.25)$
$$= .5 - P(0 < z < 2.25) = .5 - .4878 = .0122$$
(Using Table IV, Appendix A)

5.39 a. Let x = age of powerful woman. Using Table IV, Appendix A,

$$P(55 < x < 60) = P\left(\frac{55-50}{6.4} < z < \frac{60-50}{6.4}\right) = P(.78 < z < 1.56)$$
$$= .4406 - .2823 = .1583$$

b. $P(48 < x < 52) = P\left(\dfrac{48-50}{6.4} < z < \dfrac{52-50}{6.4}\right) = P(-.31 < z < .31)$
$$= .1217 + .1217 = .2434$$

c. $P(x < 35) = P\left(z < \dfrac{35-50}{6.4}\right) = P(z < -2.34) = .5000 - .4904 = .0096$

d. $P(x > 40) = P\left(z > \dfrac{40-50}{6.4}\right) = P(z > -1.56) = .5000 + .4406 = .9406$

e. We need to find the probability that a randomly selected woman is less than or equal to 25. $P(x \le 25) = P\left(z \le \dfrac{25-50}{6.4}\right) = P(z \le -3.91) \approx .5 - .5 = 0$. Since this observation would be very rare (probability almost 0), it is very unlikely that the woman is one of the 50 most powerful women.

5.41   a.   Let x = transmission delay of an RSVP linked wireless device. Using Table IV, Appendix A,

$$P(x < 57) = P\left(z \le \frac{57 - 48.5}{8.5}\right) = P(z \le 1.00) = .5000 + .3413 = .8413$$

   b. $P(40 < x < 60) = P\left(\frac{40 - 48.5}{8.5} < z < \frac{60 - 48.5}{8.5}\right) = P(-1.00 < z < 1.35)$

$$= .3413 + .4115 = .7528$$

5.43   a.   Let x = driver's head injury rating. Using Table IV, Appendix A,

$$P(500 < x < 700) = P\left(\frac{500 - 605}{185} < z < \frac{700 - 605}{185}\right) = P(-.57 < z < .51)$$

$$= .2157 + .1950 = .4107$$

   b. $P(400 < x < 500) = P\left(\frac{400 - 605}{185} < z < \frac{500 - 605}{185}\right) = P(-1.11 < z < -.57)$

$$= .3665 - .2157 = .1508$$

   c. $P(x < 850) = P\left(z < \frac{850 - 605}{185}\right) = P(z < 1.32) = .5 + .4066 = .9066$

   d. $P(x > 1,000) = P\left(z > \frac{1,000 - 605}{185}\right) = P(z > 2.14) = .5 - .4838 = .0162$

   e. $P(x > x_o) = .10 \Rightarrow P\left(z > \frac{x_o - 605}{185}\right) = P(z > z_0)$

      $A_1 = .5 - .1 = .4000$

      Looking up the area .4000 in Table IV gives $z_0 = 1.28$.

      $z_0 = 1.28 = \dfrac{x_o - 605}{185} \Rightarrow x_o = 185(1.28) + 605 = 841.8$

5.45   a.   Let x = ambulance response time at Station A. Then x has a normal distribution with $\mu = 7.5$ and $\sigma = 2.5$. Using Table IV, Appendix A,

$$P(x \le 9) = P\left(z \le \frac{9 - 7.5}{2.5}\right) = P(z \le .60) = .5 + .2257 = .7257.$$ Since this probability is

      less than the requirement of .9, the regulations are not met at EMS station A.

   b. $P(x \le 2) = P\left(z \le \frac{2 - 7.5}{2.5}\right) = P(z \le -2.20) = .5 - .4861 = .0139.$ If the call was serviced

      by Station A, the probability of a response time of 2 minutes or less would be extremely unusual. Thus, we would conclude that the call was probably not serviced by Station A.

5.47 a. Let x = height of women over 20 years old. We are given that $\mu = 64$ and that the distribution of x is normal. However, we are given no information about the standard deviation. The answers to parts a through e will depend on what each person uses as the standard deviation.

f-a. Find x_L and x_U such that $P(x_L < x < x_U) = .50$.

First, we need to find z_0 such that $P(-z_0 < z < z_0) = .50$. Since we know that the center of the z distribution is 0, half of the area or $.50/2 = .25$ will be between $-z_0$ and 0 and half will be between 0 and z_0.

Look up .25 in the body of Table IV, Appendix A to find $z_0 = .67$.

We know

$$P(x_L < x < x_U) = P\left(\frac{x_L - 64}{2.6} < z < \frac{x_U - 64}{2.6}\right) = P(-.67 < z < .67) = .50$$

Thus,

$$\frac{x_L - 64}{2.6} = -.67 \quad \text{and} \quad \frac{x_U - 64}{2.6} = .67$$

$$\Rightarrow x_L = 64 - .67(2.6) = 64 - 1.74 = 62.26$$

$$\Rightarrow x_U = 64 + .67(2.6) = 64 + 1.74 = 65.74$$

f-b. Find x_L and x_U such that $P(x_L < x < x_U) = .75$.

First, we need to find z_0 such that $P(-z_0 < z < z_0) = .75$. Since we know that the center of the z distribution is 0, half of the area or $.75/2 = .375$ will be between $-z_0$ and 0 and half will be between 0 and z_0.

Look up .375 in the body of Table IV, Appendix A to find $z_0 = 1.15$.

We know

$$P(x_L < x < x_U) = P\left(\frac{x_L - 64}{2.6} < z < \frac{x_U - 64}{2.6}\right) = P(-1.15 < z < 1.15) = .75$$

Thus,

$$\frac{x_L - 64}{2.6} = -1.15 \quad \text{and} \quad \frac{x_U - 64}{2.6} = 1.15$$

$$\Rightarrow x_L = 64 - 1.15(2.6) = 64 - 2.99 = 61.01$$

$$\Rightarrow x_U = 64 + 1.15(2.6) = 64 + 2.99 = 66.99$$

f-c. Find x_L and x_U such that $P(x_L < x < x_U) = .90$.

First, we need to find z_0 such that $P(-z_0 < z < z_0) = .90$. Since we know that the center of the z distribution is 0, half of the area or .90/2 = .45 will be between $-z_0$ and 0 and half will be between 0 and z_0.

Look up .45 in the body of Table IV, Appendix A to find $z_0 = 1.645$.

We know

$$P(x_L < x < x_U) = P\left(\frac{x_L - 64}{2.6} < z < \frac{x_U - 64}{2.6}\right) = P(-1.645 < z < 1.645) = .90$$

Thus,

$$\frac{x_L - 64}{2.6} = -1.645 \text{ and } \frac{x_U - 64}{2.6} = 1.645$$

$$\Rightarrow x_L = 64 - 1.645(2.6) = 64 - 4.28 = 59.72$$

$$\Rightarrow x_U = 64 + 1.645(2.6) = 64 + 4.28 = 68.28$$

f-d. Find x_L and x_U such that $P(x_L < x < x_U) = .95$.

First, we need to find z_0 such that $P(-z_0 < z < z_0) = .95$. Since we know that the center of the z distribution is 0, half of the area or .95/2 = .475 will be between $-z_0$ and 0 and half will be between 0 and z_0.

Look up .475 in the body of Table IV, Appendix A to find $z_0 = 1.96$.

We know

$$P(x_L < x < x_U) = P\left(\frac{x_L - 64}{2.6} < z < \frac{x_U - 64}{2.6}\right) = P(-1.96 < z < 1.96) = .95$$

Thus,

$$\frac{x_L - 64}{2.6} = -1.96 \text{ and } \frac{x_U - 64}{2.6} = 1.96$$

$$\Rightarrow x_L = 64 - 1.96(2.6) = 64 - 5.10 = 58.90$$

$$\Rightarrow x_U = 64 + 1.96(2.6) = 64 + 5.10 = 69.10$$

f-e. Find x_L and x_U such that $P(x_L < x < x_U) = .99$.

First, we need to find z_0 such that $P(-z_0 < z < z_0) = .99$. Since we know that the center of the z distribution is 0, half of the area or .99/2 = .495 will be between $-z_0$ and 0 and half will be between 0 and z_0.

We look up .495 in the body of Table IV, Appendix A to find $z_0 = 2.575$.

We know $P(x_L < x < x_U) = P\left(\dfrac{x_L - 64}{2.6} < z < \dfrac{x_U - 64}{2.6}\right) = P(-2.575 < z < 2.575) = .99$

Thus,

$$\dfrac{x_L - 64}{2.6} = -2.575 \quad \text{and} \quad \dfrac{x_U - 64}{2.6} = 2.575$$

$$\Rightarrow x_L = 64 - 2.575(2.6) = 64 - 6.70 = 57.30$$

$$\Rightarrow x_U = 64 + 2.575(2.6) = 64 + 6.70 = 70.70$$

5.49   Let x = number of additional Electoral College votes. Then x has a normal distribution with $\mu = 241.5$ and $\sigma = 49.8$. If the candidate wins the popular vote in California, then he/she needs only $270 - 55 = 215$ additional votes.

$$P(x \geq 215) = P\left(z \geq \dfrac{215 - 241.5}{49.8}\right) = P(z \geq -.53) = .5 + .2019 = .7019$$

5.51   a.   If z is a standard normal random variable,

$Q_L = z_L$ is the value of the standard normal distribution which has 25% of the data to the left and 75% to the right.

Find z_L such that $P(z < z_L) = .25$

$A_1 = .50 - .25 = .25$.

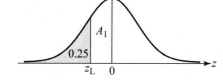

Look up the area $A_1 = .25$ in the body of Table IV of Appendix A; $z_L = -.67$ (taking the closest value). If interpolation is used, $-.675$ would be obtained.

$Q_U = z_U$ is the value of the standard normal distribution which has 75% of the data to the left and 25% to the right.

Find z_U such that $P(z < z_U) = .75$

$A_1 + A_2 = P(z \leq 0) + P(0 \leq z \leq z_U)$
$= .5 + P(0 \leq z \leq z_U) = .75$

Therefore, $P(0 \leq z \leq z_U) = .25$.

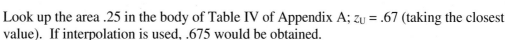

Look up the area .25 in the body of Table IV of Appendix A; $z_U = .67$ (taking the closest value). If interpolation is used, .675 would be obtained.

b. Recall that the inner fences of a box plot are located $1.5(Q_U - Q_L)$ outside the hinges (Q_L and Q_U).

To find the lower inner fence,

$$Q_L - 1.5(Q_U - Q_L) = -.67 - 1.5[.67 - (-.67)]$$
$$= -.67 - 1.5(1.34)$$
$$= -2.68 \quad (-2.70 \text{ if } z_L = -.675 \text{ and } z_U = .675)$$

The upper inner fence is:

$$Q_U + 1.5(Q_U - Q_L) = .67 + 1.5[.67 - (-.67)]$$
$$= .67 + 1.5(1.34)$$
$$= 2.68 \quad (2.70 \text{ if } z_L = -.675 \text{ and } z_U = .675)$$

c. Recall that the outer fences of a box plot are located $3(Q_U - Q_L)$ outside the hinges (Q_L and Q_U).

To find the lower outer fence,

$$Q_L - 3(Q_U - Q_L) = -.67 - 3[.67 - (-.67)]$$
$$= -.67 - 3(1.34)$$
$$= -4.69 \quad (-4.725 \text{ if } z_L = -.675 \text{ and } z_U = .675)$$

The upper outer fence is:

$$Q_U + 3(Q_U - Q_L) = .67 + 3[.67 - (-.67)]$$
$$= .67 + 3(1.34)$$
$$= 4.69 \quad (4.725 \text{ if } z_L = -.675 \text{ and } z_U = .675)$$

d. $P(z < -2.68) + P(z > 2.68)$
    $= 2P(z > 2.68)$
    $= 2(.5000 - .4963)$
    $= 2(.0037) = .0074$
    (Table IV, Appendix A)

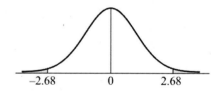

(or $2(.5000 - .4965) = .0070$ if -2.70 and 2.70 are used)

$P(z < -4.69) + P(z > 4.69)$
    $= 2P(z > 4.69)$
    $\approx 2(.5000 - .5000) \approx 0$

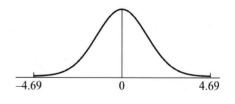

e. In a normal probability distribution, the probability of an observation being beyond the inner fences is only .0074 and the probability of an observation being beyond the outer fences is approximately zero. Since the probabilities are so small, there should not be any observations beyond the inner and outer fences. Therefore, they are probably outliers.

5.53   Several statistical techniques are based on the assumption that the population is approximately normally distributed. Thus, it is important to determine whether the sample data come from a normal population so that the techniques can be properly applied.

5.55   a.   The proportion of measurements that one would expect to fall in the interval $\mu \pm \sigma$ is about .68.

        b.   The proportion of measurements that one would expect to fall in the interval $\mu \pm 2\sigma$ is about .95.

        c.   The proportion of measurements that one would expect to fall in the interval $\mu \pm 3\sigma$ is about .997.

5.57   If the data are normally distributed, then the normal probability plot should be an approximate straight line. Of the three plots, only plot c implies that the data are normally distributed. The data points in plot c form an approximately straight line. In both plots a and b, the plots of the data points do not form a straight line.

5.59   a.   Using MINITAB, the stem-and-leaf display is:

        Stem-and-Leaf Display: Data

```
Stem-and-leaf of Data  N  = 40
Leaf Unit = 10

  14   0   00012223333344
 (10)  0   5556666889
  16   1   001334
  10   1   5689
   6   2   01
   4   2
   4   3   000
   1   3
   1   4   0
```

        The distribution from the stem-and-leaf display looks like it is skewed to the right. Thus, it does not appear that the data are normally distributed.

        b.   Using MINITAB, the descriptive statistics are:

        Descriptive Statistics: Data

```
Variable   N  Mean  StDev  Minimum   Q1  Median    Q3  Maximum
Data      40 105.1   95.8     1.00 37.5    65.0 152.8    401.0
```

        The lower quartile is $Q_L = 37.5$, the upper quartile is $Q_U = 152.8$, and $s = 95.8$.

        c.   The interquartile range is $IQR = Q_U - Q_L = 152.8 - 37.5 = 115.3$.
              $IQR / s = 115.3 / 95.8 = 1.204$. If the data are approximately normal, then this ratio should be 1.3. Since the actual ratio is 1.204 which is close to 1.3, it appears that the data could be approximately normal.

d.   Using MINITAB, the normal probability plot is:

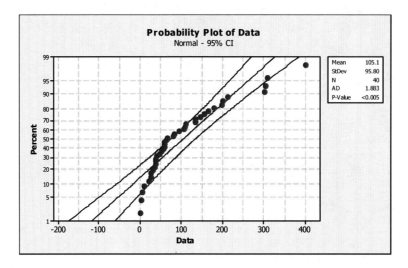

Since the data points do not fall along the straight line, the data do not appear to be normally distributed.

5.61   a. $\text{IQR} = Q_U - Q_L = 54 - 47 = 7$.

b.   From the printout, $s = 6.444$.

c.   If the data are approximately normal, then $\text{IQR} / s \approx 1.3$. For this data, $\text{IQR} / s = 7 / 6.444 = 1.09$. This is somewhat close to 1.3, so the data may be normal.

d.   Using MINITAB, a relative frequency histogram of the data is:

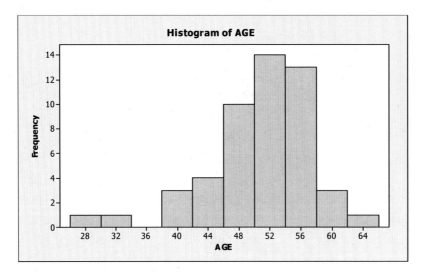

From the histogram, the data appear to be close to mound-shaped (slightly skewed to the left), so the data may be normally distributed.

5.63   The interquartile range is $\mathrm{IQR} = Q_U - Q_L = 6 - 1 = 5$. $\mathrm{IQR}/s = 5/6.09 = .82$. This is much less than the 1.3 that we would expect if the data were normal. It appears that these data are not approximately normal. In addition, $\bar{x} = 4.71$ and $s = 6.09$. Since the number of updates cannot be negative, the data must be skewed to the right since one standard deviation below the mean would be a negative number which is impossible.

5.65   a.   From the graph, the histogram of the difference between the elevation estimates is too peaked. This means that there are too many observations in the very center of the distribution and not enough observations in the tails for the distribution to be normal.

        b.   The interval $\mu \pm 2\sigma$ would be $.28 \pm 2(1.6) \Rightarrow .28 \pm 3.2 \Rightarrow (-2.92, 3.48)$. From the graph, almost all the observations are between are between -2 and 2. Thus, the interval $(-2.92, 3.48)$ will contain more that 95% of the 400 elevation differences.

5.67   **American League**

To determine if the distribution of batting averages is approximately normal, we will run through the tests. Using MINITAB, a histogram of the data with a normal curve included is:

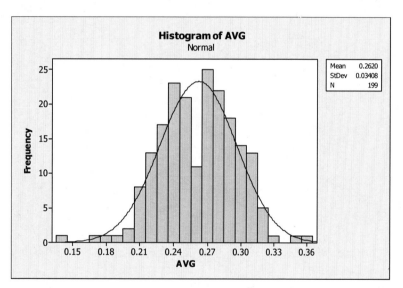

From the graph, the data appear to be approximately mound-shaped. The data may be normal.

Using MINITAB, the descriptive statistics are:

Descriptive Statistics: AVG

Variable	N	Mean	StDev	Minimum	Q1	Median	Q3	Maximum
AVG	199	0.26198	0.03408	0.14400	0.23600	0.26500	0.28600	0.36500

The interval $\bar{x} \pm s \Rightarrow .26198 \pm .03408 \Rightarrow (.22790, .29606)$ contains 133 of the 199 observations. The proportion is $133/199 = .668$. This is very close to the .68 from the Empirical Rule.

The interval $\bar{x} \pm 2s \Rightarrow .26198 \pm 2(.03408) \Rightarrow .26198 \pm .06816 \Rightarrow (.19382, .33014)$ contains 193 of the 199 observations. The proportion is $193/199 = .970$. This is somewhat larger than the .95 from the Empirical Rule.

The interval $\bar{x} \pm 3s \Rightarrow .26198 \pm 3(.03408) \Rightarrow .26198 \pm .10224 \Rightarrow (.15974, .36422)$ contains 197 of the 199 observations. The proportion is $197/199 = .990$. This is very close to the .997 from the Empirical Rule. Thus, it appears that the data may be normal.

The lower quartile is $Q_L = .236$ and the upper quartile is $Q_U = .286$. The interquartile range is $IQR = Q_U - Q_L = .286 - .236 = .050$. From the printout, $s = .03408$. $IQR/s = .050/.03408 = 1.467$. This is a fairly close to the 1.3 that we would expect if the data were normal. Thus, there is evidence that the data may be normal.

Using MINITAB, the normal probability plot is:

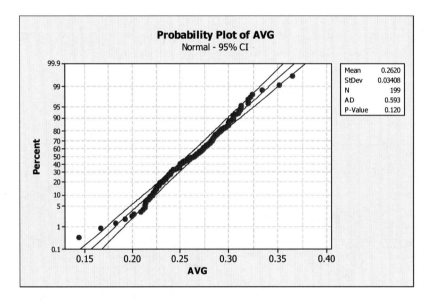

The data are very close to a straight line. Thus, it appears that the data may be normal.

From all of the indicators, it appears that the batting averages from the American League come from an approximate normal distribution.

National League

To determine if the distribution of batting averages is approximately normal, we will run through the tests. Using MINITAB, a histogram of the data with a normal curve included is:

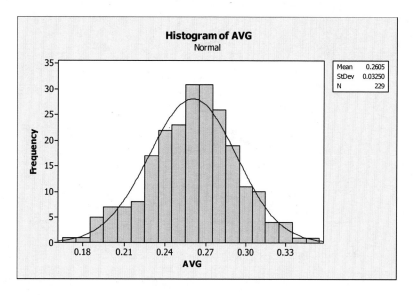

From the graph, the data appear to be approximately mound-shaped. The data may be normal.

Using MINITAB, the descriptive statistics are:

Descriptive Statistics: AVG

```
Variable    N     Mean    StDev  Minimum       Q1   Median       Q3  Maximum
AVG       229  0.26047  0.03250  0.17100  0.23850  0.26100  0.28200  0.35300
```

The interval $\bar{x} \pm s \Rightarrow .26047 \pm (.03250) \Rightarrow (.22797, .29297)$ contains 160 of the 229 observations. The proportion is $160 / 229 = .699$. This is very close to the .68 from the Empirical Rule.

The interval $\bar{x} \pm 2s \Rightarrow .26047 \pm 2(.03250) \Rightarrow .26047 \pm .06500 \Rightarrow (.19547, .32547)$ contains 216 of the 229 observations. The proportion is $216 / 229 = .943$. This is very close to the .95 from the Empirical Rule.

The interval $\bar{x} \pm 3s \Rightarrow .26047 \pm 3(.03250) \Rightarrow .26047 \pm .09750 \Rightarrow (.16297, .35797)$ contains 229 of the 229 observations. The proportion is $229 / 229 = 1.00$. This is very close to the .997 from the Empirical Rule. Thus, it appears that the data may be normal.

The lower quartile is $Q_L = .2385$ and the upper quartile is $Q_U = .282$. The interquartile range is $\text{IQR} = Q_U - Q_L = .282 - .2385 = .0435$. From the printout, $s = .0325$.
$\text{IQR} / s = .0435 / .0325 = 1.338$. This is a very close to the 1.3 that we would expect if the data were normal. Thus, there is evidence that the data may be normal.

Using MINITAB, the normal probability plot is:

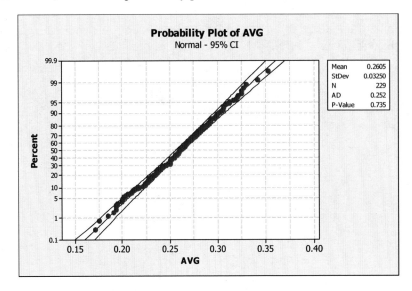

The data are very close to a straight line. Thus, it appears that the data may be normal.

From all of the indicators, it appears that the batting averages from the National League come from an approximate normal distribution.

5.69   We will look at the 4 methods for determining if the data are normal. First, we will look at a histogram of the data. Using MINITAB, the histogram of the driver's head injury ratings is:

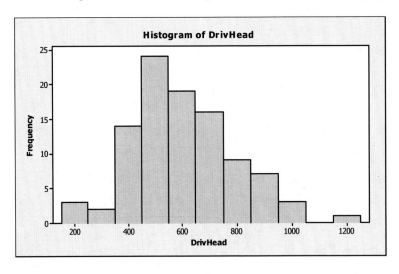

From the histogram, the data appear to be somewhat mound-shaped, but also somewhat skewed to the right. This indicates that the data may be normal.

Next, we look at the intervals $\bar{x} \pm s$, $\bar{x} \pm 2s$, $\bar{x} \pm 3s$. If the proportions of observations falling in each interval are approximately .68, .95, and 1.00, then the data are approximately normal.

Using MINITAB, the summary statistics are:

Descriptive Statistics: DrivHead

```
Variable   N  Mean   StDev  Minimum    Q1  Median     Q3  Maximum
DrivHead  98  603.7  185.4    216.0  475.0   605.0  724.3   1240.0
```

$\bar{x} \pm s \Rightarrow 603.7 \pm 185.4 \Rightarrow (418.3,\ 789.1)$ 68 of the 98 values fall in this interval. The proportion is .69. This is very close to the .68 we would expect if the data were normal.

$\bar{x} \pm 2s \Rightarrow 603.7 \pm 2(185.4) \Rightarrow 603.7 \pm 370.8 \Rightarrow (232.9,\ 974.5)$ 96 of the 98 values fall in this interval. The proportion is .98. This is a fair amount larger than the .95 we would expect if the data were normal.

$\bar{x} \pm 3s \Rightarrow 603.7 \pm 3(185.4) \Rightarrow 603.7 \pm 556.2 \Rightarrow (47.5,\ 1,159.9)$ 97 of the 98 values fall in this interval. The proportion is .99. This is fairly close to the .997 we would expect if the data were normal.

From this method, it appears that the data may be normal.

Next, we look at the ratio of the IQR to s. $IQR = Q_U - Q_L = 724.3 - 475 = 249.3$.

$\dfrac{IQR}{s} = \dfrac{249.3}{185.4} = 1.3$ This is equal to the 1.3 we would expect if the data were normal. This method indicates the data may be normal.

Finally, using MINITAB, the normal probability plot is:

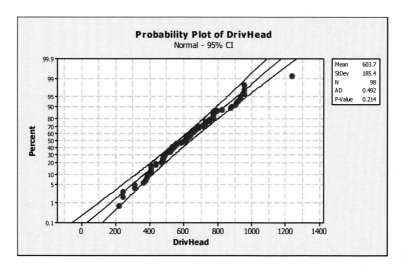

Since the data form a fairly straight line, the data may be normal.

From the 4 different methods, all indications are that the driver's head injury rating data are normal.

5.71   We are given that the mean is 13.2 and the standard deviation is 19.5. We know that the number of minutes per day each student used their laptop for taking notes cannot be negative. Since the standard deviation is larger than the mean, we know that no observations can be more than 1 standard deviation below the mean. Since the Empirical Rule indicates that approximately 16% of the observations will be more than 1 standard deviation below the mean, we know that the distribution of the number of minutes per day each student used their laptop for taking notes is unlikely to be normal.

5.73   When n is large, there often are not Binomial Tables for finding probabilities. If n is large, then finding the probabilities of events by using the formulas can be quite tedious and time consuming. The normal approximation gives very good estimates of the true probabilities.

5.75   a.   In order to approximate the binomial distribution with the normal distribution, the interval $\mu \pm 3\sigma \Rightarrow np \pm 3\sqrt{npq}$ should lie in the range 0 to n.

When $n = 25$ and $p = .4$,
$$np \pm 3\sqrt{npq} \Rightarrow 25(.4) \pm 3\sqrt{25(.4)(1-.4)}$$
$$\Rightarrow 10 \pm 3\sqrt{6} \Rightarrow 10 \pm 7.3485 \Rightarrow (2.6515, 17.3485)$$

Since the interval calculated does lie in the range 0 to 25, we can use the normal approximation.

b. $\mu = np = 25(.4) = 10$ $\sigma^2 = npq = 25(.4)(.6) = 6$

c. $P(x \geq 9) = 1 - P(x \leq 8) = 1 - .274 = .726$ (Table II, Appendix A)

d. $P(x \geq 9) \approx P\left(z \geq \dfrac{(9-.5)-10}{\sqrt{6}}\right)$

$= P(z \geq -.61) = .5000 + .2291 = .7291$
(Using Table IV in Appendix A.)

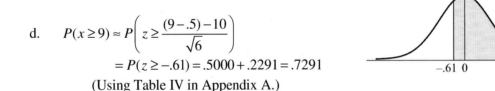

5.77   a.   Using Table II, $P(x \leq 11) = .345$
$$\mu = np = 25(.5) = 12.5 \,,\ \sigma = \sqrt{npq} = \sqrt{25(.5)(.5)} = 2.5$$

Using the normal approximation,
$$P(x \leq 11) \approx P\left(z \leq \frac{(11+.5)-12.5}{2.5}\right) = P(z \leq -.40) = .5 - .1554 = .3446$$
(from Table IV, Appendix A)

b.   Using Table II, $P(x \geq 16) = 1 - P(x \leq 15) = 1 - .885 = .115$

Using the normal approximation,
$$P(x \geq 16) \approx P\left(z \geq \frac{(16-.5)-12.5}{2.5}\right) = P(z \geq 1.2) = .5 - .3849 = .1151$$

(from Table IV, Appendix A)

c. Using Table II, $P(8 \le x \le 16) = P(x \le 16) - P(x \le 7) = .946 - .022 = .924$

Using the normal approximation,

$$P(8 \le x \le 16) \approx P\left(\frac{(8-.5)-12.5}{2.5} \le z \le \frac{(16+.5)-12.5}{2.5} \right)$$
$$= P(-2.0 \le z \le 1.6) = .4772 + .4452 = .9224$$

(from Table IV, Appendix A)

5.79 x is a binomial random variable with $n = 100$ and $p = .4$.
$$\mu \pm 3\sigma \Rightarrow np \pm 3\sqrt{npq} \Rightarrow 100(.4) \pm 3\sqrt{100(.4)(1-.4)}$$
$$\Rightarrow 40 \pm 3(4.8990) \Rightarrow (25.303, 54.697)$$

Since the interval lies in the range 0 to 100, we can use the normal approximation to approximate the probabilities.

a. $P(x \le 35) \approx P\left(z \le \frac{(35+.5)-40}{4.899} \right)$

$= P(z \le -.92)$

$= .5000 - .3212 = .1788$

(Using Table IV in Appendix A.)

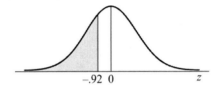

b. $P(40 \le x \le 50)$

$\approx P\left(\frac{(40-.5)-40}{4.899} \le z \le \frac{(50+.5)-40}{4.899} \right)$

$= P(-.10 \le z \le 2.14)$

$= P(-.10 \le z \le 0) + P(0 \le z \le 2.14)$

$= .0398 + .4838 = .5236$

(Using Table IV in Appendix A.)

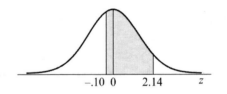

c. $P(x \ge 38) \approx P\left(z \ge \frac{(38-.5)-40}{4.899} \right)$

$= P(z \ge -.51)$

$= .5000 + .1950 = .6950$

(Using Table IV in Appendix A.)

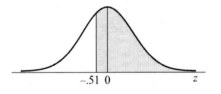

5.81 a. Let x = number of bottles that contain tap water in 65 trials. Then x is a binomial random variable with $n = 65$ and $p = .25$.

$E(x) = \mu = np = 65(.25) = 16.25$

b. $\sigma = \sqrt{npq} = \sqrt{65(.25)(.75)} = \sqrt{12.1875} = 3.4911$

c. $z = \dfrac{x-\mu}{\sigma} = \dfrac{20-16.25}{3.4911} = 1.07$

d. $\mu \pm 3\sigma \Rightarrow 16.25 \pm 3(3.4911) \Rightarrow 16.25 \pm 10.4733 \Rightarrow (5.7767,\ 26.7233)$ Since the interval lies in the range 0 to 65, we can use the normal approximation to approximate the binomial probability.

$$P(x \ge 20) = P\left(z \ge \frac{(20-.5)-16.25}{3.4911}\right) = P(z \ge .93) = .5 - .3238 = .1762$$

(Using Table IV, Appendix A)

5.83   a. $n = 10,000$ and $p = .15$

$$E(x) = \mu = np = 10,000(.15) = 1,500$$

$$\sigma^2 = npq = 10,000(.15)(.85) = 1,275$$

b. $\sigma = \sqrt{1,275} = 35.707$

Using the normal approximation to the binomial,

$$P(x > 1,600) \approx P\left(z > \frac{(1,600+.5)-1,500}{35.707}\right) = P(z > 2.81) = .5 - .4975 = .0025$$
(from Table IV, Appendix A)

c.   Using the normal approximation to the binomial,

$$P(x > 6,500) \approx P\left(z > \frac{(6,500+.5)-1,500}{35.707}\right) = P(z > 140.04) \approx .5 - .5 = 0$$
(from Table IV, Appendix A)

Since this probability is essentially 0, we would not expect to see more than 6,500 deaths in any one year.

5.85   Let x = number of guppies that survive in 300 trials. Then x is a binomial random variable with $n = 300$ and $p = .6$.

$\mu \pm 3\sigma \Rightarrow np \pm 3\sqrt{npq} \Rightarrow 300(.6) \pm 3\sqrt{300(.6)(1-.6)} \Rightarrow 180 \pm 3\sqrt{72} \Rightarrow 180 \pm 25.456$
$\Rightarrow (154.544,\ 205.456)$

Since the interval lies in the range 0 to 300, we can use the normal approximation to approximate the binomial probability.

$$P(x < 100) = P\left(z < \frac{(100-.5)-180}{\sqrt{72}}\right) = P(z < -9.49) \approx .5 - .5 = 0$$
(Using Table IV, Appendix A)

5.87 a.   For $n = 100$ and $p = .01$:

$$\mu \pm 3\sigma \Rightarrow np \pm 3\sqrt{npq} \Rightarrow 100(.01) \pm 3\sqrt{100(.01)(.99)}$$
$$\Rightarrow 1 \pm 3(.995) \Rightarrow 1 \pm 2.985 \Rightarrow (-1.985, 3.985)$$

Since the interval does not lie in the range 0 to 100, we cannot use the normal approximation to approximate the probabilities.

b.   For $n = 100$ and $p = .5$:

$$\mu \pm 3\sigma \Rightarrow np \pm 3\sqrt{npq} \Rightarrow 100(.5) \pm 3\sqrt{100(.5)(.5)}$$
$$\Rightarrow 50 \pm 3(5) \Rightarrow 50 \pm 15 \Rightarrow (35, 65)$$

Since the interval lies in the range 0 to 100, we can use the normal approximation to approximate the probabilities.

c.   For $n = 100$ and $p = .9$:

$$\mu \pm 3\sigma \Rightarrow np \pm 3\sqrt{npq} \Rightarrow 100(.9) \pm 3\sqrt{100(.9)(.1)}$$
$$\Rightarrow 90 \pm 3(3) \Rightarrow 90 \pm 9 \Rightarrow (81, 99)$$

Since the interval lies in the range 0 to 100, we can use the normal approximation to approximate the probabilities.

5.89 Let x = number of guests who participate in the conservation program. Then x is a binomial random variable with $n = 200$ and $p = .45$.

$$\mu = np = 200(.45) = 90 \quad \text{and} \quad \sigma = \sqrt{npq} = \sqrt{200(.45)(.55)} = \sqrt{49.5} = 7.0356$$

$$\mu \pm 3\sigma \Rightarrow 90 \pm 3(7.0356) \Rightarrow 90 \pm 21.1068 \Rightarrow (68.8932, 111.1068)$$

Since the above is completely contained in the interval 0 to 200, the normal approximation is valid.

$$P(x > 110) = P\left(z > \frac{(110 + .5) - 90}{7.0356}\right) = P(z > 2.91) = .5 - .4982 = .0018$$

Since this probability is so small, it is very unlikely that the director's claim is correct.

5.91 a.   The expected number of passengers detained will be:

$$E(x) = \mu = np = 1,500(.2) = 300$$

b.   For $n = 4,000$, $E(x) = \mu = np = 4,000(.2) = 800$

c. $$P(x > 600) \approx P\left(z > \frac{(600 + .5) - 800}{\sqrt{4000(.2)(.8)}}\right) = P(z > -7.89) = .5 + .5 = 1.0$$

5.93   The amount of time or distance between occurrences of random events can be described by the exponential probability distribution. This distribution has the property that the mean is equal to the standard deviation.

5.95   For $\theta = 3$, $f(x) = \dfrac{1}{3}e^{-x/3}$. Using Table V, Appendix A and a calculator:

x	f(x)
0	.333
1	.239
2	.171
3	.123
4	.088
5	.063
6	.045

For $\theta = 1$, $f(y) = e^{-y}$. Using Table V, Appendix A:

x	f(x)
0	1.000
1	.368
2	.135
3	.050
4	.018
5	.007
6	.002

Using MINITAB, the plot is:

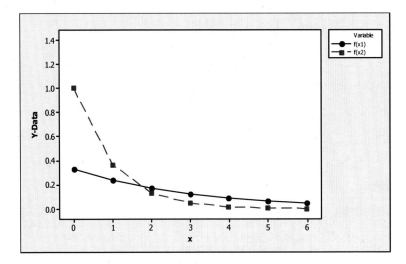

5.97 $P(x \geq a) = e^{-a/\theta} = e^{-a/1} = e^{-a}$. Using Table V, Appendix A:

a. $P(x > 1) = e^{-1} = .367879$

b. $P(x \leq 3) = 1 - P(x > 3) = 1 - e^{-3} = 1 - .049787 = .950213$

c. $P(x > 1.5) = e^{-1.5} = .223130$

d. $P(x \le 5) = 1 - P(x > 5) = 1 - e^{-5} = 1 - .006738 = .993262$

5.99 $\mu = \theta = 1$, $\sigma = \theta = 1$

$\mu \pm 2\sigma \Rightarrow 1 \pm 2(1) \Rightarrow 1 \pm 2 \Rightarrow (-1, 3)$

$P(-1 < x < 3) = P(0 < x < 3) = 1 - P(x \ge 3) = 1 - e^{-3/1} = 1 - e^{-1} = 1 - .049787 = .950213$

(from Table V, Appendix A)

5.101 a. $P(x > 2) = e^{-2/2.5} = e^{-.8} = .449329$ (using Table V, Appendix A)

b. $P(x < 5) = 1 - P(x \ge 5) = 1 - e^{-5/2.5} = 1 - e^{-2} = 1 - .135335 = .864665$

(Using Table V, Appendix A)

5.103 a. Let x = time between successive email notifications. Then x has an exponential distribution with $\mu = \theta = 95$. Two minutes is equal to 120 seconds.

$P(x \ge 120) = e^{-120/95} = e^{-1.263} = .282760$

b. Using MINITAB, the descriptive statistics are:

Descriptive Statistics: INTTIME

Variable	N	Mean	StDev	Minimum	Q1	Median	Q3	Maximum
INTTIME	267	95.52	91.54	1.86	30.59	70.88	133.34	513.52

The sample mean is 95.52 which is very close to 95. In addition, the standard deviation is 91.54, which is also close to 95, which is the standard deviation of the exponential distribution.

Using MINITAB, a histogram of the data with an exponential distribution with mean 95 superimposed on the graph is:

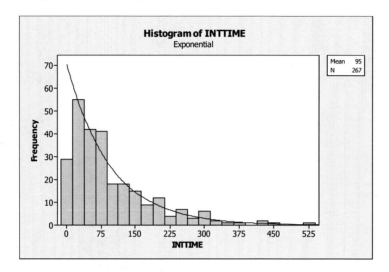

Based on the sample mean and standard deviation along with the graph, it appears that the data follow an exponential distribution with $\theta = 95$.

5.105   a.   Let x = length of time elapsed before the winning goal is scored. Then x has an exponential distribution with $\mu = \theta = 9.15$.

$$P(x \le 3) = 1 - P(x > 3) = 1 - e^{-3/9.15} = 1 - e^{-.327869} = 1 - .720457 = .279543$$

   b. $P(x > 20) = e^{-20/9.15} = e^{-2.185792} = .1123887$

5.107   a.   For $\mu = \theta = 17$. To graph the distribution, we will pick several values of x and find the value of $f(x)$, where x = time between arrivals of the smaller craft at the pier.

$$f(x) = \frac{1}{\theta} e^{-x/\theta} = \frac{1}{17} e^{-x/17}$$

$$f(1) = \frac{1}{17} e^{-1/17} = .0555$$

$$f(3) = \frac{1}{17} e^{-3/17} = .0493$$

$$f(5) = \frac{1}{17} e^{-5/17} = .0438$$

$$f(7) = \frac{1}{17} e^{-7/17} = .0390$$

$$f(10) = \frac{1}{17} e^{-10/17} = .0327$$

$$f(15) = \frac{1}{17} e^{-15/17} = .0243$$

$$f(20) = \frac{1}{17} e^{-20/17} = .0181$$

$$f(25) = \frac{1}{17}e^{-25/17} = .0135$$

The graph is:

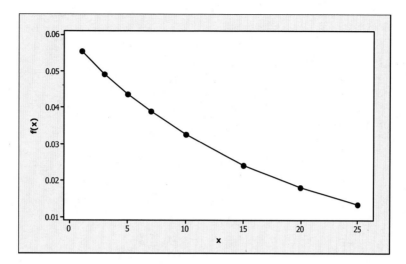

b. We want to find the probability that the time between arrivals is less than 15 minutes.

$$P(x < 15) = 1 - P(x \geq 15) = 1 - e^{-15/17} = 1 - .4138 = .5862$$

5.109 a. Let x = time till failure at the end of the product's lifetime. Then x has an exponential distribution with $\theta = 500$.

$$P(x < 700) = 1 - P(x > 700) = 1 - e^{-700/500} = 1 - e^{-1.4} = 1 - .246597 = .753403$$

b. Let y = time till failure during the product's normal lifetime. Then y has a uniform distribution with $c = 100$ and $d = 1000$.

$$P(y < 700) = \frac{700 - 100}{1000 - 100} = \frac{600}{900} = .6667$$

c. $P(x < 830) = 1 - P(x > 830) = 1 - e^{-830/500} = 1 - e^{-1.66} = 1 - .190139 = .809861$

$$P(y < 830) = \frac{830 - 100}{1000 - 100} = \frac{730}{900} = .8111$$

The two probabilities, .809861 and .8111, are almost the same.

5.111 a. For $\theta = 250$, $P(x > a) = e^{-a/250}$

For $a = 300$ and $b = 200$, show $P(x > a + b) \geq P(x > a)P(x > b)$

$$P(x > 300 + 200) = P(x > 500) = e^{-500/250} = e^{-2} = .135335$$

$$P(x > 300)P(x > 200) = e^{-300/250}e^{-200/250} = e^{-1.2}e^{-.8} = .301194(.449329) = .135335$$

Since $P(x > 300 + 200) = P(x > 300)P(x > 200)$, then
$P(x > 300 + 200) \geq P(x > 300)P(x > 200)$

Also, show $P(x > 300 + 200) \leq P(x > 300)P(x > 200)$. Since we already showed that
$P(x > 300 + 200) = P(x > 300)P(x > 200)$,
then $P(x > 300 + 200) \leq P(x > 300)P(x > 200)$.

b.   Let $a = 50$ and $b = 100$. Show $P(x > a + b) \geq P(x > a)P(x > b)$

$$P(x > 50 + 100) = P(x > 150) = e^{-150/250} = e^{-.6} = .548812$$

$$P(x > 50)P(x > 100) = e^{-50/250}e^{-100/250} = e^{-.2}e^{-.4} = .818731(.670320) = .548812$$

Since $P(x > 50 + 150) = P(x > 50)P(x > 100)$, then
$P(x > 50 + 150) \geq P(x > 50)P(x > 100)$

Also, show $P(x > 50 + 150) \leq P(x > 50)P(x > 100)$. Since we already showed that
$P(x > 50 + 150) = P(x > 50)P(x > 100)$, then $P(x > 50 + 150) \leq P(x > 50)P(x > 100)$.

c.   Show $P(x > a + b) \geq P(x > a)P(x > b)$
$$P(x > a + b) = e^{-(a+b)/250} = e^{-a/250}e^{-b/250} = P(x > a)P(x > b)$$

5.113   a.   For the probability density function, $f(x) = \dfrac{e^{-x/7}}{7}$, $x > 0$, x is an exponential random variable.

b.   For the probability density function, $f(x) = \dfrac{1}{20}$, $5 < x < 25$, x is a uniform random variable.

c.   For the probability function, $f(x) = \dfrac{e^{-.5[(x-10)/5]^2}}{5\sqrt{2\pi}}$, x is a normal random variable.

5.115   a. $P(z \leq 2.1) = A_1 + A_2 = P(z < 0) + P(0 \leq z \leq 2.1)$
$= .5 + .4821 = .9821$

b. $P(z \geq 2.1) = A_2 = .5 - A_1 = .5 - P(0 \leq z \leq 2.1)$
$= .5 - .4821 = .0179$

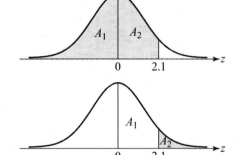

c. $P(z \geq -1.65) = A_1 + A_2 = P(-1.65 \leq z \leq 0) + .5$
$= .4505 + .5 = .9505$

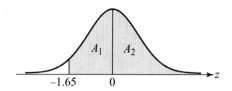

d. $P(-2.13 \leq z \leq -.41)$
$= P(-2.13 \leq z \leq 0) - P(-.41 \leq z \leq 0)$
$= .4834 - .1591 = .3243$

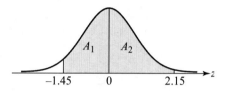

e. $P(-1.45 \leq z \leq 2.15) = A_1 + A_2$
$= P(-1.45 \leq z \leq 0) - P(0 \leq z \leq 2.15)$
$= .4265 + .4842 = .9107$

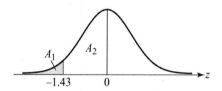

f. $P(z \leq -1.43) = A_1 = .5 - A_2$
$= .5 - P(-1.43 \leq z \leq 0)$
$= .5 - .4236 = .0764$

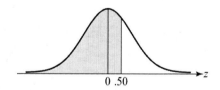

5.117 a. $P(x \leq 75) = P\left(z \leq \dfrac{75 - 70}{10}\right)$
$= P(z \leq .50)$
$= .5 + .1915 = .6915$

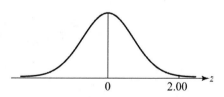

b. $P(x \geq 90) = P\left(z \geq \dfrac{90 - 70}{10}\right)$
$= P(z \geq 2.00)$
$= .5 - .4772 = .0228$

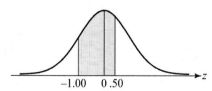

c. $P(60 \leq x \leq 75) = P\left(\dfrac{60 - 70}{10} \leq z \leq \dfrac{75 - 70}{10}\right)$
$= P(-1.00 \leq z \leq .50)$
$= .3413 + .1915 = .5328$

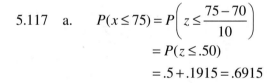

d. This is the probability of the complement of the event in part **a**. Therefore:
$P(x > 75) = 1 - P(x \leq 75) = 1 - .6915 = .3085$

e. $P(x=75)=0$ (True for any continuous random variable.)

f. $P(x \le 95) = P\left(z \le \dfrac{95-70}{10} \right)$

$= P(z \le 2.50)$

$= .5 + .4938 = .9938$

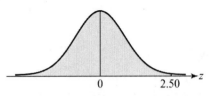

5.119 $\mu = np = 100(.5) = 50$, $\sigma = \sqrt{npq} = \sqrt{100(.5)(.5)} = 5$

a. $P(x \le 48) = P\left(z \le \dfrac{(48+.5)-50}{5} \right)$

$= P(z \le -.30)$

$= .5 - .1179 = .3821$

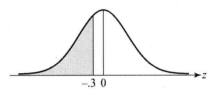

b. $P(50 \le x \le 65)$

$= P\left(\dfrac{(50-.5)-50}{5} \le z \le \dfrac{(65+.5)-50}{5} \right)$

$= P(-.10 \le z \le 3.10)$

$= .0398 + .5 = .5398$

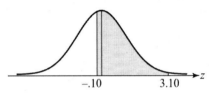

c. $P(x \ge 70) = P\left(z \ge \dfrac{(70-.5)-50}{5} \right)$

$= P(z \ge 3.90)$

$= .5 - .5 = 0$

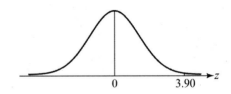

d. $P(55 \le x \le 58)$

$= P\left(\dfrac{(55-.5)-50}{5} \le z \le \dfrac{(58+.5)-50}{5} \right)$

$= P(.90 \le z \le 1.70)$

$= P(0 \le z \le 1.70) - P(0 \le z \le .90)$

$= .4554 - .3159 = .1395$

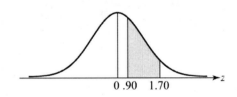

e. $P(x = 62)$

$= P\left(\dfrac{(62-.5)-50}{5} \le z \le \dfrac{(62+.5)-50}{5} \right)$

$= P(2.30 \le z \le 2.50)$

$= P(0 \le z \le 2.50) - P(0 \le z \le 2.30)$

$= .4938 - .4893 = .0045$

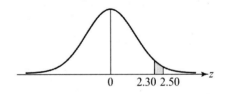

f. $P(x \leq 49 \text{ or } x \geq 72)$

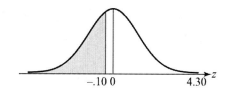

$$= P\left(z \leq \frac{(49+.5)-50}{5}\right) + P\left(z \geq \frac{(72-.5)-50}{5}\right)$$

$$= P(z \leq -.10) + P(z \geq 4.30)$$

$$= (.5 - .0398) + (.5 - .5) = .4602$$

5.121   Based only on the histogram provided, I would say that the data are probably not normally distributed. There are several observations that are very small compared to the rest of the data and some that are very large compared to the rest of the data.

5.123   Let x = score for first time students on mathematics achievement test. Then x is a normal random variable with $\mu = 77$ and $\sigma = 7.3$.

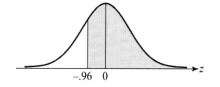

$$P(x \geq 70) = P\left(z \geq \frac{70-77}{7.3}\right)$$

$$= P(z \geq -.96)$$

$$= .3315 + .5 = .8315$$

Thus, 83.15% of students will pass the test the first time.

5.125   a.   Let x = alkalinity level of water specimens collected from the Han River.

Using Table IV, Appendix A,

$$P(x > 45) = P\left(z > \frac{45-50}{3.2}\right) = P(z > -1.56) = .5 + .4406 = .9406 \,.$$

b.   Using Table IV, Appendix A,

$$P(x < 55) = P\left(z < \frac{55-50}{3.2}\right) = P(z < 1.56) = .5 + .4406 = .9406 \,.$$

c.   Using Table IV, Appendix A,

$$P(51 < x < 52) = P\left(\frac{51-50}{3.2} < z < \frac{52-50}{3.2}\right) = P(.31 < z < .63) = .2357 - .1217 = .1140 \,.$$

5.127   Let x = number of African-Americans with sickle-cell anemia. Then x is a binomial random variable with n = 1,000 and p = .08.

$$\mu = np = 1,000(.08) = 80 \text{ and } \sigma = \sqrt{npq} = \sqrt{1,000(.08)(.92)} = 8.579$$

a. $P(x > 175) = P\left(z > \dfrac{(175+.5)-80}{8.579}\right) = P(z > 11.13) \approx .5 - .5 = 0$

b. $P(x<140)=P\left(z<\dfrac{(140-.5)-80}{8.579}\right)=P(z<6.94)=.5+.5=1.0$

5.129   For the uniform distribution,

$$f(x)=\begin{cases}\dfrac{1}{d-c}=\dfrac{1}{90-75}=\dfrac{1}{15} & 75\le x\le 90 \\ 0 & \text{otherwise}\end{cases}$$

a. $P(x>80)=(90-80)\left(\dfrac{1}{15}\right)=.667$

b. $P(80<x<85)=(85-80)\left(\dfrac{1}{15}\right)=.333$

c. $E(x)=\mu=\dfrac{c+d}{2}=\dfrac{75+90}{2}=82.5^{\circ}\text{F}$

5.131   The $\text{IQR}=Q_{U}-Q_{L}=34-14=20$. $\text{IQR}/s=20/12.8=1.5625$. If the data are normally distributed, this ratio should be close to 1.3. Since 1.5625 is fairly close to 1.3, this indicates that the data are approximately normal.

5.133   a.   Using Table IV, Appendix A, with $\mu=9.06$ and $\sigma=2.11$,

$$P(x<6)=\left(z<\dfrac{6-9.06}{2.11}\right)=P(z<-1.45)=.5-P(-1.45\le z\le 0)$$
$$=.5-.4265=.0735$$

b. $P(8<x<10)=P\left(\dfrac{8-9.06}{2.11}<z<\dfrac{10-9.06}{2.11}\right)=P(-.50<z<.45)$
$$=P(-.50<z<0)+P(0<z<.45)=.1915+.1736=.3651$$

c. $P(x<x_{0})=.2000$.

If $P(x<x_{0})=.2000$, then $P\left(z<\dfrac{x_{0}-9.06}{2.11}\right)=P(z<z_{0})=.2000$

If $P(z<z_{0})=.2000$, then $P(z_{0}<z<0)=.3000$. Looking up .3000 in Table IV, the z-score is .84. Since $z_{0}<0$, $z_{0}=-.84$. Now, we must convert z_{0} back to an x score.

$$z_{0}=\dfrac{x_{0}-9.06}{2.11}\Rightarrow-.84=\dfrac{x_{0}-9.06}{2.11}\Rightarrow x_{0}=2.11(-.84)+9.06=7.29$$

5.135 $c = 0, d = 4$

$$f(x) = \begin{cases} \dfrac{1}{d-c} = \dfrac{1}{4-0} = \dfrac{1}{4} & 0 \le x \le 4 \\ 0 & \text{otherwise} \end{cases}$$

$$\mu = \frac{c+d}{2} = \frac{0+4}{2} = 2, \ \sigma = \frac{d-c}{\sqrt{12}} = \frac{4-0}{\sqrt{12}} = 1.155$$

$$15 \text{ seconds} = \frac{15}{60} = .25 \text{ minutes}$$

It takes .25 minutes to go from floor 2 to floor 1. Thus, we must find the probability that the waiting time is less than $1.5 - .25 = 1.25$.

$$P(x < 1.25) = (1.25 - 0)\frac{1}{4} = 1.25\left(\frac{1}{4}\right) = .3125$$

5.137 Let x = velocity of a galaxy located within the galaxy cluster A2142. Then x is a normal random variable with mean $\mu = 27,117$ km/s and standard deviation $\sigma = 1,280$ km/s.

$$P(x \le 24,350) = P\left(z \le \frac{24,350 - 27,117}{1,280}\right) = P(z \le -2.16) = .5 - .4846 = .0154$$

Since the probability of observing a galaxy from the galaxy cluster A2142 with a velocity of 24,350 km/s or slower is so small ($p = .0154$), it would be very unlikely that the galaxy observed came from the galaxy cluster A2142.

5.139 Let x = number of beech trees damaged by fungi in 200 trees. Then x is a binomial random variable with $n = 200$ and $p = .25$.

$$\mu = np = 200(.25) = 50$$

$$\sigma = \sqrt{npq} = \sqrt{200(.25)(.75)} = \sqrt{37.5} = 6.124$$

$$P(x > 100) \approx P\left(z > \frac{(100 + .5) - 50}{6.124}\right) = P(z > 8.25) \approx 0$$

Since this probability is so small, it is almost impossible for more than half of the trees to have fungi damage.

5.141   First, we will graph the data. Using MINITAB, a histogram of the data with a normal
distribution superimposed on the graph is:

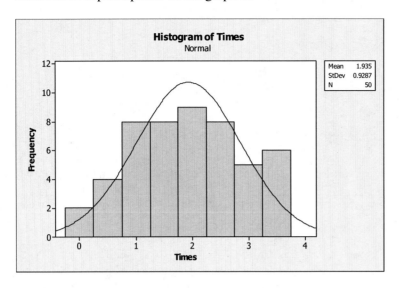

The data are somewhat mound shaped. The data could be normally distributed.

The descriptive statistics of the data are:

Descriptive Statistics: Times

Variable	N	Mean	StDev	Minimum	Q1	Median	Q3	Maximum
Times	50	1.935	0.929	0.0100	1.218	1.835	2.645	3.500

We know if the data are normally distributed, then about 68% of the observations will fall
within 1 standard deviations of the mean, about 95% of the observations will fall within 2
standard deviations of the mean, and about all of the observations will fall within 3 standard
deviations of the mean.

$\bar{x} \pm s \Rightarrow 1.935 \pm .929 \Rightarrow (1.006, 2.864)$. 33 of the 50 or 33/50 = .66 of the observations are
within 1 standard deviation of the mean. This is close to the .68 if the data are normal.

$\bar{x} \pm 2s \Rightarrow 1.935 \pm 2(.929) \Rightarrow 1.935 \pm 1.858 \Rightarrow (.077, 3.793)$. 49 of the 50 or 49/50 = .98 of the
observations are within 2 standard deviations of the mean. This is higher than the .95 if the
data are normal.

$\bar{x} \pm 3s \Rightarrow 1.935 \pm 3(.929) \Rightarrow 1.935 \pm 2.787 \Rightarrow (-.852, 4.722)$. 50 of the 50 or 59/50 = 1.00 of
the observations are within 3 standard deviations of the mean. This is equal to the 1.00 if the
data are normal.

The data appear to not be exactly normal.

$IQR = 3.5 - 1.218 = 2.282$ and $s = .929$. $\dfrac{IQR}{s} = \dfrac{2.282}{.929} = 2.456$. This is much larger than the
1.3 expected is the data are normal. This indicates that the data are probably not normal.

Finally, the probability plot of the data using MINITAB is:

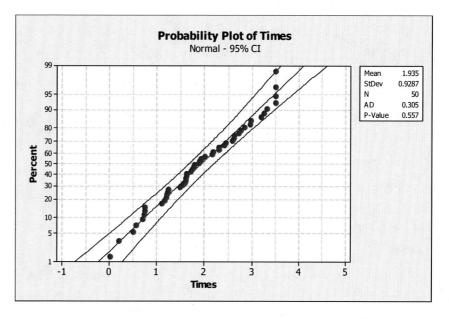

If the data are normal, then the normal probability plot should be a straight line. On the right side of the graph, the plot of the data is almost a vertical line. There is some indication that the data are not normally distributed.

Thus, it appears that the data may not be normal.

5.143   a.   Using Table IV, Appendix A, with $\mu = 50$ and $\sigma = 12$,

$$P(x > d) = .30 \Rightarrow P\left(z > \frac{d-50}{12}\right) = P(z > z_0) = .30$$

$A_1 = .5 - .30 = .2000$
Looking up the area .2000 in Table IV gives $z_0 = .52$.

$$z_0 = .52 = \frac{d-50}{12} \Rightarrow d = 12(.52) + 50 = 56.24 \text{ cm}$$

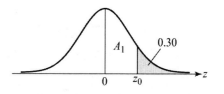

  b.   Let x = breast height diameter of the western hemlock. Then x has an exponential distribution with a mean of 30 cm.

$$P(x > 25) = e^{-25/30} = e^{-.833333} = .434598$$

5.145   a.   (i) $P(0 < z < 1.2) \approx z(4.4 - z)/10 = 1.2(4.4 - 1.2)/10 = .384$

       (ii) $P(0 < z < 2.5) \approx .49$

      (iii) $P(z > .8) = .5 - P(0 < z < .8) \approx .5 - .8(4.4 - .8)/10 = .5 - .288 = .212$

      (iv) $P(z < 1.0) = .5 + P(0 < z < 1.0) = .5 + 1(4.4 - 1)/10 = .5 + .34 = .84$

b.   Using Table IV, Appendix A:

  (i) $P(0 < z < 1.2) = .3849$

  (ii) $P(0 < z < 2.5) = .4938$

  (iii) $P(z > .8) = .5 - P(0 < z < .8) = .5 - .2881 = .2119$

  (iv) $P(z < 1.0) = .5 + P(0 < z < 1.0) = .5 + .3413 = .8413$

c.   For each part, the absolute error is:

  (i) $\text{Error} = |.384 - .3849| = .0009$

  (ii) $\text{Error} = |.49 - .4938| = .0038$

  (iii) $\text{Error} = |.212 - .2119| = .0001$

  (iv) $\text{Error} = |.84 - .8413| = .0013$

  In all cases, the absolute error is less than .0052.

5.147   Since we know for a mound-shaped distribution that approximately 95% of the data lies within two standard deviations of the mean, by symmetry approximately 2.5% would fall below 2 standard deviations below the mean and approximately 2.5% would fall more than 2 standard deviations above the mean. Therefore, the amount of time allotted should be two standard deviations above the mean. This is:

$$\mu + 2\sigma = 40 + 2(6) = 52 \text{ minutes}$$

5.149   From Table IV, Appendix A, and $\sigma = .4$:

$$P(x > 6) = .01$$

$$P(x > 6) = P\left(z > \frac{6 - \mu}{.4}\right) = P(z > z_0) = .01$$

$A_1 = .5 - .01 = .4900$
From Table IV, $z_0 = 2.33$

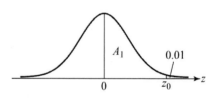

$$2.33 = \frac{6 - \mu}{.4} \Rightarrow \mu = 6 - 2.33(.4) = 5.068$$

5.151   Using Table IV, Appendix A, with $\mu = .27$ and $\sigma = .04$,

$$P(x \leq .14) = P\left(z \leq \frac{.14 - .27}{.04}\right) = P(z < -3.25) \approx .5 - .5 = .0000$$

Since the probability of observing a reading of .14 or smaller is so small ($p \approx .0000$) if the pressure is .1 MPa, it would be extremely unlikely that this reading was obtained at a pressure of .1 MPa.

5.153 a. Class I (very bright) consists of those with IQs above the 95th percentile. Thus, there would be .05 or 5% of the people in Class I.

Class II (bright) consists of those with IQs between the 75th and 95th percentile. Thus, there would be $.95 - .75 = .20$ or 20% of the people in Class II.

Class III (normal) consists of those with IQs between the 25th and 75th percentile. Thus, there would be $.75 - .25 = .50$ or 50% of the people in Class III.

Class IV (dull) consists of those with IQs between the 5th and 25th percentile. Thus, there would be $.25 - .05 = .20$ or 20% of the people in Class IV.

Class V (very dull) consists of those with IQs below the 5th percentile. Thus, there would be .05 or 5% of the people in Class V.

b. Suppose we pick a pair of percentiles and compute the corresponding z-scores. Suppose we compute the z-scores for the 50th and 55th percentiles.

The z-score corresponding to the 50th percentile is $z = 0$. The z-score corresponding to the 55th percentile is:

$$P(z < z_0) = .55 \Rightarrow z_0 = .13$$

Now, suppose we pick another pair of percentiles that are also 5 points apart. Suppose we compute the z-scores for the 94th and 99th percentiles.

The z-score corresponding to the 94th percentile is:

$$P(z < z_0) = .94 \Rightarrow z_0 = 1.555$$

The z-score corresponding to the 99th percentile is:

$$P(z < z_0) = .99 \Rightarrow z_0 = 2.33$$

The z-scores corresponding to the 50th and 55th percentiles are 0 and .13. The difference is $.13 - 0 = .13$. The z-scores corresponding to the 94th and 99th percentiles are 1.555 and 2.33. The difference is $2.33 - 1.555 = .775$. Even though the differences in the percentiles are the same, the differences in the z-scores are much different. The information in the z-scores is more informative.

c. If the distribution of the IQs is skewed to the right, then the tail to the right is much longer than the tail to the left. If the distribution of the IQs is skewed to the left, then the tail to the left is much longer than the tail to the right. However, the proportions in the 5 cognitive classes would not differ, regardless of the shape of the distribution.

Class I (very bright) consists of those with IQs above the 95th percentile. Thus, there would be .05 or 5% of the people in Class I.

Class II (bright) consists of those with IQs between the 75th and 95th percentile. Thus, there would be $.95 - .75 = .20$ or 20% of the people in Class II.

Class III (normal) consists of those with IQs between the 25th and 75th percentile. Thus, there would be $.75 - .25 = .50$ or 50% of the people in Class III.

Class IV (dull) consists of those with IQs between the 5th and 25th percentile. Thus, there would be $.25 - .05 = .20$ or 20% of the people in Class IV.

Class V (very dull) consists of those with IQs below the 5th percentile. Thus, there would be .05 or 5% of the people in Class V.

Chapter

6

Sampling Distributions

6.1   A population parameter is a numerical descriptive measure of a population. Because it is based on the observations in a population, its value is almost always unknown.

A sample statistic is a numerical descriptive measure of a sample. It is calculated from the observations in the sample.

6.3   a–b. The different samples of $n = 2$ with replacement and their means are:

Possible Samples	$\bar{x}$	Possible Samples	$\bar{x}$
0, 0	0	4, 0	2
0, 2	1	4, 2	3
0, 4	2	4, 4	4
0, 6	3	4, 6	5
2, 0	1	6, 0	3
2, 2	2	6, 2	4
2, 4	3	6, 4	5
2, 6	4	6, 6	6

c.   Since each sample is equally likely, the probability of any 1 being selected is

$$\frac{1}{4}\left(\frac{1}{4}\right) = \frac{1}{16}$$

d. $P(\bar{x}=0) = \dfrac{1}{16}$

$P(\bar{x}=1) = \dfrac{1}{16} + \dfrac{1}{16} = \dfrac{2}{16}$

$P(\bar{x}=2) = \dfrac{1}{16} + \dfrac{1}{16} + \dfrac{1}{16} = \dfrac{3}{16}$

$P(\bar{x}=3) = \dfrac{1}{16} + \dfrac{1}{16} + \dfrac{1}{16} + \dfrac{1}{16} = \dfrac{4}{16}$

$P(\bar{x}=4) = \dfrac{1}{16} + \dfrac{1}{16} + \dfrac{1}{16} = \dfrac{3}{16}$

$P(\bar{x}=5) = \dfrac{1}{16} + \dfrac{1}{16} = \dfrac{2}{16}$

$P(\bar{x}=6) = \dfrac{1}{16}$

$\bar{x}$	$p(\bar{x})$
0	1/16
1	2/16
2	3/16
3	4/16
4	3/16
5	2/16
6	1/16

e.

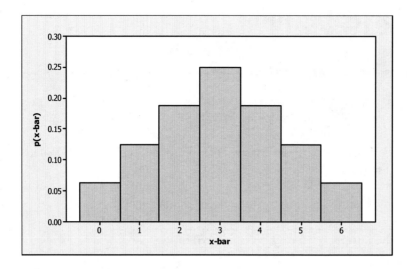

6.5   If the observations are independent of each other, then

$P(1, 1) = p(1)p(1) = .2(.2) = .04$
$P(1, 2) = p(1)p(2) = .2(.3) = .06$
$P(1, 3) = p(1)p(3) = .2(.2) = .04$
　　etc.

a.

Possible Samples	$\bar{x}$	$p(\bar{x})$	Possible Samples	$\bar{x}$	$p(\bar{x})$
1, 1	1	.04	3, 4	3.5	.04
1, 2	1.5	.06	3, 5	4	.02
1, 3	2	.04	4, 1	2.5	.04
1, 4	2.5	.04	4, 2	3	.06
1, 5	3	.02	4, 3	3.5	.04
2, 1	1.5	.06	4, 4	4	.04
2, 2	2	.09	4, 5	4.5	.02
2, 3	2.5	.06	5, 1	3	.02
2, 4	3	.06	5, 2	3.5	.03
2, 5	3.5	.03	5, 3	4	.02
3, 1	2	.04	5, 4	4.5	.02
3, 2	2.5	.06	5, 5	5	.01
3, 3	3	.04			

Summing the probabilities, the probability distribution of $\bar{x}$ is:

$\bar{x}$	$p(\bar{x})$
1	.04
1.5	.12
2	.17
2.5	.20
3	.20
3.5	.14
4	.08
4.5	.04
5	.01

b.

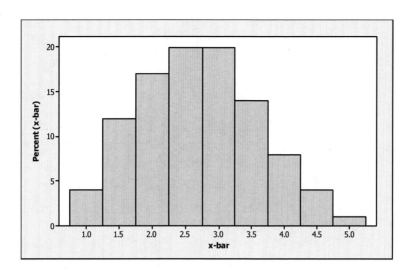

c. $P(\overline{x} \geq 4.5) = .04 + .01 = .05$

d. No. The probability of observing $\overline{x} = 4.5$ or larger is small (.05).

6.7 a. For a sample of size $n = 2$, the sample mean and sample median are exactly the same. Thus, the sampling distribution of the sample median is the same as that for the sample mean (see Exercise 6.5**a**).

b. The probability histogram for the sample median is identical to that for the sample mean (see Exercise 6.5**b**).

6.9 a. Answers will vary. MINITAB was used to generate 500 samples of size $n = 25$ observations from a uniform population from 00 to 99. The first 10 samples along with the sample means are shown in the table below:

Sample	Observations																									Mean	Var
1	29	57	34	57	82	63	57	94	59	28	29	48	29	12	88	7	93	25	67	69	84	59	28	90	48	53.44	679.76
2	83	27	46	42	18	40	52	5	81	38	58	28	30	16	57	60	45	92	82	19	30	12	42	16	13	41.28	604.79
3	68	15	75	26	46	69	46	81	97	23	38	17	7	47	32	85	98	20	69	8	28	33	7	34	61	45.2	809.75
4	60	38	42	88	85	1	58	5	74	36	4	73	68	99	17	65	13	41	49	97	15	6	95	65	9	48.12	1066.11
5	62	76	37	6	70	14	95	94	12	18	63	65	33	37	44	86	71	67	35	10	17	53	21	27	65	47.12	761.44
6	53	56	92	5	57	13	57	13	7	91	2	76	98	21	93	44	91	74	30	17	99	77	49	86	5	52.24	1178.44
7	38	54	18	62	31	70	72	85	50	57	8	44	48	8	21	50	90	75	2	86	50	1	87	43	49	47.96	741.54
8	54	29	55	23	50	90	66	46	9	30	39	29	33	67	67	93	60	46	28	2	11	63	96	61	46	47.72	631.63
9	32	30	33	92	57	53	40	88	18	60	9	3	7	34	98	62	25	13	67	3	98	96	40	66	25	45.96	976.29
10	7	24	53	4	5	48	10	21	52	23	82	51	55	76	56	44	81	32	57	27	36	84	37	22	49	41.44	582.84

Using MINTAB, the histogram of the sample means is:

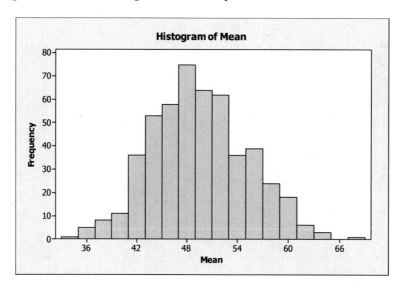

b. The sample variances for the first 10 samples are also shown in the table in part a. Using MINITAB, the histogram of the 500 sample variances is:

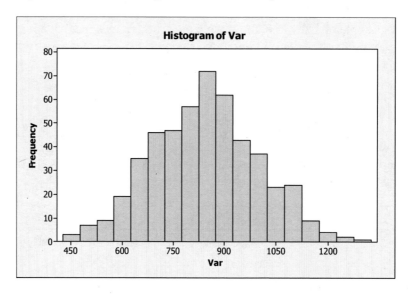

6.11   If the sampling distribution of a sample statistic has a mean equal to the population parameter the statistic is intended to estimate, the statistic is said to be an unbiased estimator of the parameter. If the mean of the sampling distribution is not equal to the parameter, the statistic is said to be a biased estimator of the parameter.

6.13   The properties of an ideal estimator are unbiased and minimum variance. Ideally, when we are estimating a parameter, we would like to use an unbiased estimator. In addition, of all unbiased estimators, we would like to choose the estimator with the minimum variance.

6.15   a. $\mu \sum xp(x) = 2\left(\dfrac{1}{3}\right) + 4\left(\dfrac{1}{3}\right) + 9\left(\dfrac{1}{3}\right) = \dfrac{15}{3} = 5$

b.   The possible samples of size $n = 3$, the sample means, and the probabilities are:

Possible Samples	$\overline{x}$	$p(\overline{x})$	m	Possible Samples	$\overline{x}$	$p(\overline{x})$	m
2, 2, 2	2	1/27	2	4, 4, 4	4	1/27	4
2, 2, 4	8/3	1/27	2	4, 4, 9	17/3	1/27	4
2, 2, 9	13/3	1/27	2	4, 9, 2	5	1/27	4
2, 4, 2	8/3	1/27	2	4, 9, 4	17/3	1/27	4
2, 4, 4	10/3	1/27	4	4, 9, 9	22/3	1/27	9
2, 4, 9	5	1/27	4	9, 2, 2	13/3	1/27	2
2, 9, 2	13/3	1/27	2	9, 2, 4	5	1/27	4
2, 9, 4	5	1/27	4	9, 2, 9	20/3	1/27	9
2, 9, 9	20/3	1/27	9	9, 4, 2	5	1/27	4
4, 2, 2	8/3	1/27	2	9, 4, 4	17/3	1/27	4
4, 2, 4	10/3	1/27	4	9, 4, 9	22/3	1/27	9
4, 2, 9	5	1/27	4	9, 9, 2	20/3	1/27	9
4, 4, 2	10/3	1/27	4	9, 9, 4	22/3	1/27	9
				9, 9, 9	9	1/27	9

The sampling distribution of $\overline{x}$ is:

$\overline{x}$	$p(\overline{x})$
2	1/27
8/3	3/27
10/3	3/27
4	1/27
13/3	3/27
5	6/27
17/3	3/27
20/3	3/27
22/3	3/27
9	1/27
	27/27

$$E(\overline{x}) = \sum \overline{x}p(\overline{x}) = 2\left(\frac{1}{27}\right) + \frac{8}{3}\left(\frac{3}{27}\right) + \frac{10}{3}\left(\frac{3}{27}\right) + 4\left(\frac{1}{27}\right) + \frac{13}{3}\left(\frac{3}{27}\right)$$
$$+ 5\left(\frac{6}{27}\right) + \frac{17}{3}\left(\frac{3}{27}\right) + \frac{20}{3}\left(\frac{3}{27}\right) + \frac{22}{3}\left(\frac{3}{27}\right) + 9\left(\frac{1}{27}\right)$$
$$= \frac{2}{27} + \frac{8}{27} + \frac{10}{27} + \frac{4}{27} + \frac{13}{27} + \frac{30}{27} + \frac{17}{27} + \frac{20}{27} + \frac{22}{27} + \frac{9}{27} = \frac{135}{27} = 5$$

Since $\mu = 5$ in part **a**, and $E(\overline{x}) = \mu = 5$, $\overline{x}$ is an unbiased estimator of μ.

c.   The median was calculated for each sample and is shown in the table in part **b**. The sampling distribution of m is:

m	$p(m)$
2	7/27
4	13/27
9	7/27
	27/27

$$E(m) = \sum mp(m) = 2\left(\frac{7}{27}\right) + 4\left(\frac{13}{27}\right) + 9\left(\frac{7}{27}\right) = \frac{14}{27} + \frac{52}{27} + \frac{63}{27} = \frac{129}{27} = 4.778$$

The $E(m) = 4.778 \neq \mu = 5$. Thus, m is a biased estimator of μ.

d.   Use the sample mean, $\bar{x}$. It is an unbiased estimator.

6.17   Answers will vary. MINITAB was used to generate 500 samples of size $n = 25$ observations from a uniform population from 1 to 50. The first 10 samples along with the sample means and medians are shown in the table below:

Sample	Observations	Mean	Median
1	28 27 11 19 50 30 47 26 9 33 50 15 21 41 31 41 35 32 32 17 6 32 39 34 21	29.08	31
2	8 4 32 32 3 45 18 9 40 3 42 21 44 50 42 14 24 10 36 6 15 47 26 48 28	25.88	26
3	6 20 27 1 50 14 21 37 46 23 1 34 42 47 24 46 8 29 18 28 40 39 49 33 23	28.24	28
4	45 12 26 13 40 17 11 43 8 35 20 8 44 48 13 46 49 17 47 27 5 45 9 21 36	27.4	26
5	40 38 25 37 47 2 17 40 32 6 22 30 23 2 18 22 14 6 22 3 43 47 16 35 35	24.88	23
6	17 8 43 27 21 5 18 45 31 15 2 38 22 18 7 9 3 35 23 45 24 39 38 35 37	24.20	23
7	40 1 22 29 6 8 22 20 36 18 45 16 29 9 6 3 49 34 24 40 27 5 49 11 30	23.16	22
8	25 3 44 34 29 6 33 32 43 6 43 24 49 14 37 8 46 44 1 12 36 18 30 25 4	25.84	29
9	7 33 36 41 30 13 17 19 14 36 20 39 41 20 15 38 12 37 14 9 19 2 37 15 8	22.88	19
10	4 46 49 49 45 49 24 3 25 22 27 28 23 17 14 6 35 5 20 34 4 41 9 15 3	23.88	23

Using MINITAB, side-by side histograms of the means and medians of the 500 samples are:

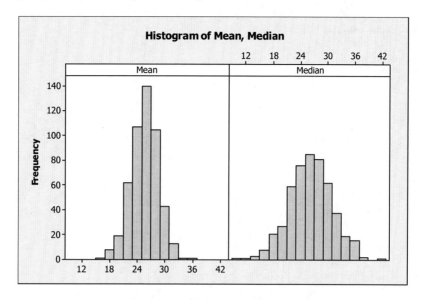

a. Yes, it appears that $\bar{x}$ and the median are unbiased estimators of the population mean. The centers of both distributions above appear to be around 25 to 26. In fact, the mean of the sampling distribution of $\bar{x}$ is 25.65 and the mean of the sampling distribution of the median is 25.73.

b. The sampling distribution of the median has greater variation because it is more spread out than the sampling distribution of $\bar{x}$.

6.19   a. Refer to the solution to Exercise 6.5. The values of s^2 and the corresponding probabilities are listed below:

$$s^2 = \frac{\sum x^2 - \dfrac{\left(\sum x\right)^2}{n}}{n-1}$$

For sample 1, 1; $s^2 = \dfrac{2 - \dfrac{2^2}{2}}{1} = 0$

For sample 1, 2: $s^2 = \dfrac{5 - \dfrac{3^2}{2}}{1} = .5$

The rest of the values are calculated and shown:

s^2	$p(s^2)$	s^2	$p(s^2)$
0.0	.04	0.5	.04
0.5	.06	2.0	.02
2.0	.04	4.5	.04
4.5	.04	2.0	.06
8.0	.02	0.5	.04
0.5	.06	0.0	.04
0.0	.09	0.5	.02
0.5	.06	8.0	.02
2.0	.06	4.5	.03
4.5	.03	2.0	.02
2.0	.04	0.5	.02
0.5	.06	0.0	.01
0.0	.04		

The sampling distribution of s^2 is:

s^2	$p(s^2)$
0.0	.22
0.5	.36
2.0	.24
4.5	.14
8.0	.04

b. $$\sigma^2 = \sum (x - \mu)^2 \, p(x) = (1 - 2.7)^2(.2) + (2 - 2.7)^2(.3) + (3 - 2.7)^2(.2)$$
$$+ (4 - 2.7)^2(.2) + (5 - 2.7)^2(.1) = 1.61$$

c. $E(s^2) = \sum s^2 p(s^2) = 0(.22) + .5(.36) + 2(.24) + 4.5(.14) + 8(.04) = 1.61$

d. The sampling distribution of s is listed below, where $s = \sqrt{s^2}$:

s	$p(s)$
0.000	.22
0.707	.36
1.414	.24
2.121	.14
2.828	.04

e. $E(s) = \sum s p(s) = 0(.22) + .707(.36) + 1.41(.24) + 2.1212(.14) + 2.828(.04) = 1.00394$

Since $E(s) = 1.00394$ is not equal to $\sigma = \sqrt{\sigma^2} = \sqrt{1.61} = 1.269$, s is a biased estimator of σ .

6.21 The symbol $\mu_{\bar{x}}$ represents the mean of the sampling distribution of $\bar{x}$. The symbol $\sigma_{\bar{x}}$ represents the standard deviation of the sampling distribution of $\bar{x}$.

6.23 The standard deviation of the sampling distribution of $\bar{x}$, $\sigma_{\bar{x}}$, is the standard deviation of the population from which the sample is selected divided by $\sqrt{n}$.

6.25 The Central Limit Theorem states: Consider a random sample of n observations selected from a population (*any* population) with mean μ and standard deviation σ. Then, when n is sufficiently large, the sampling distribution of $\bar{x}$ will be approximately a normal distribution with mean $\mu_{\bar{x}} = \mu$ and standard deviation $\sigma_{\bar{x}} = \sigma/\sqrt{n}$. The larger the sample size, the better will be the normal approximation to the sampling distribution of $\bar{x}$.

6.27 a. $\mu_{\bar{x}} = \mu = 100, \sigma_{\bar{x}} = \dfrac{\sigma}{\sqrt{n}} = \dfrac{\sqrt{100}}{\sqrt{4}} = 5$

b. $\mu_{\bar{x}} = \mu = 100, \sigma_{\bar{x}} = \dfrac{\sigma}{\sqrt{n}} = \dfrac{\sqrt{100}}{\sqrt{25}} = 2$

c. $\mu_{\bar{x}} = \mu = 100, \sigma_{\bar{x}} = \dfrac{\sigma}{\sqrt{n}} = \dfrac{\sqrt{100}}{\sqrt{100}} = 1$

d. $\mu_{\bar{x}} = \mu = 100, \sigma_{\bar{x}} = \dfrac{\sigma}{\sqrt{n}} = \dfrac{\sqrt{100}}{\sqrt{50}} = 1.414$

e. $\mu_{\bar{x}} = \mu = 100, \sigma_{\bar{x}} = \dfrac{\sigma}{\sqrt{n}} = \dfrac{\sqrt{100}}{\sqrt{500}} = .447$

f. $\mu_{\bar{x}} = \mu = 100, \sigma_{\bar{x}} = \dfrac{\sigma}{\sqrt{n}} = \dfrac{\sqrt{100}}{\sqrt{1000}} = .316$

6.29 a. $\mu = \sum xp(x) = 1(.1) + 2(.4) + 3(.4) + 8(.1) = 2.9$

$\sigma^2 = \sum (x - \mu)^2 p(x) = (1 - 2.9)^2(.1) + (2 - 2.9)^2(.4) + (3 - 2.9)^2(.4) + (8 - 2.9)^2(.1)$
$= .361 + .324 + .004 + 2.601 = 3.29$

$\sigma = \sqrt{3.29} = 1.814$

b. The possible samples, values of $\bar{x}$, and associated probabilities are listed:

Possible Samples	$\bar{x}$	$p(\bar{x})$	Possible Samples	$\bar{x}$	$p(\bar{x})$
1, 1	1	.01	3, 1	2	.04
1, 2	1.5	.04	3, 2	2.5	.16
1, 3	2	.04	3, 3	3	.16
1, 8	4.5	.01	3, 8	5.5	.04
2, 1	1.5	.04	8, 1	4.5	.01
2, 2	2	.16	8, 2	5	.04
2, 3	2.5	.16	8, 3	5.5	.04
2, 8	5	.04	8, 8	8	.01

$P(1, 1) = p(1)p(1) = .1(.1) = .01$
$P(1, 2) = p(1)p(2) = .1(.4) = .04$
$P(1, 3) = p(1)p(3) = .1(.4) = .04$
  etc.

The sampling distribution of $\bar{x}$ is:

$\bar{x}$	$p(\bar{x})$
1	.01
1.5	.08
2	.24
2.5	.32
3	.16
4.5	.02
5	.08
5.5	.08
8	.01
	1.00

c.
$$\mu_{\bar{x}} = E(\bar{x}) = \sum \bar{x}p(\bar{x}) = 1(.01) + 1.5(.08) + 2(.24) + 2.5(.32) + 3(.16) + 4.5(.02)$$
$$5(.08) + 5.5(.08) + 8(.01) = 2.9 = \mu$$

$$\sigma_{\bar{x}}^2 = \sum (\bar{x} - \mu_{\bar{x}})^2\, p(\bar{x}) = (1 - 2.9)^2(.01) + (1.5 - 2.9)^2(.08) + (2 - 2.9)^2(.24)$$
$$+ (2.5 - 2.9)^2(.32) + (3 - 2.9)^2(.16) + (4.5 - 2.9)^2(.02)$$
$$+ (5 - 2.9)^2(.08) + (5.5 - 2.9)^2(.08) + (8 - 2.9)^2(.01)$$
$$= .0361 + .1568 + .1944 + .0512 + .0016 + .0512 + .3528$$
$$+ .5408 + .2601 = 1.645$$

$$\sigma_{\bar{x}} = \sqrt{1.645} = 1.283$$
$$\sigma_{\bar{x}} = \frac{\sigma}{\sqrt{n}} = \frac{1.814}{\sqrt{2}} = 1.283$$

6.31 a. $\mu_{\bar{x}} = \mu = 30$, $\sigma_{\bar{x}} = \frac{\sigma}{\sqrt{n}} = \frac{16}{\sqrt{100}} = 1.6$

b. By the Central Limit Theorem, the sampling distribution of $\bar{x}$ is approximately normal.

c. $P(\bar{x} \geq 28) = P\left(z \geq \dfrac{28-30}{1.6}\right) = P(z \geq -1.25) = .5 + .3944 = .8944$

d. $P(22.1 \leq \bar{x} \leq 26.8) = P\left(\dfrac{22.1-30}{1.6} \leq z \leq \dfrac{26.8-30}{1.6}\right) = P(-4.94 \leq z \leq -2)$
$= .5 - .4772 = .0228$

e. $P(\bar{x} \leq 28.2) = P\left(z \leq \dfrac{28.2-30}{1.6}\right) = P(z \leq -1.125) = .5 - \left(\dfrac{.3868 + .3708}{2}\right)$
$= .5 - .3697 = .1303$

f. $P(\bar{x} \geq 27.0) = P\left(z \geq \dfrac{27.0-30}{1.6}\right) = P(z \geq -1.88) = .5 + .4699 = .9699$

6.33   Answers will vary. MINITAB was used to generate the samples. The side-by-side histograms of the distributions of $\bar{x}$ were generated by MINITAB.

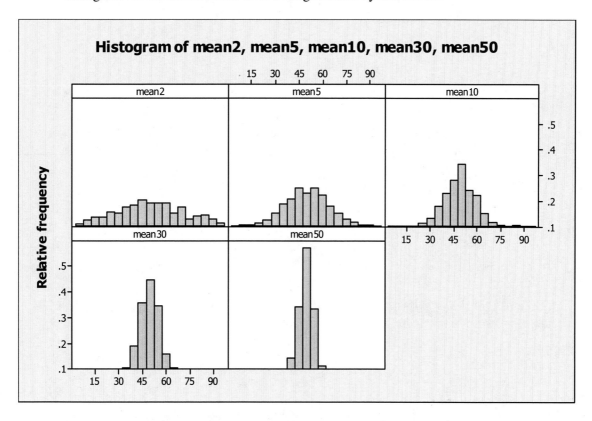

It appears that each of the distributions is centered around the same value, approximately 49.5. All of the distributions appear to be symmetric. As the sample size increases, the spread of the sampling distribution of $\bar{x}$ decreases.

6.35   a.   By the Central Limit Theorem, the sampling distribution of $\bar{x}$ is approximately normal. The mean of the $\bar{x}$ distribution is $\mu_{\bar{x}} = \mu = 320$ and the standard deviation of the $\bar{x}$ distribution is $\sigma_{\bar{x}} = \dfrac{\sigma}{\sqrt{n}} = \dfrac{100}{\sqrt{100}} = \dfrac{100}{10} = 10$.

b. $P(300 < \bar{x} < 310) = P\left(\dfrac{300 - 320}{10} < z < \dfrac{310 - 320}{10}\right)$

$= P(-2 < z < -1) = .4772 - .3413 = .1359$

(Using Table IV, Appendix A)

c. $P(\bar{x} > 360) = P\left(z > \dfrac{360 - 320}{10}\right) = P(z > 4) \approx .5 - .5 = 0$ (Using Table IV, Appendix A)

6.37   a. $\mu_{\bar{x}}$ is the mean of the sampling distribution of $\bar{x}$. $\mu_{\bar{x}} = \mu = 79$.

b. $\sigma_{\bar{x}}$ is the standard deviation of the sampling distribution of $\bar{x}$. $\sigma_{\bar{x}} = \dfrac{\sigma}{\sqrt{n}} = \dfrac{23}{\sqrt{100}} = 2.3$

c.   By the Central Limit Theorem, the sampling distribution of $\bar{x}$ is approximately normal.

d. $z = \dfrac{\bar{x} - \mu_{\bar{x}}}{\sigma_{\bar{x}}} = \dfrac{80 - 79}{2.3} = 0.43$

e. $P(\bar{x} > 80) = P(z > 0.43) = .5 - .1664 = .3336$ (Using Table IV, Appendix A.)

6.39   Let $\bar{x}$ = sample mean amount of miraculin produced. By the Central Limit Theorem, the sampling distribution of $\bar{x}$ is approximately normal with $\mu_{\bar{x}} = \mu = 105.3$ and $\sigma_{\bar{x}} = \dfrac{\sigma}{\sqrt{n}} = \dfrac{8.0}{\sqrt{64}} = 1$.

$P(\bar{x} < 103) = P\left(z < \dfrac{103 - 105.3}{1}\right) = P(z < -2.30) = .5 - .4893 = .0107$

(using Table IV, Appendix A)

Since this probability is so small, we would not expect to observe a value of $\bar{x}$ less than 103 micro-grams per gram of fresh weight.

6.41   a. $E(\bar{x}) = \mu_{\bar{x}} = \mu = .10$ and $V(\bar{x}) = \sigma_{\bar{x}}^2 = \dfrac{\sigma^2}{n} = \dfrac{.10^2}{50} = .0002$

b.   By the Central Limit Theorem, the sampling distribution of $\bar{x}$ is approximately normal if the sample size is sufficiently large. In this case, $n = 50$ is considered large.

c. $P(\bar{x} > .13) = P\left(z > \dfrac{.13 - .10}{\sqrt{.0002}}\right) = P(z > 2.12) = .5 - .4830 = .0170$

6.43 a. By the Central Limit Theorem, the sampling distribution of $\bar{x}$ is approximately normal if the sample size is sufficiently large. In this case, $n = 50$ is considered large.

$$\mu_{\bar{x}} = \mu = .53, \quad \sigma_{\bar{x}} = \frac{\sigma}{\sqrt{n}} = \frac{.193}{\sqrt{50}} = .0273$$

b. $P(\bar{x} > .58) = P\left(z > \dfrac{.58 - .53}{.0273}\right) = P(z > 1.83) = .5 - .4664 = .0336$
(Using Table IV, Appendix A)

c. $P(\bar{x} > .59 \mid \mu = .58) = P\left(z > \dfrac{.59 - .58}{.0273}\right) = P(z > .37) = .5 - .1443 = .3557$
(Using Table IV, Appendix A) The above is the probability *after tensioning and loading*.

$$P(\bar{x} > .59 \mid \mu = .53) = P\left(z > \dfrac{.59 - .53}{.0273}\right) = P(z > 2.20) = .5 - .4772 = .0228$$
(Using Table IV, Appendix A) The above is the probability *before tensioning*.

Since the probability *after tensioning* is not small, this would not be an unusual event. Since the probability before tensioning is very small, this would be an unusual event. The sample measurements were probably obtained after tensioning and loading.

6.45 a. By the Central Limit Theorem, the sampling distribution of $\bar{x}$ is approximately normal if the sample size is sufficiently large. In this case, $n = 326$ is considered large.

$$\mu_{\bar{x}} = \mu = 6, \quad \sigma_{\bar{x}} = \frac{\sigma}{\sqrt{n}} = \frac{10}{\sqrt{326}} = .5538$$

$$P(\bar{x} > 7.5) = P\left(z > \frac{7.5 - 6}{.5538}\right) = P(z > 2.71) = .5 - .4966 = .0034$$

b. $P(\bar{x} > 300) = P\left(z > \dfrac{300 - 6}{.5538}\right) = P(z > 530.88) \approx .5 - .5 = 0$

Since it is essentially impossible to get a sample mean of 300 if the population mean was 6, we would conclude that the mean PFOA in people who live near DuPont's Teflon-making facility is not 6 ppb but something much larger.

6.47 **Handrubbing**

$$\mu_{\bar{x}} = \mu = 35, \quad \sigma_{\bar{x}} = \frac{\sigma}{\sqrt{n}} = \frac{59}{\sqrt{50}} = 8.344$$

$$P(\bar{x} < 30) = P\left(z < \frac{30-35}{8.344}\right) = P(z < -.60) = .5 - .2257 = .2743$$

(Using Table IV, Appendix A.)

Handwashing

$$\mu_{\bar{x}} = \mu = 69, \quad \sigma_{\bar{x}} = \frac{\sigma}{\sqrt{n}} = \frac{106}{\sqrt{50}} = 14.991$$

$$P(\bar{x} < 30) = P\left(z < \frac{30-69}{14.991}\right) = P(z < -2.60) = .5 - .4953 = .0047$$

(Using Table IV, Appendix A.)

This sample of workers probably came from the handrubbing group. The probability of seeing a mean of 30 or smaller for this group is .2743, while the probability of observing a mean of 30 or less for the handwashers is .0047. This would be an extremely rare event if the workers were handwashers because the probability is so small. It would not be a very unusual event for the handrubbers because the probability is not that small.

6.49   The statement "The sample mean, $\bar{x}$, will always be equal to μ_x" is false. The sample mean is used to estimate the population mean, but the probability that it is equal to the population mean is 0.

6.51   The statement "The standard error of $\bar{x}$ will always be small than σ" is true. The standard error of $\bar{x}$ is $\sigma_{\bar{x}} = \dfrac{\sigma}{\sqrt{n}}$. Thus, the standard error of $\bar{x}$ will always be smaller than σ.

6.53   a.   "The sampling distribution of the sample statistic A" is the probability distribution of the variable A.

      b.   "A" is an unbiased estimator of α because the mean of the sampling distribution of A is α.

      c.   If both A and B are unbiased estimators of α, then the statistic whose standard deviation is smaller is a better estimator of α.

      d.   No. The Central Limit Theorem applies only to the sample mean. If A is the sample mean, $\bar{x}$, and n is sufficiently large, then the Central Limit Theorem will apply. However, both A and B cannot be sample means. Thus, we cannot apply the Central Limit Theorem to both A and B.

6.55   By the Central Limit Theorem, the sampling distribution of $\bar{x}$ is approximately normal.

$$\mu_{\bar{x}} = \mu = 19.6, \quad \sigma_{\bar{x}} = \frac{3.2}{\sqrt{68}} = .388$$

      a. $P(\bar{x} \le 19.6) = P\left(z \le \dfrac{19.6-19.6}{.388}\right) = P(z \le 0) = .5$ (using Table IV, Appendix A)

b. $P(\bar{x} \leq 19) = P\left(z \leq \dfrac{19-19.6}{.388}\right) = P(z \leq -1.55) = .5 - .4394 = .0606$

(using Table IV, Appendix A)

c. $P(\bar{x} \geq 20.1) = P\left(z \geq \dfrac{20.1-19.6}{.388}\right) = P(z \geq 1.29) = .5 - .4015 = .0985$

(using Table IV, Appendix A)

d. $P(19.2 \leq \bar{x} \leq 20.6) = P\left(\dfrac{19.2-19.6}{.388} < z < \dfrac{20.6-19.6}{.388}\right)$

$= P(-1.03 \leq z \leq 2.58) = .3485 + .4951 = .8436$

(using Table IV, Appendix A)

6.57   Answers will vary. One hundred samples of size $n = 2$ were selected from a uniform distribution on the interval from 0 to 10. The process was repeated for samples of size $n = 5$, $n = 10$, $n = 30$, and $n = 50$. For each sample, the value of $\bar{x}$ was computed. Using MINITAB, the histograms for each set of 100 $\bar{x}$'s were constructed:

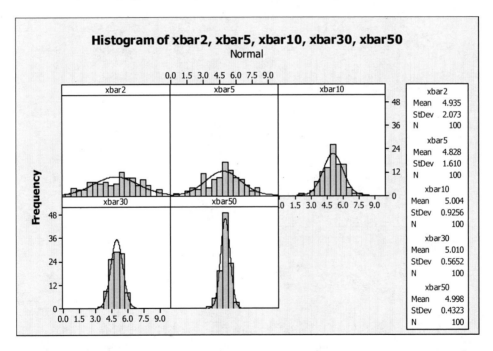

For small sizes of n, the sampling distributions of $\bar{x}$ are somewhat normal. As n increases, the sampling distributions of $\bar{x}$ become more normal.

6.59   Given: $\mu = 100$ and $\sigma = 10$

n	1	5	10	20	30	40	50
$\dfrac{\sigma}{\sqrt{n}}$	10	4.472	3.162	2.236	1.826	1.581	1.414

The graph of $\dfrac{\sigma}{\sqrt{n}}$ against n is given here:

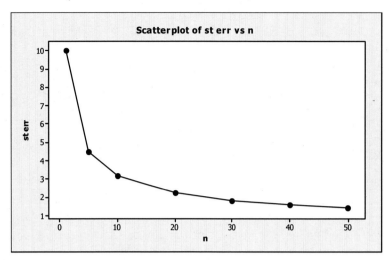

6.61   a. $\mu_{\bar{x}}$ is the mean of the sampling distribution of $\bar{x}$. $\mu_{\bar{x}} = \mu = 106$.

b. $\sigma_{\bar{x}}$ is the standard deviation of the sampling distribution of $\bar{x}$. $\sigma_{\bar{x}} = \dfrac{\sigma}{\sqrt{n}} = \dfrac{16.4}{\sqrt{36}} = 2.73$

c.   By the Central Limit Theorem, the sampling distribution of $\bar{x}$ is approximately normal.

d. $z = \dfrac{\bar{x} - \mu_{\bar{x}}}{\sigma_{\bar{x}}} = \dfrac{100 - 106}{2.73} = -2.20$

e. $P(\bar{x} < 100) = P(z < -2.20) = .5 - .4864 = .0136$ (Using Table IV, Appendix A.)

6.63   Let $\bar{x}$ = sample mean score change. By the Central Limit Theorem, the sampling distribution of $\bar{x}$ is approximately normal

with $\mu_{\bar{x}} = \mu = 19$ and $\sigma_{\bar{x}} = \dfrac{\sigma}{\sqrt{n}} = \dfrac{65}{\sqrt{100}} = 6.5$.

$$P(\bar{x} < 10) = P\left(z < \dfrac{10 - 19}{6.5}\right) = P(z < -1.38) = .5 - .4162 = .0838$$

(using Table IV, Appendix A)

6.65   From Exercise 6.64, $\sigma = .001$. We must assume the Central Limit theorem applies (n is only 25). Thus, the distribution of $\bar{x}$ is approximately normal with $\mu_{\bar{x}} = \mu = .501$

and $\sigma_{\bar{x}} = \dfrac{\sigma}{\sqrt{n}} = \dfrac{.001}{\sqrt{25}} = .0002$. Using Table IV, Appendix A,

$$P(\bar{x} < .4994) + P(\bar{x} > .5006) = P\left(z < \dfrac{.4994 - .501}{.0002}\right) + P\left(z > \dfrac{.5006 - .501}{.0002}\right)$$

$$= P(z < -8) + P(z > -2) = (.5 - .5) + (.5 + .4772) = .9772$$

6.67 For this problem, $\mu_{\bar{x}} = \mu = 4.59$ and $\sigma_{\bar{x}} = \frac{\sigma}{\sqrt{n}} = \frac{2.95}{\sqrt{50}} = .4172$

a. $P(\bar{x} \geq 6) = P\left(z \geq \frac{6-4.59}{.4172}\right) = P(z \geq 3.38) \approx .5 - .5 = 0$

(using Table IV, Appendix A)

Since the probability of observing a sample mean CAHS score of 6 or higher is so small (p is essentially 0), we would not expect to see a sample mean of 6 or higher.

b. If a psychologist actually observed $x = 6.2$, we would conclude that he has seen an extremely unusual event, or the population from which the sample was drawn probably has a mean greater than 4.59.

6.69 a. We must assume that the distribution of surface roughness scores is normally distributed because the sample size is only n = 20. Thus, the Central Limit Theorem will not be valid. If the distribution being sampled from is normal, then the sampling distribution of $\bar{x}$ is normal.

$\mu_{\bar{x}} = \mu = 1.8$, $\sigma_{\bar{x}} = \frac{\sigma}{\sqrt{n}} = \frac{.5}{\sqrt{20}} = .1118$

$P(\bar{x} > 1.85) = P\left(z > \frac{1.85-1.8}{.1118}\right) = P(z > .45) = .5 - .1736 = .3264$

b. $\bar{x} = \frac{\sum x}{n} = \frac{37.62}{20} = 1.881$

c. $P(\bar{x} > 1.881) = P\left(z > \frac{1.881-1.8}{.1118}\right) = P(z > .72) = .5 - .2642 = .2358$

Since the probability of getting a sample mean of 1.881 or anything more unusual is not small (probability is .2358), the assumptions made in part **a** appear to be valid.

6.71 Even though the number of flaws per piece of siding has a Poisson distribution, the Central Limit Theorem implies that the distribution of the sample mean will be approximately normal with $\mu_{\bar{x}} = \mu = 2.5$ and $\sigma_{\bar{x}} = \frac{\sigma}{\sqrt{n}} = \frac{\sqrt{2.5}}{\sqrt{35}} = .2673$. Therefore,

$P(\bar{x} > 2.1) = P\left(z > \frac{2.1-2.5}{\sqrt{2.5}/\sqrt{35}}\right) = P(z > -1.50) = .5 + .4332 = .9332$

6.73 a. By the Central Limit Theorem, the distribution of $\bar{x}$ is approximately normal, with $\mu_{\bar{x}} = \mu = 157$ and $\sigma_{\bar{x}} = \frac{\sigma}{\sqrt{n}} = \frac{3}{\sqrt{40}} = .474$.

The sample mean is 1.3 psi below 157 or $\bar{x} = 157 - 1.3 = 155.7$

$$P(\bar{x} \le 155.7)$$
$$= P(\bar{x} \le 155.7) = P\left(z \le \frac{155.7 - 157}{.474}\right) = P(z \le -2.74) = .5 - .4969 = .0031$$

(using Table IV, Appendix A)

If the claim is true, it is very unlikely (probability = .0031) to observe a sample mean 1.3 psi below 157 psi. Thus, the actual population mean is probably not 157 but something lower.

b. $$P(\bar{x} \le 155.7) = P\left(z \le \frac{155.7 - 156}{.474}\right) = P(z \le -.63) = .5 - .2357 = .2643$$

(using Table IV, Appendix A)

The observed sample is more likely if $\mu = 156$ rather than 157.

$$P(\bar{x} \le 155.7) = P\left(z \le \frac{155.7 - 158}{.474}\right) = P(z \le -4.85) \approx .5 - .5 = 0$$

The observed sample is less likely if $\mu = 158$ rather than 157.

c.   If $\sigma = 2$, $\sigma_{\bar{x}} = \frac{\sigma}{\sqrt{n}} = \frac{2}{\sqrt{40}} = .316$.

$$P(\bar{x} \le 155.7) = P\left(z \le \frac{155.7 - 157}{.316}\right) = P(z \le -4.11) = .5 - .5 = 0$$

(using Table IV, Appendix A)

The observed sample is less likely if $\sigma = 2$ than if $\sigma = 3$.

If $\sigma = 6$, $\sigma_{\bar{x}} = \frac{\sigma}{\sqrt{n}} = \frac{6}{\sqrt{40}} = .949$.

$$P(\bar{x} \le 155.7) = P\left(z \le \frac{155.7 - 157}{.949}\right) = P(z \le -1.37) = .5 - .4147 = .0853$$

(using Table IV, Appendix A)

The observed sample is more likely if $\sigma = 6$ than if $\sigma = 3$.

Inferences Based on a Single Sample
Estimation with Confidence Intervals

7.1 The unknown population parameter (e.g., mean or proportion) that we are interested in estimating is called the target parameter.

7.3 An interval estimator estimates μ with a range of values, while a point estimator estimates μ with a single point.

7.5 Yes. As long as the sample size is sufficiently large, the Central Limit Theorem says the distribution of $\bar{x}$ is approximately normal regardless of the original distribution.

7.7 a. For $\alpha = .10$, $\alpha/2 = .10/2 = .05$. $z_{\alpha/2} = z_{.05}$ is the z-score with .05 of the area to the right of it. The area between 0 and $z_{.05}$ is $.5 - .05 = .4500$. Using Table IV, Appendix A, $z_{.05} = 1.645$.

b. For $\alpha = .01$, $\alpha/2 = .01/2 = .005$. $z_{\alpha/2} = z_{.005}$ is the z-score with .005 of the area to the right of it. The area between 0 and $z_{.005}$ is $.5 - .005 = .4950$. Using Table IV, Appendix A, $z_{.005} = 2.58$.

c. For $\alpha = .05$, $\alpha/2 = .05/2 = .025$. $z_{\alpha/2} = z_{.025}$ is the z-score with .025 of the area to the right of it. The area between 0 and $z_{.025}$ is $.5 - .025 = .4750$. Using Table IV, Appendix A, $z_{.025} = 1.96$.

d. For $\alpha = .20$, $\alpha/2 = .20/2 = .10$. $z_{\alpha/2} = z_{.10}$ is the z-score with .10 of the area to the right of it. The area between 0 and $z_{.10}$ is $.5 - .10 = .4000$. Using Table IV, Appendix A, $z_{.10} = 1.28$.

7.9 a. For confidence coefficient .95, $\alpha = .05$ and $\alpha/2 = .05/2 = .025$. From Table IV, Appendix A, $z_{.025} = 1.96$. The confidence interval is:

$$\bar{x} \pm z_{.025} \frac{\sigma}{\sqrt{n}} \Rightarrow 28 \pm 1.96 \frac{20}{\sqrt{75}} \Rightarrow 28 \pm 4.53 \Rightarrow (23.47, 32.53)$$

b. $$\bar{x} \pm z_{.025} \frac{\sigma}{\sqrt{n}} \Rightarrow 102 \pm 1.96 \frac{20}{\sqrt{200}} \Rightarrow 102 \pm 2.77 \Rightarrow (99.23, 104.77)$$

c. $$\bar{x} \pm z_{.025} \frac{\sigma}{\sqrt{n}} \Rightarrow 15 \pm 1.96 \frac{20}{\sqrt{100}} \Rightarrow 15 \pm 3.92 \Rightarrow (11.08, 18.92)$$

d. $$\bar{x} \pm z_{.025} \frac{\sigma}{\sqrt{n}} \Rightarrow 4.05 \pm 1.96 \frac{20}{\sqrt{100}} \Rightarrow 4.05 \pm 3.92 \Rightarrow (0.13, 7.97)$$

e. No. Since the sample size in each part was large (n ranged from 75 to 200), the Central Limit Theorem indicates that the sampling distribution of $\bar{x}$ is approximately normal.

7.11   a.   For confidence coefficient .95, $\alpha = .05$ and $\alpha / 2 = .05 / 2 = .025$. From Table IV, Appendix A, $z_{.025} = 1.96$. The confidence interval is:

$$\bar{x} \pm z_{\alpha/2} \frac{s}{\sqrt{n}} \Rightarrow 83.2 \pm 1.96 \frac{6.4}{\sqrt{100}} \Rightarrow 83.2 \pm 1.25 \Rightarrow (81.95, \ 84.45)$$

b.   The confidence coefficient of .95 means that in repeated sampling, 95% of all confidence intervals constructed will include μ.

c.   For confidence coefficient .99, $\alpha = .01$ and $\alpha / 2 = .01 / 2 = .005$. From Table IV, Appendix A, $z_{.005} = 2.58$. The confidence interval is:

$$\bar{x} \pm z_{\alpha/2} \frac{s}{\sqrt{n}} \Rightarrow 83.2 \pm 2.58 \frac{6.4}{\sqrt{100}} \Rightarrow 83.2 \pm 1.65 \Rightarrow (81.55, \ 84.85)$$

d.   As the confidence coefficient increases, the width of the confidence interval also increases.

e.   Yes. Since the sample size is 100, the Central Limit Theorem applies. This ensures the distribution of $\bar{x}$ is normal, regardless of the original distribution.

7.13   a.   The point estimate for the average number of latex gloves used per week by all healthcare workers with a latex allergy is $\bar{x} = 19.3$.

b.   For confidence coefficient .95, $\alpha = .05$ and $\alpha / 2 = .05 / 2 = .025$. From Table IV, Appendix A, $z_{.025} = 1.96$. The 95% confidence interval is:

$$\bar{x} \pm z_{.025} \sigma_{\bar{x}} \Rightarrow \bar{x} \pm 1.96 \frac{\sigma}{\sqrt{n}} \Rightarrow 19.3 \pm 1.96 \frac{11.9}{\sqrt{46}} \Rightarrow 19.3 \pm 3.44 \Rightarrow (15.86, \ 22.74)$$

c.   We are 95% confident that the average number of latex gloves used per week by all healthcare workers with a latex allergy is between 15.86 and 22.74.

d.   We must assume that we have a random sample from the target population and that the sample size is sufficiently large.

7.15   For confidence coefficient .99, $\alpha = .01$ and $\alpha / 2 = .01 / 2 = .005$. From Table IV, Appendix A, $z_{.005} = 2.58$. The confidence interval is:

$$\bar{x} \pm z_{.005} \frac{s}{\sqrt{n}} \Rightarrow 39 \pm 2.58 \frac{6}{\sqrt{100}} \Rightarrow 39 \pm 1.55 \Rightarrow (37.45, \ 40.55)$$

We are 99% confident that the true mean WR score for all convicted drug dealers is between 37.45 and 40.55.

7.17   a.   Some preliminary calculations are:

$$\bar{x} = \frac{\sum x}{n} = \frac{181.56}{504} = .3602$$

$$s^2 = \frac{\sum x^2 - \dfrac{\left(\sum x\right)^2}{n}}{n-1} = \frac{74.6546 - \dfrac{181.56^2}{504}}{504 - 1} = \frac{9.24977143}{503} = .018389$$

$$s = \sqrt{.018389} = .1356$$

For confidence coefficient .90, $\alpha = .10$ and $\alpha/2 = .10/2 = .05$. From Table IV, Appendix A, $z_{.05} = 1.645$. The 90% confidence interval is:

$$\bar{x} \pm z_{.05}\sigma_{\bar{x}} \Rightarrow \bar{x} \pm 1.645\frac{\sigma}{\sqrt{n}} \Rightarrow .3602 \pm 1.645\frac{.1356}{\sqrt{504}} \Rightarrow .3602 \pm .0099 \Rightarrow (.3503, \ .3701)$$

b.   We are 90% confident that the true mean visible albedo value of all Canadian Arctic ice ponds is between .3503 and .3701.

In repeated sampling 90% of all intervals constructed in the same manner will contain the true mean and 10% will not.

c.   **First-year Ice**:

Some preliminary calculations are:

$$\bar{x} = \frac{\sum x}{n} = \frac{26.63}{88} = .3026$$

$$s^2 = \frac{\sum x^2 - \dfrac{\left(\sum x\right)^2}{n}}{n-1} = \frac{10.6923 - \dfrac{26.63^2}{88}}{88 - 1} = \frac{2.633698864}{87} = .03027$$

$$s = \sqrt{.03027} = .1740$$

The 90% confidence interval is:

$$\bar{x} \pm z_{.05}\sigma_{\bar{x}} \Rightarrow \bar{x} \pm 1.645\frac{\sigma}{\sqrt{n}} \Rightarrow .3026 \pm 1.645\frac{.1740}{\sqrt{88}} \Rightarrow .3026 \pm .0305 \Rightarrow (.2721, \ .3331)$$

We are 90% confident that the true mean visible albedo value of all First-year Canadian Arctic ice ponds is between .2721 and .3331.

Landfast Ice:

Some preliminary calculations are:

$$\bar{x} = \frac{\sum x}{n} = \frac{71.04}{196} = .3624$$

$$s^2 = \frac{\sum x^2 - \frac{\left(\sum x\right)^2}{n}}{n-1} = \frac{30.161 - \frac{71.04^2}{196}}{196-1} = \frac{4.41262449}{195} = .02263$$

$$s = \sqrt{.02263} = .1504$$

The 90% confidence interval is:

$$\bar{x} \pm z_{.05}\sigma_{\bar{x}} \Rightarrow \bar{x} \pm 1.645\frac{\sigma}{\sqrt{n}} \Rightarrow .3624 \pm 1.645\frac{.1504}{\sqrt{196}} \Rightarrow .3624 \pm .0177 \Rightarrow (.3447, \ .3801)$$

We are 90% confident that the true mean visible albedo value of all Landfast Canadian Arctic ice ponds is between .3447 and .3801.

Multi-year Ice:

Some preliminary calculations are:

$$\bar{x} = \frac{\sum x}{n} = \frac{83.89}{220} = .3813$$

$$s^2 = \frac{\sum x^2 - \frac{\left(\sum x\right)^2}{n}}{n-1} = \frac{33.8013 - \frac{83.89^2}{220}}{220-1} = \frac{1.81251773}{219} = .00827$$

$$s = \sqrt{.00827} = .0910$$

The 90% confidence interval is:

$$\bar{x} \pm z_{.05}\sigma_{\bar{x}} \Rightarrow \bar{x} \pm 1.645\frac{\sigma}{\sqrt{n}} \Rightarrow .3813 \pm 1.645\frac{.0910}{\sqrt{220}} \Rightarrow .3813 \pm .0101 \Rightarrow (.3712, \ .3914)$$

We are 90% confident that the true mean visible albedo value of all Multi-year Canadian Arctic ice ponds is between .3712 and .3914.

7.19   a.   The parameter of interest is the mean effect size for all psychological studies of personality and aggressive behavior.

b.   No, the distribution does not look normal. The data appear to be skewed to the right. The shape of this distribution is not of interest because the sample size is large, $n = 109$. By the Central Limit Theorem, the sampling distribution of $\bar{x}$ is normal regardless of the original distribution.

c.   From the printout, the 95% confidence interval for μ is (0.4786, 0.8167). We are 95% confident that the true mean effect size is between 0.4786 and 0.8167.

d. Yes. Since 0 is not contained in the 95% confidence interval, it is not a likely value for the true mean effect size. Since all the values in the 95% confident interval are above 0, the researchers are justified in concluding the true effect size mean is greater than 0 or that those who score high on a personality test are more aggressive than those who score low.

7.21 a. For confidence coefficient .99, $\alpha = .01$ and $\alpha/2 = .01/2 = .005$. From Table IV, Appendix A, $z_{.005} = 2.58$. The confidence interval is:

$$\bar{x} \pm z_{\alpha/2}\frac{s}{\sqrt{n}} \Rightarrow 1.13 \pm 2.58\frac{2.21}{\sqrt{72}} \Rightarrow 1.13 \pm .672 \Rightarrow (.458,\ 1.802)$$

We are 99% confident that the true mean number of pecks made by chickens pecking at blue string is between .458 and 1.802.

b. Yes, there is evidence that chickens are more apt to peck at white string. The mean number of pecks for white string is 7.5. Since 7.5 is not in the 99% confidence interval for the mean number of pecks at blue string, it is not a likely value for the true mean for blue string.

7.23 a. For confidence coefficient .95, $\alpha = .05$ and $\alpha/2 = .05/2 = .025$. From Table IV, Appendix A, $z_{.025} = 1.96$. The confidence interval is:

$$\bar{x} \pm z_{\alpha/2}\frac{s}{\sqrt{n}} \Rightarrow 19 \pm 1.96\frac{65}{\sqrt{265}} \Rightarrow 19 \pm 7.826 \Rightarrow (11.174,\ 26.826)$$

b. For confidence coefficient .95, $\alpha = .05$ and $\alpha/2 = .05/2 = .025$. From Table IV, Appendix A, $z_{.025} = 1.96$. The confidence interval is:

$$\bar{x} \pm z_{\alpha/2}\frac{s}{\sqrt{n}} \Rightarrow 7 \pm 1.96\frac{49}{\sqrt{265}} \Rightarrow 7 \pm 5.90 \Rightarrow (1.10,\ 12.90)$$

c. The SAT-Mathematics test would be more likely to have a mean change of 15 because 15 is in the 95% confidence interval for the mean change in SAT-Mathematics. Since 15 is in the confidence interval, it is a likely value. The value 15 is not in the 95% confidence interval for the SAT-Verbal. Thus, it is not a likely value.

7.25 a. For confidence coefficient .95, $\alpha = .05$ and $\alpha/2 = .05/2 = .025$. From Table IV, Appendix A, $z_{.025} = 1.96$.

For Males, the 95% confidence interval is:

$$\bar{x} \pm z_{.025}\sigma_{\bar{x}} \Rightarrow \bar{x} \pm 1.96\frac{\sigma}{\sqrt{n}} \Rightarrow 16.79 \pm 1.96\frac{13.57}{\sqrt{128}} \Rightarrow 16.79 \pm 2.35 \Rightarrow (14.44,\ 19.14)$$

For Females, the 95% confidence interval is:

$$\bar{x} \pm z_{.025}\sigma_{\bar{x}} \Rightarrow \bar{x} \pm 1.96\frac{\sigma}{\sqrt{n}} \Rightarrow 10.79 \pm 1.96\frac{11.53}{\sqrt{184}} \Rightarrow 10.79 \pm 1.67 \Rightarrow (9.12,\ 12.46)$$

b.   Since the two intervals are independent, the probability that at least one of the 2 confidence intervals will not contain the population mean is equal to 1 minus the probability that neither of the 2 confidence intervals will not contain the population mean. Thus the probability is:

P(at least one interval will not contain mean)
= 1 − P(neither interval will not contain the mean)
= 1 − P(both will contain the mean) = 1 − .95(.95) = 1 − .9025 = .0975.

c.   The 95% confidence interval for the males is (14.44, 19.14). The 95% confidence interval for the females is (9.12, 12.46). Since all of the values in the male interval are larger than all the values in the female interval, we can infer that the males consume the most alcohol, on average, per week.

7.27   The two problems (and corresponding solutions) with using a small sample to estimate μ are:

1.   The shape of the sampling distribution of the sample mean $\bar{x}$ now depends on the shape of the population that is sampled. The Central Limit Theorem no longer applies since the sample size is small. However, if the sampled population is normal or approximately normal, then the sampling distribution of $\bar{x}$ is always normal.

2.   The population standard deviation σ is almost always unknown. Although it is still true that $\sigma_{\bar{x}} = \dfrac{\sigma}{\sqrt{n}}$, the sample standard deviation s may provide a poor approximation for σ when the sample size is small. Instead of using $z = \dfrac{\bar{x}-\mu}{\frac{\sigma}{\sqrt{n}}}$, we use $t = \dfrac{\bar{x}-\mu}{\frac{s}{\sqrt{n}}}$ as the statistic.

7.29   a.   If x is normally distributed, the sampling distribution of $\bar{x}$ is normal, regardless of the sample size. If n is large, then the test statistic is z. If n is small, then the test statistic would be t.

b.   If nothing is known about the distribution of x, the sampling distribution of $\bar{x}$ is approximately normal if n is sufficiently large. If n is not large, the distribution of $\bar{x}$ is unknown if the distribution of x is not known.

7.31   a. $P(t \geq t_0) = .025$ where df = 10
$t_0 = 2.228$

b. $P(t \geq t_0) = .01$ where df = 17
$t_0 = 2.567$

c. $P(t \leq t_0) = .005$ where df = 6

Because of symmetry, the statement can be rewritten

$P(t \geq -t_0) = .005$ where df = 6
$t_0 = -3.707$

d. $P(t \le t_0) = .05$ where df = 13
$$t_0 = -1.771$$

7.33   First, we must compute $\bar{x}$ and s.

$$\bar{x} = \frac{\sum x}{n} = \frac{30}{6} = 5 \qquad s^2 = \frac{\sum x^2 - \frac{(\sum x)^2}{n}}{n-1} = \frac{176 - \frac{(30)^2}{6}}{6-1} = \frac{26}{5} = 5.2$$

$$s = \sqrt{5.2} = 2.2804$$

a.   For confidence coefficient .90, $\alpha = .10$ and $\alpha / 2 = .10 / 2 = .05$. From Table VI, Appendix A, with df = $n - 1 = 6 - 1 = 5$, $t_{.05} = 2.015$. The 90% confidence interval is:

$$\bar{x} \pm t_{.05} \frac{s}{\sqrt{n}} \Rightarrow 5 \pm 2.015 \frac{2.2804}{\sqrt{6}} \Rightarrow 5 \pm 1.876 \Rightarrow (3.124, \ 6.876)$$

b.   For confidence coefficient .95, $\alpha = .05$ and $\alpha / 2 = .05 / 2 = .025$. From Table VI, Appendix A, with df = $n - 1 = 6 - 1 = 5$, $t_{.025} = 2.571$. The 95% confidence interval is:

$$\bar{x} \pm t_{.025} \frac{s}{\sqrt{n}} \Rightarrow 5 \pm 2.571 \frac{2.2804}{\sqrt{6}} \Rightarrow 5 \pm 2.394 \Rightarrow (2.606, \ 7.394)$$

c.   For confidence coefficient .99, $\alpha = .01$ and $\alpha / 2 = .01 / 2 = .005$. From Table VI, Appendix A, with df = $n - 1 = 6 - 1 = 5$, $t_{.005} = 4.032$. The 99% confidence interval is:

$$\bar{x} \pm t_{.005} \frac{s}{\sqrt{n}} \Rightarrow 5 \pm 4.032 \frac{2.2804}{\sqrt{6}} \Rightarrow 5 \pm 3.754 \Rightarrow (1.246, \ 8.754)$$

d.   a)   For confidence coefficient .90, $\alpha = .10$ and $\alpha / 2 = .10 / 2 = .05$. From Table VI, Appendix A, with df = $n - 1 = 25 - 1 = 24$, $t_{.05} = 1.711$. The 90% confidence interval is:

$$\bar{x} \pm t_{.05} \frac{s}{\sqrt{n}} \Rightarrow 5 \pm 1.711 \frac{2.2804}{\sqrt{25}} \Rightarrow 5 \pm .780 \Rightarrow (4.220, \ 5.780)$$

b)   For confidence coefficient .95, $\alpha = .05$ and $\alpha / 2 = .05 / 2 = .025$. From Table VI, Appendix A, with df = $n - 1 = 25 - 1 = 24$, $t_{.025} = 2.064$. The 95% confidence interval is:

$$\bar{x} \pm t_{.025} \frac{s}{\sqrt{n}} \Rightarrow 5 \pm 2.064 \frac{2.2804}{\sqrt{25}} \Rightarrow 5 \pm .941 \Rightarrow (4.059, \ 5.941)$$

c)   For confidence coefficient .99, $\alpha = .01$ and $\alpha / 2 = .01 / 2 = .005$. From Table VI, Appendix A, with df $= n - 1 = 25 - 1 = 24$, $t_{.005} = 2.797$. The 99% confidence interval is:

$$\bar{x} \pm t_{.005} \frac{s}{\sqrt{n}} \Rightarrow 5 \pm 2.797 \frac{2.2804}{\sqrt{25}} \Rightarrow 5 \pm 1.276 \Rightarrow (3.724,\ 6.276)$$

Increasing the sample size decreases the width of the confidence interval.

7.35   a.   The target parameter is $\mu =$ the mean trap spacing for the population of red spiny lobster fishermen fishing in Baja California Sur, Mexico.

b.   Using MINITAB, the descriptive statistics are:

Descriptive Statistics: Spacing

Variable	N	Mean	StDev	Minimum	Q1	Median	Q3	Maximum
Spacing	7	89.86	11.63	70.00	82.00	93.00	99.00	105.00

The point estimate for μ is $\bar{x} = 89.86$.

c.   We do not know what the distribution of the population of trap spacings is. The Central Limit Theorem also does not apply because the sample size ($n = 7$) is too small. The sample standard deviation, s, is no longer necessarily a good estimate for the population standard deviation σ.

d.   For confidence coefficient .95, $\alpha = .05$ and $\alpha / 2 = .05 / 2 = .025$. From Table VI, Appendix A, with df $= n - 1 = 7 - 1 = 6$, $t_{.025} = 2.447$. The confidence interval is:

$$\bar{x} \pm t_{.025,6} \frac{s}{\sqrt{n}} \Rightarrow 89.86 \pm 2.447 \frac{11.63}{\sqrt{7}} \Rightarrow 89.86 \pm 10.756 \Rightarrow (79.104,\ 100.616)$$

e.   We are 95% confident that the mean trap spacing for the population of red spiny lobster fishermen fishing in Baja California Sur, Mexico is between 79.104 and 100.616 meters.

f.   We must assume that the population of trap spacings is normal and that a random sample of trapping spacings was selected.

7.37   a.   From the printout, the 95% confidence interval is (652.76, 817.40).

b.   For confidence coefficient .95, $\alpha = .05$ and $\alpha / 2 = .05 / 2 = .025$. From Table VI, Appendix A, with df $= n - 1 = 12 - 1 = 11$, $t_{.025} = 2.201$. The confidence interval is:

$$\bar{x} \pm t_{.025,11} \frac{s}{\sqrt{n}} \Rightarrow 735.08 \pm 2.201 \frac{129.565}{\sqrt{12}} \Rightarrow 735.08 \pm 82.32 \Rightarrow (652.76,\ 817.40)$$

c.   We are 95% confident that the true mean number of minutes of daylight per day in Sharon, PA is between 652.76 and 817.40 minutes.

d. If a random sample of size 12 was taken from the population, then we could see most observations in one season, like summer or winter. Since the amount of daylight is greatly affected by the time of the year, this might not be the best plan. By sampling one observation from each month, we are guaranteed that the sample will be fairly representative. However, it will not be random.

7.39 a. The point estimate for the mean amount of cesium in lichen specimens collected in Alaska is $\bar{x} = .009027$.

b. For confidence coefficient .95, $\alpha = .05$ and $\alpha / 2 = .05 / 2 = .025$. From Table VI, Appendix A, with df $= n - 1 = 9 - 1 = 8$, $t_{.025} = 2.306$.

c. The 95% confidence interval is:

$$\bar{x} \pm t_{.025} \frac{s}{\sqrt{n}} \Rightarrow .009027 \pm 2.306 \frac{.004854}{\sqrt{9}} \Rightarrow .009027 \pm .003731 \Rightarrow (.005296, \ .012758)$$

d. This interval is the same as that found on the printout.

e. We are 95% confident that the mean amount of cesium in lichen specimens collected in Alaska is between .005296 and .012758 microcuries per milliliter.

7.41 For confidence coefficient .90, $\alpha = .10$ and $\alpha / 2 = .10 / 2 = .05$. From Table VI, Appendix A, with df $= n - 1 = 17 - 1 = 16$, $t_{.05} = 1.746$. The confidence interval is:

$$\bar{x} \pm t_{.05,16} \frac{s}{\sqrt{n}} \Rightarrow .11 \pm 1.746 \frac{.19}{\sqrt{17}} \Rightarrow .11 \pm .08 \Rightarrow (.03, \ .19)$$

We are 90% confident that the true mean score difference for all amusiacs is between .03 and .19.

We must assume that the population of score differences is approximately normal and that the data were randomly selected.

7.43 a. **Both Untreated**: For confidence coefficient .90, $\alpha = .10$ and $\alpha / 2 = .10 / 2 = .05$. From Table VI, Appendix A, with df $= n - 1 = 29 - 1 = 28$, $t_{.05} = 1.701$. The 90% confidence interval is:

$$\bar{x} \pm t_{.05,28} \frac{s}{\sqrt{n}} \Rightarrow 20.9 \pm 1.701 \frac{3.34}{\sqrt{29}} \Rightarrow 20.9 \pm 1.055 \Rightarrow (19.845, \ 21.955)$$

Male Treated: For confidence coefficient .90, $\alpha = .10$ and $\alpha / 2 = .10 / 2 = .05$. From Table VI, Appendix A, with df $= n - 1 = 23 - 1 = 22$, $t_{.05} = 1.717$. The 90% confidence interval is:

$$\bar{x} \pm t_{.05,22} \frac{s}{\sqrt{n}} \Rightarrow 20.3 \pm 1.717 \frac{3.50}{\sqrt{23}} \Rightarrow 20.3 \pm 1.253 \Rightarrow (19.047, \ 21.553)$$

Female Treated: For confidence coefficient .90, $\alpha = .10$ and $\alpha / 2 = .10 / 2 = .05$. From Table VI, Appendix A, with df $= n - 1 = 18 - 1 = 17$, $t_{.05} = 1.740$. The 90% confidence interval is:

$$\bar{x} \pm t_{.05,17} \frac{s}{\sqrt{n}} \Rightarrow 22.9 \pm 1.740 \frac{4.37}{\sqrt{18}} \Rightarrow 22.9 \pm 1.792 \Rightarrow (21.108, \ 24.692)$$

Both Treated: For confidence coefficient .90, $\alpha = .10$ and $\alpha / 2 = .10 / 2 = .05$. From Table VI, Appendix A, with df $= n - 1 = 21 - 1 = 20$, $t_{.05} = 1.725$. The 90% confidence interval is:

$$\bar{x} \pm t_{.05,20} \frac{s}{\sqrt{n}} \Rightarrow 18.6 \pm 1.725 \frac{2.11}{\sqrt{21}} \Rightarrow 18.6 \pm 0.794 \Rightarrow (17.806, \ 19.394)$$

b.   The female/male pair that appears to produce the highest mean number of eggs is the **Female treated** pair. The 90% confidence interval for this pair is the highest. This interval does overlap with the **Both treated** pair and the **Males treated** pair, but just barely.

7.45   a.   For confidence coefficient .90, $\alpha = .10$ and $\alpha / 2 = .10 / 2 = .05$. From Table VI, Appendix A, with df $= n - 1 = 4 - 1 = 3$, $t_{.05} = 2.353$. The confidence interval is:

$$\bar{x} \pm t_{.05,3} \frac{s}{\sqrt{n}} \Rightarrow 1.43 \pm 2.353 \frac{.13}{\sqrt{4}} \Rightarrow 1.43 \pm .153 \Rightarrow (1.277, 1.583)$$

b.   We are 90% confident that the true mean peptide score in alleles of the antigen-produced protein is between 1.277 and 1.583.

c.   "90% confidence" means that in repeated sampling, 90% of all intervals constructed in this manner will contain the true mean.

7.47   a.   For confidence coefficient .95, $\alpha = .05$ and $\alpha / 2 = .05 / 2 = .025$. From Table VI, Appendix A, with df $= n - 1 = 15 - 1 = 14$, $t_{.025} = 2.145$. The confidence interval is:

$$\bar{x} \pm t_{.025} \frac{s}{\sqrt{n}} \Rightarrow 37.3 \pm 2.145 \frac{13.9}{\sqrt{15}} \Rightarrow 37.3 \pm 7.70 \Rightarrow (29.60, 45.00)$$

b.   We are 95% confident that the true mean adrenocorticotropin level in sleepers one-hour prior to anticipated waking is between 29.60 and 45.00

c.   Suppose we assume that the true mean adrenocorticotropin level of sleepers three hours before anticipated wake-up time is 25.5. Since the 95% confidence interval for the true mean adrenocorticotropin level in sleepers one-hour prior to anticipated waking (29.60, 45.00) does not contain 25.5, there is evidence to indicate that the true mean adrenocorticotropin level of sleepers one hour before anticipated wake-up time is greater than that of sleepers three hours prior to anticipated waking.

7.49   An unbiased estimator is one in which the mean of the sampling distribution is the parameter of interest, i.e., $E(\hat{p}) = p$.

7.51   a.   The sample size is large enough if both $n\hat{p} \geq 15$ and $n\hat{q} \geq 15$.

$n\hat{p} = 196(.64) = 125.44$ and $n\hat{q} = 196(.36) = 70.56$. Since both of these numbers are greater than or equal to 15, the sample size is sufficiently large to conclude the normal approximation is reasonable.

b.   For confidence coefficient .95, $\alpha = .05$ and $\alpha/2 = .05/2 = .025$. From Table IV, Appendix A, $z_{.025} = 1.96$. The confidence interval is:

$$\hat{p} \pm z_{.025}\sqrt{\frac{\hat{p}\hat{q}}{n}} \Rightarrow .64 \pm 1.96\sqrt{\frac{.64(.36)}{196}} \Rightarrow .64 \pm .067 \Rightarrow (.537, .707)$$

c.   We are 95% confident the true value of p is between .372 and .468.

d.   "95% confidence" means that if repeated samples of size 196 were selected from the population and 95% confidence intervals formed, 95% of all confidence intervals will contain the true value of p.

7.53   The sample size is large enough if both $n\hat{p} \geq 15$ and $n\hat{q} \geq 15$.

a. $n\hat{p} = 500(.05) = 25$ and $n\hat{q} = 500(.95) = 475$. Since both of these numbers are greater than or equal to 15, the sample size is sufficiently large to conclude the normal approximation is reasonable.

b. $n\hat{p} = 100(.05) = 5$ and $n\hat{q} = 100(.95) = 95$. Since the first number is not greater than 15, the sample size is not sufficiently large to conclude the normal approximation is reasonable.

c. $n\hat{p} = 10(.5) = 5$ and $n\hat{q} = 10(.5) = 5$. Since neither of these numbers is greater than or equal to 15, the sample size is not sufficiently large to conclude the normal approximation is reasonable.

d. $n\hat{p} = 10(.3) = 3$ and $n\hat{q} = 10(.7) = 7$. Since neither of these numbers is greater than or equal to 15, the sample size is not sufficiently large to conclude the normal approximation is reasonable.

7.55   a.   The population of interest is all American adults.

b.   The sample in the study is the 1,000 American adults surveyed.

c.   The parameter of interest is $p =$ proportion of all adults who say Starbucks coffee is overpriced.

d.   The sample size is large enough if both $n\hat{p} \geq 15$ and $n\hat{q} \geq 15$. For this problem, $\hat{p} = .73$. $n\hat{p} = 1000(.73) = 730$ and $n\hat{q} = 1000(.27) = 270$. Since both of these numbers are greater than or equal to 15, the sample size is sufficiently large to conclude the normal approximation is reasonable.

For confidence coefficient .95, $\alpha = .05$ and $\alpha/2 = .05/2 = .025$. From Table IV, Appendix A, $z_{.025} = 1.96$. The 95% confidence interval is:

$$\hat{p} \pm z_{.025}\sqrt{\frac{pq}{n}} \approx \hat{p} \pm 1.96\sqrt{\frac{\hat{p}\hat{q}}{n}} \Rightarrow .73 \pm 1.96\sqrt{\frac{.73(.27)}{1000}} \Rightarrow .73 \pm .028 \Rightarrow (.702, .758)$$

We are 95% confident that the true proportion of American adults who think Starbucks coffee is overpriced is between .702 and .758.

7.57   a.   The population of interest is the set of gun ownerships (Yes or No) of all adults in the U.S.

      b.   The parameter of interest is the true percentage or proportion, p, of all adults in the U.S. who own at least one gun.

      c.   The estimate of the population proportion is $\hat{p} = .26$. The estimate of the population percentage is $.26(100\%) = 26\%$.

      d.   The sample size is large enough if both $n\hat{p} \geq 15$ and $n\hat{q} \geq 15$.

$n\hat{p} = 2{,}770(.26) = 720.2$ and $n\hat{q} = 2{,}770(.74) = 2{,}049.8$. Since both of these numbers are greater than or equal to 15, the sample size is sufficiently large to conclude the normal approximation is reasonable.

For confidence coefficient .99, $\alpha = .01$ and $\alpha/2 = .01/2 = .005$. From Table IV, Appendix A, $z_{.005} = 2.58$. The 99% confidence interval is:

$$\hat{p} \pm z_{.005}\sqrt{\frac{pq}{n}} \Rightarrow \hat{p} \pm 2.58\sqrt{\frac{\hat{p}\hat{q}}{n}} \Rightarrow .26 \pm 2.58\sqrt{\frac{.26(.74)}{2{,}770}} \Rightarrow .26 \pm .02 \Rightarrow (.24, .28)$$

The 99% confidence interval for the true percentage is (24%, 28%).

      e.   We are 99% confident that the true percentage of adults in the U.S who own at least one gun is between 24% and 28%.

      f.   "99% confidence" means that in repeated sampling 99% of all confidence intervals constructed in this manner will contain the true percentage.

7.59   a.   The sample size is large enough if both $n\hat{p} \geq 15$ and $n\hat{q} \geq 15$.

$n\hat{p} = 1{,}000(.63) = 630$ and $n\hat{q} = 1{,}000(.37) = 370$. Since both of the numbers are greater than or equal to 15, the sample size is sufficiently large to conclude the normal approximation is reasonable.

For confidence coefficient .95, $\alpha = .05$ and $\alpha / 2 = .05 / 2 = .025$. From Table IV, Appendix A, $z_{.025} = 1.96$. The 95% confidence interval is:

$$\hat{p} \pm z_{.025}\sqrt{\frac{pq}{n}} \Rightarrow \hat{p} \pm 1.96\sqrt{\frac{\hat{p}\hat{q}}{n}} \Rightarrow .63 \pm 1.96\sqrt{\frac{.63(.37)}{1,000}} \Rightarrow .63 \pm .030 \Rightarrow (.600, \quad .660)$$

b. Yes, we would be surprised. Since .70 is not in the 95% confidence interval, it is not a likely value for the true value of the population proportion of adults who would choose to sleep when home sick.

7.61   The estimate of the true proportion of all U. S. teenagers who have used at least one informal element in a school writing assignment is $\hat{p} = \dfrac{448}{700} = .64$.

The sample size is large enough if both $n\hat{p} \geq 15$ and $n\hat{q} \geq 15$. For this problem, $\hat{p} = .64$. $n\hat{p} = 700(.64) = 448$ and $n\hat{q} = 700(.36) = 252$. Since both of these numbers are greater than or equal to 15, the sample size is sufficiently large to conclude the normal approximation is reasonable.

For confidence coefficient .99, $\alpha = .01$ and $\alpha / 2 = .01 / 2 = .005$. From Table IV, Appendix A, $z_{.005} = 2.58$. The 99% confidence interval is:

$$\hat{p} \pm z_{.005}\sqrt{\frac{pq}{n}} \approx \hat{p} \pm 2.58\sqrt{\frac{\hat{p}\hat{q}}{n}} \Rightarrow .64 \pm 2.58\sqrt{\frac{.64(.36)}{700}} \Rightarrow .64 \pm .047 \Rightarrow (.593, .687)$$

We are 99% confident that the true proportion of all U. S. teenagers who have used at least one informal element in a school writing assignment is between .593 and .687.

7.63   a. First, we compute $\hat{p}$. There were 1333 useable responses. Of these, 450 chose "Bible is the actual word of God and is to be taken literally".

$$\hat{p} = \frac{x}{n} = \frac{450}{1333} = .338$$

The sample size is large enough if both $n\hat{p} \geq 15$ and $n\hat{q} \geq 15$.

$n\hat{p} = 1,333(.338) = 450.6$ and $n\hat{q} = 1,333(.662) = 882.4$. Since both of the numbers are greater than or equal to 15, the sample size is sufficiently large to conclude the normal approximation is reasonable.

For confidence coefficient .95, $\alpha = .05$ and $\alpha / 2 = .05 / 2 = .025$. From Table IV, Appendix A, $z_{.025} = 1.96$. The 95% confidence interval is:

$$\hat{p} \pm z_{.025}\sqrt{\frac{pq}{n}} \Rightarrow \hat{p} \pm 1.96\sqrt{\frac{\hat{p}\hat{q}}{n}} \Rightarrow .338 \pm 1.96\sqrt{\frac{.338(.662)}{1,333}}$$

$$\Rightarrow .338 \pm .025 \Rightarrow (.313, \quad .363)$$

b. We are 95% confident that the true proportion of all Americans who believe that the "Bible is the actual word of God and is to be taken literally" is between .313 and .363.

c. If the responses from the survey are not representative of the population, then the validity of the results will be questionable. If the survey is sent out to a representative sample, but the response rate is low, then again, the results may be questionable.

7.65 a. The estimate of the true proportion of all mountain casualties that require a femoral shaft splint is $\hat{p} = \dfrac{1}{333} = .003$.

The sample size is large enough if both $n\hat{p} \geq 15$ and $n\hat{q} \geq 15$. For this problem, $\hat{p} = .003$. $n\hat{p} = 333(.003) = .999$ and $n\hat{q} = 333(.997) = 332.001$. Since the first of these numbers is less than to 15, the sample size is not sufficiently large to conclude the normal approximation is reasonable.

b. The Wilson adjusted sample proportion is

$$\tilde{p} = \frac{x+2}{n+4} = \frac{1+2}{333+4} = \frac{3}{337} = .009$$

For confidence coefficient .95, $\alpha = .05$ and $\alpha/2 = .05/2 = .025$. From Table IV, Appendix A, $z_{.025} = 1.96$. The Wilson adjusted 95% confidence interval is:

$$\tilde{p} \pm z_{.025}\sqrt{\frac{\tilde{p}(1-\tilde{p})}{n+4}} \Rightarrow .009 \pm 1.96\sqrt{\frac{.009(.991)}{333+4}} \Rightarrow .009 \pm .010 \Rightarrow (-.001,\ .019)$$

We are 95% confident that the true proportion of all mountain casualties that require a femoral shaft splint is between 0 and .019.

7.67 The point estimate for the proportion of health care workers with latex allergy who suspect that he/she has the allergy is $\hat{p} = \dfrac{x}{n} = \dfrac{36}{83} = .434$.

The sample size is large enough if both $n\hat{p} \geq 15$ and $n\hat{q} \geq 15$.

$n\hat{p} = 83(.434) = 36$ and $n\hat{q} = 83(.566) = 47$. Since both of the numbers are greater than or equal to 15, the sample size is sufficiently large to conclude the normal approximation is reasonable.

Since no confidence limit is given, we will use 95%. For confidence coefficient .95, $\alpha = .05$ and $\alpha/2 = .05/2 = .025$. From Table IV, Appendix A, $z_{.025} = 1.96$. The 95% confidence interval is:

$$\hat{p} \pm z_{.025}\sigma_{\hat{p}} \Rightarrow \hat{p} \pm 1.96\sqrt{\frac{\hat{p}\hat{q}}{n}} \Rightarrow .434 \pm 1.96\sqrt{\frac{.434(.566)}{83}} \Rightarrow .434 \pm .107 \Rightarrow (.327,\ .541)$$

We are 95% confident that the proportion of health care workers with latex allergy who suspect that he/she has the allergy is between .327 and .541.

7.69   The statement "For a specified sampling error SE, increasing the confidence level $(1-\alpha)$ will lead to a larger n when determining sample size" is True.

7.71   To compute the necessary sample size, use

$$n = \frac{(z_{\alpha/2})^2 \sigma^2}{(SE)^2} \text{ where } \alpha = 1 - .95 = .05 \text{ and } \alpha/2 = .05/2 = .025$$

From Table IV, Appendix A, $z_{.025} = 1.96$. Thus,

$$n = \frac{(1.96)^2 (5.4)}{.2^2} = 518.616 \approx 519$$

You would need to take 519 samples.

7.73   a.   Range = 39 − 31 = 8. $\sigma \approx \dfrac{\text{Range}}{4} = \dfrac{8}{4} = 2$

For confidence coefficient .90, $\alpha = .10$ and $\alpha/2 = .10/2 = .05$. From Table IV, Appendix A, $z_{.05} = 1.645$.

The sample size is $n = \dfrac{z_{\alpha/2}^2 \sigma^2}{(SE)^2} = \dfrac{1.645^2 (2^2)}{.15^2} = 481.07 \approx 482$

b. $\sigma \approx \dfrac{\text{Range}}{6} = \dfrac{8}{6} = 1.333$

The sample size is $n = \dfrac{z_{\alpha/2}^2 \sigma^2}{(SE)^2} = \dfrac{1.645^2 (1.333^2)}{.15^2} = 213.7 \approx 214$

7.75   For confidence coefficient .90, $\alpha = .10$ and $\alpha/2 = .10/2 = .05$. From Table IV, Appendix A, $z_{.05} = 1.645$.

We know $\hat{p}$ is in the middle of the interval, so $\hat{p} = \dfrac{.54 + .26}{2} = .4$

The confidence interval is $\hat{p} \pm z_{.05} \sqrt{\dfrac{\hat{p}\hat{q}}{n}} \Rightarrow .4 \pm 1.645 \sqrt{\dfrac{.4(.6)}{n}}$

We know $.4 - 1.645\sqrt{\dfrac{.4(.6)}{n}} = .26$

$\Rightarrow .4 - \dfrac{.8059}{\sqrt{n}} = .26$

$\Rightarrow .4 - .26 = \dfrac{.8059}{\sqrt{n}} \Rightarrow \sqrt{n} = \dfrac{.8059}{.14} = 5.756$

$\Rightarrow n = 5.756^2 = 33.1 \approx 34$

7.77   a.   The width of a confidence interval is $2(SE) = 2z_{\alpha/2}\dfrac{\sigma}{\sqrt{n}}$

For confidence coefficient .95, $\alpha = .05$ and $\alpha/2 = .05/2 = .025$. From Table IV, Appendix A, $z_{.025} = 1.96$.

For $n = 16$, $W = 2z_{\alpha/2}\dfrac{\sigma}{\sqrt{n}} = 2(1.96)\dfrac{1}{\sqrt{16}} = 0.98$

For $n = 25$, $W = 2z_{\alpha/2}\dfrac{\sigma}{\sqrt{n}} = 2(1.96)\dfrac{1}{\sqrt{25}} = 0.784$

For $n = 49$, $W = 2z_{\alpha/2}\dfrac{\sigma}{\sqrt{n}} = 2(1.96)\dfrac{1}{\sqrt{49}} = 0.56$

For $n = 100$, $W = 2z_{\alpha/2}\dfrac{\sigma}{\sqrt{n}} = 2(1.96)\dfrac{1}{\sqrt{100}} = 0.392$

For $n = 400$, $W = 2z_{\alpha/2}\dfrac{\sigma}{\sqrt{n}} = 2(1.96)\dfrac{1}{\sqrt{400}} = 0.196$

b.

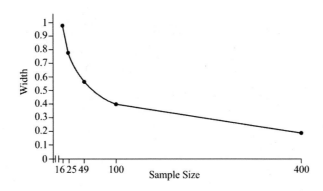

7.79 From Exercise 7.35, the sample standard deviation is $s = 11.63$. We will use this to estimate the population standard deviation. The necessary sample size is:

$$n = \frac{z_{\alpha/2}^2 \sigma^2}{(SE)^2} = \frac{1.96^2 (11.63)^2}{5^2} = 20.78 \approx 21$$

Thus, we will need to sample approximately 21 teams of fishermen.

7.81 a. The sample size may not be large enough. The sample size is large enough if both $n\hat{p} \geq 15$ and $n\hat{q} \geq 15$.

$$\hat{p} = \frac{x}{n} = \frac{52}{60} = .867$$

$n\hat{p} = 60(.867) = 52$ and $n\hat{q} = 60(.133) = 8$. Since the second number is not greater than or equal to 15, the sample size is not sufficiently large to conclude the normal approximation is reasonable.

b. For confidence coefficient .90, $\alpha = .10$ and $\alpha / 2 = .10 / 2 = .05$. From Table IV, Appendix A, $z_{.05} = 1.645$. We will use the sample proportion to estimate the population proportion:

$$\hat{p} = \frac{x}{n} = \frac{52}{60} = .867$$

$$n = \frac{z_{\alpha/2}^2 pq}{(SE)^2} = \frac{1.645^2 (.867)(.133)}{.05^2} = 124.8 \approx 125$$

7.83 For confidence coefficient .90, $\alpha = .10$ and $\alpha / 2 = .10 / 2 = .05$. From Table IV, Appendix A, $z_{.05} = 1.645$.

$$n = \frac{z_{\alpha/2}^2 \sigma^2}{(SE)^2} = \frac{1.645^2 10.9^2}{4^2} = 20.09 \approx 21$$

7.85 From Exercise 7.41, the standard deviation was $s = .19$. We will use this to estimate the population standard deviation. For confidence coefficient .90, $\alpha = .10$ and $\alpha / 2 = .10 / 2 = .05$. From Table IV, Appendix A, $z_{.05} = 1.645$. The necessary sample size is:

$$n = \frac{z_{\alpha/2}^2 \sigma^2}{(SE)^2} = \frac{1.645^2 (.19)^2}{.05^2} = 39.075 \approx 40$$

Thus, we will need to sample approximately 40 amusiacs.

7.87   For confidence coefficient .99, $\alpha = .01$ and. From Table IV, Appendix A, $z_{.005} = 2.58$. From the previous estimate, we will use $\hat{p} = .333$ to estimate p.

$$n = \frac{z_{\alpha/2}^2 pq}{(SE)^2} = \frac{2.58^2(.333)(.667)}{.01^2} = 14,784.5966 \approx 14,785$$

7.89   From Exercise 7.64, our estimate of p is $\hat{p} = .021$. We will use this to estimate the sample size.

For confidence coefficient .99, $\alpha = .01$ and $\alpha / 2 = .01 / 2 = .005$. From Table IV, Appendix A, $z_{.005} = 2.58$.

$$n = \frac{z_{\alpha/2}^2 pq}{(SE)^2} = \frac{2.58^2(.021)(.979)}{.04^2} = 85.5 \approx 86$$

We would need to sample 86 U.S. adults.

Now, the sample from Japan may not be representative because college students were used. To be extremely conservative, we could estimate the needed sample size by estimating p with .5. Now the needed sample size would be:

$$n = \frac{z_{\alpha/2}^2 pq}{(SE)^2} = \frac{2.58^2(.5)(.5)}{.04^2} = 1040.06 \approx 1041$$

This is much larger than the sample size found using the estimate from Exercise 7.64.

7.91   For confidence coefficient .90, $\alpha = .10$ and $\alpha / 2 = .10 / 2 = .05$. From Table IV, Appendix A, $z_{.05} = 1.645$.

The sample size is $n = \dfrac{\left(z_{\alpha/2}\right)^2 \sigma^2}{(SE)^2} = \dfrac{(1.645)^2(10^2)}{1^2} = 270.6 \approx 271$

7.93   The sampling distribution used to find interval estimates for σ^2 is the chi-square distribution.

7.95   The degrees of freedom associated with the chi-square sampling distribution for a sample of size n is $n - 1$.

7.97   a.   For confidence level .90, $\alpha = .10$ and $\alpha / 2 = .10 / 2 = .05$. From Table VII, Appendix A, with df $= n - 1 = 50-1 = 49$, $\chi_{.05,49}^2 \approx 67.5048$ and $\chi_{.95,49}^2 \approx 34.7642$. The 90% confidence interval is:

$$\frac{(n-1)s^2}{\chi_{.05}^2} \le \sigma^2 \le \frac{(n-1)s^2}{\chi_{.95}^2} \Rightarrow \frac{(50-1)2.5^2}{67.5048} \le \sigma^2 \le \frac{(50-1)2.5^2}{34.7642} \Rightarrow 4.537 \le \sigma^2 \le 8.809$$

b. For confidence level .90, $\alpha = .10$ and $\alpha/2 = .10/2 = .05$. From Table VII, Appendix A, with df $= n - 1 = 15\text{-}1 = 14$, $\chi^2_{.05,14} = 23.6848$ and $\chi^2_{.95,14} = 6.57063$. The 90% confidence interval is:

$$\frac{(n-1)s^2}{\chi^2_{.05}} \leq \sigma^2 \leq \frac{(n-1)s^2}{\chi^2_{.95}} \Rightarrow \frac{(15-1).02^2}{23.6848} \leq \sigma^2 \leq \frac{(15-1).02^2}{6.57063} \Rightarrow .00024 \leq \sigma^2 \leq .00085$$

c. For confidence level .90, $\alpha = .10$ and $\alpha/2 = .10/2 = .05$. From Table VII, Appendix A, with df $= n - 1 = 22\text{-}1 = 21$, $\chi^2_{.05,21} = 32.6705$ and $\chi^2_{.95,21} = 11.5913$. The 90% confidence interval is:

$$\frac{(n-1)s^2}{\chi^2_{.05}} \leq \sigma^2 \leq \frac{(n-1)s^2}{\chi^2_{.95}} \Rightarrow \frac{(22-1)31.6^2}{32.6705} \leq \sigma^2 \leq \frac{(22-1)31.6^2}{11.5913} \Rightarrow 641.86 \leq \sigma^2 \leq 1,809.09$$

d. For confidence level .90, $\alpha = .10$ and $\alpha/2 = .10/2 = .05$. From Table VII, Appendix A, with df $= n - 1 = 5\text{-}1 = 4$, $\chi^2_{.05,4} = 9.48773$ and $\chi^2_{.95,4} = .710721$. The 90% confidence interval is:

$$\frac{(n-1)s^2}{\chi^2_{.05}} \leq \sigma^2 \leq \frac{(n-1)s^2}{\chi^2_{.95}} \Rightarrow \frac{(5-1)1.5^2}{9.48773} \leq \sigma^2 \leq \frac{(5-1)1.5^2}{.710721} \Rightarrow .94859 \leq \sigma^2 \leq 12.6632$$

7.99   Using MINITAB, the descriptive statistics are:

Descriptive Statistics: x

```
Variable   N   Mean   StDev   Minimum    Q1   Median    Q3   Maximum
x          6   6.17   3.31      2.00    2.75    6.50   8.75    11.00
```

For confidence level .95, $\alpha = .05$ and $\alpha/2 = .05/2 = .025$. From Table VII, Appendix A, with df $= n - 1 = 6 - 1 = 5$, $\chi^2_{.025,5} = 12.8325$ and $\chi^2_{.975,5} = .831211$. The 95% confidence interval is:

$$\frac{(n-1)s^2}{\chi^2_{.025}} \leq \sigma^2 \leq \frac{(n-1)s^2}{\chi^2_{.975}} \Rightarrow \frac{(6-1)3.31^2}{12.8325} \leq \sigma^2 \leq \frac{(6-1)3.31^2}{.831211} \Rightarrow 4.2689 \leq \sigma^2 \leq 65.9044$$

7.101   a. For confidence level .95, $\alpha = .05$ and $\alpha/2 = .05/2 = .025$. From Table VII, Appendix A, with df $= n - 1 = 106 - 1 = 105$, $\chi^2_{.025,105} \approx 129.561$ and $\chi^2_{.975,105} = 74.2219$. The 95% confidence interval is:

$$\frac{(n-1)s^2}{\chi^2_{.025}} \leq \sigma^2 \leq \frac{(n-1)s^2}{\chi^2_{.975}} \Rightarrow \frac{(106-1)19.5^2}{129.561} \leq \sigma^2 \leq \frac{(106-1)19.5^2}{74.2219} \Rightarrow 308.166 \leq \sigma^2 \leq 537.931$$

We are 95% confident that the true variance of the times per day that laptops are used for taking notes for all middle school students is between 308.166 and 537.931.

    b.   Yes. One of the assumptions necessary for this confidence interval to be valid is that the population being sampled from is approximately normal. Since the population of times is not normal, the validity of this confidence interval is questionable.

7.103   a.   From the printout, the 95% confidence interval for the variance is: (8.6, 45.7). We are 95% confident that the true variance of rebound lengths is between 8.6 and 45.7.

    b.   From the printout, the 95% confidence interval for the standard deviation is: (2.94, 6.76). We are 95% confident that the true standard deviation of rebound lengths is between 2.94 and 6.76.

    c.   We must assume that the sample was a random sample from the target population and that the target population is approximately normal.

7.105   Using MINITAB, the descriptive statistics are:

Descriptive Statistics: Depth

```
Variable   N    Mean  StDev  Minimum      Q1  Median      Q3  Maximum
Depth      18  16.499  1.970   13.250  15.285  16.160  18.123   19.700
```

For confidence level .95, $\alpha = .05$ and $\alpha/2 = .05/2 = .025$. From Table VII, Appendix A, with df $= n - 1 = 18 - 1 = 17$, $\chi^2_{.025,17} = 30.1910$ and $\chi^2_{.975,17} = 7.56418$. The 95% confidence interval for the variance is:

$$\frac{(n-1)s^2}{\chi^2_{.025}} \le \sigma^2 \le \frac{(n-1)s^2}{\chi^2_{.975}} \Rightarrow \frac{(18-1)1.97^2}{30.1910} \le \sigma^2 \le \frac{(18-1)1.97^2}{7.56418} \Rightarrow 2.185 \le \sigma^2 \le 8.722$$

The 95% confidence interval for the standard deviation is:

$$\sqrt{2.185} \le \sigma \le \sqrt{8.722} \Rightarrow 1.478 \le \sigma \le 2.953$$

We are 95% confident that the true standard deviation of the molar depths for the population of cheek teeth in extinct primates is between 1.478 and 2.953.

We must assume that a random sample of observations was selected from a normal distribution.

Using MINITAB, a histogram of the data is:

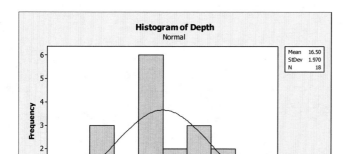

The data do not look particularly normal. The confidence interval for the standard deviation may not be valid.

7.107   From Exercise 7.35, $s = 11.63$.

For confidence level .99, $\alpha = .01$ and $\alpha / 2 = .01 / 2 = .005$. From Table VII, Appendix A, with df $= n - 1 = 7 - 1 = 6$, $\chi^2_{.005,6} = 18.5476$ and $\chi^2_{.995,6} = .675727$. The 99% confidence interval for the variance is:

$$\frac{(n-1)s^2}{\chi^2_{.005}} \leq \sigma^2 \leq \frac{(n-1)s^2}{\chi^2_{.995}} \Rightarrow \frac{(7-1)11.63^2}{18.5476} \leq \sigma^2 \leq \frac{(7-1)11.63^2}{.675727} \Rightarrow 43.755 \leq \sigma^2 \leq 1,200.990$$

We are 99% confident that the true variance of the trap spacing measurements of red spiny lobster fishermen fishing in Baja California Sur, Mexico is between 43.755 and 1,200.990.

7.109   a.   The target parameter for the average score on the SAT is μ.

b.   The target parameter for the mean time waiting at a supermarket checkout lane is μ.

c.   The target parameter for the proportion of voters in favor of legalizing marijuana is p.

b.   The target parameter for the percentage of NFL players who have ever made the Pro Bowl is p.

e.   The target parameter for the dropout rate of American college students is p.

f.   The target parameter for the variation in IQ scores of sociopaths is σ^2.

7.111   a.   For a small sample from a normal distribution with unknown standard deviation, we use the t statistic. For confidence coefficient .95, $\alpha = .05$ and $\alpha / 2 = .05 / 2 = .025$. From Table VI, Appendix A, with df $= n - 1 = 21 - 1 = 20$, $t_{.025} = 2.086$.

b.   For a large sample from a distribution with an unknown standard deviation, we can estimate the population standard deviation with s and use the z statistic. For confidence coefficient .95, $\alpha = .05$ and $\alpha / 2 = .05 / 2 = .025$. From Table IV, Appendix A, $z_{.025} = 1.96$.

c.   For a small sample from a normal distribution with known standard deviation, we use the z statistic. For confidence coefficient .95, $\alpha = .05$ and $\alpha / 2 = .05 / 2 = .025$. From Table IV, Appendix A, $z_{.025} = 1.96$.

d.   For a large sample from a distribution about which nothing is known, we can estimate the population standard deviation with s and use the z statistic. For confidence coefficient .95, $\alpha = .05$ and $\alpha / 2 = .05 / 2 = .025$. From Table IV, Appendix A, $z_{.025} = 1.96$.

e.   For a small sample from a distribution about which nothing is known, we can use neither z nor t.

7.113   a.   The point estimate for the proportion of measurements in the population with characteristic A is $\hat{p} = \dfrac{x}{n} = \dfrac{227}{400} = .5675$.

The sample size is large enough if both $n\hat{p} \geq 15$ and $n\hat{q} \geq 15$.

$n\hat{p} = 400(.5675) = 227$ and $n\hat{q} = 400(.4325) = 173$. Since both of the numbers are greater than or equal to 15, the sample size is sufficiently large to conclude the normal approximation is reasonable.

For confidence coefficient .95, $\alpha = .05$ and $\alpha / 2 = .05 / 2 = .025$. From Table IV, Appendix A, $z_{.025} = 1.96$. The 95% confidence interval is:

$$\hat{p} \pm z_{.025}\sigma_{\hat{p}} \Rightarrow \hat{p} \pm 1.96\sqrt{\dfrac{\hat{p}\hat{q}}{n}} \Rightarrow .5675 \pm 1.96\sqrt{\dfrac{.5675(.4325)}{400}}$$
$$\Rightarrow .5675 \pm .0486 \Rightarrow (.5189, \ .6161)$$

We are 95% confident that the proportion of measurements in the population with characteristic A is between .5189 and .6161.

b.   We will use $\hat{p} = .5675$ to estimate p.

For confidence coefficient .95, $\alpha = .05$ and $\alpha / 2 = .05 / 2 = .025$. From Table IV, Appendix A, $z_{.025} = 1.96$.

The sample size is $n = \dfrac{(z_{.025})^2 pq}{(SE)^2} = \dfrac{1.96^2(.5675)(.4325)}{.02^2} = 2,357.2 \approx 2,358$.

7.115   The parameters of interest for the problems are:

(1)   The question requires a categorical response. One parameter of interest might be the proportion, p, of all Americans over 18 years of age who think their health is generally very good or excellent.

(2)   A parameter of interest might be the mean number of days, μ, in the previous 30 days that all Americans over 18 years of age felt that their physical health was not good because of injury or illness.

(3)   A parameter of interest might be the mean number of days, μ, in the previous 30 days that all Americans over 18 years of age felt that their mental health was not good because of stress, depression, or problems with emotions.

(4)   A parameter of interest might be the mean number of days, μ, in the previous 30 days that all Americans over 18 years of age felt that their physical or mental health prevented them from performing their usual activities.

7.117   First, we compute $\hat{p}$: $\hat{p} = \dfrac{x}{n} = \dfrac{183}{837} = .219$

The sample size is large enough if both $n\hat{p} \geq 15$ and $n\hat{q} \geq 15$.

$n\hat{p} = 837(.219) = 183.303$ and $n\hat{q} = 837(.781) = 653.697$. Since both of the numbers are greater than or equal to 15, the sample size is sufficiently large to conclude the normal approximation is reasonable.

For confidence coefficient .90, $\alpha = .10$ and $\alpha/2 = .10/2 = .05$. From Table IV, Appendix A, $z_{.05} = 1.645$. The confidence interval is:

$$\hat{p} \pm z_{.05}\sqrt{\frac{pq}{n}} \approx \hat{p} \pm 1.645\sqrt{\frac{\hat{p}\hat{q}}{n}} \Rightarrow .219 \pm 1.645\sqrt{\frac{.219(.781)}{837}} \Rightarrow .219 \pm .024 \Rightarrow (.195,\ .243)$$

We are 95% confident that the population proportion of all pottery artifacts at Phylakopi that are painted is between .195 and .243.

7.119   a.   The point estimate of p, the true driver phone cell use rate, is $\hat{p} = \dfrac{x}{n} = \dfrac{35}{1,165} = .030$.

b.   The sample size is large enough if both $n\hat{p} \geq 15$ and $n\hat{q} \geq 15$.

$n\hat{p} = 1,165(.030) = 35$ and $n\hat{q} = 1,165(.970) = 1130$. Since both of the numbers are greater than or equal to 15, the sample size is sufficiently large to conclude the normal approximation is reasonable.

For confidence coefficient .95, $\alpha = .05$ and $\alpha/2 = .05/2 = .025$. From Table IV, Appendix A, $z_{.025} = 1.96$. The 95% confidence interval is:

$$\hat{p} \pm z_{.025}\sigma_{\hat{p}} \Rightarrow \hat{p} \pm 1.96\sqrt{\frac{\hat{p}\hat{q}}{n}} \Rightarrow .030 \pm 1.96\sqrt{\frac{.030(.970)}{1,165}} \Rightarrow .030 \pm .010 \Rightarrow (.020,\ .040)$$

c.   We are 95% confident that the true driver cell phone use rate is between .020 and .040.

d.   For confidence coefficient .95, $\alpha = .05$ and $\alpha / 2 = .05 / 2 = .025$. From Table IV, Appendix A, $z_{.025} = 1.96$. We will use $\hat{p} = .030$ from part **a** to estimate p.

$$n = \frac{z_{\alpha/2}^2 pq}{(SE)^2} = \frac{1.96^2(.03)(.97)}{.005^2} = 4,471.6 \approx 4,472$$

We would need to sample 4,472 drivers.

7.121   a.   The confidence level desired by the researchers is .90.

b.   The sampling error desired by the researchers is $SE = .05$.

c.   Since no value was given for p, we will use $\hat{p} = \dfrac{x}{n} = \dfrac{64}{106} = .604$ (from Exercise 7.120).

For confidence coefficient .90, $\alpha = .10$ and $\alpha / 2 = .10 / 2 = .05$. From Table IV, Appendix A, $z_{.05} = 1.645$.

The sample size is $n = \dfrac{(z_{.05})^2 pq}{(SE)^2} = \dfrac{1.645^2(.604)(.396)}{.05^2} = 258.9 \approx 259$.

7.123   a. $\bar{x} = \dfrac{\sum x}{n} = \dfrac{39}{21} = 1.857$ $s^2 = \dfrac{\sum x^2 - \dfrac{(\sum x)^2}{n}}{n-1} = \dfrac{101 - \dfrac{39^2}{21}}{21-1} = \dfrac{28.57143}{20} = 1.4286$

$s = \sqrt{1.4286} = 1.1952$

b.   Using MINITAB, a histogram of the data is:

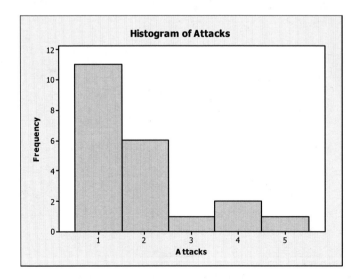

Although this is a histogram of the sample, it should reflect the distribution of the population fairly well. The population distribution appears to be skewed to the right. However, the distribution is somewhat mound-shaped. Since the sample size of $n = 21$ is somewhat close to 30, the sampling distribution of $\bar{x}$ may be approximately normal.

c. For confidence coefficient .90, $\alpha = .10$ and $\alpha / 2 = .10 / 2 = .05$. From Table VI, Appendix A, with df $= n - 1 = 21 - 1 = 20$, $t_{.05} = 1.725$. The 90% confidence interval is:

$$\bar{x} \pm t_{.05,20}\frac{s}{\sqrt{n}} \Rightarrow 1.857 \pm 1.725\frac{1.1952}{\sqrt{21}} \Rightarrow 1.857 \pm .450 \Rightarrow (1.407,\ 2.307)$$

d. We are 90% confident that the true mean number of individual suicide bombings and/or attacks per incident is between 1.407 and 2.307.

e. With repeated sampling, 90% of all confidence intervals constructed in this manner will contain the true mean and 10% will not.

7.125 Some preliminary calculations:

$$\bar{x} = \frac{\sum x}{n} = \frac{6.44}{6} = 1.073 \qquad s^2 = \frac{\sum x^2 - \dfrac{\left(\sum x\right)^2}{n}}{n-1} = \frac{7.1804 - \dfrac{6.44^2}{6}}{6-1} = .0536$$

$$s = \sqrt{.0536} = .2316$$

a. For confidence coefficient .95, $\alpha = .05$ and $\alpha / 2 = .05 / 2 = .025$. From Table VI, Appendix A, with df $= n - 1 = 6 - 1 = 5$, $t_{.025} = 2.571$. The confidence interval is:

$$\bar{x} \pm t_{\alpha/2}\frac{s}{\sqrt{n}} \Rightarrow 1.073 \pm 2.571\frac{.2316}{\sqrt{6}} \Rightarrow 1.073 \pm .243 \Rightarrow (.830,\ 1.316)$$

We are 95% confident that the true average decay rate of fine particles produced from oven cooking or toasting is between .830 and 1.316.

b. The phrase "95% confident" means that in repeated sampling, 95% of all confidence intervals constructed in this manner will contain the true mean.

c. For confidence level .95, $\alpha = .05$ and $\alpha / 2 = .05 / 2 = .025$. From Table VII, Appendix A, with df $= n - 1 = 6 - 1 = 5$, $\chi^2_{.025,5} = 12.8325$ and $\chi^2_{.975,5} = .831211$. The 95% confidence interval for the variance is:

$$\frac{(n-1)s^2}{\chi^2_{.025}} \le \sigma^2 \le \frac{(n-1)s^2}{\chi^2_{.975}} \Rightarrow \frac{(6-1).2316^2}{12.8325} \le \sigma^2 \le \frac{(6-1).2316^2}{.831211} \Rightarrow .0209 \le \sigma^2 \le .3227$$

The 95% confidence interval for the standard deviation is:

$$\sqrt{.0209} \le \sigma \le \sqrt{.3227} \Rightarrow .1446 \le \sigma \le .5681$$

We are 95% confident that the true standard deviation of the decay rate is between .1446 and .5681.

d.   In order for the inferences above to be valid, the distribution of decay rates must be normally distributed.

e.   For confidence coefficient .95, $\alpha = .05$ and $\alpha / 2 = .05 / 2 = .025$. From Table IV, Appendix A, $z_{.025} = 1.96$.

$$n = \frac{z_{\alpha/2}^2 \alpha^2}{(SE)^2} = \frac{1.96^2 (.2316)^2}{.04^2} = 128.786 \approx 129$$

We would need to sample 129 decay rates.

7.127   a.   First, we compute $\hat{p}$: $\hat{p} = \frac{x}{n} = \frac{665}{1007} = .660$

The sample size is large enough if both $n\hat{p} \ge 15$ and $n\hat{q} \ge 15$.

$n\hat{p} = 1{,}007(.660) = 665$ and $n\hat{q} = 1{,}007(.340) = 342$. Since both of the numbers are greater than or equal to 15, the sample size is sufficiently large to conclude the normal approximation is reasonable.

For confidence coefficient .95, $\alpha = .05$ and $\alpha / 2 = .05 / 2 = .025$. From Table IV, Appendix A, $z_{.025} = 1.96$. The confidence interval is:

$$\hat{p} \pm z_{.025}\sqrt{\frac{pq}{n}} \approx \hat{p} \pm 1.96\sqrt{\frac{\hat{p}\hat{q}}{n}} \Rightarrow .660 \pm 1.96\sqrt{\frac{.660(.340)}{1007}} \Rightarrow .660 \pm .029 \Rightarrow (.631,\ .689)$$

We are 95% confident that the proportion of U.S. workers who take their lunch to work is between .631 and .689.

b.   First, we compute $\hat{p}$: $\hat{p} = \frac{x}{n} = \frac{200}{665} = .301$

The sample size is large enough if both $n\hat{p} \ge 15$ and $n\hat{q} \ge 15$.

$n\hat{p} = 665(.301) = 200$ and $n\hat{q} = 665(.699) = 465$. Since both of the numbers are greater than or equal to 15, the sample size is sufficiently large to conclude the normal approximation is reasonable.

For confidence coefficient .95, $\alpha = .05$ and $\alpha/2 = .05/2 = .025$. From Table IV, Appendix A, $z_{.025} = 1.96$. The confidence interval is:

$$\hat{p} \pm z_{.025}\sqrt{\frac{pq}{n}} \approx \hat{p} \pm 1.96\sqrt{\frac{\hat{p}\hat{q}}{n}} \Rightarrow .301 \pm 1.96\sqrt{\frac{.301(.699)}{665}} \Rightarrow .301 \pm .035 \Rightarrow (.266, .336)$$

We are 95% confident that the proportion of U.S. workers who take their lunch to work who brown-bag their lunch is between .266 and .336.

7.129   a.   For confidence coefficient .99, $\alpha = .01$ and $\alpha/2 = .01/2 = .005$. From Table IV, Appendix A, $z_{.005} = 2.58$. The confidence interval is:

$$\bar{x} \pm z_{.005}\frac{s}{\sqrt{n}} \Rightarrow .044 \pm 2.58\frac{.884}{\sqrt{197}} \Rightarrow .044 \pm .162 \Rightarrow (-.118, .206)$$

We are 99% confident that the mean inbreeding coefficient for this species of wasps is between $-.118$ and .206.

   b.   A coefficient of 0 indicates that the wasp has no tendency to inbreed. Since 0 is in the 99% confidence interval, it is a likely value for the true mean inbreeding coefficient. Thus, there is no evidence to indicate that this species of wasps has a tendency to inbreed.

7.131   a.   Using MINITAB, the descriptive statistics are:

Descriptive Statistics: CPU Time

```
Variable   N    Mean   StDev  Minimum      Q1  Median      Q3  Maximum
CPU Time   52   0.812  1.505   0.0360   0.136   0.275   0.595    8.788
```

For confidence coefficient .95, $\alpha = .05$ and $\alpha/2 = .05/2 = .025$. From Table IV, Appendix A, $z_{.025} = 1.96$. The 95% confidence interval is:

$$\bar{x} \pm z_{.025}\sigma_{\bar{x}} \Rightarrow \bar{x} \pm 1.96\frac{\sigma}{\sqrt{n}} \Rightarrow .812 \pm 1.96\frac{1.505}{\sqrt{52}} \Rightarrow .812 \pm .409 \Rightarrow (.403, \ 1.221)$$

We are 95% confident that the mean solution time for the hybrid algorithm is between .403 and 1.221.

   b.   The sample size needed would be $n = \dfrac{(z_{.025})^2\sigma^2}{(SE)^2} = \dfrac{1.96^2(1.505)^2}{.25^2} = 139.2 \approx 140$.

c. For confidence level .95, $\alpha = .05$ and $\alpha / 2 = .05 / 2 = .025$. From Table VII, Appendix A, with df $= n - 1 = 52 - 1 = 51$, $\chi^2_{.025,51} \approx 71.4202$ and $\chi^2_{.975,51} \approx 32.3574$. The 95% confidence interval for the variance is:

$$\frac{(n-1)s^2}{\chi^2_{.025}} \le \sigma^2 \le \frac{(n-1)s^2}{\chi^2_{.975}} \Rightarrow \frac{(52-1)1.505^2}{71.4202} \le \sigma^2 \le \frac{(52-1)1.505^2}{32.3574} \Rightarrow 1.6174 \le \sigma^2 \le 3.5700$$

The 95% confidence interval for the standard deviation is:

$$\sqrt{1.6174} \le \sigma \le \sqrt{3.5700} \Rightarrow 1.272 \le \sigma \le 1.889$$

We are 95% confident that the true standard deviation of the solution time for the hybrid algorithm is between 1.272 and 1.889.

7.133   For all parts to this problem, we will use a 95% confidence interval. For confidence coefficient .95, $\alpha = .05$ and $\alpha / 2 = .05 / 2 = .025$. From Table IV, Appendix A, $z_{.025} = 1.96$.

a. Some preliminary calculations:

$$\hat{p} = \frac{x}{n} = \frac{14}{37} = .378$$

The sample size is large enough if both $n\hat{p} \ge 15$ and $n\hat{q} \ge 15$.

$n\hat{p} = 37(.378) = 14$ and $n\hat{q} = 37(.622) = 23$. Since the first number is not greater than or equal to 15, the sample size is not sufficiently large to conclude the normal approximation is reasonable.

Since the normal approximation is not reasonable, we will use the Wilson adjusted confidence interval.

$$\tilde{p} = \frac{x+2}{n+4} = \frac{14+2}{37+4} = \frac{16}{41} = .390$$

The Wilson adjusted 95% confidence interval is:

$$\tilde{p} \pm z_{.025}\sqrt{\frac{\tilde{p}(1-\tilde{p})}{n+4}} \Rightarrow .390 \pm 1.96\sqrt{\frac{.390(.610)}{37+4}} \Rightarrow .390 \pm .149 \Rightarrow (.241, \ .539)$$

We are 95% confident that the true proportion of suicides at the jail that are committed by inmates charged with murder/manslaughter is between .241 and .539.

b. Some preliminary calculations:

$$\hat{p} = \frac{x}{n} = \frac{26}{37} = .703$$

The sample size is large enough if both $n\hat{p} \ge 15$ and $n\hat{q} \ge 15$.

$n\hat{p} = 37(.703) = 26$ and $n\hat{q} = 37(.297) = 11$. Since the second number is not greater than or equal to 15, the sample size is not sufficiently large to conclude the normal approximation is reasonable.

Since the normal approximation is not reasonable, we will use the Wilson adjusted confidence interval.

$$\tilde{p} = \frac{x+2}{n+4} = \frac{26+2}{37+4} = \frac{28}{41} = .683$$

The Wilson adjusted 95% confidence interval is:

$$\tilde{p} \pm z_{.025}\sqrt{\frac{\tilde{p}(1-\tilde{p})}{n+4}} \Rightarrow .683 \pm 1.96\sqrt{\frac{.683(.317)}{37+4}} \Rightarrow .683 \pm .142 \Rightarrow (.541, .825)$$

We are 95% confident that the true proportion of suicides at the jail that are committed at night is between .541 and .825.

c.   Some preliminary calculations are:

$$\bar{x} = \frac{\sum x}{n} = \frac{1532}{37} = 41.405$$

$$s^2 = \frac{\sum x^2 - \frac{(\sum x)^2}{n}}{n-1} = \frac{223,606 - \frac{(1532)^2}{37}}{37-1} = 4,449.2477$$

$$s = \sqrt{s^2} = \sqrt{4,449.2477} = 66.703$$

The confidence interval is:

$$\bar{x} \pm z_{.05}\frac{s}{\sqrt{n}} \Rightarrow 41.405 \pm 1.96\frac{66.703}{\sqrt{37}} \Rightarrow 41.405 \pm 21.493 \Rightarrow (19.912, 62.898)$$

We are 95% confident that the true average length of time an inmate is in jail before committing suicide is between 19.912 and 62.898 days.

d.   Some preliminary calculations:

$$\hat{p} = \frac{x}{n} = \frac{14}{37} = .378$$

The sample size is large enough if both $n\hat{p} \geq 15$ and $n\hat{q} \geq 15$.

$n\hat{p} = 37(.378) = 14$ and $n\hat{q} = 37(.622) = 23$. Since the first number is not greater than or equal to 15, the sample size is not sufficiently large to conclude the normal approximation is reasonable.

Since the normal approximation is not reasonable, we will use the Wilson adjusted confidence interval.

$$\tilde{p} = \frac{x+2}{n+4} = \frac{14+2}{37+4} = \frac{16}{41} = .390$$

The Wilson adjusted 95% confidence interval is:

$$\tilde{p} \pm z_{.025}\sqrt{\frac{\tilde{p}(1-\tilde{p})}{n+4}} \Rightarrow .390 \pm 1.96\sqrt{\frac{.390(.610)}{37+4}} \Rightarrow .390 \pm .149 \Rightarrow (.241,\ .539)$$

We are 95% confident that the true percentage of suicides at the jail that are committed by white inmates is between 24.1% and 53.9%.

7.135 a.   The point estimate for the fraction of the entire market that refuses to purchase bars is:

$$\hat{p} = \frac{x}{n} = \frac{23}{244} = .094$$

b.   The sample size is large enough if both $n\hat{p} \geq 15$ and $n\hat{q} \geq 15$.

$n\hat{p} = 244(.094) = 23$ and $n\hat{q} = 244(.906) = 221$. Since both of the numbers are greater than or equal to 15, the sample size is sufficiently large to conclude the normal approximation is reasonable.

c.   For confidence coefficient .95, $\alpha = .05$ and $\alpha/2 = .05/2 = .025$. From Table IV, Appendix A, $z_{.025} = 1.96$. The confidence interval is:

$$\hat{p} \pm z_{.025}\sqrt{\frac{\hat{p}\hat{q}}{n}} \Rightarrow .094 \pm 1.96\sqrt{\frac{.094(.906)}{224}} \Rightarrow .094 \pm .037 \Rightarrow (.057,\ 131)$$

d.   The best estimate of the true fraction of the entire market that refuses to purchase bars six months after the poisoning is .094. We are 95% confident the true fraction of the entire market that refuses to purchase bars six months after the poisoning is between .057 and .131.

7.137 a.   For confidence coefficient .99, $\alpha = .01$ and $\alpha/2 = .01/2 = .005$. From Table VI, Appendix A, with df $= n-1 = 3-1 = 2$, $t_{.005} = 9.925$. The confidence interval is:

$$\bar{x} \pm t_{.005}\frac{s}{\sqrt{n}} \Rightarrow 49.3 \pm 9.925\frac{1.5}{\sqrt{3}} \Rightarrow 49.3 \pm 8.60 \Rightarrow (40.70,\ 57.90)$$

b.   We are 99% confident that the mean percentage of B(a)p removed from all soil specimens using the poison is between 40.70% and 57.90%.

c. We must assume that the distribution of the percentages of B(a)p removed from all soil specimens using the poison is normal.

d. For confidence coefficient .99, $\alpha = .01$ and $\alpha/2 = .01/2 = .005$. From Table IV, Appendix A, $z_{.005} = 2.575$.

$$n = \frac{z_{\alpha/2}^2 \sigma^2}{(SE)^2} = \frac{2.575^2 (1.5)^2}{.5^2} = 59.68 \approx 60$$

e. For confidence level .90, $\alpha = .10$ and $\alpha/2 = .10/2 = .05$. From Table VII, Appendix A, with df $= n - 1 = 3 - 1 = 2$, $\chi_{.05,2}^2 = 5.99147$ and $\chi_{.95,2}^2 = .102587$. The 90% confidence interval for the variance is:

$$\frac{(n-1)s^2}{\chi_{.05}^2} \leq \sigma^2 \leq \frac{(n-1)s^2}{\chi_{.95}^2} \Rightarrow \frac{(3-1)1.5^2}{5.99147} \leq \sigma^2 \leq \frac{(3-1)1.5^2}{.102587} \Rightarrow .7511 \leq \sigma^2 \leq 43.8652$$

We are 90% confidence that the true variance of the percentage of B(a)p removed is between .7511 and 43.8652.

7.139 a. From Chebyshev's Rule, we know that at least $1 - \frac{1}{k^2}$ of the observations will fall within k standard deviations of the mean. We need to find k so that $1 - \frac{1}{k^2} = .60$.

$$1 - \frac{1}{k^2} = .60 \Rightarrow .40 = \frac{1}{k^2} \Rightarrow k^2 = \frac{1}{.40} = 2.5 \Rightarrow k = \sqrt{2.5} = 1.58$$

$$s = \frac{80^{th} \text{ percentile} - 60^{th} \text{ percentile}}{2k} = \frac{73,000 - 35,100}{2(1.58)} = 11,993.67$$

For confidence coefficient .98, $\alpha = .02$ and $\alpha/2 = .02/2 = .01$. From Table IV, Appendix A, $z_{.01} = 2.33$. The sample size required is:

$$n = \frac{(z_{.01})^2 \sigma^2}{(SE)^2} = \frac{2.33^2 (11,993.67)^2}{2,000^2} = 195.23 \approx 196$$

b. See part **a**.

c. We must assume that any sample selected will be random.

7.141 a. As long as the sample is random (and thus representative), a reliable estimate of the mean weight of all the scallops can be obtained.

b. The government is using only the sample mean to make a decision. Rather than using a point estimate, they should probably use a confidence interval to estimate the true mean weight of the scallops.

c.   We will form a 95% confidence interval for the mean weight of the scallops. Using
MINITAB, the descriptive statistics are:

Descriptive Statistics: Weight

```
Variable   N    Mean    StDev  Minimum      Q1  Median    Q3   Maximum
Weight    18  0.9317  0.0753   0.8400  0.8800  0.9100  9800    1.1400
```

For confidence coefficient .95, $\alpha = .05$ and $\alpha/2 = .05/2 = .025$. From Table VI,
Appendix A, with df $= n - 1 = 18 - 1 = 17$, $t_{.025} = 2.110$.

The 95% confidence interval is:

$$\bar{x} \pm t_{.025}\frac{s}{\sqrt{n}} \Rightarrow .9317 \pm 2.110\frac{.0753}{\sqrt{18}} \Rightarrow .9317 \pm .0374 \Rightarrow (.8943, \ .9691)$$

We are 95% confident that the true mean weight of the scallops is between .8943 and
.9691. Recall that the weights have been scaled so that a mean weight of 1 corresponds
to 1/36 of a pound. Since the above confidence interval does not include 1, we have
sufficient evidence to indicate that the minimum weight restriction was violated.

Inferences Based on a Single Sample
Tests of Hypothesis

8.1 The null hypothesis is the "status quo" hypothesis, while the alternative hypothesis is the research hypothesis.

8.3 The "level of significance" of a test is α. This is the probability that the test statistic will fall in the rejection region when the null hypothesis is true.

8.5 The four possible results are:

1. Rejecting the null hypothesis when it is true. This would be a Type I error.
2. Accepting the null hypothesis when it is true. This would be a correct decision.
3. Rejecting the null hypothesis when it is false. This would be a correct decision.
4. Accepting the null hypothesis when it is false. This would be a Type II error.

8.7 When you reject the null hypothesis in favor of the alternative hypothesis, this does not prove the alternative hypothesis is correct. We are $100(1-\alpha)\%$ confident that there is sufficient evidence to conclude that the alternative hypothesis is correct.

We know the hypothesis-testing process will lead to this conclusion (reject H_0) incorrectly only $100\alpha\%$ of the time when H_0 is true.

8.9   a.   Probability of Type I error
      $= P(z > 1.96) = .5 - .4750 = .0250$
      (From Table IV, Appendix A)

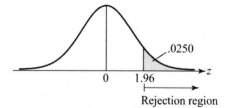

     b.   Probability of Type I error
      $= P(z > 1.645) = .5 - .4500 = .05$
      (From Table IV, Appendix A)

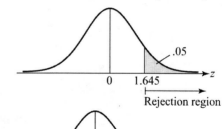

     c.   Probability of Type I error
      $= P(z > 2.575) = .5 - .4950 = .0050$
      (From Table IV, Appendix A)

d.   Probability of Type I error
$= P(z < -1.28) = .5 - .3997 = .1003$
(From Table IV, Appendix A)

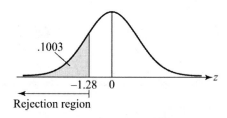

e.   Probability of Type I error
$= P(z < -1.645) + P(z > 1.645)$
$= .5 - .4500 + .5 - .4500 = .05 + .05 = .10$
(From Table IV, Appendix A)

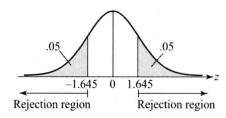

f.   Probability of Type I error
$= P(z < -2.575) + P(z > 2.575)$
$= .5 - .4950 + .5 - .4950 = .0050 + .0050$
$= .0100$ (From Table IV, Appendix A)

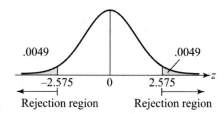

8.11   Let μ = average gain in green fees, lessons, or equipment expenditures for participating facilities.

a.   To determine if the average gain exceeds $2,400, we test:

H_0: $\mu = 2400$
H_a: $\mu > 2400$

b.   For this problem, α = the probability of concluding the average gain in green fees, lessons, or equipment expenditures for participating facilities is greater than $2,400 when, in fact, it is not.

c.   The rejection region requires $\alpha = .05$ in the upper tail of the z distribution. From Table IV, Appendix A, $z_{.05} = 1.645$. The rejection region is $z > 1.645$.

8.13   Let p = proportion of liars correctly detected by the new thermal imaging camera. To determine if the camera can correctly detect liars 75% of the time, we test:

H_0: $p = .75$

8.15   Let p = error rate of the DNA-reading device. To determine if the error rate of the DNA-reading device is less than 5%, we test:

H_0: $p = .05$
H_a: $p < .05$

8.17   a.   To determine if the average level of mercury uptake in wading birds in the Everglades today is less than 15 parts per million, we test:

$$H_0: \mu = 15$$
$$H_a: \mu < 15$$

   b.   A Type I error is rejecting H_0 when H_0 is true. In terms of this problem, we would be concluding that the average level of mercury uptake in wading birds in the Everglades today is less than 15 parts per million, when in fact, the average level of mercury uptake in wading birds in the Everglades today is equal to 15 parts per million.

   c.   A Type II error is accepting H_0 when H_0 is false. In terms of this problem, we would be concluding that the average level of mercury uptake in wading birds in the Everglades today is equal to 15 parts per million, when in fact, the average level of mercury uptake in wading birds in the Everglades today is less than 15 parts per million.

8.19   a.   The null hypothesis is: H_0: No intrusion occurs

   b.   The alternative hypothesis is: H_a: Intrusion occurs

   c. α = Probability of a Type I error = Probability of rejecting H_o when it is true =
      P(system
         provides warning when no intrusion occurs) = 1 / 1000 = .001.

      β = Probability of a Type II error = Probability of accepting H_o when it is false =
         P(system does not provide warning when an intrusion occurs) = 500 / 1000 = .5.

8.21   There are 2 conditions required for a valid large-sample hypothesis test for μ. They are:

   1.   A random sample is selected from a target population.
   2.   The sample size n is large, i.e., $n \geq 30$. (Due to the Central Limit Theorem, this condition guarantees that the test statistic will be approximately normal regardless of the shape of the underlying probability distribution of the population.)

8.23   a. H_0: $\mu = 100$
            H_a: $\mu > 100$

      The test statistic is $z = \dfrac{\bar{x} - \mu_0}{\sigma_{\bar{x}}} = \dfrac{\bar{x} - \mu_0}{\sigma / \sqrt{n}} = \dfrac{110 - 100}{60 / \sqrt{100}} = 1.67$

      The rejection region requires $\alpha = .05$ in the upper tail of the z distribution. From Table IV, Appendix A, $z_{.05} = 1.645$. The rejection region is $z > 1.645$.

      Since the observed value of the test statistic falls in the rejection region, ($z = 1.67 > 1.645$), H_0 is rejected. There is sufficient evidence to indicate the true population mean is greater than 100 at $\alpha = .05$.

b. H_0: $\mu = 100$

H_a: $\mu \neq 100$

The test statistic is $z = \dfrac{\bar{x} - \mu_0}{\sigma_{\bar{x}}} = \dfrac{110 - 100}{60 / \sqrt{100}} = 1.67$

The rejection region requires $\alpha / 2 = .05 / 2 = .025$ in each tail of the z distribution. From Table IV, Appendix A, $z_{.025} = 1.96$. The rejection region is $z < -1.96$ or $z > 1.96$.

Since the observed value of the test statistic does not fall in the rejection region, ($z = 1.67 \not> 1.96$), H_0 is not rejected. There is insufficient evidence to indicate μ differs from 100 at $\alpha = .05$.

c.   In part **a**, we rejected H_0 and concluded the mean was greater than 100. In part **b**, we did not reject H_0. There was insufficient evidence to conclude the mean was different from 100. Because the alternative hypothesis in part **a** is more specific than the one in **b**, it is easier to reject H_0.

8.25   a. $z = \dfrac{\bar{x} - \mu}{\sigma_{\bar{x}}} = \dfrac{215 - 200}{80 / \sqrt{100}} = 1.88$

The decision rule is: Reject H_0 if $z > 1.88$.

b. $\alpha = P(z > 1.88) = .5 - .4699 = .0301$ (using Table IV, Appendix A)

8.27   a.   A Type I error would be to conclude that the mean response for the population of all New York City public school children is not 3 when, in fact, the mean response is equal to 3.

A Type II error would be to conclude that the mean response for the population of all New York City public school children is equal to 3 when, in fact, the mean response is not equal to 3.

b.   To determine if the mean response for the population of all New York City public school children is different from 3, we test:

H_0: $\mu = 3$

H_a: $\mu \neq 3$

The test statistic is $z = \dfrac{\bar{x} - \mu_o}{\sigma_{\bar{x}}} = \dfrac{2.15 - 3}{\dfrac{1.05}{\sqrt{11,160}}} = -85.52$

The rejection region requires $\alpha / 2 = .05 / 2 = .025$ in each tail of the z distribution. From Table IV, Appendix A, $z_{.025} = 1.96$. The rejection region is $z < -1.96$ or $z > 1.96$.

Since the observed value of the test statistic falls in the rejection region ($z = -85.52 < -1.96$), H_0 is rejected. There is sufficient evidence to indicate the mean response for the population of all New York City public school children different from 3 at $\alpha = .05$.

   c.   To determine if the mean response for the population of all New York City public school children is different from 3, we test:

H_0: $\mu = 3$
H_a: $\mu \neq 3$

The test statistic is $z = \dfrac{\bar{x} - \mu_o}{\sigma_{\bar{x}}} = \dfrac{2.15 - 3}{\frac{1.05}{\sqrt{11,160}}} = -85.52$

The rejection region requires $\alpha / 2 = .10 / 2 = .05$ in each tail of the z distribution. From Table IV, Appendix A, $z_{.05} = 1.645$. The rejection region is $z < -1.645$ or $z > 1.645$.

Since the observed value of the test statistic falls in the rejection region ($z = -85.52 < -1.645$), H_0 is rejected. There is sufficient evidence to indicate the mean response for the population of all New York City public school children different from 3 at $\alpha = .10$.

8.29   Let μ = mean Mach rating score of all purchasing managers.

   a.   To determine if the true mean Mach rating score of all purchasing managers is different from 85, we test,

H_0: $\mu = 85$
H_a: $\mu \neq 85$

   b.   A Type I error would be concluding that the true mean Mach rating score is different from 85 when, in fact, it is equal to 85.

   c.   By definition, α = probability of committing a Type I error. With a value of $\alpha = .10$, this indicates that in repeated sampling, we will reject H_0 when it is true about 10% of the time.

   d.   The rejection region requires $\alpha / 2 = .10 / 2 = .05$ in each tail of the z distribution. From Table IV, Appendix A, $z_{.05} = 1.645$. The rejection region is $z > 1.645$ or $z < -1.645$.

   e.   The test statistic is $z = \dfrac{\bar{x} - \mu}{\sigma_{\bar{x}}} \approx \dfrac{99.6 - 85}{12.6 / \sqrt{122}} = 12.80.$

Since the observed value of the test statistic falls in the rejection region ($z = 12.80 > 1.645$), H_0 is rejected. There is sufficient evidence to indicate the true mean Mach rating score is different from 85 at $\alpha = .10$.

f.   No. Because the sample size is sufficiently large ($n = 122$), the Central Limit Theorem says that the distribution of $\bar{x}$ will be approximately normal.

8.31   Let μ = mean disclosure score. To determine if the true mean disclosure score exceeds 3, we test:

H_0: $\mu = 3$
H_a: $\mu > 3$

The test statistic is $z = \dfrac{\bar{x} - \mu_o}{\sigma_{\bar{x}}} \approx \dfrac{3.26 - 3}{\frac{.93}{\sqrt{222}}} = 4.17$

The rejection region requires $\alpha = .01$ in the upper tail of the z distribution. From Table IV, Appendix A, $z_{.01} = 2.33$. The rejection region is $z > 2.33$.

Since the observed value of the test statistic falls in the rejection region ($z = 4.17 > 2.33$), H_0 is rejected. There is sufficient evidence to indicate the true mean disclosure score exceeds 3 at $\alpha = .01$.

8.33   a.   Using MINITAB, the descriptive statistics are:

Descriptive Statistics: Bones

```
Variable    N   Mean    StDev  Minimum    Q1  Median     Q3  Maximum
Bones      41  9.258 8  1.204    6.230  8.615  9.200  9.935   12.000
```

To determine if the population mean ratio of all bones of this particular species differs from 8.5, we test:

H_0: $\mu = 8.5$
H_a: $\mu \neq 8.5$

The test statistic is $z = \dfrac{\bar{x} - \mu_0}{\sigma_{\bar{x}}} = \dfrac{9.258 - 8.5}{1.204 / \sqrt{41}} = 4.03$

The rejection region requires $\alpha / 2 = .01 / 2 = .005$ in each tail of the z distribution. From Table IV, Appendix A, $z_{.005} = 2.58$. The rejection region is $z < -2.58$ or $z > 2.58$.

Since the observed value of the test statistic falls in the rejection region ($z = 4.03 > 2.58$), H_0 is rejected. There is sufficient evidence to indicate that the true population mean ratio of all bones of this particular species differs from 8.5 at $\alpha = .01$.

b.   The practical implications of the test in part **a** is that the species from which these particular bones came from is probably not species A.

8.35   a.   Since the hypothesized value of μ_M ($60,000) falls in the 95% confidence interval, it is not an unusual value. There is no evidence to reject it. There is no evidence to indicate that the mean salary of all males with post-graduate degrees differs from $60,000.

b. To determine if the mean salary of all males with post-graduate degrees differs from $60,000, we test:

H_0: $\mu_M = 60,000$

H_a: $\mu_M \neq 60,000$

The test statistic is $z = \dfrac{\bar{x} - \mu_0}{\sigma_{\bar{x}}} = \dfrac{61,340 - 60,000}{2,185} = 0.61$

The rejection region requires $\alpha / 2 = .05 / 2 = .025$ in each tail of the z distribution. From Table IV, Appendix A, $z_{.025} = 1.96$. The rejection region is $z < -1.96$ or $z > 1.96$.

Since the observed value of the test statistic does not fall in the rejection region ($z = 0.61 \not> 1.96$), H_0 is not rejected. There is insufficient evidence to indicate the mean salary of all males with post-graduate degrees differs from $60,000 at $\alpha = .05$.

c. The inferences in parts **a** and **b** agree because the same values are used for both. The z value used in the 95% confidence interval is the same z used for the rejection region. The value of $\bar{x}$ and $s_{\bar{x}}$ are the same in both the confidence interval and the test statistic.

d. Since the hypothesized value of μ_F ($33,000) falls in the 95% confidence interval, it is not an unusual value. There is no evidence to reject it. There is no evidence to indicate that the mean salary of all females with post-graduate degrees differs from $33,000.

e. To determine if the mean salary of all females with post-graduate degrees differs from $33,000, we test:

H_0: $\mu_M = 33,000$

H_a: $\mu_M \neq 33,000$

The test statistic is $z = \dfrac{\bar{x} - \mu_0}{\sigma_{\bar{x}}} = \dfrac{32,227 - 33,000}{932} = -0.83$

The rejection region requires $\alpha / 2 = .05 / 2 = .025$ in each tail of the z distribution. From Table IV, Appendix A, $z_{.025} = 1.96$. The rejection region is $z < -1.96$ or $z > 1.96$.

Since the observed value of the test statistic does not fall in the rejection region ($z = -0.83 \not< -1.96$), H_0 is not rejected. There is insufficient evidence to indicate the mean salary of all females with post-graduate degrees differs from $33,000 at $\alpha = .05$.

f. The inferences in parts **d** and **e** agree because the same values are used for both. The z value used in the 95% confidence interval is the same z used for the rejection region. The value of $\bar{x}$ and $s_{\bar{x}}$ are the same in both the confidence interval and the test statistic.

8.37   a.   To determine if the mean social interaction score of all Connecticut mental health patients differs from 3, we test:

H_0: $\mu = 3$

H_a: $\mu \neq 3$

The test statistic is $z = \dfrac{\bar{x} - \mu_0}{\sigma_{\bar{x}}} = \dfrac{2.95 - 3}{1.10 / \sqrt{6,681}} = -3.72$

The rejection region requires $\alpha / 2 = .01 / 2 = .005$ in each tail of the z distribution. From Table IV, Appendix A, $z_{.005} = 2.58$. The rejection region is $z < -2.58$ or $z > 2.58$.

Since the observed value of the test statistic falls in the rejection region ($z = -3.72 < -2.58$), H_0 is rejected. There is sufficient evidence to indicate that the mean social interaction score of all Connecticut mental health patients differs from 3 at $\alpha = .01$.

   b.   From the test in part a, we found that the mean social interaction score was statistically different from 3. However, the sample mean score was 2.95. Practically speaking, 2.95 is very similar to 3.0. The very large sample size, $n = 6,681$, makes it very easy to find statistical significance, even when no practical significance exists.

   c.   Because the variable of interest is measured on a 5-point scale, it is very unlikely that the population of the ratings will be normal. However, because the sample size was extremely large, ($n = 6,681$), the Central Limit Theorem will apply. Thus, the distribution of $\bar{x}$ will be normal, regardless of the distribution of x. Thus, the analysis used above is appropriate.

8.39   In general, small p-values support the alternative hypothesis, H_a. The p-value is the probability of observing your test statistic or anything more unusual, given H_0 is true. If the p-value is small, it would be very unusual to observe your test statistic, given H_0 is true. This would indicate that H_0 is probably false and H_a is probably true.

8.41   a.   Since the p-value $= .10$ is greater than $\alpha = .05$, H_0 is not rejected.

   b.   Since the p-value $= .05$ is less than $\alpha = .10$, H_0 is rejected.

   c.   Since the p-value $= .001$ is less than $\alpha = .01$, H_0 is rejected.

   d.   Since the p-value $= .05$ is greater than $\alpha = .025$, H_0 is not rejected.

   e.   Since the p-value $= .45$ is greater than $\alpha = .10$, H_0 is not rejected.

8.43 p-value $= P(z \geq 2.17) = .5 - P(0 < z < 2.17) = .5 - .4850 = .0150$

(using Table IV, Appendix A)

8.45   The test statistic is $z = \dfrac{\bar{x} - \mu_0}{\sigma_{\bar{x}}} = \dfrac{49.4 - 50}{4.1 / \sqrt{100}} = -1.46$

The *p*-value $= p = P(z \geq -1.46) = .5 + .4279 = .9279$

Since the *p*-value is so large, there is no evidence to reject H_0 for any reasonable value of α.

8.47   a.   The *p*-value reported by SPSS is for a two-tailed test. Thus, $P(z \leq -1.63) + P(z \geq 1.63)$ $= .1032$. For this one-tailed test, the *p*-value $= P(z \leq -1.63) = .1032/2 = .0516$.

Since the *p*-value $= .0516 > \alpha = .05$, H_0 is not rejected. There is insufficient evidence to indicate μ is less than 75 at $\alpha = .05$.

b.   For this one-tailed test, the *p*-value $= P(z \leq 1.63)$. Since $P(z \leq -1.63) = .1032/2 = .0516$, $P(z \leq 1.63) = 1 - .0516 = .9484$.

Since the *p*-value $= .9484 > \alpha = .10$, H_0 is not rejected. There is insufficient evidence to indicate μ is less than 75 at $\alpha = .10$.

c.   For this one-tailed test, the *p*-value $= P(z \geq 1.63) = .1032/2 = .0516$.

Since the *p*-value $= p = .0516 < \alpha = .10$, H_0 is rejected. There is sufficient evidence to indicate μ is greater than 75 at $\alpha = .10$.

d.   For this two-tailed test, the *p*-value $= .1032$.

Since the *p*-value $= .1032 > \alpha = .01$, H_0 is not rejected. There is insufficient evidence to indicate μ differs from 75 at $\alpha = .01$.

8.49   a.   From Exercise 8.27, $z = -85.52$. The *p*-value is $p = P(z \leq -85.52) + P(z \geq 85.52)$ $= (.5 - .5) + (.5 - .5) = 0$ (using Table IV, Appendix A).

b.   The *p*-value is $p = 0$. Since the *p*-value is less than $\alpha = .10$, H_0 is rejected. There is sufficient evidence to indicate the mean response for the population of all New York City public school children different from 3 at $\alpha = .10$.

8.51   From the printout, the *p*-value $< .0001$. Since the *p*-value $< .0001 < \alpha = .01$, H_0 is rejected. There is sufficient evidence to indicate that the true population mean ratio of all bones of this particular species differs from 8.5 at $\alpha = .01$.

8.53   a.   Let $\mu =$ mean forecast error for buy-side analysts. To determine if the mean forecast error for buy-side analysts is positive, we test:

H_0: $\mu = 0$
H_a: $\mu > 0$

The test statistic is $z = \dfrac{\bar{x} - \mu_o}{\sigma_{\bar{x}}} \approx \dfrac{.85 - 0}{\dfrac{1.93}{\sqrt{3,526}}} = 26.15$

The p-value is $p = P(z \geq 26.51) \approx .5 - .5 = 0$.

Since the p-value is less than $\alpha = .01$ ($p = 0 < .01$), H_0 is rejected. There is sufficient evidence to indicate the mean forecast error for buy-side analysts is positive at $\alpha = .01$.

b.   Let μ = mean forecast error for sell-side analysts. To determine if the mean forecast error for sell-side analysts is negative, we test:

$H_0 : \mu = 0$
$H_a : \mu < 0$

The test statistic is $z = \dfrac{\bar{x} - \mu_o}{\sigma_{\bar{x}}} \approx \dfrac{-.05 - 0}{\dfrac{.85}{\sqrt{58,562}}} = -14.24$

The p-value is $p = P(z \leq -14.24) \approx .5 - .5 = 0$.

Since the p-value is less than $\alpha = .01$ ($p = 0 < .01$), H_0 is rejected. There is sufficient evidence to indicate the mean forecast error for sell-side analysts is negative at $\alpha = .01$.

8.55   a.   The test statistic is $z = \dfrac{\bar{x} - \mu_0}{\sigma_{\bar{x}}} = \dfrac{10.2 - 0}{31.3 / \sqrt{50}} = 2.3$

b.   To determine if the mean level of feminization differs from 0%, we test:

$H_0: \ \mu = 0$
$H_a: \ \mu \neq 0$

Since the alternative hypothesis contains $\neq$, this is a two-tailed test. The p-value is $p = P(z \leq -2.3) + P(z \geq 2.3) = .5 - .4893 + .5 - .4893 = .0214$.
(Using Table IV, Appendix A)

Since the p-value is so small, there is evidence to reject H_0. There is sufficient evidence to indicate the mean level of feminization is different from 0% for any value of $\alpha < .0214$.

c.   The test statistic is $z = \dfrac{\bar{x} - \mu_0}{\sigma_{\bar{x}}} = \dfrac{15 - 0}{25.1 / \sqrt{50}} = 4.23$.

    Since the alternative hypothesis contains $\neq$, this is a two-tailed test. The p-value is $p = P(z \leq -4.23) + P(z \geq 4.23) \approx .5 - .5 + .5 - .5 = 0$.

    Since the p-value is so small, there is evidence to reject H_0. There is sufficient evidence to indicate the mean level of feminization is different from 0% for any value of α.

8.57   The z and t distributions are alike in that they are both symmetric, bell-shaped distributions centered at 0. They differ in variability, though. The z distribution always has $\sigma^2 = 1$, while the variance of the t distribution is always larger and depends on the degrees of freedom.

8.59 a. $P(t > 1.440) = .10$
        (Using Table VI, Appendix A, with df = 6)

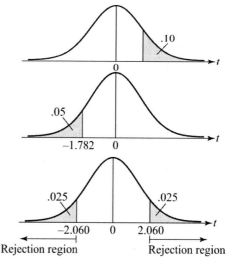

    b. $P(t < -1.782) = .05$
        (Using Table VI, Appendix A, with df = 12)

    c. $P(t < -2.060) = P(t > 2.060) = .025$
        (Using Table VI, Appendix A, with df = 25)

8.61 a. The rejection region requires $\alpha / 2 = .05 / 2 = .025$ in each tail of the t distribution with df $= n - 1 = 14 - 1 = 13$. From Table VI, Appendix A, $t_{.025} = 2.160$. The rejection region is $t < -2.160$ or $t > 2.160$.

    b. The rejection region requires $\alpha = .01$ in the upper tail of the t distribution with df $= n - 1 = 24 - 1 = 23$. From Table VI, Appendix A, $t_{.01} = 2.500$. The rejection region is $t > 2.500$.

    c. The rejection region requires $\alpha = .10$ in the upper tail of the t distribution with df $= n - 1 = 9 - 1 = 8$. From Table VI, Appendix A, $t_{.10} = 1.397$. The rejection region is $t > 1.397$.

    d. The rejection region requires $\alpha = .01$ in the lower tail of the t distribution with df $= n - 1 = 12 - 1 = 11$. From Table VI, Appendix A, $t_{.01} = 2.718$. The rejection region is $t < -2.718$.

    e. The rejection region requires $\alpha / 2 = .10 / 2 = .05$ in each tail of the t distribution with df $= n - 1 = 20 - 1 = 19$. From Table VI, Appendix A, $t_{.05} = 1.729$. The rejection region is $t < -1.729$ or $t > 1.729$.

Copyright © 2013 Pearson Education, Inc.

f.   The rejection region requires $\alpha = .05$ in the lower tail of the t distribution with df $=$ $n - 1 = 4 - 1 = 3$. From Table VI, Appendix A, $t_{.05} = 2.353$. The rejection region is $t < -2.353$.

8.63   a. H_0: $\mu = 6$

            H_a: $\mu < 6$

            The test statistic is $t = \dfrac{\bar{x} - \mu_0}{s / \sqrt{n}} = \dfrac{4.8 - 6}{1.3 / \sqrt{5}} = -2.064$.

            The rejection region requires $\alpha = .05$ in the lower tail of the t distribution with df $= n - 1$ $= 5 - 1 = 4$. From Table VI, Appendix A, $t_{.05} = 2.132$. The rejection region is $t < -2.132$.

            Since the observed value of the test statistic does not fall in the rejection region $\left(t = -2.064 \not< -2.132\right)$, H_0 is not rejected. There is insufficient evidence to indicate the mean is less than 6 at $\alpha = .05$.

   b. H_0: $\mu = 6$

            H_a: $\mu = 6$

            The test statistic is $t = -2.064$ (from **a**).

            The rejection region requires $\alpha / 2 = .05 / 2 = .025$ in each tail of the t distribution with df $= n - 1 = 5 - 1 = 4$. From Table VI, Appendix A, $t_{.025} = 2.776$. The rejection region is $t < -2.776$ or $t > 2.776$.

            Since the observed value of the test statistic does not fall in the rejection region ($t = -2.064 \not< -2.776$), H_0 is not rejected. There is insufficient evidence to indicate the mean is different from 6 at $\alpha = .05$.

   c.   For part **a**, the p-value $= P(t \leq -2.064)$.

            From Table VI, with df $= 4$, $.05 < P(t \leq -2.064) < .10$ or $.05 < p$-value $< .10$. Using MINITAB, the p-value is $p = .0539809$.

            For part **b**, the p-value $= P(t \leq -2.064) + P(t \geq 2.064)$.

            From Table VI, with df $= 4$, $2(.05) < p$-value $< 2(.10)$ or $.10 < p$-value $< .20$. Using MINITAB, the p-value is $p = 2(.0539809) = .1079618$.

8.65   a.   Let $\mu =$ average trap spacing measurement. To determine if the true average trap spacing measurement is different from 95, we test:

            $H_0 : \mu = 95$

            $H_a : \mu \neq 95$

b. For this example, the sample mean is 89.9. If another sample of size 7 was selected, chances are, the sample mean would not be 89.9. The next sample could possibly have a sample mean greater than 95. We need to determine how unusual a value 89.9 is for $\bar{x}$ if the true mean is 95. Also, we need to take into account the variability in the data.

c. The test statistic is $t = \dfrac{\bar{x} - \mu_0}{\dfrac{s}{\sqrt{n}}} = \dfrac{89.9 - 95}{\dfrac{11.6}{\sqrt{7}}} = -1.163$

d. To find the p-value, we need to find $p = P(t \le -1.163) + P(t \ge 1.163)$. From Table VI, Appendix A, with df $= n - 1 = 7 - 1 = 6$, $P(t \ge 1.440) = .10$. Thus, $P(t \ge 1.163) > .10$. The *p*-value is

$p = P(t \le -1.163) + P(t \ge 1.163) > .10 + .10 > .20.$

Using MINITAB, the exact *p*-value is .289.

e. Let $\alpha = .10$. Thus, the probability of concluding that the average strap spacing measurement is different from 95 when, in fact, it is equal to 95 is .10.

f. Since the *p*-value is greater than $\alpha = .10$ ($p = .289 > .10$), H_0 is not rejected. There is insufficient evidence to indicate the average trap spacing measurement is different from 95 at $\alpha = .10$.

g. We must assume that a random sample was drawn from a normal population of trap spacing measurements.

h. From Exercise 7.35, the 95% confidence interval is (79.095, 100.625). Since 95 falls in the interval, it is a likely value for μ. We would not reject it. This agrees with the results in part f.

8.67 To determine if the mean Dental Anxiety Scale score differs from 11, we test:

H_0: $\mu = 11$
H_a: $\mu \ne 11$

The test statistic is $t = \dfrac{\bar{x} - \mu_0}{s/\sqrt{n}} = \dfrac{10.7 - 11}{3.6/\sqrt{15}} = -.32$

The rejection region requires $\alpha / 2 = .05 / 2 = .025$ in each tail of the t distribution with df $= n - 1 = 15 - 1 = 14$. From Table VI, Appendix A, $t_{.025} = 2.145$. The rejection region is $t < -2.145$ or $t > 2.145$.

Since the observed value of the test statistic does not fall in the rejection region ($t = -.32 \not< -2.145$), H_0 is not rejected. There is insufficient evidence to indicate the mean Dental Anxiety Scale score differs from 11 at $\alpha = .05$.

8.69   Let μ = mean dentary depth of molars. To determine if the sample of 18 cheek teeth come from some other extinct primate species, we test:

H_0: $\mu = 15$
H_a: $\mu \neq 15$

The test statistic is $t = \dfrac{\overline{x} - \mu_0}{\frac{s}{\sqrt{n}}} = 3.229$ (From the printout.)

From the printout, the p-value is $p = .005$. Since the p-value is so small, we would reject H_0 for any reasonable value of α.   There is sufficient evidence to indicate that the sample of 18 cheek teeth come from some other extinct primate species other than species A.

8.71   To determine if the mean amount of cesium in lichen specimens differs from .003, we test:

H_0: $\mu = .003$
H_a: $\mu \neq .003$

From the printout, the test statistic is $t = 3.725$ and the p-value is $p = .0058$.

Since the p-value is less than α ($p = .0058 < \alpha = .10$), H_0 is rejected. There is sufficient evidence to indicate the mean amount of cesium in lichen specimens differs from .003 at $\alpha = .10$.

8.73   From Exercise 7.44, $\overline{x} = 358.45$ and s = 117.8172. To determine if the mean skidding distance is less than 425 meters, we test:

H_0: $\mu = 425$
H_a: $\mu < 425$

The test statistic is $t = \dfrac{\overline{x} - \mu_o}{s/\sqrt{n}} = \dfrac{358.45 - 425}{\frac{117.8172}{\sqrt{20}}} = -2.53$

The rejection region requires $\alpha = .10$ in the lower tail of the t-distribution with df = $n - 1$ = 20 – 1 = 19. From Table VI, Appendix A, $t_{.10} = 1.328$. The rejection region is $t < -1.328$.

Since the observed value of the test statistic falls in the rejection region ($t = -2.53 < -1.328$), H_0 is rejected. There is sufficient evidence to indicate the mean skidding distance is less than 425 meters at $\alpha = .10$. Thus, there is enough evidence to refute the claim.

8.75   a.   To determine if the true mean bias differs from 0, we test:

$$H_0: \mu = 0$$
$$H_a: \mu \neq 0$$

From the printout, the test statistics is $t = .21$ and the *p*-value is $p = .840$.

Since the *p*-value is larger than $\alpha = .10$, ($p = .840 > .10$), H_0 is not rejected. There is insufficient evidence to indicate that the true mean bias differs from 0.

   b.   The scores given are the average scores of different people. There are several people with averages scores that are positive and several with average scores which are negative. These scores will tend to cancel each other out. Also, each individual could travel in circles to the right and circles to the left. Again, these would tend to cancel each other out.

8.77   Typically, qualitative data are associated with making inferences about a population proportion.

8.79   The sample size is large enough if both np_o and nq_o are greater than or equal to 15.

   a. $np_o = 500(.05) = 25$ and $nq_o = 500(1 - .05) = 500(.95) = 475$. Since both of these values are greater than 15, the sample size is large enough to use the normal approximation.

   b. $np_o = 100(.99) = 99$ and $nq_o = 100(1 - .99) = 100(.01) = 1$. Since the second number is less than 15, the sample size is not large enough to use the normal approximation.

   c. $np_o = 50(.2) = 10$ and $nq_o = 50(1 - .2) = 50(.8) = 40$. Since the first number is less than 15, the sample size is not large enough to use the normal approximation.

   d. $np_o = 20(.2) = 4$ and $nq_o = 20(1 - .2) = 20(.8) = 16$. Since the first number is less than 15, the sample size is not large enough to use the normal approximation.

   e. $np_o = 10(.4) = 4$ and $nq_o = 10(1 - .4) = 10(.6) = 6$. Since both numbers are less than 15, the sample size is not large enough to use the normal approximation.

8.81   a. $z = \dfrac{\hat{p} - p_0}{\sqrt{\dfrac{p_0 q_0}{n}}} = \dfrac{.84 - .9}{\sqrt{\dfrac{.9(.1)}{100}}} = -2.00$

   b.   The denominator in Exercise 8.80 is $\sqrt{\dfrac{.75(.25)}{100}} = .0433$ as compared to $\sqrt{\dfrac{.9(.1)}{100}} = .03$

in part **a**. Since the denominator in this problem is smaller, the absolute value of z is larger.

   c.   The rejection region requires $\alpha = .05$ in the lower tail of the z distribution. From Table IV, Appendix A, $z_{.05} = 1.645$. The rejection region is $z < -1.645$.

Since the observed value of the test statistic falls in the rejection region ($z = -2.00 < -1.645$), H_0 is rejected. There is sufficient evidence to indicate the population proportion is less than .9 at $\alpha = .05$.

d.   The p-value = $P(z \leq -2.00) = .5 - .4772 = .0228$ (from Table IV, Appendix A).

8.83   From Exercise 7.54, $n = 50$ and since p is the proportion of consumers who do not like the snack food, $\hat{p}$ will be:

$$\hat{p} = \frac{\text{Number of 0's in sample}}{n} = \frac{29}{50} = .58$$

In order for the inference to be valid, the sample size must be large enough. The sample size is large enough if both np_o and nq_o are greater than or equal to 15.

$np_o = 50(.5) = 25$ and $nq_o = 50(1 - .5) = 50(.5) = 25$. Since both of these values are greater than 15, the sample size is large enough to use the normal approximation.

a. H_0: $p = .5$
      H_a: $p > .5$

The test statistic is $z = \dfrac{\hat{p} - p_0}{\sigma_{\hat{p}}} = \dfrac{\hat{p} - p_0}{\sqrt{\dfrac{p_0 q_0}{n}}} = \dfrac{.58 - .5}{\sqrt{\dfrac{.5(1 - .5)}{50}}} = 1.13$

The rejection region requires $\alpha = .10$ in the upper tail of the z distribution. From Table IV, Appendix A, $z_{.10} = 1.28$. The rejection region is $z > 1.28$.

Since the observed value of the test statistic does not fall in the rejection region ($z = 1.13 \not> 1.28$), H_0 is not rejected. There is insufficient evidence to indicate the proportion of customers who do not like the snack food is greater than .5 at $\alpha = .10$.

b. p-value = $P(z \geq 1.13) = .5 - .3708 = .1292$ (Using Table IV, Appendix A.)

8.85   a.   The population parameter of interest is p, the proportion of items that had the wrong price when scanned through the electronic checkout scanner at California Wal-Mart stores.

b.   To determine if the true proportion of items incorrectly scanned at California Wal-Mart stores exceeds the 2% NIST standard, we test:

H_0: $p = .02$
H_a: $p > .02$

c.   The test statistic is $z = \dfrac{\hat{p} - p_o}{\sqrt{\dfrac{p_o q_o}{n}}} = \dfrac{.083 - .02}{\sqrt{\dfrac{.02(.98)}{1000}}} = 14.23$

The rejection region requires $\alpha = .05$ in the upper tail of the z distribution. From Table IV, Appendix A, $z_{.05} = 1.645$. The rejection region is $z > 1.645$.

d. Since the observed value of the test statistic falls in the rejection region ($z = 14.23 > 1.645$), H_o is rejected. There is sufficient evidence to indicate the true proportion of items incorrectly scanned at California Wal-Mart stores exceeds the 2% NIST standard at $\alpha = .05$.

e. In order for the inference to be valid, the sample size must be large enough. The sample size is large enough if both np_o and nq_o are greater than or equal to 15.

    $np_o = 1000(.02) = 20$ and $nq_o = 1000(1 - .02) = 1000(.98) = 980$. Since both of these values are greater than 15, the sample size is large enough to use the normal approximation.

8.87   Some preliminary calculations:

$$\hat{p} = \frac{x}{n} = \frac{401}{835} = .48$$

In order for the inference to be valid, the sample size must be large enough. The sample size is large enough if both np_o and nq_o are greater than or equal to 15.

$np_o = 835(.45) = 375.75$ and $nq_o = 835(1 - .45) = 835(.55) = 459.25$. Since both of these values are greater than 15, the sample size is large enough to use the normal approximation.

To determine if more than 45% of male youths are raised in a single-parent family, we test:

H_0: $p = .45$
H_a: $p > .45$

The test statistic is $z = \dfrac{\hat{p} - p_0}{\sqrt{\dfrac{p_0 q_0}{n}}} = \dfrac{.48 - .45}{\sqrt{\dfrac{.45(.55)}{835}}} = 1.74$

The rejection region requires $\alpha = .05$ in the upper tail of the z distribution. From Table IV, Appendix A, $z_{.05} = 1.645$. The rejection region is $z > 1.645$.

Since the observed value of the test statistic falls in the rejection region ($z = 1.74 > 1.645$), H_0 is rejected. There is sufficient evidence that more than 45% of male youths are raised in a single-parent family at $\alpha = .05$.

8.89   Let p = true proportion of students who prefer the new, computerized method of identifying common conifers.

The sample proportion is $\hat{p} = \dfrac{138}{171} = .81$.

In order for the inference to be valid, the sample size must be large enough. The sample size is large enough if both np_o and nq_o are greater than or equal to 15.

$np_o = 171(.7) = 119.7$ and $nq_o = 171(1 - .7) = 171(.3) = 51.3$. Since both of these values are greater than 15, the sample size is large enough to use the normal approximation.

To determine if more than 70% of all students prefer the new, computerized method of identifying common conifers, we test:

H_0: $p = .70$
H_a: $p > .70$

The test statistic is $z = \dfrac{\hat{p} - p_0}{\sigma_{\hat{p}}} = \dfrac{\hat{p} - p_0}{\sqrt{\dfrac{p_0 q_0}{n}}} = \dfrac{.81 - .70}{\sqrt{\dfrac{.70(1 - .70)}{171}}} = 3.14$

The p-value for the test is $p = P(z \geq 3.14) \approx .5 - .5 = 0$. (From Table IV, Appendix A.) Since the p-value is so small, we would reject H_0 for any reasonable value of α. There is sufficient evidence to indicate that more than 70% of all students prefer the new, computerized method of identifying common conifers. Thus, Confir ID should be added to the curriculum at SRU.

8.91   First, we calculate the estimate for p: $\hat{p} = \dfrac{x}{n} = \dfrac{37}{148} = .25$

In order for the inference to be valid, the sample size must be large enough. The sample size is large enough if both np_o and nq_o are greater than or equal to 15.

$np_o = 148(.20) = 29.6$ and $nq_o = 148(1 - .20) = 148(.80) = 118.4$. Since both of these values are greater than 15, the sample size is large enough to use the normal approximation.

To determine if more than 20% of all freshman college students believe in the Big Bang Theory, we test:

H_0: $p = .20$
H_a: $p > .20$

The test statistic is $z = \dfrac{\hat{p} - p_o}{\sqrt{\dfrac{p_o q_o}{n}}} = \dfrac{.25 - .20}{\sqrt{\dfrac{.20(.80)}{148}}} = 1.52$

Since no value of α was given, we will choose $\alpha = .05$. The rejection region requires $\alpha = .05$ in the upper tail of the z-distribution. From Table IV, Appendix A, $z_{.05} = 1.645$. The rejection region is $z > 1.645$.

Since the observed value of the test statistic does not fall in the rejection region ($z = 1.52 \not> 1.645$), H_0 is not rejected. There is insufficient evidence to indicate that more than 20% of all freshman college students believe in the Big Bang Theory at $\alpha = .05$.

Thus, we would not be willing to state that more than 20% of all freshman college students believe in the Big Bang Theory. We are 95% confident of this decision.

8.93   a.   The point estimate for p is $\hat{p} = \dfrac{x}{n} = \dfrac{24}{33} = .727$

In order for the inference to be valid, the sample size must be large enough. The sample size is large enough if both np_0 and nq_0 are greater than or equal to 15.

$np_0 = 33(.60) = 19.8$ and $nq_0 = 33(1 - .60) = 33(.40) = 13.2$. Since the second value is not greater than 15, the sample size is may not be large enough to use the normal approximation.

To determine if the cream will improve the skin of more than 60% of middle-aged women, we test:

H_0: $p = .60$
H_a: $p > .60$

The test statistic is $z = \dfrac{\hat{p} - p_0}{\sqrt{\dfrac{p_0 q_0}{n}}} = \dfrac{.727 - .60}{\sqrt{\dfrac{.60(.40)}{33}}} = 1.49$

The rejection region requires $\alpha = .05$ in the upper tail of the z distribution. From Table IV, Appendix A, $z_{.05} = 1.645$. The rejection region is $z > 1.645$.

Since the observed value of the test statistic does not fall in the rejection region ($z = 1.49 \not> 1.645$), H_0 is not rejected. There is insufficient evidence to indicate the cream will improve the skin of more than 60% of middle-aged women at $\alpha = .05$.

b.   The p-value of the test is $p = P(z \geq 1.49) = .5 - .4319 = .0681$.

Since the p-value is not less than α ($p = .0681 \not< .05$), H_0 is not rejected. There is insufficient evidence to indicate the cream will improve the skin of more than 60% of middle-aged women at $\alpha = .05$.

8.95   We know that the miscarriage rate for pregnant women expecting a boy in the United States is 3 in 1,000 or .003. Let p = miscarriage rate for pregnant women expecting a boy in the United States during September 2001.

To determine if the male fetal death rate in September, 2001 was higher than expected, we would test:

H_0: $p = .003$
H_a: $p > .003$

The rejection region requires $\alpha = .05$ in the upper tail of the z distribution. From Table IV, Appendix A, $z_{.05} = 1.645$. The rejection region is $z > 1.645$.

The test statistic would be $z = \dfrac{\hat{p} - p_o}{\sqrt{\dfrac{p_o q_o}{n}}} = \dfrac{\hat{p} - .003}{\sqrt{\dfrac{.003(.997)}{2000}}} = 1.645$.

We now need to solve this for $\hat{p}$.

$$\frac{\hat{p} - .003}{\sqrt{\dfrac{.003(.997)}{2000}}} = 1.645 \Rightarrow \frac{\hat{p} - .003}{.001223} = 1.645 \Rightarrow \hat{p} - .003 = .002 \Rightarrow \hat{p} = .005$$

Thus, there would need to be $2000(.005) = 10$ miscarriages in the 2,000 pregnant women to support the claim.

8.97   The power of a test is equal to $1 - \beta$. As β increases, the power decreases.

8.99   a.

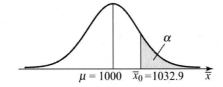

b. $z = \dfrac{\overline{x}_0 - \mu_0}{\sigma_{\overline{x}}} \Rightarrow \overline{x}_0 = \mu_0 + z_\alpha \sigma_{\overline{x}} = \mu_0 + z_\alpha \dfrac{\sigma}{\sqrt{n}}$ where $z_\alpha = z_{.05} = 1.645$ from Table IV, Appendix A.

Thus, $\overline{x}_0 = 1000 + 1.645 \dfrac{120}{\sqrt{36}} = 1032.9$

c.

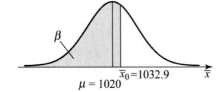

d. $\beta = P(\bar{x}_0 < 1032.9 \text{ when } \mu = 1020) = P\left(z < \dfrac{1032.9 - 1020}{120/\sqrt{36}}\right)$

$= P(z < .65) = .5 + .2422 = .7422$

e. $Power = 1 - \beta = 1 - .7422 = .2578$

8.101 a. The sampling distribution of $\bar{x}$ will be approximately normal (by the Central Limit Theorem) with $\mu_{\bar{x}} = \mu = 50$ and $\sigma_{\bar{x}} = \dfrac{\sigma}{\sqrt{n}} = \dfrac{20}{\sqrt{64}} = 2.5$.

b. The sampling distribution of $\bar{x}$ will be approximately normal (by the Central Limit Theorem) with $\mu_{\bar{x}} = \mu = 45$ and $\sigma_{\bar{x}} = \dfrac{\sigma}{\sqrt{n}} = \dfrac{20}{\sqrt{64}} = 2.5$.

c. First, find $\bar{x}_0$ so that $P(\bar{x} < \bar{x}_0) = .10$ under H_0.

$$P(\bar{x} < \bar{x}_0) = P\left(z < \dfrac{\bar{x}_0 - 50}{20/\sqrt{64}}\right) = P\left(z < \dfrac{\bar{x}_0 - 50}{2.5}\right) = P(z < z_0) = .10$$

From Table IV, Appendix A, $z_{.10} = -1.28$

$$z_0 = \dfrac{\bar{x}_0 - 50}{2.5} \Rightarrow -1.28 = \dfrac{\bar{x}_0 - 50}{2.5} \Rightarrow \bar{x}_0 = 46.8$$

Now, find

$$\beta = P(\bar{x} > 46.8 \text{ when } \mu = 45) = P\left(z > \dfrac{46.8 - 45}{20/\sqrt{64}}\right) = P(z > .72) = .5 - .2642 = .2358$$

d. $Power = 1 - \beta = 1 - .2358 = .7642$

8.103 a. The sampling distribution of $\hat{p}$ will be approximately normal (by the Central Limit Theorem) with $\mu_{\hat{p}} = p_o = .7$ and $\sigma_{\hat{p}} = \sqrt{\dfrac{p_o q_o}{n}} = \sqrt{\dfrac{.7(.3)}{100}} = .0458$.

b. The sampling distribution of $\hat{p}$ will be approximately normal (by the Central Limit Theorem) with $\mu_{\hat{p}} = p_o = .65$ and $\sigma_{\hat{p}} = \sqrt{\dfrac{p_o q_o}{n}} = \sqrt{\dfrac{.65(.35)}{100}} = .0477$.

c. First, find $\hat{p}_{o,L}$ and $\hat{p}_{o,U}$ such that $P(\hat{p} < \hat{p}_{o,L}) = P(\hat{p} > \hat{p}_{o,U}) = .025$, assuming H_0 is true.

$$P(\hat{p} < \hat{p}_{o,L}) = P\left(z < \dfrac{\hat{p}_{o,L} - .7}{.0458}\right) = P(z < z_{o,L}) = .025$$

From Table IV, Appendix A, $z_{o,L} = -1.96$

Thus, $z_{o,L} = \dfrac{\hat{p}_{o,L} - .7}{.0458} \Rightarrow \hat{p}_{o,L} = -1.96(.0458) + .7 = .6102$

$P(\hat{p} > \hat{p}_{o,U}) = P\left(z > \dfrac{\hat{p}_{o,U} - .7}{.0458}\right) = P(z > z_{o,U}) = .025$

From Table IV, Appendix A, $z_{o,U} = 1.96$

Thus, $z_{o,U} = \dfrac{\hat{p}_{o,U} - .7}{.0458} \Rightarrow \hat{p}_{o,U} = 1.96(.0458) + .7 = .7898$

Now, find

$\beta = P(.6102 < \hat{p} < .7898$ when $\mu_{\hat{p}} = .65) = P\left(\dfrac{.6102 - .65}{.0477} < z < \dfrac{.7898 - .65}{.0477}\right)$

$= P(-.83 < z < 2.93) = .2967 + .4983 = .7950$

(Using Table IV, Appendix A)

d. $\beta = P(.6102 < \hat{p} < .7898$ when $\mu_{\hat{p}} = .71) = P\left(\dfrac{.6102 - .71}{.0477} < z < \dfrac{.7898 - .71}{.0477}\right)$

$= P(-2.09 < z < 1.67) = .4817 + .4525 = .9342$

8.105   a.   First, we find $\bar{x}_o$ such that $P(\bar{x} < \bar{x}_o) = .01$, under H_0.

From Exercise 8.104, we know that $z_o = -2.33$. Thus,

$z_o = \dfrac{\bar{x}_o - \mu_o}{\dfrac{\sigma}{\sqrt{n}}} = \dfrac{\bar{x}_o - 20}{\dfrac{11.9}{\sqrt{100}}} = \dfrac{\bar{x}_o - 20}{1.19} = -2.33$

$\Rightarrow \bar{x}_o = -2.33(1.19) + 20 = 17.227$

For $\mu = 19$,

$Power = P(\bar{x} < 17.227$ when $\mu = 19) = P\left(z < \dfrac{17.227 - 19}{1.19}\right)$

$= P(z < -1.49) = .5 - .4319 = .0681$

For $\mu = 18$,

$Power = P(\bar{x} < 17.227$ when $\mu = 18) = P\left(z < \dfrac{17.227 - 18}{1.19}\right)$

$= P(z < -0.65) = .5 - .2422 = .2578$

For $\mu = 16$,

$$Power = P(\bar{x} < 17.227 \text{ when } \mu = 16) = P\left(z < \frac{17.227 - 16}{1.19} \right)$$

$$= P(z < 1.03) = .5 + .3485 = .8485$$

For $\mu = 14$,

$$Power = P(\bar{x} < 17.227 \text{ when } \mu = 14) = P\left(z < \frac{17.227 - 14}{1.19} \right)$$

$$= P(z < 2.71) = .5 + .4966 = .9966$$

For $\mu = 12$,

$$Power = P(\bar{x} < 17.227 \text{ when } \mu = 12) = P\left(z < \frac{17.227 - 12}{1.19} \right)$$

$$= P(z < 4.39) \approx .5 + .5 = 1$$

Using MINITAB, the plot is:

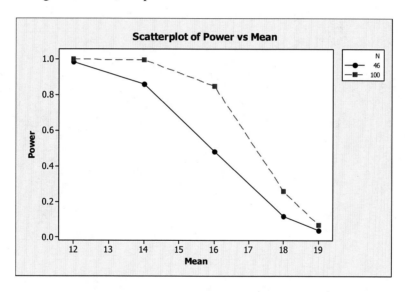

The power is much greater for $n = 100$ than for $n = 46$.

8.107 First, we must compute the value of $\bar{x}$ corresponding to the border of the rejection region. From Exercise 8.32,

H_0: $\mu = 10,000$
H_a: $\mu > 10,000$

The rejection region is $z > 1.645$ with $\alpha = .05$. The border value for the one-sided test is:

$$\bar{x}_0 = \mu_0 + z_\alpha \sigma_{\bar{x}} = \mu_0 + z_\alpha \frac{\sigma}{\sqrt{n}} = 10,000 + 1.645 \frac{1600}{\sqrt{67}} = 10,000 + 321.55 = 10,321.55$$

Next, we convert this border value to a z-score using the alternative distribution with mean $\mu_a = 10,500$. The z-score is:

$$z = \frac{\bar{x}_0 - \mu_a}{\sigma_{\bar{x}}} = \frac{10,321.55 - 10,500}{\frac{1,600}{\sqrt{67}}} = -.91$$

The probability of making a Type II error is:

$$\beta = P(\bar{x} \le 10,321.55 \,|\, \mu_a = 10,500) = P(z \le -.91) = .5 - .3186 = .1814$$

8.109   First, we find $\hat{p}_{oL}$ such that $P(\hat{p} < \hat{p}_{oL}) = .005$ and $\hat{p}_{oU}$ such that $P(\hat{p} > \hat{p}_{oU}) = .005$

From Exercise 8.86, we know that $z_{oL} = -2.575$ and $z_{oU} = 2.58$. Thus,

$$z_{oL} = \frac{\hat{p}_{oL} - p_o}{\sqrt{\frac{p_o q_o}{n}}} = \frac{\hat{p}_{oL} - .5}{\sqrt{\frac{.5(.5)}{121}}} = \frac{\hat{p}_{oL} - .5}{.0455} = -2.58$$

$$\Rightarrow \hat{p}_{oL} = -2.58(.0455) + .5 = .383$$

and

$$z_{oU} = \frac{\hat{p}_{oU} - p_o}{\sqrt{\frac{p_o q_o}{n}}} = \frac{\hat{p}_{oU} - .5}{\sqrt{\frac{.5(.5)}{121}}} = \frac{\hat{p}_{oU} - .5}{.0455} = 2.58$$

$$\Rightarrow \hat{p}_{oU} = 2.58(.0455) + .5 = .617$$

$$Power = P(\hat{p}_{oL} < .383 \text{ when } p = .65) + P(\hat{p}_{oH} > .617 \text{ when } p = .65)$$

$$= P\left(z < \frac{.383 - .65}{\sqrt{\frac{.65(.35)}{121}}} \right) + P\left(z > \frac{.617 - .65}{\sqrt{\frac{.65(.35)}{121}}} \right)$$

$$= P(z < -6.16) + P(z > -.76) = 0 + .5 + .2764 = .7764$$

8.111   The conditions required for a valid test for σ^2 are:

1.   A random sample is selected from the target population.
2.   The population from which the sample is selected has a distribution that is approximately normal.

8.113   The statement "When the sample size n is large, no assumptions about the population are necessary to test the population variance σ^2" is false. The population from which the sample is drawn must be approximately normal.

8.115   a. $df = n - 1 = 16 - 1 = 15$; reject H_0 if $\chi^2 < 6.26214$ or $\chi^2 > 27.4884$

b. $df = n - 1 = 23 - 1 = 22$; reject H_0 if $\chi^2 > 40.2894$

c. $df = n - 1 = 15 - 1 = 14$; reject H_0 if $\chi^2 > 21.0642$

d. $df = n - 1 = 13 - 1 = 12$; reject H_0 if $\chi^2 < 3.57056$

e. $df = n - 1 = 7 - 1 = 6$; reject H_0 if $\chi^2 < 1.63539$ or $\chi^2 > 12.5916$

f. $df = n - 1 = 25 - 1 = 24$; reject H_0 if $\chi^2 < 13.8484$

8.117   a. H_0: $\sigma^2 = 1$
H_a: $\sigma^2 > 1$

The test statistic is $\chi^2 = \dfrac{(n-1)s^2}{\sigma_0^2} = \dfrac{(100-1)(1.84)}{1} = 182.16$

The rejection region requires $\alpha = .05$ in the upper tail of the χ^2 distribution with df = $n - 1 = 100 - 1 = 99$. From Table VII, Appendix A, $\chi^2_{.05} \approx 124.324$. The rejection region is $\chi^2 > 124.324$.

Since the observed value of the test statistic falls in the rejection region ($\chi^2 = 182.16 > 124.324$), H_0 is rejected. There is sufficient evidence to indicate the variance is larger than 1 at $\alpha = .05$.

b.   In part **b** of Exercise 8.116, the test statistic was $\chi^2 = 11.04$. In Exercise 8.116, we did not reject H_0, but we did in this Exercise. As the sample size increases, it becomes easier to reject H_0.

8.119   a.   The rejection region requires $\alpha / 2 = .01 / 2 = .005$ in each tail of the χ^2 distribution with df = $n - 1 = 46 - 1 = 45$. From Table VII, Appendix A, $\chi^2_{.005} \approx 66.7659$ and $\chi^2_{.995} \approx 20.7065$. The rejection region is $\chi^2 < 20.7065$ or $\chi^2 > 66.7659$.

b.   The test statistic is $\chi^2 = \dfrac{(n-1)s^2}{\sigma_o^2} = \dfrac{(46-1)11.9^2}{100} = 63.7245$

c.   Since the observed value of the test statistic does not fall in the rejection region ($\chi^2 = 63.7245 \not> 66.7659$ and $\chi^2 = 63.7245 \not< 20.7065$), H_0 is not rejected. There is insufficient evidence to indicate the variance is different from 100 at $\alpha = .01$.

8.121   a.   The parameter of interest is σ^2, the variance of the population of rock bounces.

b.   To determine if the variance differs from 10 m^2, we test

$$H_0: \ \sigma^2 = 10$$
$$H_a: \ \sigma^2 \neq 10$$

c.   The test statistic is $\chi^2 = \dfrac{(n-1)s^2}{\sigma_0^2} = \dfrac{(13-1)(4.0947291)^2}{10} = 20.12$

d.   The rejection region requires $\alpha = .10/2 = .05$ in each tail of the χ^2 distribution with df = $n - 1 = 13 - 1 = 12$. From Table VII, Appendix A, $\chi_{.05}^2 = 21.0261$ and $\chi_{.95}^2 = 5.22603$. The rejection region is $\chi^2 > 21.0261$ or $\chi^2 < 5.22603$.

e.   Since the observed value of the test statistic does not fall in the rejection region ($\chi^2 = 20.12 \ngtr 21.0261$ and $\chi^2 = 20.12 \nless 5.22603$), H_0 is not rejected. There is insufficient evidence to indicate the variance differs from 10 m^2 at $\alpha = .10$.

f.   We must assume that the population of rock bounces is normally distributed.

8.123   To determine if the true standard deviation of the point-spread errors exceed 15 (variance exceeds 225), we test:

$$H_0: \ \sigma^2 = 225$$
$$H_a: \ \sigma^2 > 225$$

The test statistic is $\chi^2 = \dfrac{(n-1)s^2}{\sigma_0^2} = \dfrac{(240-1)13.3^2}{225} = 187.896$

The rejection region requires $\alpha = .05$ in the upper tail of the χ^2 distribution with df = $n - 1 = 240 - 1 = 239$. The maximum value of df in Table VII is 100. Thus, we cannot find the rejection region using Table VII. Using a statistical package, $\chi_{.05}^2 = 276.062$. The rejection region is $\chi^2 > 276.062$.

Since the observed value of the test statistics does not fall in the rejection region ($\chi^2 = 187.896 \ngtr 276.062$), H_0 is not rejected. There is insufficient evidence to indicate that the true standard deviation of the point-spread errors exceed 15 at $\alpha = .05$.

(Since the observed variance (or standard deviation) is less than the hypothesized value of the variance (or standard deviation) under H_0, there is no way H_0 will be rejected for any reasonable value of α.)

8.125   To determine if the variance of birthweights of babies delivered by cocaine-dependent women is less than 200,000, we test:

H_0: $\sigma^2 = 200,000$

H_a: $\sigma^2 < 200,000$

The test statistic is $\chi^2 = \dfrac{(n-1)s^2}{\sigma_0^2} = \dfrac{(16-1)410^2}{200,000} = 12.6075$

The rejection region requires $\alpha = .01$ in the lower tail of the χ^2 distribution with df = $n - 1 =$ 16 – 1 = 15. From Table VII, Appendix A, $\chi_{.99}^2 = 5.22935$. The rejection region is $\chi^2 < 5.22935$.

Since the observed value of the test statistic does not fall in the rejection region ($\chi^2 = 12.6075 \nless 5.22935$), H_0 is not rejected. There is insufficient evidence to indicate that the variance of birthweights of babies delivered by cocaine-dependent women is less than 200,000 at $\alpha = .01$.

8.127 To determine if the true conduction time standard deviation is less than 7 nanoseconds, we test:

H_0: $\sigma^2 = 7^2 = 49$

H_a: $\sigma^2 < 49$

The test statistic is $\chi^2 = \dfrac{(n-1)s^2}{\sigma_o^2} = \dfrac{(18-1)6.3^2}{7^2} = 13.77$

The rejection region requires $\alpha = .01$ in the lower tail of the χ^2 distribution with df = $n - 1 = 18 - 1 = 17$. From Table VII, Appendix A, $\chi_{.99}^2 = 6.40776$. The rejection region is $\chi^2 < 6.40776$.

Since the observed value of the test statistic does not fall in the rejection region ($\chi^2 = 13.77 \nless 6.40776$), H_0 is not rejected. There is insufficient evidence to indicate the true conduction time standard deviation is less than 7 nanoseconds at $\alpha = .01$.

8.129 The smaller the *p*-value associated with a test of hypothesis, the stronger the support for the **alternative** hypothesis. The *p*-value is the probability of observing your test statistic or anything more unusual, given the null hypothesis is true. If this value is small, it would be very unusual to observe this test statistic if the null hypothesis were true. Thus, it would indicate the alternative hypothesis is true.

8.131 The elements of the test of hypothesis that should be specified prior to analyzing the data are: null hypothesis, alternative hypothesis, and significance level.

8.133 The larger the *p*-value associated with a test of hypothesis, the stronger the support for the **null** hypothesis. The *p*-value is the probability of observing your test statistic or anything more unusual, given the null hypothesis is true. If this value is large, it would not be very unusual to observe this test statistic if the null hypothesis were true. Thus, it would lend support that the null hypothesis is true.

8.135   a. H_a: $p = .35$
H_a: $p < .35$

The test statistic is $z = \dfrac{\hat{p} - p_0}{\sqrt{\dfrac{p_0 q_0}{n}}} = \dfrac{.29 - .35}{\sqrt{\dfrac{.35(.65)}{200}}} = -1.78$

The rejection region requires $\alpha = .05$ in the lower tail of the z distribution. From Table IV, Appendix A, $z_{.05} = 1.645$. The rejection region is $z < -1.645$.

Since the observed value of the test statistic falls in the rejection region ($z = -1.78 < -1.645$), H_0 is rejected. There is sufficient evidence to indicate $p < .35$ at $\alpha = .05$.

b. H_0: $p = .35$
H_a: $p \neq .35$

The test statistic is $z = -1.78$ (from **a**).

The rejection region requires $\alpha / 2 = .05 / 2 = .025$ in each tail of the z distribution. From Table IV, Appendix A, $t_{.025} = 1.96$. The rejection region is $z < -1.96$ or $z > 1.96$.

Since the observed value of the test statistic does not fall in the rejection region ($z = -1.78 \not< -1.96$), H_0 is not rejected. There is insufficient evidence to indicate p is different from .35 at $\alpha = .05$.

c.   For confidence coefficient .95, $\alpha = .05$ and $\alpha / 2 = .05 / 2 = .025$. From Table IV, Appendix A, $z_{.025} = 1.96$. The confidence interval is:

$$\hat{p} \pm z_{.025} \sqrt{\frac{\hat{p}\hat{q}}{n}} \Rightarrow .29 \pm 1.96 \sqrt{\frac{.29(.71)}{200}} \Rightarrow .29 \pm .063 \Rightarrow (.227, .353)$$

d.   For confidence coefficient .99, $\alpha = .01$ and $\alpha / 2 = .01 / 2 = .005$. From Table IV, Appendix A, $z_{.005} = 2.58$. The confidence interval is:

$$\hat{p} \pm z_{.005} \sqrt{\frac{\hat{p}\hat{q}}{n}} \Rightarrow .29 \pm 2.58 \sqrt{\frac{.29(.71)}{200}} \Rightarrow .29 \pm .083 \Rightarrow (.207, .373)$$

e. $n = \dfrac{\left(z_{\alpha/2}\right)^2 pq}{B^2} = \dfrac{2.58^2(.29)(.71)}{.05^2} = 548.2 \approx 549$

(We used $\hat{p} = .29$ to estimate the value of p.)

8.137 a. H_0: $\sigma^2 = 30$
H_a: $\sigma^2 > 30$

The test statistic is $\chi^2 = \dfrac{(n-1)s^2}{\sigma_0^2} = \dfrac{(41-1)(6.9)^2}{30} = 63.48$

The rejection region requires $\alpha = .05$ in the upper tail of the χ^2 distribution with df $= n - 1 = 40$. From Table VII, Appendix A, $\chi^2_{.05} = 55.7585$. The rejection region is $\chi^2 > 55.7585$.

Since the observed value of the test statistic falls in the rejection region ($\chi^2 = 63.48 >$ 55.7585), H_0 is rejected. There is sufficient evidence to indicate the variance is larger than 30 at $\alpha = .05$.

     b. H_0: $\sigma^2 = 30$
H_a: $\sigma^2 \neq 30$

The test statistic is $\chi^2 = 63.48$ (from part **a**).

The rejection region requires $\alpha / 2 = .05 / 2 = .025$ in each tail of the χ^2 distribution with df $= n - 1 = 40$. From Table VII, Appendix A, $\chi^2_{.025} = 59.3417$ and $\chi^2_{.975} = 24.4331$. The rejection region is $\chi^2 < 24.4331$ or $\chi^2 > 59.3417$.

Since the observed value of the test statistic falls in the rejection region $\left(\chi^2 = 63.48 > 59.3417 \right)$, H_0 is rejected. There is sufficient evidence to indicate the variance is not 30 at $\alpha = .05$.

8.139 a.   The null hypothesis would be: H_0: $p = .45$

     b.   The null hypothesis would be: H_0: $\mu = 2.5$

8.141 a.   To determine if the true mean score of all sleep-deprived subjects is less than 80, we test:

H_0: $\mu = 80$
H_a: $\mu < 80$

The test statistic is $t = \dfrac{\bar{x} - \mu_0}{s / \sqrt{n}} = \dfrac{63 - 80}{17 / \sqrt{12}} = -3.46$

The rejection region requires $\alpha = .05$ in the lower tail of the t distribution with df $= n - 1$ $= 12 - 1 = 11$. From Table VI, Appendix A, $t_{.05} = 1.796$. The rejection region is $t < -1.796$.

Since the observed value of the test statistic falls in the rejection region ($t = -3.46 <$ -1.796), H_0 is rejected. There is sufficient evidence to indicate that the true mean score of all sleep-deprived subjects is less than 80 at $\alpha = .05$.

b.   We must assume that the overall test scores of sleep-deprived students are normally distributed.

8.143 a.   To determine if the mean number of suicide bombings for all Al Qaeda attacks against the U.S. differs from 2.5, we test:

H_0: $\mu = 2.5$
H_a: $\mu \neq 2.5$

The test statistic is $t = -2.46$ and the p-value is $p = 0.023$. (from the printout)

Since the p-value (0.023) is less than $\alpha = .10$, H_0 is rejected. There is sufficient evidence to indicate that the mean number of suicide bombings for all Al Qaeda attacks against the U.S. differs from 2.5 at $\alpha = .10$.

b.   The 90% confidence interval is (1.407, 2.307). Since the hypothesized value of the mean (2.5) does not fall in the 90% confidence interval, it is not a likely value. Thus, we would reject H_0 and conclude that the true mean number of suicide bombings for all Al Qaeda attacks against the U.S. differs from 2.5 at $\alpha = .10$.

c.   Yes, the two inferences agree. Since the same level of confidence was used for each and the sample statistics are the same, the results must agree.

d.   We must assume that the distribution that we are sampling from is approximately normal.

e.   Using MINITAB, a histogram of the data is:

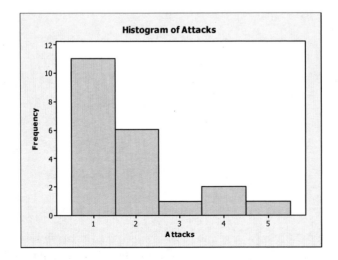

Although this is a histogram of the sample, it should reflect the distribution of the population fairly well. The population distribution appears to be skewed to the right. Thus, the assumption of normality is probably not valid.

8.145   a.   Let p = proportion of the subjects who felt that the masculinization of face shape decreased attractiveness of the male face. If there is no preference for the unaltered or morphed male face, then $p = .5$.

For this problem, $\hat{p} = x/n = 58/67 = .866$.

In order for the inference to be valid, the sample size must be large enough. The sample size is large enough if both np_0 and nq_0 are greater than or equal to 15.

$np_0 = 67(.5) = 33.5$ and $nq_0 = 67(1 - .5) = 67(.5) = 33.5$. Since both of these values are greater than 15, the sample size is large enough to use the normal approximation.

To determine if the subjects showed a preference for either the unaltered or morphed face, we test:

H_0: $p = .5$
H_a: $p \neq .5$

b.   The test statistic is $z = \dfrac{\hat{p} - p_0}{\sqrt{\dfrac{p_0 q_0}{n}}} = \dfrac{.866 - .50}{\sqrt{\dfrac{.50(.50)}{67}}} = 5.99$

c.   By definition, the p-value is the probability of observing our test statistic or anything more unusual, given H_0 is true. For this problem, $p = P(z \leq -5.99) + P(z \geq 5.99)$ $\approx .5 - .5 + .5 - .5 = 0$. This corresponds to what is listed in the Exercise.

d.   The rejection region requires $\alpha/2 = .01/2 = .005$ in each tail of the z distribution. From Table IV, Appendix A, $z_{.005} = 2.58$. The rejection region is $z < -2.58$ or $z > 2.58$.

Since the observed value of the test statistic falls in the rejection region ($z = 5.99 > 2.58$), H_0 is rejected. There is sufficient evidence to indicate that the subjects showed a preference for either the unaltered or morphed face at $\alpha = .01$.

8.147   To determine if the mean alkalinity level of water in the tributary exceeds 50 mpl, we test:

H_0: $\mu = 50$
H_a: $\mu > 50$

The test statistic is $z = \dfrac{\overline{x} - \mu_0}{\sigma_{\overline{x}}} = \dfrac{67.8 - 50}{14.4/\sqrt{100}} = 12.36$

The rejection region requires $\alpha = .01$ in the upper tail of the z distribution. From Table IV, Appendix A, $z_{.01} = 2.33$. The rejection region is $z > 2.33$.

Since the observed value of the test statistic falls in the rejection region ($z = 12.36 > 2.33$), H_0 is rejected. There is sufficient evidence to indicate that the mean alkalinity level of water in the tributary exceeds 50 mpl at $\alpha = .01$.

8.149   To determine if the true mean PTSD score of all World War II aviator POWs is less than 16, we test:

H_0: $\mu = 16$

H_a: $\mu < 16$

The test statistic is $z = \dfrac{\bar{x} - \mu_o}{\sigma_{\bar{x}}} = \dfrac{9.00 - 16}{9.32/\sqrt{33}} = -4.31$

The rejection region requires $\alpha = .10$ in the lower tail of the z distribution. From Table IV, Appendix A, $z_{.10} = 1.28$. The rejection region is $z < -1.28$.

Since the observed value of the test statistic falls in the rejection region ($z = -4.31 < -1.28$), H_0 is rejected. There is sufficient evidence to indicate that the true mean PTSD score of all World War II aviator POWs is less than 16 at $\alpha = .10$.

8.151   a.   To determine if the true mean inbreeding coefficient μ for this species of wasp exceeds 0, we test:

H_0: $\mu = 0$

H_a: $\mu > 0$

The test statistic is $z = \dfrac{\bar{x} - \mu_0}{\sigma_{\bar{x}}} = \dfrac{.044 - 0}{.884/\sqrt{197}} = 0.70$

The rejection region requires $\alpha = .05$ in the upper tail of the z distribution. From Table IV, Appendix A, $z_{.05} = 1.645$. The rejection region is $z > 1.645$.

Since the observed value of the test statistic does not fall in the rejection region ($z = 0.70 \not> 1.645$), H_0 is not rejected. There is insufficient evidence to indicate that the true mean inbreeding coefficient μ for this species of wasp exceeds 0 at $\alpha = .05$.

b.   This result agrees with that of Exercise 7.129. The confidence interval in Exercise 7.129 was $(-.118, .206)$. Since this interval contains 0, there is no evidence to indicate that the mean is different from 0. This is the same conclusion that was reached in part **a**.

8.153   a.   To determine if the production process should be halted, we test:

H_0: $\mu = 3$

H_a: $\mu > 3$

where μ = mean amount of PCB in the effluent.

The test statistic is $z = \dfrac{\bar{x} - \mu_0}{\sigma_{\bar{x}}} = \dfrac{3.1 - 3}{.5/\sqrt{50}} = 1.41$

The rejection region requires $\alpha = .01$ in the upper tail of the z distribution. From Table IV, Appendix A, $z_{.01} = 2.33$. The rejection region is $z > 2.33$.

Since the observed value of the test statistic does not fall in the rejection region, ($z =$ 1.41 $\not> 2.33$), H_0 is not rejected. There is insufficient evidence to indicate the mean amount of PCB in the effluent is more than 3 parts per million at $\alpha = .01$. Do not halt the manufacturing process.

b.   As plant manager, I do not want to shut down the plant unnecessarily. Therefore, I want $\alpha = P$(shut down plant when $\mu = 3$) to be small.

8.155   a.   No, it increases the risk of falsely rejecting H_0, i.e., closing the plant unnecessarily.

b.   First, find $\bar{x}_0$ such that $P(\bar{x} > \bar{x}_0) = P(z > z_0) = .05$.

From Table IV, Appendix A, $z_0 = 1.645$

$$z = \frac{\bar{x} - \mu_0}{\sigma / \sqrt{n}} \Rightarrow 1.645 = \frac{\bar{x}_0 - 3}{.5 / \sqrt{50}} \Rightarrow \bar{x}_0 = 3.116$$

Then, compute

$$\beta = P(\bar{x}_0 \leq 3.116 \text{ when } \mu = 3.1) = P\left(z \leq \frac{3.116 - 3.1}{.5 / \sqrt{50}} \right) = P(z \leq .23) = .5 + .0910 = .5910$$

$$Power = 1 - \beta = 1 - .5910 = .4090$$

c.   The power of the test increases as α increases.

8.157   The point estimate for p is $\hat{p} = \frac{x}{n} = \frac{85}{124} = .685$

In order for the inference to be valid, the sample size must be large enough. The sample size is large enough if both np_0 and nq_0 are greater than or equal to 15.

$np_0 = 124(.167) = 20.708$ and $nq_0 = 124\,(1 - .167) = 124\,(.833) = 103.292$. Since both of these values are greater than 15, the sample size is large enough to use the normal approximation.

To determine if the students will tend to choose the three-grill display so that Grill #2 is a compromise between a more and a less desirable grill ($p > .167$), we test:

H_0: $p = .167$
H_a: $p > .167$

The test statistic is $z = \dfrac{\hat{p} - p_0}{\sqrt{\dfrac{p_0 q_0}{n}}} = \dfrac{.685 - .167}{\sqrt{\dfrac{.167(.822)}{124}}} = 15.46$

The rejection region requires $\alpha = .05$ in the upper tail of the z distribution. From Table IV, Appendix A, $z_{.05} = 1.645$. The rejection region is $z > 1.645$.

Since the observed value of the test statistic falls in the rejection region ($z = 15.46 > 1.645$), H_0 is rejected. There is sufficient evidence to indicate the students will tend to choose the three-grill display so that Grill #2 is a compromise between a more and a less desirable grill ($p > .167$) at $\alpha = .05$.

8.159   a.   A Type I error is rejecting the null hypothesis when it is true. In this problem, we would be concluding that the individual is a liar when, in fact, the individual is telling the truth.

A Type II error is accepting the null hypothesis when it is false. In this problem, we would be concluding that the individual is telling the truth when, in fact, the individual is a liar.

b.   The probability of a Type I error would be the probability of concluding the individual is a liar when he/she is telling the truth. From the problem, it stated that the polygraph would indicate that of 500 individuals who were telling the truth, 185 would be liars. Thus, an estimate of the probability of a Type I error would be 185/500 = .37.

The probability of a Type II error would be the probability of concluding the individual is telling the truth when he/she is a liar. From the problem, it stated that the polygraph would indicate that of 500 individuals who were liars, 120 would be telling the truth. Thus, an estimate of the probability of a Type II error would be 120/500 = .24.

8.161   a.   Using MINITAB, the descriptive statistics are:

Descriptive Statistics: PCB

```
Variable   N  Mean   StDev    Minimum     Q1  Median      Q3  Maximum
PCB       48 9.292   2.103  0.000000000  9.000   9.000  10.000   13.000
```

To determine if the data can refute the manufacturer's claim, we test:

H_0: $\mu = 10$
H_a: $\mu < 10$

The test statistic is $z = \dfrac{\bar{x} - \mu_0}{\sigma_{\bar{x}}} = \dfrac{9.292 - 10}{2.103/\sqrt{48}} = -2.33$

The rejection region requires $\alpha = .05$ in the lower tail of the z distribution. From Table IV, Appendix A, $z_{.05} = 1.645$. The rejection region is $z < -1.645$.

Since the observed value of the test statistic falls in the rejection region ($z = -2.33 < -1.645$), H_0 is rejected. There is sufficient evidence to indicate that the true average number of inspections is less than 10 at $\alpha = .05$.

We agree with the potential buyer who doubts the manufacturer's claim.

b. First, we must compute the value of $\bar{x}$ corresponding to the border of the rejection region. The rejection region is $z < -1.645$ with $\alpha = .05$. The border value for the one-sided test is:

$$\bar{x}_0 = \mu_0 - z_\alpha \sigma_{\bar{x}} = \mu_0 - z_\alpha \frac{\sigma}{\sqrt{n}} = 10 - 1.645\frac{1.2}{\sqrt{48}} = 10 - .285 = 9.715$$

Next, we convert this border value to a z-score using the alternative distribution with mean $\mu_a = 9.5$. The z-score is:

$$z = \frac{\bar{x}_0 - \mu_a}{\sigma_{\bar{x}}} = \frac{9.715 - 9.5}{\frac{1.2}{\sqrt{48}}} = 1.24$$

The probability of rejecting H_0 is:

$$Power = P(\bar{x} \leq 9.715 \mid \mu_a = 9.5) = P(z \leq 1.24) = .5 + .3925 = .8925$$

8.163 a. We want to show that more than 17,000 of the 18,200 signatures are valid. Thus, we want to show that the proportion of valid signatures is greater than $17{,}000/18{,}200 = .934$.

From the Exercise, $\hat{p} = 98/100 = .98$.

In order for the inference to be valid, the sample size must be large enough. The sample size is large enough if both np_0 and nq_0 are greater than or equal to 15.

$np_0 = 100(.934) = 93.4$ and $nq_0 = 100(1 - .934) = 100(.066) = 6.6$. Since the second value is not greater than 15, the sample size may not be large enough to use the normal approximation.

To determine if more than .934 of the signatures are valid, we test:

H_0: $p = .934$
H_a: $p > .934$

The test statistic is $z = \dfrac{\hat{p} - p_0}{\sqrt{\dfrac{p_0 q_0}{n}}} = \dfrac{.98 - .934}{\sqrt{\dfrac{.934(.066)}{100}}} = 1.85$

Since no α is given in the problem, we will choose $\alpha = .05$. The rejection region requires $\alpha = .05$ in the upper tail of the z distribution. From Table IV, Appendix A, $z_{.05} = 1.645$. The rejection region is $z > 1.645$.

Since the observed value of the test statistic falls in the rejection region ($z = 1.85 >$ 1.645), H_0 is rejected. There is sufficient evidence to indicate that the true proportion of valid signatures is greater than .934 at $\alpha = .05$. This indicates that more than 17,000 of the 18,200 signatures are valid.

b.   We want to show that more than 16,000 of the 18,200 signatures are valid. Thus, we want to show that the proportion of valid signatures is greater than $16,000/18,200 = .879$.

Again, from the Exercise, $\hat{p} = 98/100 = .98$.

In order for the inference to be valid, the sample size must be large enough. The sample size is large enough if both np_o and nq_o are greater than or equal to 15.

$np_o = 100(.879) = 87.9$ and $nq_o = 100(1 - .879) = 100(.121) = 12.1$. Since the second value is not greater than 15, the sample size may not be large enough to use the normal approximation.

To determine if more than .879 of the signatures are valid, we test:

H_0: $p = .879$
H_a: $p > .879$

The test statistic is $z = \dfrac{\hat{p} - p_0}{\sqrt{\dfrac{p_0 q_0}{n}}} = \dfrac{.98 - .879}{\sqrt{\dfrac{.879(.121)}{100}}} = 3.10$

Since no α is given in the problem, we will choose $\alpha = .05$. The rejection region requires $\alpha = .05$ in the upper tail of the z distribution. From Table IV, Appendix A, $z_{.05} = 1.645$. The rejection region is $z > 1.645$.

Since the observed value of the test statistic falls in the rejection region ($z = 3.10 >$ 1.645), H_0 is rejected. There is sufficient evidence to indicate that the true proportion of valid signatures is greater than .879 at $\alpha = .05$. This indicates that more than 16,000 of the 18,200 signatures are valid.

Inferences Based on Two Samples
Confidence Intervals and Tests of Hypotheses

9.1   When the samples are large, the sampling distribution of $(\bar{x}_1 - \bar{x}_2)$ is approximately normal by the Central Limit Theorem. The mean of the sampling distribution is $(\mu_1 - \mu_2)$ and the standard deviation is $\sqrt{\dfrac{\sigma_1^2}{n_1} + \dfrac{\sigma_2^2}{n_2}}$.

9.3   a.   No. Both populations must be normal.

   b.   No. Both population variances must be equal.

   c.   No. Both populations must be normal.

   d.   Yes.

   e.   No. Both populations must be normal.

9.5   The confidence interval for $(\mu_1 - \mu_2)$ is $(-10, -4)$. The correct inference is **b**: $\mu_1 < \mu_2$. Since all the values in the interval are negative, there is evidence that $\mu_1 < \mu_2$.

9.7   a. $\mu_{\bar{x}_1} = \mu_1 = 14$

   $\sigma_{\bar{x}} = \dfrac{\sigma_1}{\sqrt{n_1}} = \dfrac{4}{\sqrt{100}} = .4$

   b. $\mu_{\bar{x}_2} = \mu_2 = 10$

   $\sigma_{\bar{x}_2} = \dfrac{\sigma_2}{\sqrt{n_2}} = \dfrac{3}{\sqrt{100}} = .3$

   c. $\mu_{\bar{x}_1 - \bar{x}_2} = \mu_1 - \mu_2 = 14 - 10 = 4$

   $\sigma_{\bar{x}_1 - \bar{x}_2} = \sqrt{\dfrac{\sigma_1^2}{n_1} + \dfrac{\sigma_2^2}{n_2}} = \sqrt{\dfrac{4^2}{100} + \dfrac{3^2}{100}} = \sqrt{\dfrac{25}{100}} = .5$

   d.   Since $n_1 \geq 30$ and $n_2 \geq 30$, the sampling distribution of $\bar{x}_1 - \bar{x}_2$ is approximately normal by the Central Limit Theorem.

9.9   Some preliminary calculations are:

$$\bar{x}_1 = \frac{\sum x_1}{n_1} = \frac{11.8}{5} = 2.36 \qquad s_1^2 = \frac{\sum x_1^2 - \frac{\left(\sum x_1\right)^2}{n_1}}{n_1 - 1} = \frac{30.78 - \frac{(11.8)^2}{5}}{5 - 1} = .733$$

$$\bar{x}_2 = \frac{\sum x_2}{n_2} = \frac{14.4}{4} = 3.6 \qquad s_2^2 = \frac{\sum x_2^2 - \frac{\left(\sum x_2\right)^2}{n_2}}{n_2 - 1} = \frac{53.1 - \frac{(14.4)^2}{4}}{4 - 1} = .42$$

a. $$s_p^2 = \frac{(n_1 - 1)s_1^2 + (n_2 - 1)s_2^2}{n_1 + n_2 - 2} = \frac{(5 - 1).773 + (4 - 1).42}{5 + 4 - 2} = \frac{4.192}{7} = .5989$$

b. H_0: $\mu_1 - \mu_2 = 0$

    H_a: $\mu_1 - \mu_2 < 0$

The test statistic is $t = \dfrac{(\bar{x}_1 - \bar{x}_2) - D_0}{\sqrt{s_p^2\left(\dfrac{1}{n_1} + \dfrac{1}{n_2}\right)}} = \dfrac{(2.36 - 3.6) - 0}{\sqrt{.5989\left(\dfrac{1}{5} + \dfrac{1}{4}\right)}} = \dfrac{-1.24}{.5191} = -2.39$

The rejection region requires $\alpha = .10$ in the lower tail of the t distribution with df = $n_1 + n_2 - 2 = 5 + 4 - 2 = 7$. From Table VI, Appendix A, $t_{.10} = 1.415$. The rejection region is $t < -1.415$.

Since the test statistic falls in the rejection region ($t = -2.39 < -1.415$), H_0 is rejected. There is sufficient evidence to indicate that $\mu_2 > \mu_1$ at $\alpha = .10$.

c.   A small sample confidence interval is needed because $n_1 = 5 < 30$ and $n_2 = 4 < 30$.

For confidence coefficient .90, $\alpha = .10$ and $\alpha / 2 = .05$. From Table VI, Appendix A, with df = $n_1 + n_2 - 2 = 5 + 4 - 2 = 7$, $t_{.05} = 1.895$. The 90% confidence interval for ($\mu_1 - \mu_2$) is:

$$(\bar{x}_1 - \bar{x}_2) \pm t_{.05}\sqrt{s_p^2\left(\frac{1}{n_1} + \frac{1}{n_2}\right)} \Rightarrow (2.36 - 3.6) \pm 1.895\sqrt{.5989\left(\frac{1}{5} + \frac{1}{4}\right)}$$

$$\Rightarrow -1.24 \pm .98 \Rightarrow (-2.22, -0.26)$$

d.   The confidence interval in part **c** provides more information about ($\mu_1 - \mu_2$) than the test of hypothesis in part **b**. The test in part **b** only tells us that μ_2 is greater than μ_1. However, the confidence interval estimates what the difference is between μ_1 and μ_2.

9.11   a.   The p-value is $p = .115$.

Since the p-value is not small, there is no evidence to reject H_0. There is no evidence to indicate that the population means are different for $\alpha \le .10$.

b.   The *p*-value would be half of the *p*-value in part **a**. The *p*-value = *p* = .1150/2 = .0575.

There is no evidence to reject H_0 for $\alpha = .05$. There is no evidence to indicate that the mean for population 1 is less than the mean for population 2 for $\alpha = .05$.

There is evidence to reject H_0 for $\alpha > .0575$. There is evidence to indicate that the mean for population 1 is less than the mean for population 2 for $\alpha > .0575$.

9.13   a.   Let μ_1 = mean test score of students in classrooms using software products and μ_2 = mean test score of students in classrooms not using software products. The parameter of interest is $\mu_1 - \mu_2$.

b.   To determine if the mean test score of students in classrooms using software products is higher than the mean test score of students in classrooms not using software products, we test:

$$H_0\text{: } \mu_1 - \mu_2 = 0$$
$$H_a\text{: } \mu_1 - \mu_2 > 0$$

c.   Since the *p*-value is so large (*p* = .62), H_0 is not rejected for any reasonable value of α. There is insufficient evidence to indicate the mean test score of students in classrooms using software products is higher than the mean test score of students in classrooms not using software products. This agrees with the conclusion of the DOE.

9.15   a.   Let μ_1 = mean recall of those who received only visual aspects of the ad and μ_2 = mean recall of those who received an audiovisual presentation. To determine if the mean recall of ad information by those who receive an audiovisual presentation is different from those who receive only the visual aspects of the ad, we test:

$$H_0\text{: } \mu_1 - \mu_2 = 0$$
$$H_a\text{: } \mu_1 - \mu_2 \neq 0$$

b.   Some preliminary calculations are:

$$s_p^2 = \frac{(n_1 - 1)s_1^2 + (n_2 - 1)s_2^2}{n_1 + n_2 - 2} = \frac{(20 - 1)1.98^2 + (20 - 1)2.13^2}{20 + 20 - 2} = 4.22865$$

The test statistic is $t = \dfrac{\overline{x}_1 - \overline{x}_2}{\sqrt{s_p^2 \left(\dfrac{1}{n_1} + \dfrac{1}{n_2} \right)}} = \dfrac{3.70 - 3.30}{\sqrt{4.22865 \left(\dfrac{1}{20} + \dfrac{1}{20} \right)}} = .62$

c.   The rejection region requires $\alpha / 2 = .10 / 2 = .05$ in each tail of the *t* distribution with df = $n_1 + n_2 - 2 = 20 + 20 - 2 = 38$. From Table VI, Appendix A, $t_{.05} \approx 1.684$. The rejection region is $t > 1.684$ or $t < -1.684$.

d. Since the observed value of the test statistic does not fall in the rejection region
($t = .62 \not> 1.684$), H_0 is not rejected. There is insufficient evidence to indicate the mean
recall of ad information by those who receive an audiovisual presentation is different
from those who receive only the visual aspects of the ad at $\alpha = .10$.

Although we cannot prove the researchers' theory is correct (Accept H_o), there is
evidence that supports the theory.

e. A p-value of .62 is the probability of observing our test statistic or anything more
unusual. For this problem, the p-value = $P(t > .62) + P(t < -.62) = .62$. Since the p-
value is so large, H_o will not be rejected for any reasonable value of α.

f. In order for the inference to be valid, the following conditions must be met:

1. Both populations being sampled from are normal
2. Both population variances must be the same
3. The samples must be random and independent.

9.17 a. Let μ_1 = mean number of books read by students who earned an "A" grade and μ_2 =
mean number of books read by students who earned a "B" or "C" grade. The parameter
of interest for this problem is $\mu_1 - \mu_2$, the difference in the mean number of books read
between students who received an "A" grade and those who received a "B or C" grade.

b. Using MINITAB, the descriptive statistics are:

Descriptive Statistics: A, B-C

```
Variable  N   Mean   StDev  Variance  Minimum  Median  Maximum
A         8   37.00  8.70   75.71     24.00    36.50   53.00
B-C       6   24.50  8.53   72.70     16.00    21.50   40.00
```

Some preliminary calculations are:

$$s_p^2 = \frac{(n_1-1)s_1^2+(n_2-1)s_2^2}{n_1+n_2-2} = \frac{(8-1)75.71+(6-1)72.70}{8+6-2} = 74.456$$

For confidence coefficient .95, $\alpha = .05$ and $\alpha/2 = .05/2 = .025$. From Table VI,
Appendix A, with $df = n_1 + n_2 - 2 = 8 + 6 - 2 = 12$, $t_{.025} = 2.179$. The confidence
interval is:

$$(\bar{x}_1 - \bar{x}_2) \pm t_{.025}\sqrt{s_p^2\left(\frac{1}{n_1}+\frac{1}{n_2}\right)} \Rightarrow (37-24.5) \pm 2.179\sqrt{74.456\left(\frac{1}{8}+\frac{1}{6}\right)}$$

$$\Rightarrow 12.5 \pm 10.154 \Rightarrow (2.346, 22.654)$$

c. We are 95% confident that the true difference in the mean number of books read
between the students who received an "A" grade and those who received a "B or C"
grade is between 2.346 and 22.654.

d.   In Exercise 2.33, the stem-and-leaf display indicated that the students who earned A's tended to read more books than those who earned B's and C's. This agrees with the confidence interval in part b. Because all the values in the confidence interval are positive, the mean number of books read by the students earning A's is larger than the mean number read by students earning B's or C's.

9.19   a.   From the printout, the 95% confidence interval is (-0.60, 7.95). We are 95% confident that the difference in mean FNE scores for bulimic and normal female students is between -0.60 and 7.95.

b.   We must assume that the distribution of FNE scores for bulimic female students and the distribution of the FNE scores for the normal female students are normally distributed. We must also assume that the variances of the two populations are equal and the samples are random and independent.

We must assume that the distribution of FNE scores for bulimic female students and the distribution of the FNE scores for the normal female students are normally distributed. We must also assume that the variances of the two populations are equal and the samples are random and independent.

Both sample distributions look somewhat mound-shaped and the sample variances are fairly close in value. Thus, both assumptions appear to be reasonably satisfied.

9.21   Let μ_1 = mean drug concentration for site 1 and μ_2 = mean drug concentration for site 2. To determine if there is a difference in the mean drug concentration of the 2 sites, we test:

$$H_0: \ \mu_1 - \mu_2 = 0$$
$$H_a: \ \mu_1 - \mu_2 \neq 0$$

From the printout, the test statistic is $t = .57$ and the p-value is $p = .573$. Since the *p*-value is so large, H_0 is not rejected. There is insufficient evidence to indicate a difference in the mean drug concentration of the 2 sites for any reasonable value of α.

9.23   Let μ_1 = mean total rating for group 1 (strengthen favored position) and μ_2 = mean total rating for group 2 (weaken opposing position).

Some preliminary calculations are:

$$s_p^2 = \frac{(n_1 - 1)s_1^2 + (n_2 - 1)s_2^2}{n_1 + n_2 - 2} = \frac{(26-1)12.5^2 + (26-1)12.2^2}{26 + 26 - 2} = \frac{7,627.25}{50} = 152.545$$

To determine if the difference in the mean total ratings for the two groups differ, we test:

$$H_0: \ \mu_1 - \mu_2 = 0$$
$$H_a: \ \mu_1 - \mu_2 \neq 0$$

The test statistic is $t = \dfrac{(\bar{x}_1 - \bar{x}_2) - 0}{\sqrt{s_p^2\left(\dfrac{1}{n_1} + \dfrac{1}{n_2}\right)}} = \dfrac{(28.6 - 24.9) - 0}{\sqrt{152.545\left(\dfrac{1}{26} + \dfrac{1}{26}\right)}} = 1.08$

The rejection region requires $\alpha/2 = .05/2 = .025$ in each tail of the t distribution. From Table VI, Appendix A, with df $= n_1 + n_2 - 2 = 26 + 26 - 2 = 50$, $t_{.025} \approx 2.010$. The rejection region is $t < -2.010$ or $t > 2.010$.

Since the observed value of the test statistic does not fall in the rejection region ($t = 1.08 \ngtr 2.010$), H_0 is not rejected. There is insufficient evidence to indicate there is a difference in the mean total ratings for the two groups at $\alpha = .05$.

9.25   Using MINITAB, the descriptive statistics are:

Descriptive Statistics: Species

```
Variable  Region        N   Mean  StDev Minimum    Q1  Median     Q3  Maximum
Species   Dry Steppe    5  14.00  21.31    3.00  3.00    5.00  29.50    52.00
          Gobi Desert   6  11.83  18.21    4.00  4.00    4.50  16.00    49.00

Variable  Region           Q3  Maximum
Species   Dry Steppe    29.50    52.00
          Gobi Desert   16.00    49.00
```

Let μ_1 = average number of ant species found in the Dry Steppe and μ_2 = average number of ant species found in the Gobi Desert.

$$s_p^2 = \frac{(n_1-1)s_1^2 + (n_2-1)s_2^2}{n_1+n_2-2} = \frac{(5-1)21.31^2 + (6-1)18.21^2}{5+6-2} = 386.0539$$

To determine if there is a difference in the average number of ant species found at sites in the two regions, we test:

H_0: $\mu_1 - \mu_2 = 0$
H_a: $\mu_1 - \mu_2 \neq 0$

The test statistic is $t = \dfrac{(\bar{x}_1 - \bar{x}_2) - D_o}{\sqrt{s_p^2\left(\frac{1}{n_1}+\frac{1}{n_2}\right)}} = \dfrac{(14-11.83)-0}{\sqrt{386.0539\left(\frac{1}{5}+\frac{1}{6}\right)}} = 0.18$

The rejection region requires $\alpha/2 = .05/2 = .025$ in each tail of the t-distribution with df $= n_1 + n_2 - 2 = 5 + 6 - 2 = 9$. From Table VI, Appendix B, $t_{.025} = 2.262$. The rejection region is $t < -2.262$ or $t > 2.262$.

Since the observed value of the test statistic does not fall in the rejection region ($t = 0.18 \ngtr 2.262$), H_0 is not rejected. There is insufficient evidence to indicate a difference in the average number of ant species found at sites in the two regions at $\alpha = .05$.

9.27   a.   Let μ_1 = mean MFS score for the violent event group and μ_2 = mean MFS score for the avoided-violent event group. The target parameter of interest is $\mu_1 - \mu_2$ = difference in the mean MFS score between the 2 groups.

b. We are given the sample mean scores for the two groups. However, in order to determine if there is a difference in the 2 population means, we also need to know the population standard deviations for the 2 groups or the variances for the 2 groups. This information was not provided.

c. The *p*-value of the test is

$$p = P(z \leq -1.21) + P(z \geq 1.21) = (.5 - .3869) + (.5 - .3869) = .1131 + .1131 = .2262$$

d. Since the observed p-value is greater than α ($p = .2262 > \alpha = .10$), H_0 is not rejected. There is insufficient evidence to indicate a difference in the mean MFS score between the violent event group and the avoided-violent event group at $\alpha = .10$.

9.29   a. From the information given, we have no idea what the standard deviations are. Whether the population means are different or not depends on how variable the data are.

   b. Let μ_1 = mean pain intensity rating for blacks and μ_2 = mean pain intensity rating for whites.

   To determine if blacks, on average, have a higher pain intensity rating than whites, we test:

   H_0: $\mu_1 - \mu_2 = 0$
   H_a: $\mu_1 - \mu_2 > 0$

   The rejection region requires $\alpha = .05$ in the upper tail of the *z* distribution. From Table IV, Appendix A, $z_{.05} = 1.645$. Thus, in order to reject H_0, the test statistic would have to be greater than 1.645.

   Substituting known values into the test statistic, we can solve for s_p.

   $$z = \frac{\bar{x}_1 - \bar{x}_2}{\sqrt{s_p^2 \left(\frac{1}{n_1} + \frac{1}{n_2}\right)}} > 1.645 \quad \Rightarrow \quad \frac{8.2 - 6.9}{\sqrt{s_p^2 \left(\frac{1}{55} + \frac{1}{159}\right)}} > 1.645$$

   $$\Rightarrow 1.3 > 1.645(s_p)\sqrt{\left(\frac{1}{55} + \frac{1}{159}\right)} \quad \Rightarrow \quad \frac{1.3}{1.645(.1564)} > s_p \quad \Rightarrow 5.053 > s_p$$

   Thus, if both s_1 and s_2 were less than 5.053, then we would conclude that blacks, on average would have a higher pain intensity than whites. (It is possible that either s_1 or s_2 could be greater than 5.053 and we would still reject H_0. As long as $s_p < 5.053$, we would reject H_0.)

   c. From part **b**, we know we would reject H_0 if $s_p < 5.053$. Thus, we would not reject H_0 if $s_p \geq 5.053$. This could happen if both s_1 and s_2 are greater than 5.053. (It is possible that either s_1 or s_2 could be less than 5.053 and we would still reject H_0. As long as $s_p \geq 5.053$, we would reject H_0.)

9.31   In a paired difference experiment, the observations should be paired before collecting the data.

9.33   a.   The rejection region requires $\alpha = .05$ in the upper tail of the t distribution with df $= n_d - 1 = 10 - 1 = 9$. From Table VI, Appendix A, $t_{.05} = 1.833$. The rejection region is $t > 1.833$.

b.   From Table VI, with df $= n_d - 1 = 20 - 1 = 19$, $t_{.10} = 1.328$. The rejection region is $t > 1.328$.

c.   From Table VI, with df $= n_d - 1 = 5 - 1 = 4$, $t_{.025} = 2.776$. The rejection region is $t > 2.776$.

d.   From Table VI, with df $= n_d - 1 = 9 - 1 = 8$, $t_{.01} = 2.896$. The rejection region is $t > 2.896$.

9.35   a.

Pair	Difference
1	3
2	2
3	2
4	4
5	0
6	1

$$\overline{x}_d = \frac{\sum x_d}{n_d} = \frac{12}{6} = 2 \qquad s_d^2 = \frac{\sum x_d^2 - \dfrac{\left(\sum x_d\right)^2}{n_d}}{n_d - 1} = \frac{\left(34 - \dfrac{(12)^2}{6}\right)}{5} = 2$$

b. $\mu_d = \mu_1 - \mu_2$

c.   For confidence coefficient .95, $\alpha = .05$ and $\alpha / 2 = .025$. From Table VI, Appendix A, with df $= n_d - 1 = 6 - 1 = 5$, $t_{.025} = 2.571$. The confidence interval is:

$$\overline{x}_d \pm t_{\alpha/2} \frac{s_d}{\sqrt{n_d}} \Rightarrow 2 \pm 2.571 \frac{\sqrt{2}}{\sqrt{6}} \Rightarrow 2 \pm 1.484 \Rightarrow (.516, 3.484)$$

d. H_0: $\mu_d = 0$
H_a: $\mu_d \neq 0$

The test statistic is $t = \dfrac{\overline{x}_d - 0}{s_d / \sqrt{n_d}} = \dfrac{2}{\sqrt{2} / \sqrt{6}} = 3.46$

The rejection region requires $\alpha / 2 = .05 / 2 = .025$ in each tail of the t distribution with df $= n_d - 1 = 6 - 1 = 5$. From Table VI, Appendix A, $t_{.025} = 2.571$. The rejection region is $t < -2.571$ or $t > 2.571$.

Since the observed value of the test statistic falls in the rejection region ($t = 3.46 > 2.571$), H_0 is rejected. There is sufficient evidence to indicate that the mean difference is different from 0 at $\alpha = .05$.

9.37   a. H_0: $\mu_d = 10$

H_a: $\mu_d \neq 10$

The test statistic is $z = \dfrac{\bar{x}_d - 10}{s_d / \sqrt{n_d}} = \dfrac{11.7 - 10}{6 / \sqrt{40}} = 1.79$

The rejection region requires $\alpha / 2 = .05 / 2 = .025$ in each tail of the z distribution. From Table IV, Appendix A, $z_{.025} = 1.96$. The rejection region is $z < -1.96$ or $z > 1.96$.

Since the observed value of the test statistic does not fall in the rejection region ($z = 1.79 \not> 1.96$), H_0 is not rejected. There is insufficient evidence to indicate the difference in the population means is different from 10 at $\alpha = .05$.

.   b.   The *p*-value is

$p = P(z \leq -1.79) + P(z \geq 1.79) = (.5 - .4633) + (.5 - .4633) = .0367 + .0367 = .0734$

9.39   a.   The health status is measured on all patients two times, both before and after handling museum objects. Thus, the two sets of measurements are not independent. The data need to be analyzed as paired differences.

     b.   The differences between the "before" and "after" measurements are:

Session	Difference	Session	Difference
1	-7	17	0
2	-12	18	-11
3	-9	19	-11
4	-9	20	-10
5	1	21	-12
6	-13	22	-7
7	-16	23	-4
8	-3	24	-2
9	-7	25	-8
10	-2	26	-1
11	-6	27	4
12	-10	28	-17
13	-9	29	-11
14	-9	30	-12
15	1	31	-5
16	-14	32	-13

c.   Using MINITAB, the descriptive statistics are:

Descriptive Statistics: Before, After, Diff

Variable	N	Mean	StDev	Minimum	Q1	Median	Q3	Maximum
Before	32	51.66	10.09	30.00	43.00	51.00	59.00	73.00
After	32	59.28	9.60	42.00	52.25	59.00	64.75	83.00
Diff	32	-7.625	5.272	-17.000	-11.750	-9.000	-3.250	4.000

The mean of the differences is $\mu_d = -7.625$ and the standard deviation of the differences is $s_d = 5.272$.

d.   For confidence coefficient .90, $\alpha = .10$ and $\alpha / 2 = .05$. From Table IV, Appendix A, $z_{.05} = 1.645$. The confidence interval is:

$$\bar{x}_d \pm z_{.05} \frac{s_d}{\sqrt{n_d}} \Rightarrow -7.625 \pm 1.645 \frac{5.272}{\sqrt{32}} \Rightarrow -7.625 \pm 1.533 \Rightarrow (-9.158, -6.092)$$

e.   We are 90% confident that the true difference in mean health status before and after handling museum objects is between -9.158 and -6.092. Since the interval is entirely negative, there is evidence to indicate the mean after handling the museum objects is higher than the mean before handling the museum objects. Thus, handling museum objects has a positive impact on a sick patient's well-being.

9.41   a.   Let μ_1 = mean BOLD score for the first session and μ_2 = mean BOLD score when the placebo was applied. The target parameter is $\mu_d = \mu_1 - \mu_2$, the difference in the mean BOLD scores between the first session and the session involving the placebo.

b.   A paired difference design was used to collect the data. Each experimental person had two observations measured on it.

c.   To determine if there was a placebo effect, we test:

H_0: $\mu_d = 0$
H_a: $\mu_d > 0$

d.   The test statistic is $t = \dfrac{\bar{x}_d - 0}{s_d / \sqrt{n_d}} = \dfrac{0.21 - 0}{.47 / \sqrt{24}} = 2.19$

Since no α value was given, we will use $\alpha = .05$. The rejection region requires $\alpha = .05$ in the upper tail of the t distribution with df = $n_d - 1 = 24 - 1 = 23$. From Table VI, Appendix A, $t_{.05} = 1.714$. The rejection region is $t > 1.714$.

Since the observed value of the test statistic falls in the rejection region $(t = 2.19 > 1.714)$, H_0 is rejected. There is sufficient evidence to indicate that there is a placebo effect at $\alpha = .05$.

e. Since the *p*-value is less than α ($p = .02 < \alpha = .05$), H_0 is rejected. There is sufficient evidence to indicate that there is a placebo effect at $\alpha = .05$.

9.43 Using MINITAB, the descriptive statistics are:

Descriptive Statistics: Initial, Final, Diff

Variable	N	Mean	StDev	Minimum	Q1	Median	Q3	Maximum
Initial	3	5.640	1.075	4.560	4.560	5.650	6.710	6.710
Final	3	5.453	1.125	4.270	4.270	5.580	6.510	6.510
Diff	3	0.1867	0.1106	0.0700	0.0700	0.2000	0.2900	0.2900

Let μ_1 = mean initial pH and μ_2 = mean final pH. The parameter of interest is $\mu_d = \mu_1 - \mu_2$. To determine if the mean initial pH differs significantly from the mean pH after 30 days, we test:

$$H_0: \ \mu_d = 0$$
$$H_a: \ \mu_d \neq 0$$

The test statistics is $t = \dfrac{\bar{x}_d - 0}{\dfrac{s_d}{\sqrt{n_d}}} = \dfrac{.1867}{\dfrac{.1106}{\sqrt{3}}} = 2.92$

The rejection region requires $\alpha / 2 = .05 / 2 = .025$ in each tail of the t distribution. From Table VI, Appendix A, with $df = n_d - 1 = 3 - 1 = 2$, $t_{.025} = 4.303$. The rejection region is $t < -4.303$ or $t > 4.303$.

Since the observed value of the test statistic does not fall in the rejection region ($t = 2.92 \not> 4.303$), H_0 is not rejected. There is insufficient evidence to indicate the mean initial pH differs significantly from the mean pH after 30 days at $\alpha = .05$.

9.45 a. Since all the data were collected from twins, the observations from each pair are not independent of each other. Thus, the data were collected as paired data. If the data are collected as paired data, they must be analyzed as paired data.

b. Let μ_1 = mean level of overall leisure activity for the control group and μ_2 = mean level of overall leisure activity for the demented group. The target parameter is $\mu_d = \mu_1 - \mu_2$, the difference in the mean level of overall leisure activity for the control group and the demented group.

Using MINITAB, the descriptive statistics are:

Descriptive Statistics: Diff

Variable	N	Mean	StDev	Minimum	Q1	Median	Q3	Maximum
Diff	107	1.953	10.082	-28.000	-4.000	2.000	9.000	29.000

We will use a 95% confidence interval to estimate the difference in the mean level of leisure activity between the control group and the demented group. For confidence coefficient .95, $\alpha = .05$ and $\alpha / 2 = .05 / 2 = .025$. From Table IV, appendix A, $z_{.025} = 1.96$.

The 95% confidence interval is:

$$\bar{x}_d \pm z_{.025}\frac{s_d}{\sqrt{n_d}} \Rightarrow 1.953 \pm 1.96\frac{10.082}{\sqrt{107}} \Rightarrow 1.953 \pm 1.910 \Rightarrow (0.043,\ 3.863)$$

We are 95% confident that the difference in the mean level of leisure activity between the control group and the demented group is between 0.043 and 3.863. Since 0 is not contained in this interval, there is evidence of a difference in the mean level of leisure activity between the control group and the demented group at $\alpha = .05$. Since all the values are positive, there is evidence that the mean level of leisure activity for the control group is larger than that for the demented group.

9.47   Using MINITAB, the descriptive statistics are:

Descriptive Statistics: Before, After, Diff

```
Variable    N    Mean   StDev      Minimum       Q1   Median       Q3   Maximum
Before     13   2.513   1.976        0.270    0.805    2.400    3.405     7.350
After      13   1.506   1.448  0.000000000    0.260    1.360    2.380     4.920
Diff       13   1.007   1.209       -0.850    0.265    0.560    2.335     2.780
```

Let μ_1 = mean number of crashes before the camera and μ_2 = mean number of crashes after the camera. The parameter of interest is $\mu_d = \mu_1 - \mu_2$. To determine if the mean number of crashes after the camera is smaller than before the camera, we test:

H_0: $\mu_d = 0$

H_a: $\mu_d > 0$

The test statistics is $t = \dfrac{\bar{x}_d - 0}{\dfrac{s_d}{\sqrt{n_d}}} = \dfrac{1.007}{\dfrac{1.209}{\sqrt{13}}} = 3.00$

Since no α is given, we will use $\alpha = .05$. The rejection region requires $\alpha = .05$ in the upper tail of the t distribution. From Table VI, Appendix A, with $df = n_d - 1 = 13 - 1 = 12$, $t_{.05} = 1.782$. The rejection region is $t > 1.782$.

Since the observed value of the test statistic falls in the rejection region ($t = 3.00 > 1.782$), H_0 is rejected. There is sufficient evidence to indicate the mean number of crashes after the camera is smaller than before the camera at $\alpha = .05$.

9.49   Some preliminary calculations are:

Patient	Time 1	Time 2	Difference
1	5	5	0
2	1	3	−2
3	0	0	0
4	1	1	0
5	0	1	−1
6	2	1	1
7	5	6	−1
8	1	2	−1
9	0	9	−9
10	5	8	−3

Patient	Time 1	Time 2	Difference
11	7	10	−3
12	0	3	−3
13	3	9	−6
14	5	8	−3
15	7	12	−5
16	10	16	−6
17	5	5	0
18	6	3	3
19	9	6	3
20	11	8	3

$$\bar{x}_d = \frac{\sum x_d}{n_d} = \frac{-33}{20} = -1.65$$

$$s_d^2 = \frac{\sum x_d^2 - \frac{\left(\sum x_d\right)^2}{n_d}}{n_d - 1} = \frac{249 - \frac{(-33)^2}{20}}{20 - 1} = \frac{194.55}{19} = 10.23947 \qquad s_d = \sqrt{10.23947} = 3.1999$$

Let μ_1 = mean homophone confusion errors for time 1 and μ_2 = mean homophone confusion errors for time 2. Then $\mu_d = \mu_1 - \mu_2$.

To determine if Alzheimer's patients show a significant increase in mean homophone confusion errors over time, we test:

H_0: $\mu_d = 0$

H_a: $\mu_d < 0$

The test statistic is $t = \dfrac{\bar{x}_d - 0}{s_d / \sqrt{n_d}} = \dfrac{-1.65 - 0}{3.1999 / \sqrt{20}} = -2.306$

Since no α was given we will use $\alpha = .05$. The rejection region requires $\alpha = .05$ in the lower tail of the t distribution with df $= n_d = 20 - 1 = 19$. From Table VI, Appendix A, $t_{.05} = 1.729$. The rejection region is $t < -1.729$.

Since the observed value of the test statistic falls in the rejection region ($t = -2.306 < -1.729$), H_0 is rejected. There is sufficient evidence to indicate Alzheimer's patients show a significant increase in mean homophone confusion errors over time at $\alpha = .05$.

We must assume that the population of differences is normally distributed and that the sample was randomly selected. A stem-and-leaf display of the data indicates that the data are mound-shaped. It appears that these assumptions are valid.

9.51   The conditions required for a large-sample inference about $(p_1 - p_2)$ are:

      1.   The two samples are randomly selected in an independent manner from the two target populations.

      2.   The sample sizes, n_1 and n_2, are both large so that the sampling distribution of $(\hat{p}_1 - \hat{p}_2)$ will be approximately normal. (This condition will be satisfied if all of the following are true: $n_1\hat{p}_1 \geq 15$, $n_1\hat{q}_1 \geq 15$ and $n_2\hat{p}_2 \geq 15$, $n_2\hat{q}_2 \geq 15$.)

9.53   a.   The distribution of x_1 is binomial with n_1 trials and the probability of success, p_1. The distribution of x_2 is binomial with n_2 trials and the probability of success, p_2.

      b.   For large samples, the sampling distribution of $(\hat{p}_1 - \hat{p}_2)$ is approximately normal with mean $(p_1 - p_2)$, and standard deviation $\sigma_{\hat{p}_1 - \hat{p}_2} = \sqrt{\dfrac{p_1 q_1}{n_1} + \dfrac{p_2 q_2}{n_2}}$.

9.55   For confidence coefficient .95, $\alpha = 1 - .95 = .05$ and $\alpha/2 = .05/2 = .025$. From Table IV, Appendix A, $z_{.025} = 1.96$. The 95% confidence interval for $p_1 - p_2$ is approximately:

    a. $(\hat{p}_1 - \hat{p}_2) \pm z_{\alpha/2}\sqrt{\dfrac{\hat{p}_1\hat{q}_1}{n_1} + \dfrac{\hat{p}_2\hat{q}_2}{n_2}}$

$$\Rightarrow (.65 - .58) \pm 1.96\sqrt{\dfrac{.65(1 - .65)}{400} + \dfrac{.58(1 - .58)}{400}} \Rightarrow .07 \pm .067 \Rightarrow (.003, .137)$$

    b. $(\hat{p}_1 - \hat{p}_2) \pm z_{\alpha/2}\sqrt{\dfrac{\hat{p}_1\hat{q}_1}{n_1} + \dfrac{\hat{p}_2\hat{q}_2}{n_2}}$

$$\Rightarrow (.31 - .25) \pm 1.96\sqrt{\dfrac{.31(1 - .31)}{180} + \dfrac{.25(1 - .25)}{250}} \Rightarrow .06 \pm .086 \Rightarrow (-.026, .146)$$

    c. $(\hat{p}_1 - \hat{p}_2) \pm z_{\alpha/2}\sqrt{\dfrac{\hat{p}_1\hat{q}_1}{n_1} + \dfrac{\hat{p}_2\hat{q}_2}{n_2}}$

$$\Rightarrow (.46 - .61) \pm 1.96\sqrt{\dfrac{.46(1 - .46)}{100} + \dfrac{.61(1 - .61)}{120}} \Rightarrow -.15 \pm .131 \Rightarrow (-.281, -.019)$$

9.57 H_0: $(p_1 - p_2) = .1$
        H_a: $(p_1 - p_2) > .1$

    Since D_0 is not equal to 0, the test statistic is:

$$z = \dfrac{(\hat{p}_1 - \hat{p}_2) - D_0}{\sqrt{\dfrac{p_1 q_1}{n_1} + \dfrac{p_2 q_2}{n_2}}} = \dfrac{(\hat{p}_1 - \hat{p}_2) - D_0}{\sqrt{\dfrac{\hat{p}_1\hat{q}_1}{n_1} + \dfrac{\hat{p}_2\hat{q}_2}{n_2}}} = \dfrac{(.4 - .2) - .1}{\sqrt{\dfrac{.4(1 - .4)}{50} + \dfrac{.2(1 - .2)}{60}}} = \dfrac{1}{.0864} = 1.16$$

The rejection region requires $\alpha = .05$ in the upper tail of the z distribution. From Table IV, Appendix A, $z_{.05} = 1.645$. The rejection region is $z > 1.645$.

Since the observed value of the test statistic does not fall in the rejection region ($z = 1.16 \not> 1.645$), H_0 is not rejected. There is insufficient evidence to show $(p_1 - p_2) > .1$ at $\alpha = .05$.

9.59 a. $\hat{p}_1 = \dfrac{x_1}{n_1} = \dfrac{746}{1358} = .549$

b. $\hat{p}_2 = \dfrac{x_2}{n_2} = \dfrac{967}{1379} = .701$

c. For confidence coefficient .90, $\alpha = .10$ and $\alpha / 2 = .10 / 2 = .05$. From Table IV, Appendix A, $z_{.05} = 1.645$. The 90% confidence interval is:

$$(\hat{p}_1 - \hat{p}_2) \pm z_{.05} \sqrt{\frac{\hat{p}_1 \hat{q}_1}{n_1} + \frac{\hat{p}_2 \hat{q}_2}{n_2}} \Rightarrow (.549 - .701) \pm 1.645 \sqrt{\frac{.549(.451)}{1358} + \frac{.701(.299)}{1379}}$$

$$\Rightarrow -.152 \pm .030 \Rightarrow (-.182, \ -.122)$$

d. We are 90% confident that the true value of the difference in the proportion of Dutch boys and Dutch girls who have never bullied another student is between $-.182$ and $-.122$.

e. Since the 90% confidence interval contains only negative numbers, the proportion of girls who do not bully is larger than that for boys. Thus, the proportion of Dutch boys who bully is significantly greater than the proportion of Dutch girls who bully.

9.61 a. The observed proportion of St. John's wort patients who were in remission is:

$$\hat{p}_1 = \frac{x_1}{n_1} = \frac{14}{98} = .143$$

b. The observed proportion of placebo patients who were in remission is:

$$\hat{p}_2 = \frac{x_2}{n_2} = \frac{5}{102} = .049$$

c. Some preliminary calculations are:

$$\hat{p} = \frac{x_1 + x_2}{n_1 + n_2} = \frac{14 + 5}{98 + 102} = \frac{19}{200} = .095 \qquad \hat{q} = 1 - \hat{p} = 1 - .095 = .905$$

To determine if the proportion of St. John's wort patients in remission exceeds the proportion of placebo patients in remission, we test:

H_0: $p_1 - p_2 = 0$
H_a: $p_1 - p_2 > 0$

The test statistic is $z = \dfrac{\hat{p}_1 - \hat{p}_2}{\sqrt{\hat{p}\hat{q}\left(\dfrac{1}{n_1} + \dfrac{1}{n_2}\right)}} = \dfrac{.143 - .049}{\sqrt{.095(.905)\left(\dfrac{1}{98} + \dfrac{1}{102}\right)}} = 2.27$

The rejection region requires $\alpha = .01$ in the upper tail of the z distribution. From Table IV, Appendix A, $z_{.01} = 2.33$. The rejection region is $z > 2.33$.

Since the observed value of the test statistic does not fall in the rejection region ($z = 2.27 \not> 2.33$), H_0 is not rejected. There is insufficient evidence to indicate the proportion of St. John's wort patients in remission exceeds the proportion of placebo patients in remission at $\alpha = .01$.

d. The hypotheses and test statistic are the same as in part **c**.

The rejection region requires $\alpha = .10$ in the upper tail of the z distribution. From Table IV, Appendix A, $z_{.10} = 1.28$. The rejection region is $z > 1.28$.

Since the observed value of the test statistic falls in the rejection region ($z = 2.27 > 1.28$), H_0 is rejected. There is sufficient evidence to indicate the proportion of St. John's wort patients in remission exceeds the proportion of placebo patients in remission $\alpha = .10$.

e. By changing the value of α in this problem, the conclusion changes. When α is small ($\alpha = .01$), we did not reject H_0. When α increases ($\alpha = .10$), we rejected H_0. As α increases, the chance of making a Type I error increases, but the chance of a Type II error decreases.

9.63   a. Let p_1 = proportion of African-American drivers who were searched and p_2 = proportion of White drivers who were searched. The parameter of interest is $p_1 - p_2$, or the difference in the proportions of drivers searched between the two ethnic groups.

Some preliminary calculations are:

$\hat{p}_1 = \dfrac{x_1}{n_1} = \dfrac{12,016}{61,688} = .195$ $\qquad$ $\hat{p}_2 = \dfrac{x_2}{n_2} = \dfrac{5,312}{106,892} = .050$

$\hat{p} = \dfrac{x_1 + x_2}{n_1 + n_2} = \dfrac{12,016 + 5,312}{61,688 + 106,892} = .103$

To determine if there is a difference in the proportions of African-American and White drivers who are searched by the LA police, we test:

H_0: $p_1 - p_2 = 0$
H_a: $p_1 - p_2 \neq 0$

The test statistic is $z = \dfrac{(\hat{p}_1 - \hat{p}_2) - 0}{\sqrt{\hat{p}\hat{q}\left(\dfrac{1}{n_1} + \dfrac{1}{n_2}\right)}} = \dfrac{(.195 - .050) - 0}{\sqrt{.103(.897)\left(\dfrac{1}{61,688} + \dfrac{1}{106,892}\right)}} = 94.35$

The rejection region requires $\alpha/2 = .05/2 = .025$ in each tail of the z distribution. From Table IV, Appendix A, $z_{.025} = 1.96$. The rejection region is $z < -1.96$ or $z > 1.96$.

Since the observed value of the test statistic falls in the rejection region ($z = 94.35 > 1.96$), H_0 is rejected. There is sufficient evidence to indicate a difference in the proportions of African-American and White drivers who are searched by the LA police at $\alpha = .05$.

b. Let $p_1 =$ "hit rate" of African-American drivers and $p_2 =$ "hit rate" of White drivers. The parameter of interest is $p_1 - p_2$, or the difference in the "hit rates" for the two ethnic groups.

Some preliminary calculations are:

$$\hat{p}_1 = \frac{x_1}{n_1} = \frac{5,134}{12,016} = .427 \qquad\qquad \hat{p}_2 = \frac{x_2}{n_2} = \frac{3,006}{5,312} = .566$$

For confidence coefficient .95, $\alpha = .05$ and $\alpha/2 = .05/2 = .025$. From Table IV, Appendix A, $z_{.025} = 1.96$. The 95% confidence interval is:

$$(.427 - .566) \pm 1.96\sqrt{\frac{.427(.573)}{12,016} + \frac{.566(.434)}{5,312}} \Rightarrow -.139 \pm .016 \Rightarrow (-.155, \ -.123)$$

We are 95% confident that the difference in "hit rates" between African-American and White drivers searched by the LA police is between $-.155$ and $-.123$. Since both end points of the interval are negative, there is evidence that the "hit rate" for African-American drives is less than that for White drivers.

9.65 Let $p_1 =$ proportion of rice weevils found dead after 4 days of exposure to nitrogen gas and $p_2 =$ proportion of rice weevils found dead after 3.5 days of exposure to nitrogen gas. The parameter of interest is $p_1 - p_2$, or the difference in the proportions of rice weevils found dead between the 2 different exposures to nitrogen gas.

Some preliminary calculations are:

$$\hat{p}_1 = \frac{x_1}{n_1} = \frac{31,386}{31,421} = .999 \qquad\qquad \hat{p}_2 = \frac{x_2}{n_2} = \frac{23,516}{23,676} = .993$$

$$\hat{p} = \frac{x_1 + x_2}{n_1 + n_2} = \frac{31,386 + 23,516}{31,421 + 23,676} = .996$$

To determine if there is a difference in the proportions of rice weevils found dead between the 2 exposure times, we test:

H_0: $p_1 - p_2 = 0$
H_a: $p_1 - p_2 \neq 0$

The test statistic is $z = \dfrac{(\hat{p}_1 - \hat{p}_2) - 0}{\sqrt{\hat{p}\hat{q}\left(\frac{1}{n_1} + \frac{1}{n_2}\right)}} = \dfrac{(.999 - .993) - 0}{\sqrt{.996(.004)\left(\frac{1}{31,421} + \frac{1}{23,676}\right)}} = 11.05$

The rejection region requires $\alpha/2 = .10/2 = .05$ in each tail of the z distribution. From Table IV, Appendix A, $z_{.05} = 1.645$. The rejection region is $z < -1.645$ or $z > 1.645$.

Since the observed value of the test statistic falls in the rejection region ($z = 11.05 > 1.645$), H_0 is rejected. There is sufficient evidence to indicate a difference in the proportions of rice weevils found dead between the 2 exposure times at $\alpha = .10$.

9.67   Let p_1 = recall rate of the 60 to 72 year-old seniors and p_2 = recall rate of the 73 to 83 year-old seniors.

Some preliminary calculations are:

$$\hat{p}_1 = \frac{x_1}{n_1} = \frac{31}{40} = .775 \qquad \hat{p}_2 = \frac{x_2}{n_2} = \frac{22}{40} = .55 \qquad \hat{p} = \frac{x_1 + x_2}{n_1 + n_2} = \frac{31 + 22}{40 + 40} = .6625$$

To determine if the recall rates of the two groups differ, we test:

H_0: $p_1 - p_2 = 0$
H_a: $p_1 - p_2 \neq 0$

The test statistic is $z = \dfrac{(\hat{p}_1 - \hat{p}_2) - 0}{\sqrt{\hat{p}\hat{q}\left(\dfrac{1}{n_1} + \dfrac{1}{n_2}\right)}} = \dfrac{(.775 - .55) - 0}{\sqrt{.6625(.3375)\left(\dfrac{1}{40} + \dfrac{1}{40}\right)}} = 2.13$

The rejection region requires $\alpha/2 = .05/2 = .025$ in each tail of the z distribution. From Table IV, Appendix A, $z_{.025} = 1.96$. The rejection region is $z < -1.96$ or $z > 1.96$.

Since the observed value of the test statistic falls in the rejection region ($z = 2.13 > 1.96$), H_0 is rejected. There is sufficient evidence to indicate the recall rates of the two groups differ at $\alpha = .05$.

9.69   Let p_1 = proportion of volunteers who figured out the third rule in the group that slept and p_2 = proportion of volunteers who figured out the third rule in the group that stayed awake all night. The parameter of interest is $p_1 - p_2$, or the difference in the proportions of volunteers who figured out the third rule between the 2 groups.

Some preliminary calculations are:

$$\hat{p}_1 = \frac{x_1}{n_1} = \frac{39}{50} = .78 \qquad\qquad \hat{p}_2 = \frac{x_2}{n_2} = \frac{15}{50} = .30$$

For confidence coefficient .90, $\alpha = .10$ and $\alpha/2 = .10/2 = .05$. From Table IV, Appendix A, $z_{.05} = 1.645$. The 90% confidence interval is:

$$(.78 - .30) \pm 1.645\sqrt{\frac{.78(.22)}{50} + \frac{.30(.70)}{50}} \Rightarrow .48 \pm .144 \Rightarrow (.336, \ .624)$$

We are 90% confident that the difference in the proportions of volunteers who figured out the third rule between the 2 groups is between .336 and .624. Since both end points are greater than 0, there is evidence that the proportion of those who slept who figured out the third rule is greater than the proportion of those who stayed awake all night who figured out the third rule.

9.71   Let p_1 = proportion of Yayoi farmers with LEH defects, p_2 = proportion of Eastern Jomon foragers with LEH defects, and p_3 = proportion of Western Jomon foragers with LEH defects.

First, we will test Theory 1.

Some preliminary calculations are:

$$\hat{p} = \frac{n_1\hat{p}_1 + n_2\hat{p}_2}{n_1 + n_2} = \frac{182(.631) + 164(.482)}{182 + 164} = .560$$

To determine if foragers with a broad-based economy (Eastern Jomon) have a lower prevalence of LEH defects than early agriculturists (Yayoi), we test:

H_0: $p_1 - p_2 = 0$
H_a: $p_1 - p_2 > 0$

The test statistic is $z = \dfrac{\hat{p}_1 - \hat{p}_2}{\sqrt{\hat{p}\hat{q}\left(\frac{1}{n_1} + \frac{1}{n_2}\right)}} = \dfrac{.631 - .482}{\sqrt{.560(.440)\left(\frac{1}{182} + \frac{1}{164}\right)}} = 2.79$

The rejection region requires $\alpha = .01$ in the upper tail of the z-distribution. From Table IV, Appendix A, $z_{.01} = 2.33$. The rejection region is $z > 2.33$.

Since the observed value of the test statistic falls in the rejection region ($z = 2.79 > 2.33$), H_0 is rejected. There is sufficient evidence to indicate that foragers with a broad-based economy (Eastern Jomon) have a lower prevalence of LEH defects than early agriculturists (Yayoi) at $\alpha = .01$.

Next, we test Theory 2.

Some preliminary calculations are:

$$\hat{p} = \frac{n_1\hat{p}_1 + n_3\hat{p}_3}{n_1 + n_3} = \frac{182(.631) + 122(.648)}{182 + 122} = .638$$

To determine if foragers with a wet rice economy (Western Jomon) have a different prevalence of LEH defects than early agriculturists (Yayoi), we test:

H_0: $p_1 - p_2 = 0$
H_a: $p_1 - p_2 \neq 0$

The test statistic is $z = \dfrac{\hat{p}_1 - \hat{p}_2}{\sqrt{\hat{p}\hat{q}\left(\dfrac{1}{n_1} + \dfrac{1}{n_2}\right)}} = \dfrac{.631 - .648}{\sqrt{.638(.362)\left(\dfrac{1}{182} + \dfrac{1}{122}\right)}} = -.30$

The rejection region requires $\alpha = .01/2 = .005$ in each tail of the z-distribution. From Table IV, Appendix A, $z_{.005} = 2.58$. The rejection region is $z > 2.58$ or $z < -2.58$.

Since the observed value of the test statistic does not fall in the rejection region ($z = -.30 \not< -2.58$), H_0 is not rejected. There is insufficient evidence to indicate that foragers with a wet rice economy (Western Jomon) have a different prevalence of LEH defects than early agriculturists (Yayoi) at $\alpha = .01$.

9.73   We can estimate the values of p_1 and p_2 based on prior samples, by using an educated guess of the values, or most conservatively, estimating p_1 and p_2 with .5.

9.75 $n_1 = n_2 = \dfrac{\left(z_{\alpha/2}\right)^2 \left(\sigma_1^2 + \sigma_2^2\right)}{(SE)^2}$

For confidence coefficient .95, $\alpha = 1 - .95 = .05$ and $\alpha/2 = .05/2 = .025$. From Table IV, Appendix A, $z_{.025} = 1.96$.

$n_1 = n_2 = \dfrac{1.96^2(15+15)}{2.2^2} = 23.8 \approx 24$

9.77   a.   For confidence coefficient .99, $\alpha = 1 - .99 = .01$ and $\alpha/2 = .01/2 = .005$. From Table IV, Appendix A, $z_{.005} = 2.58$.

$n_1 = n_2 = \dfrac{\left(z_{\alpha/2}\right)^2 (p_1q_1 + p_2q_2)}{(SE)^2} = \dfrac{2.58^2\left(.4(1-.4) + .7(1-.7)\right)}{.01^2}$

$= \dfrac{2.99538}{.0001} = 29,953.8 \approx 29,954$

b.   For confidence coefficient .90, $\alpha = 1 - .90 = .10$ and $\alpha/2 = .10/2 = .05$. From Table IV, Appendix A, $z_{.05} = 1.645$. Since we have no prior information about the proportions, we use $p_1 = p_2 = .5$ to get a conservative estimate. For a width of .05, the standard error is .5.

$n_1 = n_2 = \dfrac{\left(z_{\alpha/2}\right)^2 (p_1q_1 + p_2q_2)}{(SE)^2} = \dfrac{(1.645)^2\left(.5(1-.5) + .5(1-.5)\right)}{.025^2} = 2,164.82 \approx 2,165$

c.   From part **b**, $z_{.05} = 1.645$.

$n_1 = n_2 = \dfrac{\left(z_{\alpha/2}\right)^2 (p_1q_1 + p_2q_2)}{(SE)^2} = \dfrac{1.645^2\left(.2(1-.2) + .3(1-.3)\right)}{.03^2}$

$= \dfrac{1.00123}{.0009} = 1,112.48 \approx 1,113$

9.79   For confidence coefficient .95, $\alpha = .05$ and $\alpha / 2 = .05 / 2 = .025$. From Table IV, Appendix A, $z_{.025} = 1.96$.

$$n_1 = n_2 = \frac{\left(z_{\alpha/2}\right)^2 \left(\sigma_B^2 + \sigma_N^2\right)}{(SE)} = \frac{1.96^2 (25 + 25)}{2^2} = 48.02 \approx 49$$

9.81   a.   For confidence coefficient .95, $\alpha = 1 - .95 = .05$ and $\alpha / 2 = .05 / 2 = .025$. From Table IV, Appendix A, $z_{.05} = 1.96$. The standard error is $SE = .015$. From Exercise 9.64, $\hat{p}_1 = .184$ and $\hat{p}_2 = .177$.

$$n_1 = n_2 = \frac{\left(z_{\alpha/2}\right)^2 (p_1 q_1 + p_2 q_2)}{(SE)^2} = \frac{1.96^2 (.184(.816) + .177(.823))}{.015^2} = 5050.7 \approx 5051$$

In order to estimate the difference in proportions using a 95% confidence interval to within .015, we would need to sample 5,051 observations from each population.

b.   It would be extremely difficult to obtain information on over 10,000 patients. In addition, the cost would be extremely high.

c.   A difference of .015 is very small. This difference would be a difference of only 15 out of every 1,000 patients. In practice, this difference would be almost meaningless.

9.83   For confidence coefficient .95, $\alpha = 1 - .95 = .05$ and $\alpha / 2 = .05 / 2 = .025$. From Table IV, Appendix A, $z_{.025} = 1.96$. The standard error is 5.

$$n_1 = n_2 = \frac{\left(z_{\alpha/2}\right)^2 \left(\sigma_1^2 + \sigma_2^2\right)}{(SE)^2} = \frac{(1.96)^2 (9^2 + 9^2)}{5^2} = 24.9 \approx 25$$

We would need to sample 25 subjects from each group.

9.85   For confidence coefficient .90, $\alpha = 1 - .90 = .10$ and $\alpha / 2 = .10 / 2 = .05$. From Table IV, Appendix A, $z_{.05} = 1.645$.

$$n_1 = n_2 = \frac{\left(z_{\alpha/2}\right)^2 \left(\sigma_1^2 + \sigma_2^2\right)}{(SE)^2} = \frac{1.645^2 \left(5^2 + 5^2\right)}{1^2} = 135.3 \Rightarrow 136$$

9.87   The conditions required for valid inferences about σ_1^2 / σ_2^2 are:

1.   The two sampled populations are normally distributed.
2.   The samples are randomly and independently selected from their respective populations.

9.89   The statement "H_0: $\sigma_1^2 = \sigma_2^2$ is equivalent to H_0: $\sigma_1^2 / \sigma_2^2 = 0$" is false.

The correct statement is: H_0: $\sigma_1^2 = \sigma_2^2$ is equivalent to H_0: $\sigma_1^2 / \sigma_2^2 = 1$.

9.91   a.   With $v_1 = 2$ and $v_2 = 30$, $P(F \geq 4.18) = .025$ (Table X, Appendix A)

b.   With $v_1 = 24$ and $v_2 = 10$, $P(F \geq 1.94) = .10$ (Table VIII, Appendix A)

Thus, $P(F < 1.94) = 1 - P(F \geq 1.94) = 1 - .10 = .90$

c.   With $v_1 = 9$ and $v_2 = 1$, $P(F \geq 6022) = .01$ (Table XI, Appendix A)

Thus, $P(F < 6022) = 1 - P(F \geq 6022) = 1 - .01 = .99$

d.   With $v_1 = 30$ and $v_2 = 30$, $P(F \geq 1.84) = .05$ (Table IX, Appendix A)

9.93   The test statistic for each of these is:

$$F = \frac{\text{Larger sample variance}}{\text{Smaller sample variance}}$$

a.   The rejection region requires $\alpha / 2 = .20 / 2 = .10$ in the upper tail of the F distribution. If $s_1^2 > s_2^2$, numerator df $= v_1 = 8$ and denominator df $= v_2 = 40$. From Table VIII, Appendix A, $F_{.10} = 1.83$. The rejection region is $F > 1.83$. If $s_1^2 < s_2^2$, numerator df $= v_1 = 40$ and denominator df $v_2 = 8$. From Table VIII, Appendix A, $F_{.10} = 2.36$. The rejection region is $F > 2.36$.

b.   The rejection region requires $\alpha / 2 = .10 / 2 = .05$ in the upper tail of the F distribution. If $s_1^2 > s_2^2$, numerator df $= v_1 = 8$ and denominator df $= v_2 = 40$. From Table IX, Appendix A, $F_{.05} = 2.18$. The rejection region is $F > 2.18$. If $s_1^2 < s_2^2$, numerator df $= v_1 = 40$ and denominator df $= v_2 = 8$. From Table IX, Appendix A, $F_{.05} = 3.04$. The rejection region is $F > 3.04$.

c.   The rejection region requires $\alpha / 2 = .05 / 2 = .025$ in the upper tail of the F distribution. If $s_1^2 > s_2^2$, numerator df $= v_1 = 8$ and denominator df $= v_2 = 40$. From Table X, Appendix A, $F_{.025} = 2.53$. The rejection region is $F > 2.53$. If $s_1^2 < s_2^2$, numerator df $= v_1 = 40$ and denominator df $= v_2 = 8$. From Table X, Appendix A, $F_{.025} = 3.84$. The rejection region is $F > 3.84$.

d.   The rejection region requires $\alpha / 2 = .02 / 2 = .01$ in the upper tail of the F distribution. If $s_1^2 > s_2^2$, numerator df $= v_1 = 8$ and denominator df $= v_2 = 40$. From Table XI, Appendix A, $F_{.01} = 2.99$. The rejection region is $F > 2.99$. If $s_1^2 < s_2^2$, numerator df $= v_1 = 40$ and denominator df $= v_2 = 8$. From Table XI, Appendix A, $F_{.01} = 5.12$. The rejection region is $F > 5.12$.

9.95   a. H_0: $\sigma_1^2 = \sigma_2^2$

H_a: $\sigma_1^2 \neq \sigma_2^2$

The test statistic is $F = \dfrac{\text{Larger sample variance}}{\text{Smaller sample variance}} = \dfrac{s_2^2}{s_1^2} = \dfrac{9.85}{2.87} = 3.43$

The rejection region requires $\alpha / 2 = .05 / 2 = .025$ in the upper tail of the F distribution with numerator df $v_1 = n_2 - 1 = 25 - 1 = 24$ and denominator df $v_2 = n_1 - 1 = 16 - 1 = 15$. From Table X, Appendix A, $F_{.025} = 2.70$. The rejection region is $F > 2.70$.

Since the observed value of the test statistic falls in the rejection region ($F = 3.43 > 2.70$), H_0 is rejected. There is sufficient evidence to indicate $\sigma_1^2 \neq \sigma_2^2$ at $\alpha = .05$.

b.   The *p*-value for the test is $p = 2P(F \geq 3.43)$. Using Table XI, Appendix A, with df $v_1 = n_2 - 1 = 25 - 1 = 24$ and df $v_2 = n_1 - 1 = 16 - 1 = 15$, $P(F > 3.29) = .01$. Thus, the *p*-value is $2P(F \geq 3.43) < 2(.01) = .02$ or $p < .02$.

9.97   a.   To determine if the variances of the two groups are different, we test:

H_0: $\sigma_1^2 = \sigma_2^2$

H_a: $\sigma_1^2 \neq \sigma_2^2$

b.   The test statistic is $F = \dfrac{\text{Larger sample variance}}{\text{Smaller sample variance}} = \dfrac{s_2^2}{s_1^2} = \dfrac{2.13^2}{1.98^2} = 1.157$

c.   The rejection region requires $\alpha / 2 = .10 / 2 = .05$ in the upper tail of the F distribution with numerator df $v_1 = n_2 - 1 = 20 - 1 = 19$ and denominator df $v_2 = n_1 - 1 = 20 - 1 = 19$. From Table IX, Appendix A, $F_{.05} \approx 2.16$. The rejection region is $F > 2.16$.

d.   Since the observed value of the test statistic does not fall in the rejection region ($F = 1.157 \not> 2.16$), H_0 is not rejected. There is insufficient evidence to indicate that the two variances are different at $\alpha = .10$. It appears that the assumption of equal variances is valid.

e.   It appears that the inference derived in Exercise 9.15 is valid.

9.99   To determine if the variances of the 2 groups are different, we test:

H_0: $\sigma_1^2 = \sigma_2^2$

H_a: $\sigma_1^2 \neq \sigma_2^2$

The test statistic is $F = \dfrac{\text{Larger sample variance}}{\text{Smaller sample variance}} = \dfrac{s_1^2}{s_2^2} = \dfrac{12.5^2}{12.2^2} = 1.050$

The rejection region requires $\alpha / 2 = .05 / 2 = .025$ in the upper tail of the F distribution with numerator df $v_1 = n_1 - 1 = 26 - 1 = 25$ and denominator df $v_2 = n_2 - 1 = 26 - 1 = 25$. From Table X, Appendix A, $F_{.025} \approx 2.24$. The rejection region is $F > 2.24$.

Since the observed value of the test statistic does not fall in the rejection region ($F = 1.050 \not> 2.24$), H_0 is not rejected. There is insufficient evidence to indicate a difference in the population variances between the two groups at $\alpha = .05$. Thus, the assumption appears to be valid.

9.101   Let σ_1^2 = variance of the time for the Trail Making Test for schizophrenics and σ_2^2 = variance of the time for the Trail Making Test for normal individuals.

To determine if schizophrenics have a wider range in time on the Trail Making Test than normal subjects, we test:

$$H_0: \ \sigma_1^2 = \sigma_2^2$$
$$H_a: \ \sigma_1^2 > \sigma_2^2$$

The test statistic is $F = \dfrac{\text{Larger sample variance}}{\text{Smaller sample variance}} = \dfrac{s_1^2}{s_2^2} = \dfrac{45.45^2}{16.34^2} = 7.737$

The rejection region requires $\alpha = .01$ in the upper tail of the F distribution with numerator df $v_1 = n_1 - 1 = 41 - 1 = 40$ and denominator df $v_2 = n_2 - 1 = 49 - 1 = 48$. From Table XI, Appendix A, $F_{.01} \approx 2.11$. The rejection region is $F > 2.11$.

Since the observed value of the test statistic falls in the rejection region ($F = 7.737 > 2.11$), H_0 is rejected. There is sufficient evidence to indicate that schizophrenics have a wider range in time on the Trail Making Test than normal subjects at $\alpha = .01$.

9.103   Using MINITAB, the descriptive statistics are:

Descriptive Statistics: Honey, DM

Variable	N	Mean	StDev	Variance	Minimum	Median	Maximum
Honey	35	10.714	2.855	8.151	4.000	11.000	16.000
DM	33	8.333	3.256	10.604	3.000	9.000	15.000

Let σ_1^2 = variance of the coughing improvement scores for DM dosage group and σ_2^2 = variance of the coughing improvement scores for Honey dosage group.

To determine if the variances of the coughing improvement scores differ for the two groups, we test:

$$H_0: \ \sigma_1^2 = \sigma_2^2$$
$$H_a: \ \sigma_1^2 \ne \sigma_2^2$$

The test statistic is $F = \dfrac{\text{Larger sample variance}}{\text{Smaller sample variance}} = \dfrac{s_1^2}{s_2^2} = \dfrac{10.604}{8.151} = 1.301$.

The rejection region requires $\alpha / 2 = .10 / 2 = .05$ in the upper tail of the F distribution with numerator $df\ v_1 = n_1 - 1 = 33 - 1 = 32$ and denominator $df\ v_2 = n_2 - 1 = 35 - 1 = 34$. From Table IX, Appendix A, $F_{.05} \approx 1.84$. The rejection region is $F > 1.84$.

Since the observed value of the test statistic does not fall in the rejection region ($F = 1.301 \ngtr$ 1.84), H_0 is not rejected. There is insufficient evidence to indicate that the variances of the coughing improvement scores differ for the two groups at $\alpha = .10$. Since there is no difference in the variances for the two groups, neither treatment is preferred over the other.

9.105   Some preliminary calculations:

$$s_1^2 = \frac{\sum x_1^2 - \dfrac{\left(\sum x_1\right)^2}{n_1}}{n_1 - 1} = \frac{6620.96 - \dfrac{227^2}{9}}{9 - 1} = 110.6778$$

$$s_2^2 = \frac{\sum x_2^2 - \dfrac{\left(\sum x_2\right)^2}{n_2}}{n_2 - 1} = \frac{596.68 - \dfrac{66.4^2}{9}}{9 - 1} = 13.3494$$

a.   Let σ_1^2 = variance of the order-to-delivery times for the Persian Gulf and σ_2^2 = variance of the order-to-delivery times for Bosnia. To determine if the variances of the order-to-delivery times for the Persian Gulf and Bosnia differ, we test:

H_0: $\sigma_1^2 = \sigma_2^2$

H_a: $\sigma_1^2 \neq \sigma_2^2$

The test statistic is $F = \dfrac{\text{Larger sample variance}}{\text{Smaller sample variance}} = \dfrac{s_1^2}{s_2^2} = \dfrac{110.6778}{13.3494} = 8.29$

The rejection region requires $\alpha / 2 = .05 / 2 = .025$ in the upper tail of the F distribution with numerator df $v_1 = n_1 - 1 = 9 - 1 = 8$ denominator df $v_2 = n_2 - 1 = 9 - 1 = 8$. From Table X, Appendix A, $F_{.025} = 4.03$. The rejection region is $F > 4.03$.

Since the observed value of the test statistic falls in the rejection region ($F = 8.29 >$ 4.03), H_0 is rejected. There is sufficient evidence to indicate the variances of the order-to-delivery times for Persian Gulf and Bosnia differ at $\alpha = .05$.

b.   No. Since both sample sizes are less than 30, we must use the two-sample t-test to test for differences in means. However, one of the assumptions for the two-sample t-test is that the variances of the two populations being sampled from are equal. Since we rejected H_0 in part a, there is evidence to indicate the variances of the two populations are not the same.

9.107 a. To compare average SAT scores of males and females, the target parameter is $\mu_1 - \mu_2$.

b. To compare the difference between mean waiting times at two supermarket checkout lines, the target parameter is $\mu_1 - \mu_2$.

c. To compare proportions of Democrats and Republicans who favor legalization of marijuana, the target parameter is $p_1 - p_2$.

d. To compare variation in salaries of NBA players picked in the 1st round and the 2nd round, the target parameter is σ_1^2 / σ_2^2.

e. To compare the difference in dropout rates of college student-athletes and regular students, the target parameter is $p_1 - p_2$.

9.109 a. $s_p^2 = \dfrac{(n_1-1)s_1^2 + (n_1-1)s_2^2}{n_1 + n_2 - 2} = \dfrac{11(74.2) + 13(60.5)}{12 + 14 - 2} = 66.7792$

H_0: $\mu_1 - \mu_2 = 0$
H_a: $\mu_1 - \mu_2 > 0$

The test statistic is $t = \dfrac{(\bar{x}_1 - \bar{x}_2) - 0}{\sqrt{s_p^2\left(\dfrac{1}{n_1} + \dfrac{1}{n_2}\right)}} = \dfrac{(17.8 - 15.3) - 0}{\sqrt{66.7792\left(\dfrac{1}{12} + \dfrac{1}{14}\right)}} = .78$

The rejection region requires $\alpha = .05$ in the upper tail of the t distribution with df $= n_1 + n_2 - 2 = 12 + 14 - 2 = 24$. From Table VI, Appendix A, for df $= 24$, $t_{.05} = 1.711$. The rejection region is $t > 1.711$.

Since the observed value of the test statistic does not fall in the rejection region ($t = 0.78 \ngtr 1.711$), H_0 is not rejected. There is insufficient evidence to indicate that $\mu_1 > \mu_2$ at $\alpha = .05$.

b. For confidence coefficient .99, $\alpha = .01$ and $\alpha/2 = .01/2 = .005$. From Table VI, Appendix A, with df $= n_1 + n_2 - 2 = 12 + 14 - 2 = 24$, $t_{.005} = 2.797$. The confidence interval is:

$$(\bar{x}_1 - \bar{x}_2) \pm t_{.005}\sqrt{s_p^2\left(\dfrac{1}{n_1} + \dfrac{1}{n_2}\right)} \Rightarrow (17.8 - 15.3) \pm 2.797\sqrt{66.7792\left(\dfrac{1}{12} + \dfrac{1}{14}\right)}$$
$$\Rightarrow 2.50 \pm 8.99 \Rightarrow (-6.49, 11.49)$$

c. For confidence coefficient .99, $\alpha = .01$ and $\alpha/2 = .01/2 = .005$. From Table IV, Appendix A, $z_{.005} = 2.58$.

$$n_1 = n_2 = \dfrac{(z_{\alpha/2})^2(\sigma_1^2 + \sigma_2^2)}{(SE)^2} = \dfrac{(2.58)^2(74.2 + 60.5)}{2^2} = 224.15 \approx 225$$

9.111   a.   For confidence coefficient .90, $\alpha = .10$ and $\alpha / 2 = .05$. From Table IV, Appendix A, $z_{.05} = 1.645$. The confidence interval is:

$$(\bar{x}_1 - \bar{x}_2) \pm z_{.05} \sqrt{\frac{s_1^2}{n_1} + \frac{s_2^2}{n_2}} \Rightarrow (12.2 - 8.3) \pm 1.645 \sqrt{\frac{2.1}{135} + \frac{3.0}{148}} \Rightarrow 3.90 \pm .31 \Rightarrow (3.59, 4.21)$$

   b.
$$H_0: \ \mu_1 - \mu_2 = 0$$
$$H_a: \ \mu_1 - \mu_2 \neq 0$$

The test statistic is $z = \dfrac{(\bar{x}_1 - \bar{x}_2) - 0}{\sqrt{\dfrac{s_1^2}{n_1} + \dfrac{s_2^2}{n_2}}} = \dfrac{(12.2 - 8.3) - 0}{\sqrt{\dfrac{2.1}{135} + \dfrac{3.0}{148}}} = 20.60$

The rejection region requires $\alpha / 2 = .01 / 2 = .005$ in each tail of the z distribution. From Table IV, Appendix A, $z_{.005} = 2.58$. The rejection region is $z < -2.58$ or $z > 2.58$.

Since the observed value of the test statistic falls in the rejection region ($z = 20.60 > 2.58$), H_0 is rejected. There is sufficient evidence to indicate that $\mu_1 \neq \mu_2$ at $\alpha = .01$.

   c.   For confidence coefficient .90, $\alpha = .10$ and $\alpha / 2 = .05$. From Table IV, Appendix A, $z_{.05} = 1.645$.

$$n_1 = n_2 = \frac{(z_{\alpha/2})^2 (\sigma_1^2 + \sigma_2^2)}{(SE)^2} = \frac{(1.645)^2 (2.1 + 3.0)}{.2^2} = 345.02 \approx 346$$

9.113   From the printout, the 95% confidence interval for the difference in mean seabird densities of oiled and unoiled transects is $(-2.93, 2.49)$. We are 95% confident that the true difference in mean seabird densities of oiled and unoiled transects is between -2.93 and 2.49. Since 0 is contained in the interval, there is no evidence to indicate that the mean seabird densities are different for the oiled and unoiled transects.

9.115   a.   The variable measured for this experiment is the time needed to match the picture with the sentence.

   b.   The experimental units are the climbers.

   c.   Since the same climbers were timed at the base camp and at a camp 5 miles above sea level, the data should be analyzed as a paired difference experiment.

9.117   a.   Let μ_1 = mean height of Australian boys who repeated a grade and μ_2 = mean height of Australian boys who never repeated a grade.

To determine if the average height of Australian boys who repeated a grade is less than the average height of boys who never repeated, we test:

$$H_0: \ \mu_1 - \mu_2 = 0$$
$$H_a: \ \mu_1 - \mu_2 < 0$$

b.   The test statistic is $z = \dfrac{\bar{x}_1 - \bar{x}_2}{\sqrt{\dfrac{s_1^2}{n_1} + \dfrac{s_2^2}{n_2}}} = \dfrac{-.04 - .30}{\sqrt{\dfrac{1.17^2}{86} + \dfrac{.97^2}{1346}}} = -2.64$

The rejection region requires $\alpha = .05$ in the lower tail of the z distribution. From Table IV, Appendix A, $z_{.05} = 1.645$. The rejection region is $z < -1.645$.

Since the observed value of the test statistic falls in the rejection region ($z = -2.64 < -1.645$), H_0 is rejected. There is sufficient evidence to indicate that the average height of Australian boys who repeated a grade is less than the average height of boys who never repeated at $\alpha = .05$.

c.   Let μ_1 = mean height of Australian girls who repeated a grade and μ_2 = mean height of Australian girls who never repeated a grade.

To determine if the average height of Australian girls who repeated a grade is less than the average height of girls who never repeated, we test:

H_0: $\mu_1 - \mu_2 = 0$
H_a: $\mu_1 - \mu_2 < 0$

The test statistic is $z = \dfrac{\bar{x}_1 - \bar{x}_2}{\sqrt{\dfrac{s_1^2}{n_1} + \dfrac{s_2^2}{n_2}}} = \dfrac{.26 - .22}{\sqrt{\dfrac{.94^2}{43} + \dfrac{1.04^2}{1366}}} = .27$

The rejection region requires $\alpha = .05$ in the lower tail of the z distribution. From Table IV, Appendix A, $z_{.05} = 1.645$. The rejection region is $z < -1.645$.

Since the observed value of the test statistic does not fall in the rejection region ($z = .27 \not< -1.645$), H_0 is not rejected. There is insufficient evidence to indicate that the average height of Australian girls who repeated a grade is less than the average height of girls who never repeated at $\alpha = .05$.

9.119 a. $\hat{p}_1 = \dfrac{x_1}{n_1} = \dfrac{29}{189} = .153$

b. $\hat{p}_2 = \dfrac{x_2}{n_2} = \dfrac{32}{149} = .215$

c.   For confidence coefficient .90, $\alpha = .10$ and $\alpha / 2 = .10 / 2 = .05$. From Table IV, Appendix A, $z_{.05} = 1.645$. The 90% confidence interval is:

$(\hat{p}_1 - \hat{p}_2) \pm z_{.05} \sqrt{\dfrac{\hat{p}_1 \hat{q}_1}{n_1} + \dfrac{\hat{p}_2 \hat{q}_2}{n_2}} \Rightarrow (.153 - .215) \pm 1.645 \sqrt{\dfrac{.153(.847)}{189} + \dfrac{.215(.785)}{149}}$

$\Rightarrow -.062 \pm .070 \Rightarrow (-.132, .008)$

d.   We are 90% confident that the difference in the true proportion of super-experienced bidders who fall prey to the winner's curse and the true proportion of less-experienced bidders who fall prey to the winner's curse is between −0.132 and 0.008. Since 0 is contained in the interval, there is no evidence to indicate that experience in bidding affects the likelihood of the winner's curse occurring.

9.121   For confidence coefficient .95, $\alpha = .05$ and $\alpha/2 = .05/2 = .025$. From Table VI, Appendix A, with df $= n_d - 1 = 26 - 1 = 25$, $t_{.025} = 2.060$. The 95% confidence interval is:

$$\bar{x}_d \pm t_{.025}\frac{s_d}{\sqrt{n_d}} \Rightarrow 10.5 \pm 2.060\frac{7.6}{\sqrt{26}} \Rightarrow 10.5 \pm 3.07 \Rightarrow (7.43,\ 13.57)$$

We are 95% confident that the difference in the mean anxiety levels between before and after the visit is between 7.43 and 13.57. Since all of these values are positive, we can conclude that mean anxiety level after the visit is significantly less than the anxiety level before the visit.

9.123   a.   Let μ_1 = mean reading response times for the tongue-twister list and μ_2 = mean reading response times for the control list. Then $\mu_d = \mu_1 - \mu_2$.

To compare the mean reading response times for the tongue-twister and control lists, we test:

$H_0: \mu_d = 0$
$H_a: \mu_d \neq 0$

b.   The test statistic is $z = \dfrac{\bar{x}_d - 0}{s_d/\sqrt{n_d}} = \dfrac{.25 - 0}{.78/\sqrt{42}} = 2.08$

The *p*-value is $P(z \leq -2.08) + P(z \geq 2.08) = (.5000 - .4812) + (.5000 - .4812) = .0188 + .0188 = .0376$. (Table IV, Appendix A.)

c.   Since the observed *p*-value is less than α ($p = .0376 < .05$), H_0 is rejected. There is sufficient evidence to conclude that the mean reading response times differ for the tongue-twister and control lists at $\alpha = .05$.

9.125   Let μ_1 = mean DIQ score for SLI children and μ_2 = mean DIQ score for YND children.

Some preliminary calculations are:

$$\bar{x}_1 = \frac{\sum x_1}{n_1} = \frac{936}{10} = 93.6 \qquad s_1^2 = \frac{\sum x_1^2 - \dfrac{(\sum x_1)^2}{n_1}}{n_1 - 1} = \frac{88352 - \dfrac{(936)^2}{10}}{10-1} = 82.4889$$

$$\bar{x}_2 = \frac{\sum x_2}{n_2} = \frac{953}{10} = 95.3 \qquad s_2^2 = \frac{\sum x_2^2 - \dfrac{(\sum x_2)^2}{n_2}}{n_1 - 1} = \frac{91329 - \dfrac{(953)^2}{10}}{10-1} = 56.4556$$

$$s_p^2 = \frac{(n_1-1)s_1^2 + (n_2-1)^2 s_2^2}{n_1+n_2-2} = \frac{(10-1)82.4889 + (10-1)56.4556}{10+10-2} = 69.4723$$

To determine if the mean DIQ scores differ for the two groups, we test:

H_0: $\mu_1 - \mu_2 = 0$
H_a: $\mu_1 - \mu_2 \neq 0$

The test statistic is $t = \dfrac{\bar{x}_1 - \bar{x}_2 - 0}{\sqrt{s_p^2\left(\dfrac{1}{n_1}+\dfrac{1}{n_2}\right)}} = \dfrac{(93.6-95.3)-0}{\sqrt{69.4723\left(\dfrac{1}{10}+\dfrac{1}{10}\right)}} = -0.46$

The rejection region requires $\alpha/2 = .10/2 = .05$ in each tail of the t distribution with df = $n_1 + n_2 - 2 = 10 + 10 - 2 = 18$. From Table VI, Appendix A, $t_{.05} = 1.734$. The rejection region is $t < -1.734$ or $t > 1.734$.

Since the observed value of the test statistic does not fall in the rejection region ($t = -.46 \not< -1.734$), H_0 is not rejected. There is insufficient evidence to indicate that the mean DIQ scores differ for the two groups at $\alpha = .10$.

9.127   Some preliminary calculations:

Patient	Placebo	Inositol	D = Pl − In	Patient	Placebo	Inositol	D = Pl − In
1	0	0	0	12	0	2	−2
2	2	1	1	13	3	1	2
3	0	3	−3	14	3	1	2
4	1	2	−1	15	3	4	−1
5	0	0	0	16	4	2	2
6	10	5	5	17	6	4	2
7	2	0	2	18	15	21	−6
8	6	4	2	19	28	8	20
9	1	1	0	20	30	0	30
10	1	0	1	21	13	0	13
11	1	3	−2				

(Panic Attacks per Week)

$$\bar{x}_d = \frac{\sum x_d}{n_d} = \frac{67}{21} = 3.19 \qquad s_d^2 = \frac{\sum x_d^2 - \dfrac{\left(\sum x_d\right)^2}{n_d}}{n_d-1} = \frac{1{,}575 - \dfrac{67^2}{21}}{21-1} = 68.0619$$

To determine if there is a difference in the mean number of panic attacks between the placebo and the drug Inositol, we test:

H_0: $\mu_d = 0$
H_a: $\mu_d \neq 0$

The test statistic is $t = \dfrac{\bar{x}_d - 0}{s_d / \sqrt{n_d}} = \dfrac{3.19 - 0}{\sqrt{68.0619} / \sqrt{21}} = 1.77$

Since no α was given, we will use $\alpha = .05$. The rejection region requires $\alpha / 2 = .05 / 2 = .025$ in each tail of the t distribution with df $= n_d - 1 = 21 - 1 = 20$. From Table VI, Appendix A, $t_{.025} = 2.086$. The rejection region is $t < -2.086$ or $t > 2.086$.

Since the observed value of the test statistic does not fall in the rejection region ($t = 1.77 \not> 2.086$), H_0 is not rejected. There is insufficient evidence to indicate that there is a difference in the mean number of panic attacks between the placebo and the drug Inositol at $\alpha = .05$. There is no evidence that the drug is effective.

9.129 a. Let $p_1 =$ proportion of female students who switched due to loss of interest in SME and $p_2 =$ proportion of male students who switched due to lack of interest in SME.

Some preliminary calculations are:

$$\hat{p}_1 = \frac{x_1}{n_1} = \frac{74}{172} = .430; \quad \hat{p}_2 = \frac{x_2}{n_2} = \frac{72}{163} = .442; \quad \hat{p} = \frac{x_1 + x_2}{n_1 + n_2} = \frac{74 + 72}{172 + 163} = .436$$

To determine if the proportion of female students who switch due to lack of interest in SME differs from the proportion of males who switch due to a lack of interest, we test:

H_0: $p_1 - p_2 = 0$
H_a: $p_1 - p_2 \neq 0$

The test statistic is $z = \dfrac{(\hat{p}_1 - \hat{p}_2) - 0}{\sqrt{\hat{p}\hat{q}\left(\dfrac{1}{n_1} + \dfrac{1}{n_2}\right)}} = \dfrac{(.430 - .442) - 0}{\sqrt{.436(.564)\left(\dfrac{1}{172} + \dfrac{1}{163}\right)}} = -0.22$

The rejection region requires $\alpha / 2 = .10 / 2 = .05$ in each tail of the z distribution. From Table IV, Appendix A, $z_{.05} = 1.645$. The rejection region is $z < -1.645$ or $z > 1.645$.

Since the observed value of the test statistic does not fall in the rejection region ($z = -0.22 \not< -1.645$), H_0 is not rejected. There is insufficient evidence to indicate the proportion of female students who switch due to lack of interest in SME differs from the proportion of males who switch due to a lack of interest in SME at $\alpha = .10$.

b. Let $p_1 =$ proportion of female students who switched due to low grades in SME and $p_2 =$ proportion of male students who switched due to low grades in SME.

Some preliminary calculations are:

$$\hat{p}_1 = \frac{x_1}{n_1} = \frac{33}{172} = .192; \quad \hat{p}_2 = \frac{x_2}{n_2} = \frac{44}{163} = .270$$

For confidence coefficient .90, $\alpha = .10$ and $\alpha/2 = .10/2 = .05$. From Table IV, Appendix A, $z_{.05} = 1.645$. The confidence interval is:

$$(\hat{p}_1 - \hat{p}_2) \pm z_{.05}\sqrt{\frac{\hat{p}_1\hat{q}_1}{n_1} + \frac{\hat{p}_2\hat{q}_2}{n_2}} \Rightarrow (.192 - .270) \pm 1.645\sqrt{\frac{.192(.808)}{172} + \frac{.270(.730)}{163}}$$

$$\Rightarrow -.078 \pm .076 \Rightarrow (-.154, -.002)$$

We are 90% confident that the difference between the proportions of female and male switchers who lost confidence due to low grades in SME is between −.154 and −.002. Since the interval does not include 0, there is evidence to indicate the proportion of female switchers due to low grades is less than the proportion of male switchers due to low grades.

9.131   Let μ_1 = mean rating for music majors and μ_2 = mean rating of nonmusic majors.

For confidence coefficient .95, $\alpha = .05$ and $\alpha/2 = .05/2 = .025$. From Table IV, Appendix A, $z_{.025} = 1.96$. The confidence interval is:

$$(\bar{x}_1 - \bar{x}_2) \pm z_{.025}\sqrt{\frac{s_1^2}{n_1} + \frac{s_2^2}{n_2}} \Rightarrow (4.26 - 4.59) \pm 1.96\sqrt{\frac{.81^2}{100} + \frac{.78^2}{100}}$$

$$\Rightarrow -.33 \pm .220 \Rightarrow (-.550, -.110)$$

We are 95% confident that the difference in the mean rating between music majors and nonmusic majors is between −.55 and −.11.

9.133   a.   Some preliminary calculations are:

$$s_1^2 = \frac{\sum x_1^2 - \frac{(\sum x_1)^2}{n_1}}{n_1 - 1} = \frac{4209 - \frac{(145)^2}{5}}{5 - 1} = 1$$

$$s_2^2 = \frac{\sum x_2^2 - \frac{(\sum x_2)^2}{n_2}}{n_2 - 1} = \frac{4540 - \frac{(150)^2}{5}}{5 - 1} = 10$$

Let σ_1^2 = variance of measurements for instrument 1 and σ_2^2 = variance of measurements for instrument 2.

To determine if the variances of the measurements for the two instruments differ, we test:

H_0: $\sigma_1^2 = \sigma_2^2$
H_a: $\sigma_1^2 \neq \sigma_2^2$

The test statistic is $F = \dfrac{\text{Larger sample variance}}{\text{Smaller sample variance}} = \dfrac{s_2^2}{s_1^2} = \dfrac{10}{1} = 10$

Since no α was given, we will use $\alpha = .05$. The rejection region requires $\alpha / 2 = .05 / 2 = .025$ in the upper tail of the F distribution with numerator df $v_1 = n_2 - 1 = 5 - 1 = 4$ and denominator df $v_2 = n_1 - 1 = 5 - 1 = 4$. From Table X, Appendix A, $F_{.025} = 9.60$. The rejection region is $F > 9.60$.

Since the observed value of the test statistic falls in the rejection reigon ($F = 10 > 9.60$), H_0 is rejected. There is sufficient evidence to indicate the variances of the measurements for the two instruments differ at $\alpha = .05$.

b.   We must assume that the samples are randomly and independently selected from populations that are normally distributed.

9.135   Let p_1 = proportion of patients receiving Zyban who were not smoking one year later and p_2 = proportion of patients not receiving Zyban who were not smoking one year later.

Some preliminary calculations are:

$$\hat{p}_1 = \frac{x_1}{n_1} = \frac{71}{309} = .230 \qquad\qquad \hat{p}_2 = \frac{x_2}{n_2} = \frac{37}{306} = .121$$

$$\hat{p} = \frac{x_1 + x_2}{n_1 + n_2} = \frac{71 + 37}{309 + 306} = .176 \qquad \hat{q} = 1 = \hat{p} = 1 - .176 = .824$$

To determine if the antidepressant drug Zyban helped cigarette smokers kick their habit, we test:

H_0: $p_1 - p_2 = 0$
H_a: $p_1 - p_2 > 0$

The test statistic is $z = \dfrac{(\hat{p}_1 - \hat{p}_2) - 0}{\sqrt{\hat{p}\hat{q}\left(\dfrac{1}{n_1} + \dfrac{1}{n_2}\right)}} = \dfrac{(.230 - .121) - 0}{\sqrt{.176(.824)\left(\dfrac{1}{309} + \dfrac{1}{306}\right)}} = 3.55$

The rejection region requires $\alpha = .05$ in the upper tail of the z distribution. From Table IV, Appendix A, $z_{.05} = 1.645$. The rejection region is $z > 1.645$.

Since the observed value of the test statistic falls in the rejection region ($z = 3.55 > 1.645$), H_0 is rejected. There is sufficient evidence to indicate that the antidepressant drub Zyban helped cigarette smokers kick their habit at $\alpha = .05$.

9.137   a.   Let p_1 = proportion of 9th grade boys who gambled weekly or daily on any game in 1992 and p_2 = proportion of 9th grade boys who gambled weekly or daily on any game in 1998.

Some preliminary calculations are:

$$\hat{p}_1 = \frac{x_1}{n_1} = \frac{4,684}{21,484} = .218 \qquad \hat{p}_2 = \frac{x_2}{n_2} = \frac{5,313}{23,199} = .229$$

$$\hat{p} = \frac{x_1 + x_2}{n_1 + n_2} = \frac{4,684 + 5,313}{21,484 + 23,199} = .224$$

To determine if the proportions of 9th grade boys who gambled weekly or daily on any game in 1992 and 1998 differ, we test:

$H_0: p_1 - p_2 = 0$
$H_a: p_1 - p_2 \neq 0$

The test statistic is $z = \dfrac{(\hat{p}_1 - \hat{p}_2) - 0}{\sqrt{\hat{p}\hat{q}\left(\dfrac{1}{n_1} + \dfrac{1}{n_2}\right)}} = \dfrac{(.218 - .229) - 0}{\sqrt{.224(.776)\left(\dfrac{1}{21,484} + \dfrac{1}{23,199}\right)}} = -2.79$

The rejection region requires $\alpha / 2 = .01 / 2 = .005$ in each tail of the z distribution. From Table IV, Appendix A, $z_{.005} = 2.58$. The rejection region is $z < -2.58$ or $z > 2.58$.

Since the observed value of the test statistic falls in the rejection region ($z = -2.79 < -2.58$), H_0 is rejected. There is sufficient evidence to indicate the proportions of 9th grade boys who gambled weekly or daily on any game in 1992 and 1998 differ at $\alpha = .01$.

b.   Yes. If the sample sizes are large enough, any difference can be found to be statistically different.

To measure the magnitude of the difference in the proportions of 9th grade boys who gambled weekly or daily on any game between 1992 and 1998, we will construct a 95% confidence interval. For confidence coefficient .95, $\alpha = .05$ and $\alpha / 2 = .05 / 2 = .025$. From Table IV, Appendix A, $z_{.025} = 1.96$. The 95% confidence interval is:

$$(\hat{p}_1 - \hat{p}_2) \pm z_{.025}\sqrt{\left(\frac{\hat{p}_1\hat{q}_1}{n_1} + \frac{\hat{p}_2\hat{q}_2}{n_2}\right)} \Rightarrow (.218 - .229) \pm 1.96\sqrt{\frac{.218(.782)}{21,484} + \frac{.229(.771)}{23,199}}$$

$$\Rightarrow -.011 \pm .0077 \Rightarrow (-.0187, \ -.0033)$$

We are 95% confident that the difference in the proportions of 9th grade boys who gambled weekly or daily on any game between 1992 and 1998 is between $-.0187$ and $-.0033$.

9.139   Using MINITAB, the descriptive statistics are:

Descriptive Statistics: CREATIVE, INFO, DECPERS, SKILLS, TASKID, AGE, EDYRS

Variable	GROUP	N	Mean	StDev	Minimum	Q1	Median	Q3	Maximum
CREATIVE	NOSPILL	67	4.4478	0.5304	3.0000	4.0000	4.0000	5.0000	5.0000
	SPILLOV	47	5.2553	0.4408	5.0000	5.0000	5.0000	6.0000	6.0000
INFO	NOSPILL	67	4.657	2.226	1.000	3.000	5.000	7.000	7.000
	SPILLOV	47	5.213	1.719	1.000	4.000	5.000	7.000	7.000
DECPERS	NOSPILL	67	2.746	1.964	1.000	1.000	2.000	5.000	7.000
	SPILLOV	47	3.319	2.023	1.000	1.000	3.000	5.000	7.000
SKILLS	NOSPILL	67	4.821	1.100	2.000	4.000	5.000	5.000	7.000
	SPILLOV	47	5.851	1.161	2.000	5.000	6.000	7.000	7.000
TASKID	NOSPILL	67	4.239	1.156	1.000	3.000	4.000	5.000	7.000
	SPILLOV	47	4.809	2.028	1.000	3.000	5.000	7.000	7.000
AGE	NOSPILL	67	45.343	7.971	21.000	41.000	45.000	50.000	63.000
	SPILLOV	47	46.298	5.767	30.000	44.000	47.000	51.000	54.000
EDYRS	NOSPILL	67	13.224	1.312	12.000	12.000	13.000	14.000	18.000
	SPILLOV	47	13.085	1.060	12.000	12.000	13.000	14.000	16.000

Let μ_1 = mean characteristic for the workers with positive spillover and μ_2 = mean characteristic for the workers with no spillover. To determine if there is a difference in the mean characteristic between the positive spillover group and the no spillover group, we test:

H_0: $\mu_1 - \mu_2 = 0$
H_a: $\mu_1 - \mu_2 \neq 0$

The test statistic is $z = \dfrac{\bar{x}_1 - \bar{x}_2}{\sqrt{\left(\dfrac{s_1^2}{n_1} + \dfrac{s_2^2}{n_2}\right)}}$

Since no α was given, we will run all tests at $\alpha = .05$. The rejection region requires $\alpha/2 = .05/2 = .025$ in each tail of the z distribution. From Table IV, Appendix A, $z_{.025} = 1.96$. The rejection region is $z < -1.96$ or $z > 1.96$.

Use of Creative Ideas:

The test statistic is $z = \dfrac{\bar{x}_1 - \bar{x}_2}{\sqrt{\left(\dfrac{s_1^2}{n_1} + \dfrac{s_2^2}{n_2}\right)}} = \dfrac{5.2553 - 4.4478}{\sqrt{\dfrac{.4408^2}{47} + \dfrac{.5304^2}{67}}} = 8.85$

Since the observed value of the test statistic falls in the rejection region ($z = 8.85 > 1.96$), H_0 is rejected. There is sufficient evidence to indicate a difference in the mean use of creative ideas between the positive spillover group and the no spillover group at $\alpha = .05$.

Utilization of Information:

The test statistic is $z = \dfrac{\bar{x}_1 - \bar{x}_2}{\sqrt{\left(\dfrac{s_1^2}{n_1} + \dfrac{s_2^2}{n_2}\right)}} = \dfrac{5.213 - 4.657}{\sqrt{\dfrac{1.719^2}{47} + \dfrac{2.226^2}{67}}} = 1.50$

Since the observed value of the test statistic does not fall in the rejection region ($z = 1.50 \not> 1.96$), H_o is not rejected. There is insufficient evidence to indicate a difference in the mean utilization of information between the positive spillover group and the no spillover group at $\alpha = .05$.

Decisions Regarding Personnel Matters:

The test statistic is $z = \dfrac{\bar{x}_1 - \bar{x}_2}{\sqrt{\left(\dfrac{s_1^2}{n_1} + \dfrac{s_2^2}{n_2}\right)}} = \dfrac{3.319 - 2.746}{\sqrt{\dfrac{2.023^2}{47} + \dfrac{1.964^2}{67}}} = 1.51$

Since the observed value of the test statistic does not fall in the rejection region ($z = 1.51 \not> 1.96$), H_0 is not rejected. There is insufficient evidence to indicate a difference in the mean participation in decisions regarding personnel matters between the positive spillover group and the no spillover group at $\alpha = .05$.

Good Use of Job Skills:

The test statistic is $z = \dfrac{\bar{x}_1 - \bar{x}_2}{\sqrt{\left(\dfrac{s_1^2}{n_1} + \dfrac{s_2^2}{n_2}\right)}} = \dfrac{5.851 - 4.821}{\sqrt{\dfrac{1.161^2}{47} + \dfrac{1.100^2}{67}}} = 4.76$

Since the observed value of the test statistic falls in the rejection region ($z = 4.76 > 1.96$), H_0 is rejected. There is sufficient evidence to indicate a difference in the mean good use of skills between the positive spillover group and the no spillover group at $\alpha = .05$.

Task Identity:

The test statistic is $z = \dfrac{\bar{x}_1 - \bar{x}_2}{\sqrt{\left(\dfrac{s_1^2}{n_1} + \dfrac{s_2^2}{n_2}\right)}} = \dfrac{4.809 - 4.239}{\sqrt{\dfrac{2.028^2}{47} + \dfrac{1.156^2}{67}}} = 1.74$

Since the observed value of the test statistic does not fall in the rejection region ($z = 1.74 \not> 1.96$), H_0 is not rejected. There is insufficient evidence to indicate a difference in the mean task identity between the positive spillover group and the no spillover group at $\alpha = .05$.

Of the 5 job-related characteristics, there are only 2 that have significantly different means for the 2 groups. They are "use of creative ideas" and "good use of skills". For both of these characteristics, the mean for the positive spillover group was significantly higher than the mean for the no spillover group.

Age:

The test statistic is $z = \dfrac{\bar{x}_1 - \bar{x}_2}{\sqrt{\left(\dfrac{s_1^2}{n_1} + \dfrac{s_2^2}{n_2}\right)}} = \dfrac{46.298 - 45.343}{\sqrt{\dfrac{5.767^2}{47} + \dfrac{7.971^2}{67}}} = 0.74$

Since the observed value of the test statistic does not fall in the rejection region ($z = 0.74 \not> 1.96$), H_0 is not rejected. There is insufficient evidence to indicate a difference in the mean age between the positive spillover group and the no spillover group at $\alpha = .05$.

Education:

The test statistic is $z = \dfrac{\bar{x}_1 - \bar{x}_2}{\sqrt{\left(\dfrac{s_1^2}{n_1} + \dfrac{s_2^2}{n_2}\right)}} = \dfrac{13.085 - 13.224}{\sqrt{\dfrac{1.060^2}{47} + \dfrac{1.312^2}{67}}} = -0.62$

Since the observed value of the test statistic does not fall in the rejection region ($z = -0.62 \not> 1.96$), H_0 is not rejected. There is insufficient evidence to indicate a difference in the mean years of education between the positive spillover group and the no spillover group at $\alpha = .05$.

Analysis of Variance
Comparing More than Two Means

10.1   Since only one factor is utilized, the treatments are the four levels (A, B, C, D) of the qualitative factor.

10.3   One has no control over the levels of the factors in an observational experiment. One does have control of the levels of the factors in a designed experiment.

10.5   a.   This is an observational study. The economist has no control over the factor levels or unemployment rates.

       b.   This is a designed study. The psychologist selects the feedback programs of interest and randomly assigns five rats to each program.

       c.   This is a designed study. The marketer has control of the selection of the national publications.

       d.   This is an observational study. The load on the generators is not controlled by the utility.

       e.   This is an observational study. One has no control over the distance of the haul, the goods hauled, or the price of diesel fuel.

       f.   This is an observational study. The student does not control which state is assigned to a portion of the country.

10.7   a.   The experimental unit is a patient.

       b.   The response variable is the score on the Hamilton Depression Rating Scale.

       c.   The factor is the drug combination group.

       d.   There are 4 factor levels. Group 1 received daily doses of the antidepressant drug floxetine and a placebo; group 2 received the antidepressant drug mirtazapine plus floxetine; group 3 received mirtazapine plus veniafaxine; and group 4 received mirtazapine plus bupropion.

10.9   a.   The experimental unit is a healthy adult.

       b.   The response variable is the postural index.

       c.   There are 2 factors – gender and strength knowledge.

       d.   There are 2 levels of gender – male and female. There are 2 levels of strength knowledge – either provided or not provided.

       e.   There are a total of 2 x 2 = 4 treatments: male, knowledge provided; male, knowledge not provided; female knowledge provided; and female, knowledge not provided.

10.11 a. The experimental units are the cockatiels.

      b. This experiment is a designed experiment. The birds were randomly divided into 3 groups and each group received a different treatment.

      c. There is one factor in this study. The factor is the group.

      d. There are three levels of the group variable – Group 1 received purified water, Group 2 received purified water and liquid sucrose, and Group 3 received purified water and liquid sodium chloride.

      e. There are 3 treatments in the study. Because there is only one factor, the treatments are the same as the factor levels.

      f. The response variable is the liquid consumption.

10.13 a. There are 2 factors in this experiment – Temperature and Type of yeast. Temperature has 4 levels – 45, 48, 51, and 54°C. Type of yeast has 2 levels – Baker's and Brewer's.

      b. The response variable is the autolysis yield (recorded as a percentage).

      c. There are a total of 4 x 2 = 8 treatments for this experiment.

      d. This is a designed experiment. The levels of temperature and type of yeast are controlled by the researcher.

10.15 For a completely randomized design, independent random samples are selected for each treatment, or treatments are randomly assigned to experimental units.

10.17 The conditions required for a valid ANOVA F-test in a completely randomized design are:

      1. The samples are randomly selected in an independent manner from k treatment populations.

      2. All k sampled populations have distributions that are approximately normal.

      3. The k population variances are equal (i.e., $\sigma_1^2 = \sigma_2^2 = \ldots = \sigma_k^2$)

10.19 a. Using Table IX, $F_{.05} = 6.59$ with $v_1 = 3$, $v_2 = 4$.

      b. Using Table XI, $F_{.01} = 16.69$ with $v_1 = 3$, $v_2 = 4$.

      c. Using Table VIII, $F_{.10} = 1.61$ with $v_1 = 20$, $v_2 = 40$.

      d. Using Table X, $F_{.025} = 3.87$ with $v_1 = 12$, $v_2 = 9$.

10.21 a. In the second dot diagram **B**, the difference between the sample means is small relative to the variability within the sample observations. In the first dot diagram **A**, the values in each of the samples are grouped together with a range of 4, while in the second diagram **B**, the range of values is 8.

b. For diagram **A**,

$$\bar{x}_1 = \frac{\sum x_1}{n} = \frac{7+8+9+9+10+11}{6} = \frac{54}{6} = 9$$

$$\bar{x}_2 = \frac{\sum x_2}{n} = \frac{12+13+14+14+15+16}{6} = \frac{84}{6} = 14$$

For diagram **B**,

$$\bar{x}_1 = \frac{\sum x_1}{n} = \frac{5+5+7+11+13+13}{6} = \frac{54}{6} = 9$$

$$\bar{x}_2 = \frac{\sum x_2}{n} = \frac{10+10+12+16+18+18}{6} = \frac{84}{6} = 14$$

c. For diagram **A**,

$$\text{SST} = \sum_{i=1}^{2} n_i (\bar{x}_i - \bar{x})^2 = 6(9 - 11.5)^2 + 6(14 - 11.5)^2 + = 75$$

$$\left(\bar{x} = \frac{\sum x}{n} = \frac{54 + 84}{12} = 11.5 \right)$$

For diagram **B**,

$$\text{SST} = \sum_{i=1}^{2} n_i (\bar{x}_i - \bar{x})^2 = 6(6 - 11.5)^2 + 6(14 - 11.5)^2 = 75$$

d. For diagram **A**,

$$s_1^2 = \frac{\sum x_1^2 - \dfrac{\left(\sum x_1\right)^2}{n_1}}{n_1 - 1} = \frac{496 - \dfrac{54^2}{6}}{6 - 1} = 2$$

$$s_2^2 = \frac{\sum x_2^2 - \dfrac{\left(\sum x_2\right)^2}{n_2}}{n_2 - 1} = \frac{1186 - \dfrac{84^2}{6}}{6 - 1} = 2$$

$$\text{SSE} = (n_1 - 1)s_1^2 + (n_2 - 1)s_2^2 = (6-1)^2(2) + (6-1)^2(2) = 20$$

For diagram **B**,

$$s_1^2 = \frac{\sum x_1^2 - \dfrac{\left(\sum x_1\right)^2}{n_1}}{n_1 - 1} = \frac{558 - \dfrac{54^2}{6}}{6 - 1} = 14.4$$

$$s_2^2 = \frac{\sum x_2^2 - \frac{\left(\sum x_2\right)^2}{n_2}}{n_2 - 1} = \frac{1248 - \frac{84^2}{6}}{6 - 1} = 14.4$$

$$SSE = (n_1 - 1)s_1^2 + (n_2 - 1)s_2^2 = (6-1)14.4 + (6-1)14.4 = 144$$

e. For diagram **A**, SS(Total) = SST + SSE = 75 + 20 = 95

$$\text{SST is } \frac{SST}{SS(Total)} \times 100\% = \frac{75}{95} \times 100\% = 78.95\% \text{ of SS(Total)}$$

For diagram **B**, SS(Total) = SST + SSE = 75 + 144 = 219

$$\text{SST is } \frac{SST}{SS(Total)} \times 100\% = \frac{75}{219} \times 100\% = 34.25\% \text{ of SS(Total)}$$

f. For diagram **A**, $MST = \dfrac{SST}{k-1} = \dfrac{75}{2-1} = 75$

$$MSE = \frac{SSE}{n-k} = \frac{20}{12-2} = 2 \qquad F = \frac{MST}{MSE} = \frac{75}{2} = 37.5$$

For diagram **B**, $MST = \dfrac{SST}{k-1} = \dfrac{75}{2-1} = 75$

$$MSE = \frac{SSE}{n-k} = \frac{144}{12-2} = 14.4 \qquad F = \frac{MST}{MSE} = \frac{75}{14.4} = 5.21$$

g. The rejection region for both diagrams requires $\alpha = .05$ in the upper tail of the F distribution with numerator df $= k - 1 = 2 - 1 = 1$ and denominator df $= n - k = 12 - 2 = 10$. From Table IX, Appendix A, $F_{.05} = 4.96$. The rejection region is $F > 4.96$.

For diagram **A**, since the observed value of the test statistic falls in the rejection region ($F = 37.5 > 4.96$), H_0 is rejected. There is sufficient evidence to indicate the samples were drawn from populations with different means at $\alpha = .05$.

For diagram **B**, since the observed value of the test statistic falls in the rejection region ($F = 5.21 > 4.96$), H_0 is rejected. There is sufficient evidence to indicate the samples were drawn from populations with different means at $\alpha = .05$.

h. We must assume both populations are normally distributed with common variances.

10.23 Refer to Exercises 10.21 and 10.22, the ANOVA table is:

For diagram **A**:

Source	df	SS	MS	F
Treatment	1	75	75	37.5
Error	10	20	2	
Total	11	95		

For diagram **B**:

Source	df	SS	MS	F
Treatment	1	75	75	5.21
Error	10	144	14.4	
Total	11	219		

10.25   For all parts, the hypotheses are:

H_0: $\mu_1 = \mu_2 = \mu_3 = \mu_4 = \mu_5$
H_a: At least two treatment means differ

The rejection region for all parts is the same.

The rejection region requires $\alpha = .10$ in the upper tail of the F distribution with $v_1 = k - 1 = 5 - 1 = 4$ and $v_2 = n - k = 30 - 5 = 25$. From Table VIII, Appendix A, $F_{.10} = 2.18$. The rejection region is $F > 2.18$.

a. $SST = .2(500) = 100$ $SSE = SS(Total) - SST = 500 - 100 = 400$

$$MST = \frac{SST}{k-1} = \frac{100}{5-1} = 25 \qquad MSE = \frac{SSE}{n-k} = \frac{400}{30-5} = 16$$

$$F = \frac{MST}{MSE} = \frac{25}{16} = 1.5625$$

Since the observed value of the test statistic does not fall in the rejection region ($F = 1.5625 \not> 2.18$), H_0 is not rejected. There is insufficient evidence to indicate differences among the treatment means at $\alpha = .10$.

b. $SST = .5(500) = 250$ $SSE = SS(Total) - SST = 500 - 250 = 250$

$$MST = \frac{SST}{k-1} = \frac{250}{5-1} = 62.5 \qquad MSE = \frac{SSE}{n-k} = \frac{250}{30-5} = 10$$

$$F = \frac{MST}{MSE} = \frac{62.5}{10} = 6.25$$

Since the observed value of the test statistic falls in the rejection region ($F = 6.25 > 2.18$), H_0 is rejected. There is sufficient evidence to indicate differences among the treatment means at $\alpha = .10$.

c. $SST = .8(500) = 400$ $SSE = SS(Total) - SST = 500 - 400 = 100$

$$MST = \frac{SST}{k-1} = \frac{400}{5-1} = 100 \qquad MSE = \frac{SSE}{n-k} = \frac{100}{30-5} = 4$$

$$F = \frac{MST}{MSE} = \frac{100}{4} = 25$$

Since the observed value of the test statistic falls in the rejection region, ($F = 25 > 2.18$), H_0 is rejected. There is sufficient evidence to indicate differences among the treatment means at $\alpha = .10$.

d.   The F ratio increases as the treatment sum of squares increases.

10.27   a.   Answers will vary. Using MINITAB, the numbers selected to receive the usual follow-up care are:

2, 3, 7, 9, 11, 12, 13, 14, 15, 16, 17, 18, 19, 20, 22, 23, 25, 26, 27, 31, 32, 33, 35, 42, 44, 45, 47, 48, 49, 51, 52, 56, 57, 63, 64, 65, 66, 67, 69, 70, 71, 72, 75, 78, 79, 80, 82, 83, 86, 87, 88, 89, 93, 94, 95, 96, 99, 101, 102, 103, 104, 105, 106, 107, 108, 110, 111, 112, 114, 115, 116, 117, 118, 120, 121, 123, 125, 126, 128, 130, 138, 140, 142, 143, 144, 145, 147, 152, 154, 155, 156, 159, 161, 163, 165, 166, 168, 170, 171, 173, 174, 175, 176, 177, 179, 181, 182, 183, 186, 188, 190, 191, 192, 193, 194, 196, 198, 199, 202, 204, 207, 208, 244, 212, 218, 219, 221, 224, 228, 229, 231, 232, 233, 234, 235, 240, 243, 244, 246, 252, 259, 262, 264, 267, 269, 270, 273, 274, 275, 276, 277, 278, 279, 283, 285, 286, 288, 292, 298, 300, 302, 306, 311, 314, 315, 317, 321, 323, 329, 330, 332, 334, 335, 338, 340, 342, 344, 345, 347, 354, 356, 359, 360, 362, 363, 369, 370, 371, 372, 373, 375, 376, 377, 378, 381, 383, 384, 385, 386, 391, 392, 394, 401, 402, 405

All other numbers will be assigned to the group attending yoga classes.

b.   The results from this assignment would not be valid. Those patients with the most severe cancer would probably be more fatigued and sleepy than those with less severe cancer, regardless of which group they were assigned. To assign all the most severe cancer patients to one group will skew the results.

10.29   a.   The experimental units are the NCAA tennis coaches. The dependent variable is the rating on a 7-point scale of how important the coaches think the web site was. There is one factor and it is the division of the college/university that employs the tennis coach. Since there is only one factor, the treatments are the different levels of the factor. The factor has 3 levels and thus, 3 treatments which are Division I, Division II and Division III.

b.   To determine if the mean ratings of the web sites are different for the different levels of coaches, we test:

H_0: $\mu_I = \mu_{II} = \mu_{III}$
H_a: At least one mean differs

c.   Since the p-value is less than $\alpha = .05$ ($p < .003 < .05$), H_0 is rejected. There is sufficient evidence to indicate a difference in mean responses among the different divisions of the colleges and universities at $\alpha = .05$.

10.31   a.   The experimental design used is a completely randomized design.

b.   There are 4 treatments in this experiment. The four treatments are the 4 "colonies" to which the robots were assigned – 3, 6, 9, or 12 robots per colony. The dependent variable is the energy expended per robot.

c. To determine if the mean energy expended (per robot) of the four different colony sizes differed, we test:

H_0: $\mu_1 = \mu_2 = \mu_3 = \mu_4$
H_a: At least two treatment means differ

d. The test statistic is $F = 7.70$ and the p-value is $p < .001$. Since the p-value is less than $\alpha = .05$, H_0 is rejected. There is sufficient evidence to indicate that mean energy expended (per robot) in the four different colony sizes differed at $\alpha = .05$.

10.33 a. The experimental units are the participants in the study.

b. The dependent variable is the brand recall score.

c. There is one factor in this study – TV viewing group. Since there is only one factor, the treatments correspond to the factor levels of this variable. Thus, the treatments are the same as the three levels of TV viewer group. These 3 levels are violent content code, sex content code, and neutral TV.

d. The means given are only sample means. If new samples were selected and sample means computed, the values and order of the sample means could change. In addition, the variances are not taken into account.

e. The test statistic is $F = 20.45$ and the p-value is p-value = 0.000.

f. Since the p-value is less than α (p = 0.000 < .01), H_o is rejected. There is sufficient evidence to indicate differences in the mean recall scores among the three viewing groups at $\alpha = .01$. The researchers can conclude that the content of the TV show affects the recall of imbedded commercials.

10.35 a. This experiment is a completely randomized design. The 3 treatments are honey dosage, DM dosage, and no dosage.

b. Using MINITAB, the ANOVA is:

One-way ANOVA: Honey, DM, Control

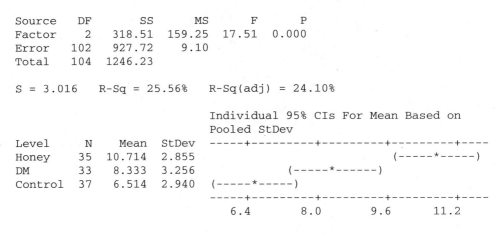

```
Source    DF        SS       MS       F       P
Factor     2    318.51   159.25   17.51   0.000
Error    102    927.72     9.10
Total    104   1246.23

S = 3.016    R-Sq = 25.56%    R-Sq(adj) = 24.10%

                                Individual 95% CIs For Mean Based on
                                Pooled StDev
Level      N     Mean   StDev   -----+---------+---------+---------+----
Honey     35   10.714   2.855                            (-----*-----)
DM        33    8.333   3.256                 (-----*------)
Control   37    6.514   2.940   (-----*-----)
                                -----+---------+---------+---------+----
                                   6.4       8.0       9.6      11.2

Pooled StDev = 3.016
```

To determine if there are differences in the mean improvement scores among the 3 groups, we test:

H_0: $\mu_1 = \mu_2 = \mu_3$
H_a: At least two treatment means differ

From the printout, the test statistic is $F = 17.51$ and the p-value is $p = 0.000$.

Since the p-value is so small $(p = 0.000)$, H_0 is rejected. There is sufficient evidence to indicate a difference in mean improvement scores among the 3 groups for any reasonable value of α.

10.37   Using MINITAB, the results of the analysis are:

One-way ANOVA: UMRB-1, UMRB-2, UMBR-3, SWRA, SD

```
Source   DF      SS      MS      F      P
Factor    4   5.836   1.459   7.25  0.001
Error    21   4.225   0.201
Total    25  10.061

S = 0.4486    R-Sq = 58.00%   R-Sq(adj) = 50.00%


                                   Individual 95% CIs For Mean Based on
                                   Pooled StDev
Level    N    Mean    StDev    -----+---------+---------+---------+----
UMRB-1   7  3.4971   0.3640                    (-----*-----)
UMRB-2   6  4.0167   0.5726                        (-----*-----)
UMRB-3   7  3.7614   0.4832                     (-----*-----)
SWRA     3  2.6433   0.3580    (--------*--------)
SD       3  2.7800   0.2587      (--------*--------)
                               -----+---------+---------+---------+----
                                 2.40      3.00      3.60      4.20
Pooled StDev = 0.4486
```

To determine if there are differences among the mean Al/Be ratios for the 5 boreholes, we test:

H_0: $\mu_1 = \mu_2 = \mu_3 = \mu_4 = \mu_5$
H_a: At least 2 means differ

From the printout, the test statistic is $F = 7.25$ and the p-value is $p = 0.001$. Since the p-value is less than $\alpha = .10$ $(p = .001 < .10)$, H_0 is rejected. There is sufficient evidence to indicate a difference in the mean Al/Be ratios among the 5 boreholes at $\alpha = .10$.

10.39   The experimentwise error rate is the probability of declaring at least one pair of means different, given all of the means are the same.

10.41   a.   The confidence interval $(-10, 5)$ for $(\mu_1 - \mu_2)$ means that we are confident that the difference between $\mu_1 - \mu_2$ is between -10 and 5. Since this interval spans both positive and negative numbers, we cannot conclude that either mean is significantly larger.

b. The confidence interval $(-10, -5)$ for $(\mu_1 - \mu_2)$ means that we are confident that the difference between $\mu_1 - \mu_2$ is between -10 and -5. Since this interval spans only negative numbers, we can conclude that μ_2 is significantly larger than μ_1.

c. The confidence interval $(5, 10)$ for $(\mu_1 - \mu_2)$ means that we are confident that the difference between $\mu_1 - \mu_2$ is between 5 and 10. Since this interval spans only positive numbers, we can conclude that μ_1 is significantly larger than μ_2.

10.43   The number of pairwise comparisons is equal to $c = k(k - 1)/2$.

a. For $k = 3$, the number of comparisons is $c = 3(3 - 1)/2 = 3$.

b. For $k = 5$, the number of comparisons is $c = 5(5 - 1)/2 = 10$.

c. For $k = 4$, the number of comparisons is $c = 4(4 - 1)/2 = 6$.

d. For $k = 10$, the number of comparisons is $c = 10(10 - 1)/2 = 45$.

10.45 $(\mu_1 - \mu_2)$: $(2, 15)$ Since all values in the interval are positive, μ_1 is significantly greater than μ_2.

$(\mu_1 - \mu_3)$: $(4, 7)$ Since all values in the interval are positive, μ_1 is significantly greater than μ_3.

$(\mu_1 - \mu_4)$: $(-10, 3)$ Since 0 is in the interval, μ_1 is not significantly different from μ_4. However, since the center of the interval is less than 0, μ_4 is larger than μ_1.

$(\mu_2 - \mu_3)$: $(-5, 11)$ Since 0 is in the interval, μ_2 is not significantly different from μ_3. However, since the center of the interval is greater than 0, μ_2 is larger than μ_3

$(\mu_2 - \mu_4)$: $(-12, -6)$ Since all values in the interval are negative, μ_4 is significantly greater than μ_2.

$(\mu_3 - \mu_4)$: $(-8, -5)$ Since all values in the interval are negative, μ_4 is significantly greater than μ_3.

Thus, the largest mean is μ_4 followed by μ_1, μ_2, and μ_3.

10.47   a. This ANOVA table is used to test whether the mean number of alternatives listed differs among the 3 groups. Since the p-value is so small ($p = .001$), H_0 is rejected. There is sufficient evidence of a difference in the mean number of alternatives listed among the 3 groups.

b. The .05 means that the probability of concluding at least two means differ when, in fact, none are different is .05.

c. From the results, we can conclude that the mean number of alternatives listed for the guilt group is significantly greater than the mean number of alternatives listed for the other two groups. There is no significant difference in the mean number of alternatives listed between the angry group and the neutral group.

10.49   a.   The total number of pairwise comparisons made in the Bonferroni analysis is $c = k(k-1)/2 = 4(4-1)/2 = 6$.

   b.   The Sourdough treatment yielded the significantly highest mean soluble magnesium level. There is no significant difference in the mean soluble magnesium level between the Control and the Yeast groups. These two treatments yielded the lowest mean soluble magnesium level.

   c.   The experimentwise error rate is .05. This means that the probability of declaring at least one pair significantly different when, in fact, none are different is .05.

10.51   a.   There are $c = k(k-1)/2 = 4(4-1)/2 = 6$ pairwise comparisons.

   b.   There are no significant differences in the mean energy expended among the colony sizes of 3, 6, and 9. However, the mean energy expended for the colony containing 12 robots was significantly less than that for all other colony sizes.

10.53   a.   To determine if there are differences in the mean dental fear scores among the three groups, we test:

H_0: $\mu_1 = \mu_2 = \mu_3$
H_a: At least two treatment means differ

The test statistic is $F = 4.43$ and the p-value is $p < .05$. If we use $\alpha = .05$, then the rejection region will be p-value $< .05$. Since the p-value is less than .05, we reject H_0. There is sufficient evidence to indicate there are differences in the mean dental fear scores among the three groups at $\alpha = .05$.

   b.   First, we arrange the means in order from the largest to the smallest. Then we draw a line between the means that are not significantly different. The results are:

Mean: 53.8   43.1   41.8
Group: Questionnaire Slide   Control

10.55   Using MINITAB, the Tukey's multiple comparison procedure yielded:

```
Tukey 90% Simultaneous Confidence Intervals
All Pairwise Comparisons

Individual confidence level = 98.44%


UMRB-1 subtracted from:

           Lower    Center    Upper    --------+---------+---------+---------+-
UMRB-2   -0.1369    0.5195   1.1760                    (----*-----)
UMRB-3   -0.3664    0.2643   0.8950                  (----*----)
SWRA     -1.6680   -0.8538  -0.0396         (------*------)
SD       -1.5314   -0.7171   0.0971           (------*------)
                                     --------+---------+---------+---------+-
                                         -1.2       0.0       1.2       2.4


UMRB-2 subtracted from:

           Lower    Center    Upper    --------+---------+---------+---------+-
UMRB-3   -0.9117   -0.2552   0.4012                 (-----*----)
SWRA     -2.2077   -1.3733  -0.5390      (------*------)
SD       -2.0710   -1.2367  -0.4023       (------*------)
                                     --------+---------+---------+---------+-
                                         -1.2       0.0       1.2       2.4


UMRB-3 subtracted from:

          Lower    Center    Upper    --------+---------+---------+---------+-
SWRA    -1.9323   -1.1181  -0.3039     (------*-----)
SD      -1.7957   -0.9814  -0.1672     (------*------)
                                    --------+---------+---------+---------+-
                                        -1.2       0.0       1.2       2.4


SWRA subtracted from:

         Lower   Center   Upper    --------+---------+---------+---------+-
SD     -0.8267   0.1367  1.1001            (-------*-------)
                                  --------+---------+---------+---------+-
                                      -1.2       0.0       1.2       2.4
```

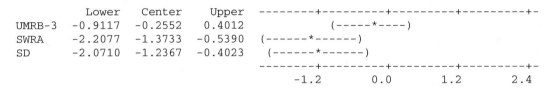

Mean:	4.0167	3.7614	3.4971	2.7800	2.6433
Group:	UMRB-2	UMRB-3	UMRB-1	SD	SWRA

The mean Al/Be ratios for UMRB-2 and UMRB-3 are significantly greater than the mean Al/Be ratios for SD and SWRA. The mean Al/Be ratio for UMRB-1 is significantly greater than the mean Al/Be ratio for SWRA. No other differences exist.

10.57 Using MINITAB, the results of Tukey's multiple comparison procedure are:

```
Tukey 95% Simultaneous Confidence Intervals
All Pairwise Comparisons

Individual confidence level = 98.94%

AR subtracted from:

        Lower    Center   Upper    ---------+---------+---------+---------+
AC    -0.3751  -0.1745   0.0260              (------*------)
A     -0.5769  -0.3764  -0.1758    (-----*------)
P     -0.2405  -0.0400   0.1605                (------*-----)
                                   ---------+---------+---------+---------+
                                        -0.30      0.00      0.30      0.60

AC subtracted from:

        Lower    Center   Upper    ---------+---------+---------+---------+
A     -0.4023  -0.2018  -0.0013              (-----*------)
P     -0.0660   0.1345   0.3351                       (-----*------)
                                   ---------+---------+---------+---------+
                                        -0.30      0.00      0.30      0.60

A subtracted from:

        Lower    Center   Upper    ---------+---------+---------+---------+
P      0.1358   0.3364   0.5369                        (-----*------)
                                   ---------+---------+---------+---------+
                                        -0.30      0.00      0.30      0.60
```

Mean:	0.4400	0.4000	0.2655	0.0636
Group:	AR	P	AC	A

The mean for the Alcohol only group is significantly less than the means for the other 3 groups. No other differences exist. The researchers' theories were supported. The Alcohol only group performed the worst. There were no significant differences between the Placebo group and the Alcohol plus caffeine group, between the Placebo group and the Alcohol plus reward group, and between the alcohol plus caffeine group and the alcohol plus reward group.

10.59 A paired difference experiment is a randomized block design with only 2 experimental units per block. In the paired difference design, the pairs are the blocks. However, a randomized block design is not limited to 2 experimental units per block.

10.61 The conditions required for a valid ANOVA F-test in a randomized block design are:

1. The *b* blocks are randomly selected and all *k* treatments are applied (in random order) to each block.
2. The distributions of observations corresponding to all *bk* block-treatment combinations are approximately normal.
3. The *bk* block-treatment distributions have equal variances.

10.63   a.   Treatment $df = k - 1 = 3 - 1 = 2$
      Block $df = b - 1 = 3 - 1 = 2$
      Error $df = n - k - b + 1 = 9 \ 3 - 3 + 1 = 4$
      Total $df = n - 1 = 9 - 1 = 8$

$$SSB = \sum_{i=1}^{b} \frac{B_i^2}{k} - CM \quad \text{from Appendix C}$$

$$\text{where } CM = \frac{\left(\sum x_i\right)^2}{n} = \frac{49^2}{9} = 266.7778$$

$$SSB = \frac{17^2}{3} + \frac{15^2}{3} + \frac{17^2}{3} - 266.7778 = .8889$$

$$SSE = SS(Total) - SST - SSB = 30.2222 - 21.5555 - .8889 = 7.7778$$

$$MST = \frac{SST}{k-1} = \frac{21.5555}{2} = 10.7778 \qquad MSB = \frac{SSB}{b-1} = \frac{.8889}{2} = .4445$$

$$MSE = \frac{SSE}{n-k-b+1} = \frac{7.7778}{4} = 1.9445$$

$$F_T = \frac{MST}{MSE} = \frac{10.7778}{1.9445} = 5.54 \qquad F_B = \frac{MSB}{MSE} = \frac{.4445}{1.9445} = .23$$

The ANOVA table is:

Source	df	SS	MS	F
Treatment	2	21.5555	10.7778	5.54
Block	2	.8889	.4445	.23
Error	4	7.7778	1.9445	
Total	8	30.2222		

b.   To determine if differences exist among the treatment means, we test:

H_0: $\mu_1 = \mu_2 = \mu_3$
H_a: At least two treatment means differ

c.   The test statistic is $F = \dfrac{MST}{MSE} = 5.54$

d.   A Type I error would be concluding at least two treatment means differ when they do not.

A Type II error would be concluding all the treatment means are the same when at least two differ.

e.   The rejection region requires $\alpha = .05$ in the upper tail of the F distribution with $v_1 = k - 1 = 3 - 1 = 2$ and $v_2 = n - k - b + 1 = 9 - 3 - 3 + 1 = 4$. From Table IX, Appendix A, $F_{.05} = 6.94$. The rejection region is $F > 6.94$.

Since the observed value of the test statistic does not fall in the rejection region ($F = 5.54 \not> 6.94$), H_0 is not rejected. There is insufficient evidence to indicate at least two of the treatment means differ at $\alpha = .05$.

10.65   a.   The ANOVA Table is as follows:

Source	df	SS	MS	F	p
Treatment	2	12.032	6.016	50.96	0.000
Block	3	71.749	23.916	202.59	0.000
Error	6	0.708	.118		
Total	11	84.489			

b.   To determine if the treatment means differ, we test:

H_0: $\mu_A = \mu_B = \mu_C$
H_a: At least two treatment means differ

The test statistic is $F = \dfrac{\text{MST}}{\text{MSE}} = 50.96$

The rejection region requires $\alpha = .05$ in the upper tail of the F distribution with $v_1 = k - 1 = 3 - 1 = 2$ and $v_2 = n - k - b + 1 = 12 - 3 - 4 + 1 = 6$. From Table IX, Appendix A, $F_{.05} = 5.14$. The rejection region is $F > 5.14$.

Since the observed value of the test statistic falls in the rejection region ($F = 50.96 > 5.14$), H_0 is rejected. There is sufficient evidence to indicate that the treatment means differ at $\alpha = .05$.

c.   To see if the blocking was effective, we test:

H_0: $\mu_1 = \mu_2 = \mu_3 = \mu_4$
H_a: At least two block means differ

The test statistic is $F = \dfrac{\text{MSB}}{\text{MSE}} = 202.59$

The rejection region requires $\alpha = .05$ in the upper tail of the F distribution with $v_1 = b - 1 = 4 - 1 = 3$ and $v_2 = n - k - b + 1 = 12 - 3 - 4 + 1 = 6$. From Table IX, Appendix A, $F_{.05} = 4.76$. The rejection region is $F > 4.76$.

Since the observed value of the test statistic falls in the rejection region ($F = 202.59 > 4.76$), H_0 is rejected. There is sufficient evidence to indicate that blocking was effective in reducing the experimental error at $\alpha = .05$.

d.   From the printout, we are given the sample means: $\bar{x}_B = 5.7$, $\bar{x}_A = 4.5$, and $\bar{x}_C = 3.2$. The confidence interval comparing the means for treatments C and B is (-3.196, -1.704). Since the endpoints are both negative, it indicates that the mean for treatment B is significantly greater than the mean from treatment C.

The confidence interval comparing the means for treatments B and A is (.379, 1.8706). Since the endpoints are both positive, it indicates that the mean for treatment B is significantly greater than the mean from treatment A.

The confidence interval comparing the means for treatments C and A is (-2.071, -0.5794). Since the endpoints are both negative, it indicates that the mean for treatment A is significantly greater than the mean from treatment C.

Thus, the rankings of the treatment means are: $\mu_B > \mu_A > \mu_C$

e.   The assumptions necessary to assure the validity of the inferences above are:

1.   The probability distributions of observations corresponding to all bk block-treatment combinations are normal.
2.   The variances of all the bk probability distributions are equal.
3.   The b blocks are randomly selected and all k treatments are applied (in random order) to each block.

10.67   a.   To determine if there is a significant difference in the mean TT index across the 15 weeks of the college football season, we test:

H_0: $\mu_1 = \mu_2 = \mu_3 = \cdots = \mu_{15}$
H_a: At least two treatment means differ

The test statistic is $F = 2.57$ and the p-value is $p = .0044$. Since the p-value is less than $\alpha = .01$ ($p = .0044 < .01$), H_0 is rejected. There is sufficient evidence of a difference in the mean TT index among the 15 weeks of the college football season at $\alpha = .01$.

b.   To determine if blocking on seasons was effective in removing an extraneous source of variation in the data, we test:

H_0: $\mu_1 = \mu_2 = \mu_3 = \cdots = \mu_6$
H_a: At least two blocking means differ

The test statistic is $F = 5.94$ and the p-value is $p = .0001$. Since the p-value is less than $\alpha = .01$ ($p = .0001 < .01$), H_0 is rejected. There is sufficient evidence that blocking on seasons was effective in removing an extraneous source of variation in the data at $\alpha = .01$.

c.   The number of pairwise comparisons is $c = k(k - 1)/2 = 15(15 - 1)/2 = 105$.

d.   There is only one significant difference in means among the 105 pairs compared. This difference is between the means of weeks 6 and 14. Thus, the mean TT score for week 6 is not significantly greater than the mean TT scores for all weeks except week 14. Similarly, the mean TT score for week 14 is not significantly less than the mean TT scores for all weeks except week 6. We can only say that week 6 is more topsy-turvy than week 14.

10.69   a.   This is a randomized block design. The blocks are the 12 plots of land. The treatments are the three methods used on the shrubs: fire, clipping, and control. The response variable is the mean number of flowers produced. The experimental units are the 36 shrubs.

b.

Treatment

		Fire	Clipping	Control
	1	Shrub 2	Shrub 1	Shrub 3
Plot	2	Shrub 3	Shrub 2	Shrub 1
	.	.	.	.
	.	.	.	.
	.	.	.	.
	12	Shrub 1	Shrub 3	Shrub 2

c.   To determine if there is a difference in the mean number of flowers produced among the three treatments, we test:

H_0: $\mu_1 = \mu_2 = \mu_3$

H_a: The mean number of flowers produced differ for at least two of the methods.

The test statistic is $F = 5.42$ and $p = .009$. We can reject the null hypothesis at the $\alpha > .009$ level of significance. At least two of the methods differ with respect to the mean number of flowers produced by pawpaws for $\alpha > .009$.

d.   The mean number of flowers produced by the Control is significantly less than the mean number of flowers produced by the Clipping method and the Burning method. The mean number of flowers produced by the Clipping method does not differ from the mean number of flowers produced by the Burning method.

10.71   a.   The treatments are the 4 pre-slaughter phases and the blocks are the cows.

b.   Using MINITAB, the ANOVA is:

Two-way ANOVA: Y versus Phase, Cow

```
Source   DF        SS        MS      F       P
Phase     3    521.13   173.708   3.63   0.030
Cow       7   1922.88   274.696   5.74   0.001
Error    21   1004.88    47.851
Total    31   3448.88

S = 6.917   R-Sq = 70.86%   R-Sq(adj) = 56.99%
```

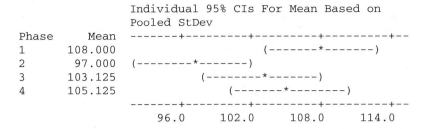

```
                    Individual 95% CIs For Mean Based on
                    Pooled StDev
Phase     Mean      -------+---------+---------+---------+--
1        108.000                          (-------*-------)
2         97.000    (--------*-------)
3        103.125            (--------*-------)
4        105.125             (-------*--------)
                    -------+---------+---------+---------+--
                       96.0      102.0     108.0     114.0
```

c.   To determine if there are differences in mean heart rates among the four pre-slaughter phases, we test:

H_0: $\mu_1 = \mu_2 = \mu_3 = \mu_4$
H_a: At least two blocking means differ

The test statistic is $F = 3.63$ and the p-value is $p = .030$. Since the p-value is less than $\alpha = .05$ ($p = .030 < .05$), H_0 is rejected. There is sufficient evidence to indicate differences in mean heart rates exist among the four pre-slaughter phases at $\alpha = .05$.

d.   Using MINITAB, the results of Tukey's multiple comparison procedure are:

```
Tukey 95.0% Simultaneous Confidence Intervals
Response Variable Y
All Pairwise Comparisons among Levels of Phase
Phase = 1  subtracted from:

Phase   Lower   Center   Upper   -------+---------+---------+---------
2      -20.64   -11.00   -1.364  (--------*-------)
3      -14.51    -4.87    4.761      (-------*-------)
4      -12.51    -2.88    6.761        (-------*-------)
                                 -------+---------+---------+---------
                                      -12        0        12


Phase = 2  subtracted from:

Phase   Lower   Center   Upper   -------+---------+---------+---------
3      -3.511    6.125   15.76                (-------*-------)
4      -1.511    8.125   17.76                 (-------*-------)
                                 -------+---------+---------+---------
                                      -12        0        12


Phase = 3  subtracted from:

Phase   Lower   Center   Upper   -------+---------+---------+---------
4      -7.636    2.000   11.64                (-------*-------)
                                 -------+---------+---------+---------
                                      -12        0        12
```

If we rank the means in order from the smallest to the larges we get:

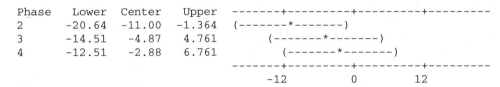

Means:	Phase 1	Phase 4	Phase 3	Phase 2
Treatments:	108.000	105.125	103.125	97.000

Comparing phase 1 mean to phase 2 mean, the confidence interval is (-20.64, -1.364). Since both end points are negative, the mean heart rate at phase 1 is significantly greater than that at phase 2.

Comparing phase 1 mean to phase 3 mean, the confidence interval is (-14.51, 4.761). Since 0 is in the confidence interval, there is no evidence that the mean heart rate at phase 1 is greater than that at phase 3.

Since there is no difference in the means for phases 1 and 3, there will not be a difference in the means for phases 1 and 4.

Comparing phase 4 mean to phase 2 mean, the confidence interval is (-1.511, 17.76). Since 0 is in the confidence interval, there is no evidence that the mean heart rate at phase 4 is greater than that at phase 2.

Since there is no difference in the means for phases 4 and 2, there will not be a difference in the means for phases 4 and 3. There will also not be a difference in the means for phases 2 and 3.

Thus, the only difference in mean heart rates is between phases 1 and 2, with the mean heart rate for phase 1 significantly greater than that for phase 2.

10.73   Using MINITAB, the ANOVA results are:

Two-way ANOVA: Temp versus Plant, Student

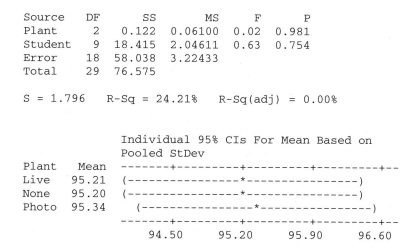

```
Source    DF      SS       MS      F      P
Plant      2   0.122   0.06100   0.02   0.981
Student    9  18.415   2.04611   0.63   0.754
Error     18  58.038   3.22433
Total     29  76.575

S = 1.796   R-Sq = 24.21%   R-Sq(adj) = 0.00%


                    Individual 95% CIs For Mean Based on
                    Pooled StDev
Plant    Mean    -------+---------+---------+---------+--
Live    95.21    (----------------*----------------)
None    95.20    (----------------*----------------)
Photo   95.34       (----------------*----------------)
                 -------+---------+---------+---------+--
                    94.50     95.20     95.90     96.60
```

To determine if there are differences among the mean temperatures among the three treatments, we test:

H_0: $\mu_1 = \mu_2 = \mu_3$
H_a: At least two treatment means differ

The test statistic is $F = 0.02$ and the associated p-value is $p = .981$. Since the p-value is very large, H_0 is not rejected. There is no evidence of a difference in mean temperature among the three treatments for any reasonable value of α. Since there is no difference, we do not need to compare the means. It appears that the presence of plants or pictures of plants does not reduce stress.

10.75   a.   The data were collected using a randomized complete block design and must be analyzed as a randomized complete block. In this case, each treatment was measured on each gene. Within each gene, the observations are not independent.

b.   To determine if the mean growth differs among the 3 light/dark conditions, we test:

H_0: $\mu_1 = \mu_2 = \mu_3$
H_a: At least two treatment means differ

c.   Using MINITAB, the results are:

Two-way ANOVA: Response versus Treatment, Block

```
Source        DF       SS       MS      F       P
Treatment      2     9.093   4.54660  5.33   0.006
Block        102    50.715   0.49720  0.58   0.999
Error        204   174.138   0.85362
Total        308   233.946

S = 0.9239     R-Sq = 25.56%    R-Sq(adj) = 0.00%


                           Individual 95% CIs For Mean Based on
                           Pooled StDev
Treatment      Mean     -----+---------+---------+---------+----
1           -0.318164   (--------*--------)
2            0.101547                        (--------*--------)
3           -0.090780                 (--------*--------)
                        -----+---------+---------+---------+----
                        -0.40     -0.20     -0.00      0.20
```

The test statistic is $F = 5.33$ and the p-value is $p = 0.006$. Since the p-value is less than $\alpha = .05$ $(p = .006 < .05)$, H_o is rejected. There is sufficient evidence to indicate a difference in mean growth among the 3 light/dark conditions at $\alpha = .05$.

d.   We will use the Bonferroni method to compare the treatment means.

$$\alpha^* = \frac{2\alpha}{k(k-1)} = \frac{2(.05)}{3(3-1)} = .0167$$

$$t_{\alpha^*/2} = t_{.0167/2} = t_{.0083} \approx 2.39 \quad \text{with } df = 204$$

FULL-DARK and TR-LIGHT:

$$\left(\overline{x}_i - \overline{x}_j\right) \pm t_{\alpha^*/2}(s)\sqrt{\frac{1}{n_i} + \frac{1}{n_j}} \Rightarrow (-.318164 - .101547) \pm 2.39(.9239)\sqrt{\frac{1}{103} + \frac{1}{103}}$$

$$\Rightarrow -.419711 \pm .307694 \Rightarrow (-.727405, \ -.112017)$$

FULL-DARK and TR-DARK:

$$(-.318164 - (-.090780)) \pm 2.39(.9239)\sqrt{\frac{1}{103} + \frac{1}{103}} \Rightarrow -.2273841 \pm .307694$$

$$\Rightarrow (-.535078, \ .080310)$$

TR-LIGHT and TR-DARK:

$$(.101547 - (-.090780)) \pm 2.39(.9239)\sqrt{\frac{1}{103} + \frac{1}{103}} \Rightarrow .192327 \pm .307694$$

$$\Rightarrow (-.115367, \ .500021)$$

A convenient summary of the results is below, where the treatment means are listed from highest to lowest with a solid line connecting those means that are not significantly different.

Means:	.101547	-.090780	-.318164
Treatments:	TR-LIGHT	TR-DARK	FULL-DARK

The mean growth for TR-LIGHT is significantly greater than the mean growth for FULL-DARK.

10.77   In a complete factorial experiment, the treatments are formed by taking every combination of all levels of both factors. The total number of treatments is equal to the total number of factor-level combinations.

10.79   The conditions that are required for valid inferences for a factorial ANOVA are:

1.   The response distribution for each factor-level combination (treatment) is normal.
2.   The response variance is constant for all treatments.
3.   Random and independent samples of experimental units are associated with each treatment.

10.81   a.   There are two factors.

b.   No, we cannot tell whether the factors are qualitative or quantitative.

c.   Yes. There are 3 levels of factor A and 5 levels of factor B.

d.   A treatment would consist of a combination of one level of factor A and one level of factor B. There are a total of $3 \times 5 = 15$ treatments.

e.   One problem with only one replicate is there are no degrees of freedom for error. This is overcome by having at least two replicates.

10.83   Some preliminary calculations are:

df for A is $a - 1 = 3 - 1 = 2$
df for B is $b - 1 = 4 - 1 = 3$
df for AB is $(a - 1)(b - 1) = 2(3) = 6$
df for Error is $n - ab = 24 - 3(4) = 12$
df for Total is $n - 1 = 24 - 1 = 23$

$$SSE = SS(Total) - SSA - SSB - SSAB = 18.1 - .8 - 5.3 - 9.6 = 2.4$$

$$MSA = \frac{SSA}{a-1} = \frac{.8}{3-1} = .40 \qquad MSB = \frac{SSB}{b-1} = \frac{5.3}{4-1} = 1.7667$$

$$MSAB = \frac{SSAB}{(a-1)(b-1)} = \frac{9.6}{(3-1)(4-1)} = 1.60$$

$$MSE = \frac{MSE}{n-ab} = \frac{2.4}{24-3(4)} = .2000$$

$$F_A = \frac{MSA}{MSE} = \frac{.4000}{.2000} = 2.00 \qquad F_B = \frac{MSB}{MSE} = \frac{1.7667}{.2000} = 8.83$$

$$F_{AB} = \frac{MSAB}{MSE} = \frac{1.6000}{.2000} = 8.00$$

The ANOVA table is:

Source	df	SS	MS	F
A	2	.8	0.4000	2.00
B	3	5.3	1.7667	8.83
AB	6	9.6	1.6000	8.00
Error	12	2.4	.2000	
Total	23	18.1		

b.   Sum of Squares for Treatment = SSA + SSB + SSAB = .8 + 5.3 + 9.6 = 15.7.

$$MST = \frac{SST}{ab-1} = \frac{15.7}{3(4)-1} = 1.4273 \qquad F_T = \frac{MST}{MSE} = \frac{1.4273}{.2000} = 7.14$$

To determine if the treatment means differ, we test:

H_0: $\mu_1 = \mu_2 = \cdots = \mu_{12}$
H_a: At least two treatment means differ

The test statistic is $F = 7.14$.

The rejection region requires $\alpha = .05$ in the upper tail of the F distribution with $v_1 = ab - 1 = 3(4) - 1 = 11$ and $v_2 = n - ab = 24 - 3(4) = 12$. From Table IX, Appendix A, $F_{.05} \approx 2.75$. The rejection region is $F > 2.75$.

Since the observed value of the test statistic falls in the rejection region ($F = 7.14 > 2.75$), H_0 is rejected. There is sufficient evidence to indicate the treatment means differ at $\alpha = .05$.

c. Yes. We need to partition the Treatment Sum of Squares into the Main Effects and Interaction Sum of Squares. Then we test whether factors A and B interact. Depending on the conclusion of the test for interaction, we either test for main effects or compare the treatment means.

d. Two factors are said to interact if the effects of one factor on the dependent variable are not the same at different levels of the second factor. If the factors interact, then tests for main effects are not necessary. We need to compare the treatment means for one factor at each level of the second.

e. To determine if the factors interact, we test:

H_0: Factors A and B do not interact to affect the response mean
H_a: Factors A and B do interact to affect the response mean

The test statistic is $F = \dfrac{\text{MSAB}}{\text{MSE}} = 8.00$

The rejection region requires $\alpha = .05$ in the upper tail of the F distribution with $v_1 = (a-1)(b-1) = (3-1)(4-1) = 6$ and $v_2 = n - ab = 24 - 3(4) = 12$. From Table IX, Appendix A, $F_{.05} = 3.00$. The rejection region is $F > 3.00$.

Since the observed value of the test statistic falls in the rejection region ($F = 8.00 > 3.00$), H_0 is rejected. There is sufficient evidence to indicate the two factors interact to affect the response mean at $\alpha = .05$.

f. No. Testing for main effects is not warranted because the interaction was significant. Instead, we compare the treatment means of one factor at each level of the second factor.

10.85 a. $\text{SSA} = .2(1000) = 200,\quad \text{SSB} = .1(1000) = 100,\quad \text{SSAB} = .1(1000) = 100$

$\text{SSE} = \text{SS(Total)} - \text{SSA} - \text{SSB} - \text{SSAB} = 1000 - 200 - 100 - 100 = 600$

$\text{SST} = \text{SSA} + \text{SSB} + \text{SSAB} = 200 + 100 + 100 = 400$

$\text{MSA} = \dfrac{\text{SSA}}{a-1} = \dfrac{200}{3-1} = 100 \qquad\qquad \text{MSB} = \dfrac{\text{SSB}}{b-1} = \dfrac{100}{3-1} = 50$

$\text{MSAB} = \dfrac{\text{SSAB}}{(a-1)(b-1)} = \dfrac{100}{(3-1)(3-1)} = 25 \qquad \text{MSE} = \dfrac{\text{SSE}}{n-ab} = \dfrac{600}{27-3(3)} = 33.333$

$\text{MST} = \dfrac{\text{SST}}{ab-1} = \dfrac{400}{3(3)-1} = 50$

$$F_A = \frac{\text{MSA}}{\text{MSE}} = \frac{100}{33.333} = 3.00 \qquad\qquad F_B = \frac{\text{MSB}}{\text{MSE}} = \frac{50}{33.333} = 1.50$$

$$F_{AB} = \frac{\text{MSAB}}{\text{MSE}} = \frac{25}{33.333} = .75 \qquad\qquad F_T = \frac{\text{MST}}{\text{MSE}} = \frac{50}{33.333} = 1.50$$

Source	df	SS	MS	F
A	2	200	100	3.00
B	2	100	50	1.50
AB	4	100	25	.75
Error	18	600	33.333	
Total	26	1000		

To determine whether the treatment means differ, we test:

$H_0:\ \mu_1 = \mu_2 = \cdots = \mu_9$
$H_a:$ At least two treatment means differ

The test statistic is $F_T = \dfrac{\text{MST}}{\text{MSE}} = 1.50$

Suppose $\alpha = .05$. The rejection region requires $\alpha = .05$ in the upper tail of the F distribution with $v_1 = ab - 1 = 3(3) - 1 = 8$ and $v_2 = n - ab = 27 - 3(3) = 18$. From Table IX, Appendix A, $F_{.05} = 2.51$. The rejection region is $F > 2.51$.

Since the observed value of the test statistic does not fall in the rejection region ($F = 1.50 \not> 2.51$), H_0 is not rejected. There is insufficient evidence to indicate the treatment means differ at $\alpha = .05$. Since there are no treatment mean differences, we have nothing more to do.

b.   SSA = .1(1000) = 100,   SSB = .1(1000) = 100,   SSAB = .5(1000) = 500

SSE = SS(Total) − SSA − SSB − SSAB = 1000 − 100 − 100 − 500 = 300

SST = SSA + SSB + SSAB = 100 + 100 + 500 = 700

$$\text{MSA} = \frac{\text{SSA}}{a-1} = \frac{100}{3-1} = 50 \qquad\qquad \text{MSB} = \frac{\text{SSB}}{b-1} = \frac{100}{3-1} = 50$$

$$\text{MSAB} = \frac{\text{SSAB}}{(a-1)(b-1)} = \frac{500}{(3-1)(3-1)} = 125 \qquad \text{MSE} = \frac{\text{SSE}}{n-ab} = \frac{300}{27-3(3)} = 16.667$$

$$\text{MST} = \frac{\text{SST}}{ab-1} = \frac{700}{9-1} = 87.5$$

$$F_A = \frac{MSA}{MSE} = \frac{50}{16.667} = 3.00 \qquad F_B = \frac{MSB}{MSE} = \frac{50}{16.667} = 3.00$$

$$F_{AB} = \frac{MSAB}{MSE} = \frac{125}{16.667} = 7.50 \qquad F_T = \frac{MST}{MSE} = \frac{87.5}{16.667} = 5.25$$

Source	df	SS	MS	F
A	2	100	50	3.00
B	2	100	50	3.00
AB	4	500	125	7.50
Error	18	300	16.667	
Total	26	1000		

To determine if the treatment means differ, we test:

H_0: $\mu_1 = \mu_2 = \cdots = \mu_9$
H_a: At least two treatment means differ

The test statistic is $F = \dfrac{MST}{MSE} = 5.25$

The rejection region requires $\alpha = .05$ in the upper tail of the F distribution with $v_1 = ab - 1 = 3(3) - 1 = 8$ and $v_2 = n - ab = 27 - 3(3) = 18$. From Table IX, Appendix A, $F_{.05} = 2.51$. The rejection region is $F > 2.51$.

Since the observed value of the test statistic falls in the rejection region ($F = 5.25 > 2.51$), H_0 is rejected. There is sufficient evidence to indicate the treatment means differ at $\alpha = .05$.

Since the treatment means differ, we next test for interaction between factors A and B. To determine if factors A and B interact, we test:

H_0: Factors A and B do not interact to affect the mean response
H_a: Factors A and B do interact to affect the mean response

The test statistic is $F = \dfrac{MSAB}{MSE} = 7.50$

The rejection region requires $\alpha = .05$ in the upper tail of the F distribution with $v_1 = (a-1)(b-1) = (3-1)(3-1) = 4$ and $v_2 = n - ab = 27 - 3(3) = 18$. From Table IX, Appendix A, $F_{.05} = 2.93$. The rejection region is $F > 2.93$.

Since the observed value of the test statistic falls in the rejection region ($F = 7.50 > 2.93$), H_0 is rejected. There is sufficient evidence to indicate the factors A and B interact at $\alpha = .05$. Since interaction is present, no tests for main effects are necessary.

c. $SSA = .4(1000) = 400, \quad SSB = .1(1000) = 100, \quad SSAB = .2(1000) = 200$

$SSE = SS(Total) - SSA - SSB - SSAB = 1000 - 400 - 100 - 200 = 300$

$$SST = SSA + SSB + SSAB = 400 + 100 + 200 = 700$$

$$MSA = \frac{SSA}{a-1} = \frac{400}{3-1} = 200 \qquad\qquad MSB = \frac{SSB}{b-1} = \frac{100}{3-1} = 50$$

$$MSAB = \frac{SSAB}{(a-1)(b-1)} = \frac{200}{(3-1)(3-1)} = 50 \qquad MSE = \frac{SSE}{n-ab} = \frac{300}{27-3(3)} = 16.667$$

$$MST = \frac{SST}{ab-1} = \frac{700}{3(3)-1} = 87.5$$

$$F_A = \frac{MSA}{MSE} = \frac{200}{16.667} = 12.00 \qquad\qquad F_B = \frac{MSB}{MSE} = \frac{50}{16.667} = 3.00$$

$$F_{AB} = \frac{MSAB}{MSE} = \frac{50}{16.667} = 3.00 \qquad\qquad F_T = \frac{MST}{MSE} = \frac{87.5}{16.667} = 5.25$$

Source	df	SS	MS	F
A	2	400	200	12.00
B	2	100	50	3.00
AB	4	200	50	3.00
Error	18	300	16.667	
Total	26	1000		

To determine if the treatment means differ, we test:

H_0: $\mu_1 = \mu_2 = \cdots = \mu_9$
H_a: At least two treatment means differ

The test statistic is $F = \dfrac{MST}{MSE} = 5.25$

The rejection region requires $\alpha = .05$ in the upper tail of the F distribution with $v_1 = ab - 1 = 3(3) - 1 = 8$ and $v_2 = n - ab = 27 - 3(3) = 18$. From Table IX, Appendix A, $F_{.05} = 2.51$. The rejection region is $F > 2.51$.

Since the observed value of the test statistic falls in the rejection region ($F = 5.25 > 2.51$), H_0 is rejected. There is sufficient evidence to indicate the treatment means differ at $\alpha = .05$.

Since the treatment means differ, we next test for interaction between factors A and B. To determine if factors A and B interact, we test:

H_0: Factors A and B do not interact to affect the mean response
H_a: Factors A and B do interact to affect the mean response

The test statistic is $F = \dfrac{MSAB}{MSE} = 3.00$

The rejection region requires $\alpha = .05$ in the upper tail of the F distribution with $v_1 = (a-1)(b-1) = (3-1)(3-1) = 4$ and $v_2 = n - ab = 27 - 3(3) = 18$. From Table IX, Appendix A, $F_{.05} = 2.93$. The rejection region is $F > 2.93$.

Since the observed value of the test statistic falls in the rejection region ($F = 3.00 > 2.93$), H_0 is rejected. There is sufficient evidence to indicate the factors A and B interact at $\alpha = .05$. Since interaction is present, no tests for main effects are necessary.

d. $SSA = .4(1000) = 400, \quad SSB = .4(1000) = 400, \quad SSAB = .1(1000) = 100$

$SSE = SS(Total) - SSA - SSB - SSAB = 1000 - 400 - 400 - 100 = 100$

$SST = SSA + SSB + SSAB = 400 + 400 + 100 = 900$

$$MSA = \frac{SSA}{a-1} = \frac{400}{3-1} = 200 \qquad\qquad MSB = \frac{SSB}{b-1} = \frac{400}{3-1} = 200$$

$$MSAB = \frac{SSAB}{(a-1)(b-1)} = \frac{100}{(3-1)(3-1)} = 25 \qquad MSE = \frac{SSE}{n-ab} = \frac{100}{27-3(3)} = 5.556$$

$$MST = \frac{SST}{ab-1} = \frac{900}{3(3)-1} = 112.5$$

$$F_A = \frac{MSA}{MSE} = \frac{200}{5.556} = 36.00 \qquad\qquad F_B = \frac{MSB}{MSE} = \frac{200}{5.556} = 36.00$$

$$F_{AB} = \frac{MSAB}{MSE} = \frac{25}{5.556} = 4.50 \qquad\qquad F_T = \frac{MST}{MSE} = \frac{112.5}{5.556} = 20.25$$

Source	df	SS	MS	F
A	2	400	200	36.00
B	2	400	200	36.00
AB	4	100	25	4.50
Error	18	100	5.556	
Total	26	1000		

To determine if the treatment means differ, we test:

H_0: $\mu_1 = \mu_2 = \cdots = \mu_9$
H_a: At least two treatment means differ

The test statistic is $F = \dfrac{MST}{MSE} = 20.25$

The rejection region requires $\alpha = .05$ in the upper tail of the F distribution with $v_1 = ab - 1 = 3(3) - 1 = 8$ and $v_2 = n - ab = 27 - 3(3) = 18$. From Table IX, Appendix A, $F_{.05} = 2.51$. The rejection region is $F > 2.51$.

Since the observed value of the test statistic falls in the rejection region ($F = 20.25 >$ 2.51), H_0 is rejected. There is sufficient evidence to indicate the treatment means differ at $\alpha = .05$.

Since the treatment means differ, we next test for interaction between factors A and B. To determine if factors A and B interact, we test:

H_0: Factors A and B do not interact to affect the mean response
H_a: Factors A and B do interact to affect the mean response

The test statistic is $F = \dfrac{MSAB}{MSE} = 4.50$

The rejection region requires $\alpha = .05$ in the upper tail of the F distribution with $v_1 = (a-1)(b-1) = (3-1)(3-1) = 4$ and $v_2 = n - ab = 27 - 3(3) = 18$. From Table IX, Appendix A, $F_{.05} = 2.93$. The rejection region is $F > 2.93$.

Since the observed value of the test statistic falls in the rejection region ($F = 4.50 >$ 2.93), H_0 is rejected. There is sufficient evidence to indicate the factors A and B interact at $\alpha = .05$. Since interaction is present, no tests for main effects are necessary.

10.87 a. This is a complete factorial experiment.

b. There are 2 factors – age and diet. Age has 2 levels – young and old. Diet has 2 levels – fine limestone and coarse limestone. There are a total of 2 x 2 = 4 treatments – young, fine limestone; young, coarse limestone; old, fine limestone; and old, coarse limestone.

c. The experimental unit is the hen.

d. The dependent variable is shell thickness.

e. The factors do not interact. This means that the effect of diet on shell thickness is the same for each age group of hens.

f. The shell thickness is not affected by the age of the hens.

g. The shell thickness is affected by the diet. The mean shell thickness of eggs produced by hens on the coarse limestone diet is significantly greater than the shell thickness of eggs produced by hens on the fine limestone diet.

10.89 a. There are 2 factors in this experiment. The first factor is wash-up event. There are 3 levels of this factor. The second factor is strata. There are 4 levels of strata: coarse-branching, medium-branching, fine-branching, and hydroid algae.

b. There are a total of $3 \times 4 = 12$ treatments in the experiment.

c. There are 2 replicates for each treatment in the experiment.

d. The total sample size is $12 \times 2 = 24$.

e. The response variable is the mussel density (percent per square centimeter).

f. The first ANOVA *F*-test is the test for differences among the treatments. If this test is significant, then the test for interaction follows.

Since not enough information is provided to test for differences among the treatment means, we will test for interaction.

To determine if Event and Strata interact to affect the mean mussel density, we test:

H_0: Event and Strata do not interact
H_a: Event and Strata interact

The test statistic is $F = 1.91$.

The *p*-value is $p > .05$. Since the *p*-value ($p > .05$) is not less than $\alpha = .05$, H_0 is not rejected. There is insufficient evidence to indicate Event and Strata interact at $\alpha = .05$.

g. Since the interaction is not significant, we will test for the main effects.

Event:

To determine if differences exist in the mean mussel densities among the three events, we test:

H_0: $\mu_1 = \mu_2 = \mu_3$
H_a: At least two event means differ

The test statistic is $F = 0.35$.

The *p*-value is $p > .05$. Since the *p*-value ($p > .05$) is not less than $\alpha = .05$, H_0 is not rejected. There is insufficient evidence to indicate differences exist in the mean mussel densities among the three events at $\alpha = .05$.

Strata:

To determine if differences exist in the mean mussel densities among the four strata, we test:

H_0: $\mu_1 = \mu_2 = \mu_3 = \mu_4$
H_a: At least two strata means differ
The test statistic is $F = 217.33$.

The *p*-value is $p < .05$. Since the *p*-value ($p < .05$) is less than $\alpha = .05$, H_0 is rejected. There is sufficient evidence to indicate differences exist in the mean mussel densities among the four strata at $\alpha = .05$.

h. The mean density for Strata Hydroid is significantly larger than the mean densities of the other 3 strata. The mean density for Strata Fine is significantly larger than the mean densities of Strata Medium and Strata Coarse. There is no difference in the mean densities of Strata Medium and Strata Coarse.

10.91   a.   A 2×2 factorial design was used for this study. The two factors are color and type of questions, each at 2 levels. There are $2 \times 2 = 4$ treatments. The 4 treatments are "blue, difficult", "blue, simple", "red, difficult", and "red, simple".

b.   Since the p-value is small ($p < .03$), H_0 is rejected. There is sufficient evidence to indicate that color and type of questions interact to affect mean exam scores. This means that the effect of color on the mean exam scores depends on the difficulty of the questions.

c.   Using MINITAB, the graph is:

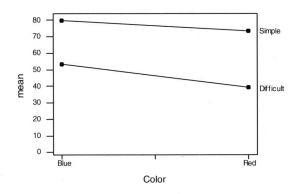

Looking at the graph, the lines are not parallel. The difference between the Blue and Red "simple" questions is not as great as the difference between the Blue and Red "difficult" questions. The effect of color on the exam scores depends on the level of difficulty.

10.93   a.

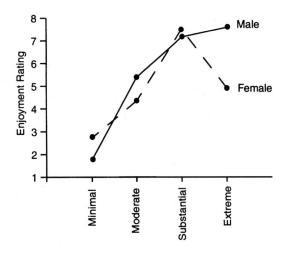

The pattern in the graph suggests interaction between suspense and gender. The difference between the mean ratings of the two genders varies depending on the suspense category. Notice that the lines cross three times.

b. To determine if interaction exists between suspense and gender, we test:

H_0: Gender and Suspense do not interact
H_a: Gender and Suspense interact

The test statistic is $F = 4.42$ and the p-value is $p = .007$. Since the p-value is so small, we will reject H_0. There is sufficient evidence to conclude that Gender and Suspense interact to affect the mean rating for any value of $\alpha > .007$.

c. Because the factors interact, there is evidence to indicate the difference between the mean enjoyment levels of males and females are different for different suspense levels.

10.95   a.   The ANOVA table is:

Source	df	F	p-value
Age	$3 - 1 = 2$	11.93	$< .001$
Book	$3 - 1 = 2$	23.64	$< .001$
Age x Book	$(3 - 1)(3 - 1) = 4$	2.99	$< .05$
Error	$108 - 3(3) = 99$		
Total	107		

b. The total number of treatments is 3 x 3 = 9. The treatments are:

18, Photos	24, Photos	30, Photos
18, Drawings	24, Drawings	30, Drawings
18, Control	24, Control	30, Control

c. To determine if the interaction between Age and Book exists, we test:

H_0: Age and Book do not interact to affect reenactment scores
H_a: Age and Book do interact to affect reenactment scores

The test statistic is $F = 2.99$ and the p-value is $p < .05$. Since the p-value is less than $\alpha = .05$, H_0 is rejected. There is sufficient evidence to indicate that Age and Book interact to affect target action scores at $\alpha = .05$. This means that the effect of Age on the reenactment scores depends on the level of Book.

d. We do not need to conduct the main effect tests because the interaction between Book and Age is significant. If the interaction is significant, it is possible that it could cover up any main effects.

e. Using MINITAB, a graph of the results is:

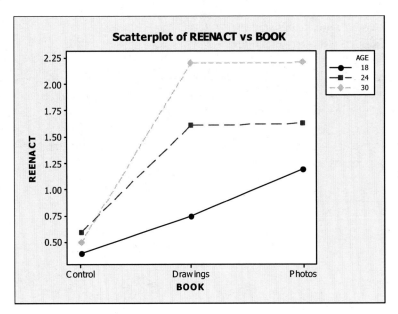

From the multiple comparisons, we can conclude:

Children at age 18 months: The mean reenactment score for children using books with Photos is significantly greater than the mean reenactment score for children using books with no photos or drawings. There are no other differences.

Children at age 24 months: The mean reenactment scores for children using books with Photos and books with Drawings are significantly greater than the mean reenactment score for children using books with no photos or drawings. There is no difference between the mean reenactment scores for children using books with Photos and children using books with Drawings.

Children at age 30 months: The mean reenactment scores for children using books with Photos and books with Drawings are significantly greater than the mean reenactment score for children using books with no photos or drawings. There is no difference between the mean reenactment scores for children using books with Photos and children using books with Drawings.

10.97 Using MINITAB, the results of the ANOVA are:

General Linear Model: NUMBER versus GROUP, SET

```
Factor   Type    Levels  Values
GROUP    fixed        3  3, 6, 12
SET      fixed        3  FIRST, LAST, MIDDLE


Analysis of Variance for NUMBER, using Adjusted SS for Tests

Source      DF   Seq SS  Adj SS  Adj MS      F      P
GROUP        2   15.267  15.267   7.633   7.59  0.001
SET          2   62.600  62.600  31.300  31.11  0.000
GROUP*SET    4    7.133   7.133   1.783   1.77  0.142
Error       81   81.500  81.500   1.006
Total       89  166.500


S = 1.00308   R-Sq = 51.05%   R-Sq(adj) = 46.22%


Unusual Observations for NUMBER

Obs   NUMBER     Fit   SE Fit  Residual  St Resid
 37  0.00000  2.30000  0.31720  -2.30000    -2.42 R
 66  5.00000  2.60000  0.31720   2.40000     2.52 R

R denotes an observation with a large standardized residual.

        Means

SET      N  NUMBER
FIRST   30  3.0000
LAST    30  1.1000
MIDDLE  30  1.4000


GROUP    N  NUMBER
 3      30  2.4000
 6      30  1.4333
12      30  1.6667


Tukey 95.0% Simultaneous Confidence Intervals
Response Variable NUMBER
All Pairwise Comparisons among Levels of GROUP
GROUP =  3  subtracted from:

GROUP  Lower   Center   Upper    ---+---------+---------+---------+---
 6     -1.586  -0.9667  -0.3477  (--------*--------)
12     -1.352  -0.7333  -0.1143      (--------*-------)
                                 ---+---------+---------+---------+---
                                 -1.40     -0.70      0.00      0.70


GROUP =  6  subtracted from:

GROUP  Lower   Center  Upper    ---+---------+---------+---------+---
12     -0.3857  0.2333  0.8523                  (--------*--------)
                                ---+---------+---------+---------+---
                                -1.40     -0.70      0.00      0.70
```

```
Tukey 95.0% Simultaneous Confidence Intervals
Response Variable NUMBER
All Pairwise Comparisons among Levels of SET
SET = FIRST   subtracted from:

SET      Lower   Center   Upper   -----+---------+---------+---------+-
LAST    -2.519  -1.900  -1.281   (-----*-----)
MIDDLE  -2.219  -1.600  -0.981     (-----*-----)
                                  -----+---------+---------+---------+-
                                    -2.0      -1.0      0.0       1.0


SET = LAST   subtracted from:

SET       Lower   Center   Upper   -----+---------+---------+---------+-
MIDDLE  -0.3190  0.3000  0.9190                        (-----*-----)
                                   -----+---------+---------+---------+-
                                     -2.0      -1.0      0.0       1.0
```

To determine if group size and photo set interact to affect the number of selections, we test:

H_0: Group size and Photo set do not interact to affect the number of selections
H_a: Group size and Photo set interact to affect the number of selections

The test statistic is $F = 1.77$ and the p-value is $p = .142$. Since the p-value is not small, H_0 is not rejected. There is insufficient evidence to indicate that group size and photo set interact to affect the number of selections for any reasonable level of α.

Since there is no evidence of an interaction, we will next test for the main effects. To determine if group size had an affect on the mean number of selections, we test:

H_0: $\mu_1 = \mu_2 = \mu_3$
H_a: At least two group size means differ

The test statistic is $F = 7.59$ and the p-value is $p = .001$. Since the p-value is so small, H_0 is rejected. There is sufficient evidence to indicate that group size has an affect on the mean number of selections for any level of α greater than .001.

To determine if photo set had an affect on the mean number of selections, we test:

H_0: $\mu_1 = \mu_2 = \mu_3$
H_a: At least two photo set means differ

The test statistic is $F = 31.11$ and the p-value is $p = .000$. Since the p-value is so small, H_0 is rejected. There is sufficient evidence to indicate that photo set has an affect the mean number of selections for any level of α greater than .000.

Since both main effects are significant, we will run Tukey's multiple comparison procedure on each main effect to find where the differences exist.

The mean number of selections made for the different group sizes are:

Means:	2.400	1.433	1.667
Groups:	3	6	12

The confidence interval comparing size 3 to size 6 is (-1.586, -.3477). Since both endpoints of the interval are negative, the mean number of selections for size 3 is significantly greater than the mean number of selections for size 6. The confidence interval comparing size 3 to size 12 is (-1.352, -.1143). Since both endpoints of the interval are negative, the mean number of selections for size 3 is significantly greater than the mean number of selections for size 12. The confidence interval comparing size 6 to size 12 is (-.3857, .8523). Since 0 is contained in the interval, there is no difference in the mean number of selections between sizes 6 and 12. Thus, there are significantly more selections made for group size 3 than for the other two sizes.

The mean number of selections made for the different photo sets are:

Means:	3.00	1.40	1.10
Groups:	First	Middle	Last

The confidence interval comparing the first photo set to the last photo set is (-2.519, -1.281). Since both endpoints of the interval are negative, the mean number of selections for the first photo set is significantly greater than the mean number of selections for the last photo set. The confidence interval comparing the first photo set to the middle photo set is (-2.219, -.981). Since both endpoints of the interval are negative, the mean number of selections for the first photo set is significantly greater than the mean number of selections for the middle photo set. The confidence interval comparing the middle photo set to the last photo set is (-.3190, .9190). Since 0 is contained in the interval, there is no difference in the mean number of selections between the last photo set and the middle photo set. Thus, there are significantly more selections made for the first photo set than for the other two photo sets.

10.99   a.   The treatment totals for the 4 treatments are:

Low Load, Ambiguous: $25(18.0) = 450$

Low Load, Common: $25(7.8) = 195$

High Load, Ambiguous: $25(6.1) = 152.5$

High Load, Common: $25(6.3) = 157.5$

b. $$CM = \frac{\left(\sum y_i\right)^2}{n} = \frac{(450 + 152.5 + 195 + 157.5)^2}{25 + 25 + 25 + 25} = \frac{955^2}{100} = 9,120.25$$

c. $SS(Load) = \dfrac{\sum A_i^2}{br} - CM = \dfrac{(450+195)^2 + (152.5+157.5)^2}{2(25)} - 9{,}120.25$

$= \dfrac{645^2 + 310^2}{50} - 9{,}120.25 \;=\; \dfrac{512{,}125}{50} - 9{,}120.25$

$= 10{,}242.5 - 9{,}120.25 = 1{,}122.25$

$SS(Name) = \dfrac{\sum B_i^2}{ar} - CM = \dfrac{(450+152.5)^2 + (195+157.5)^2}{2(25)} - 9{,}120.25$

$= \dfrac{602.5^2 + 352.5^2}{50} - 9{,}120.25 \;=\; \dfrac{487{,}262.5}{50} - 9{,}120.25$

$= 9{,}745.25 - 9{,}120.25 = 625$

$SS(Load \times Name) = \dfrac{\sum\sum AB_{ij}^2}{r} - SS(Name) - SS(Load) - CM$

$= \dfrac{450^2 + 195^2 + 152.5^2 + 157.5^2}{25} - 625 - 1{,}122.25 - 9{,}120.25$

$= 11{,}543.5 - 625 - 1{,}122.25 - 9{,}120.25 = 676$

d. The treatment variances and sums of squares are:

Low Load, Ambiguous: $s_{LA}^2 = 15^2 = 225$,
$\sum(x_i - \bar{x}_{LA})^2 = (n_{LA}-1)s_{LA}^2 = (25-1)225 = 5{,}400$

Low Load, Common: $s_{LC}^2 = 9.5^2 = 90.25$,
$\sum(x_i - \bar{x}_{LC})^2 = (n_{LC}-1)s_{LC}^2 = (25-1)90.25 = 2{,}166$

High Load, Ambiguous: $s_{HA}^2 = 9.5^2 = 90.25$,
$\sum(x_i - \bar{x}_{HA})^2 = (n_{HA}-1)s_{HA}^2 = (25-1)90.25 = 2{,}166$

High Load, Common: $s_{HC}^2 = 10^2 = 100$,
$\sum(x_i - \bar{x}_{HC})^2 = (n_{HC}-1)s_{HC}^2 = (25-1)100 = 2{,}400$

e. SSE = 5,400 + 2,166 + 2,166 + 2,400 = 12,132

f. SS(Total) = SS(Load) + SS(Name) + SS(Load x Name) + SSE

$= 1{,}122.25 + 625 + 676 + 12{,}132 = 14{,}555.25$

g.　The ANOVA table is:

Source	df	SS	MSE	F
Load	$2-1=1$	1,122.25	1,122.25	8.88
Name	$2-1=1$	625	625	4.95
Load x Name	$1 \times 1 = 1$	676	676	5.35
Error	$100 - 4 = 96$	12,132	126.375	
Total	99	14,555.25		

h.　The researchers calculated the F value to be $F = 5.34$. We calculated the F to be $F = 5.35$. The difference could be due to some round-off error.

i.　To determine if Load and Name interact to affect the number of jelly beans taken, we test:

H_0: Load and Name do not interact to affect the number of jelly beans taken
H_a: Load and Name do interact to affect the number of jelly beans taken

The test statistic is $F = 5.35$.

The rejection region requires $\alpha = .05$ in the upper tail of the F distribution with numerator df $v_1 = (a-1)(b-1) = (2-1)(2-1) = 1$ and denominator df $v_2 = n - ab = 100 - 2(2) = 96$. From Table IX, Appendix A, $F_{.05} \approx 3.96$. The rejection region is $F > 3.96$.

Since the observed value of the test statistic falls in the rejection region ($F = 5.35 > 3.96$), H_0 is rejected. There is sufficient evidence to indicate that Load and Name interact to affect the number of jelly beans taken at $\alpha = .05$.

Since the interaction is significant, there is no need to run the main effects test. Using MINITAB, a graphical display of the results is:

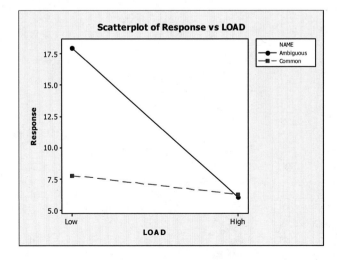

From the graph, it appears that when the Name of the jelly beans is Ambiguous, there is a significant difference in the mean number of jelly beans taken for the two loads. When Load is Low, there is a significantly greater mean number of jelly beans taken than when Load is high. When the Name of the jelly beans is Common, there appears to be no significant difference in the mean number of jelly beans taken between the 2 levels of load.

j.   In order for the techniques to be valid, the following assumptions are necessary:

1. We must assume that the response distribution for each factor-level combination (treatment) is normal
2. We must assume that the response variance is constant for all treatments.
3. We must assume that we have random and independent samples of experimental units associated with each treatment.

10.101 In a completely randomized design, independent random selection of treatments to be assigned to experimental units is required; in a randomized block design, sets (blocks) of matched experimental units are employed, with an experimental unit in each block randomly assigned a treatment.

10.103 When the overall level of significance of a multiple comparisons procedure is α, the level of significance for each comparison is less than α.

10.105 a. $\text{SST} = \text{SS(Total)} - \text{SS(Block)} - \text{SSE} = 22.31 - 10.688 - .288 = 11.334$

$$\text{MST} = \frac{\text{SST}}{k-1} = \frac{11.334}{4-1} = 3.778, \quad \text{df} = k - 1 = 4 - 1 = 3$$

$$\text{MS(Block)} = \frac{\text{SS(Block)}}{b-1} = \frac{10.688}{5-1} = 2.672, \quad \text{df} = b - 1 = 5 - 1 = 4$$

$$\text{MSE} = \frac{\text{SSE}}{n-k-b+1} = \frac{.288}{20-4-5+1} = .024, \quad \text{df} = n - k - b + 1 = 20 - 4 - 5 + 1 = 12$$

$$F_T = \frac{\text{MST}}{\text{MSE}} = \frac{3.778}{.024} = 157.42 \qquad F_{Block} = \frac{\text{MS(Block)}}{\text{MSE}} = \frac{2.672}{.024} = 111.33$$

The ANOVA Table is:

Source	df	SS	MS	F
Treatment	3	11.334	3.778	157.42
Block	4	10.688	2.672	111.33
Error	12	.288	.024	
Total	19	22.310		

b.   To determine if there is a difference among the treatment means, we test:

H_0: $\mu_A = \mu_B = \mu_C = \mu_D$
H_a: At least two treatment means differ

The test statistic is $F = \dfrac{\text{MST}}{\text{MSE}} = 157.42$

The rejection region requires $\alpha = .05$ in the upper tail of the F distribution with $v_1 = k - 1 = 4 - 1 = 3$ and $v_2 = n - k - b + 1 = 20 - 4 - 5 + 1 = 12$. From Table IX, Appendix A, $F_{.05} = 3.49$. The rejection region is $F > 3.49$.

Since the observed value of the test statistic falls in the rejection region ($F = 157.42 > 3.49$), H_0 is rejected. There is sufficient evidence to indicate a difference among the treatment means at $\alpha = .05$.

c. Since there is evidence of differences among the treatment means, we need to compare the treatment means. The number of pairwise comparisons is $c = k(k - 1)/2 = 4(4 - 1)/2 = 6$.

d. To determine if there are differences among the block means, we test:

H_0: All the block means are the same
H_a: At least two block means differ

The test statistic is $F = \dfrac{\text{MS(Block)}}{\text{MSE}} = 111.33$

The rejection region requires $\alpha = .05$ in the upper tail of the F distribution with $v_1 = b - 1 = 5 - 1 = 4$ and $v_2 = n - k - b + 1 = 20 - 4 - 5 + 1 = 12$. From Table IX, Appendix A, $F_{.05} = 3.26$. The rejection region is $F > 3.26$.

Since the observed value of the test statistic falls in the rejection region ($F = 111.33 > 3.26$), H_0 is rejected. There is sufficient evidence that the block means differ at $\alpha = .05$.

10.107 a. For this study, the experimental units are accountants.

b. The response variable is income.

c. There are 2 factors for this study: Mach rating and gender.

d. There are 3 levels of Mach rating: high, moderate, and low. There are 2 levels od gender: male and female.

e. There are a total of 3 x 2 = 6 treatments. They are: high male; high female; moderate male; moderate female; low male; and low female.

10.109 a. The data are collected as a completely randomized design because twenty boxes of each size were randomly selected and tested.

b. Yes. The confidence intervals surrounding each of the means do not overlap. This would indicate that there is a difference in the means for the two sizes.

c.   No. Several of the confidence intervals overlap. This would indicate that the mean compression strengths of the sizes that have intervals that overlap are not significantly different.

10.111 a.   To determine if the average heights of male Australian school children differ among the age groups, we test:

H_0: $\mu_1 = \mu_2 = \mu_3$
H_a: At least two treatment means differ

b.   The test statistic is $F = 4.57$ and the p-value is $p = .01$. Since the p-value is less than α ($p = .01 < .05$), H_0 is rejected. There is sufficient evidence to indicate the average heights of male Australian school children differ among the age groups at $\alpha = .05$.

c.   To determine if the average heights of female Australian school children differ among the age groups, we test:

H_0: $\mu_1 = \mu_2 = \mu_3$
H_a: At least two treatment means differ

The test statistic is $F = 0.85$ and the p-value is $p = .43$. Since the p-value is not less than α ($p = .43 > .05$), H_0 is not rejected. There is insufficient evidence to indicate the average heights of female Australian school children differ among the age groups at $\alpha = .05$.

d.   For the boys, differences were found in the mean standardized heights among the three tertiles. For the girls, no differences were found in the mean standardized heights among the three tertiles.

e.   From the table, there is no difference in the mean height between the youngest tertile and the middle tertile. However, the mean height for the oldest tertile has a mean height that is significantly smaller than the mean heights for the other two tertiles.

f.   The experimentwise error rate for the inference made in part **e** is $\alpha = .05$. The probability of concluding at least two means are different when none are different is .05.

g.   No differences were found among the means of the three groups of girls. Therefore, we do not need to find where the differences are.

10.113 a.   There are two factors. The factor measuring luckiness has three levels: lucky, unlucky, and uncertain. The factor measuring competition has two levels: competitive and noncompetitive.

b.   There are three tests:

To determine if interaction between Luckiness and Competition exists, we test:

H_0: The factors Luckiness and Competition do not
H_a: Luckiness and Competition interact to affect mean response

The test statistic is $F = 0.72$ and the p-value is $p = .72$. Since the p-value is so large, H_0 is not rejected. There is insufficient evidence to indicate Luckiness and Competition interact for any reasonable value of α.

To determine if the mean percentage guessed correctly differs among the three levels of luckiness, we test:

H_0: $\mu_1 = \mu_2 = \mu_3$
H_a: At least two Luckiness mean percentages differ

The test statistic is $F = 1.39$ and the p-value is $p = .26$. Since the p-value is so large, H_0 is not rejected. There is insufficient evidence to indicate differences in the mean percentages guessed correctly among the three levels of Luckiness for any level of $\alpha < .26$.

To determine if differences exist in the mean percentages guessed correctly between the two competition levels, we test:

H_0: $\mu_1 = \mu_2$
H_a: The two competition means differ.

The test statistic is $F = 2.84$ and the p-value is $p = .10$. Since the p-value is not small, H_0 is not rejected. There is insufficient evidence to indicate the mean percentages guessed correctly differ between the two levels of Competition for $\alpha < .10$.

10.115 a. The experimenters expected there to be much variation in the number of participants from week to week (more participants at the beginning and fewer as time goes on). Thus, by blocking on weeks, this extraneous source of variation can be controlled.

b. Using MINITAB, the ANOVA results are:

Two-way ANOVA: Walkers versus Week, Group

```
Source  DF      SS      MS      F       P
Group    4  1185.0  296.25  39.87  0.000
Week     5   386.4   77.28  10.40  0.000
Error   20   148.6    7.43
Total   29  1720.0

S = 2.726   R-Sq = 91.36%   R-Sq(adj) = 87.47%


                  Individual 95% CIs For Mean Based on
                  Pooled StDev
Group    Mean    --------+---------+---------+---------+
1      2.6667    (--*---)
2     17.0000                        (---*---)
3     20.6667                             (--*---)
4     10.5000               (---*--)
5      9.1667              (---*---)
                  --------+---------+---------+---------+
                      6.0      12.0      18.0      24.0
```

The ANOVA table is:

Source	df	SS	MS	F	p
Prompt	4	1185.0	296.25	39.87	0.000
Week	5	386.4	77.28	10.40	0.000
Error	20	148.6	7.43		
Total	29	1720.0			

c.   To determine if differences exist in the mean number of walkers per week among the five walker groups, we test:

H_0: $\mu_1 = \mu_2 = \mu_3 = \mu_4 = \mu_5$
H_a: At least two treatment means differ

where μ_i represents the mean number of walkers in group i.

The test statistic is $F = 39.87$.

The rejection region requires $\alpha = .05$ in the upper tail of the F distribution with $v_1 = k - 1 = 5 - 1 = 4$ and $v_2 = n - k - b + 1 = 30 - 5 - 6 + 1 = 20$. From Table IX, Appendix A, $F_{.05} = 2.87$. The rejection region is $F > 2.87$.

Since the observed value of the test statistic falls in the rejection region ($F = 39.87 > 2.87$), H_0 is rejected. There is sufficient evidence to indicate differences exist among the mean number of walkers per week among the 5 walker groups at $\alpha = .05$.

d.   Using SAS, the Tukey multiple comparison results are:

```
        Tukey's Studentized Range (HSD) Test for walkers

NOTE: This test controls the Type I experimentwise error rate, but it generally has a
                higher Type II error rate than REGWQ.


            Alpha                                   0.05
            Error Degrees of Freedom                  20
            Error Mean Square                       7.43
            Critical Value of Studentized Range  4.23186
            Minimum Significant Difference        4.7092


        Means with the same letter are not significantly different.


    Tukey Grouping         Mean      N    group

                A        20.667      6    FREQ/HIGH
                A
                A        17.000      6    FREQ/LOW

                B        10.500      6    INFREQ/LOW
                B
                B         9.167      6    INFREQ/HIGH

                C         2.667      6    CONTROL
```

There is no difference in the mean number of walkers between the Freq/High group and the Freq/Low group. There is no difference in the mean number of walkers between the Infreq/Low group and the Infreq/High group.

The mean number of walkers in the Control group is significantly less than the means for the other 4 groups. The mean number of walkers in the Infreq/Low and Infre/High groups is significantly less than the mean number of walkers in the Freq/Low and Freq/High groups.

e.   In order for the above inferences to be valid, the following assumptions must hold:

1)   The probability distributions of observations corresponding to all bk block-treatment conditions are normal.
2)   The variances of all the bk probability distributions are equal.
3)   The b blocks are randomly selected and all k treatments are applied (in random order) to each block.

10.117   To determine if there are differences in mean shell thicknesses among the four housing systems, we test:

H_o: $\mu_1 = \mu_2 = \mu_3 = \mu_4$
H_a: At least two means differ

The test statistics is $F = 11.74$ and the p-value is $p = .000$. Since the p-value is so small, H_0 is rejected. There is sufficient evidence to indicate there are differences in mean shell thicknesses among the four housing systems for any reasonable value of α.

To determine if there are differences in mean percent overruns among the four housing systems, we test:

H_o: $\mu_1 = \mu_2 = \mu_3 = \mu_4$
H_a: At least two means differ

The test statistics is $F = 31.36$ and the p-value is $p = .000$. Since the p-value is so small, H_0 is rejected. There is sufficient evidence to indicate there are differences in mean percent overruns among the four housing systems for any reasonable value of α.

To determine if there are differences in mean penetration strengths among the four housing systems, we test:

H_o: $\mu_1 = \mu_2 = \mu_3 = \mu_4$
H_a: At least two means differ

The test statistics is $F = 1.70$ and the p-value is $p = .193$. Since the p-value is not small, H_0 is not rejected. There is insufficient evidence to indicate there are differences in mean penetration strengths among the four housing systems for any $\alpha < .193$.

10.119 a.   Using MINITAB, the ANOVA table is:

One-way ANOVA: y versus FACES

```
Source  DF    SS    MS    F      P
FACES    5  23.09  4.62  3.96  0.007
Error   30  34.99  1.17
Total   35  58.07

S = 1.080   R-Sq = 39.75%   R-Sq(adj) = 29.71%
```

To determine if differences exist among the mean dominance ratings of the six facial expressions, we test:

H_0: $\mu_1 = \mu_2 = \mu_3 = \mu_4 = \mu_5 = \mu_6$
H_a: At least two treatment means differ

where μ_i represents the mean dominance rating for facial expression i.

The test statistic is $F = 3.96$.

The rejection region requires $\alpha = .10$ in the upper tail of the F distribution with $v_1 = k - 1 = 6 - 1 = 5$ and $v_2 = n - k = 36 - 6 = 30$. Using Table VIII, Appendix A, $F_{.10} = 2.05$. The rejection region is $F > 2.05$.

Since the observed value of the test statistic falls in the rejection region ($F = 3.96 > 2.05$), H_0 is rejected. There is sufficient evidence to indicate that differences exist among the mean dominance ratings among the facial expressions at $\alpha = .10$.

b.   Using MINITAB, the results of Tukey's multiple comparison procedure using $\alpha = .05$ is:

```
Tukey 95% Simultaneous Confidence Intervals
All Pairwise Comparisons among Levels of FACES

Individual confidence level = 99.51%


FACES = 1 subtracted from:

FACES   Lower   Center   Upper     +---------+---------+---------+---------
2      -2.144  -0.248   1.647              (---------*--------)
3      -3.487  -1.592   0.304      (--------*---------)
4      -1.731   0.165   2.061              (--------*--------)
5      -3.794  -1.898  -0.003      (---------*--------)
6      -3.069  -1.173   0.722          (--------*---------)
                                  +---------+---------+---------+---------
                                    -4.0      -2.0      0.0       2.0
```

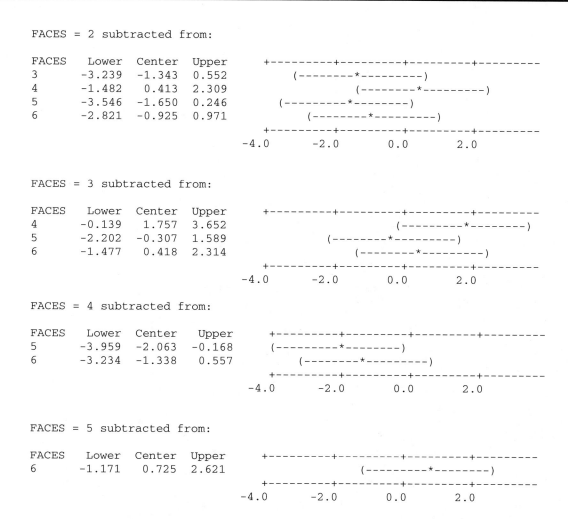

```
FACES = 2 subtracted from:

FACES   Lower   Center   Upper      +---------+---------+---------+---------
3      -3.239  -1.343   0.552             (--------*---------)
4      -1.482   0.413   2.309                     (--------*---------)
5      -3.546  -1.650   0.246          (---------*-------)
6      -2.821  -0.925   0.971             (--------*---------)
                                   +---------+---------+---------+---------
                                 -4.0      -2.0       0.0       2.0


FACES = 3 subtracted from:

FACES   Lower   Center   Upper      +---------+---------+---------+---------
4      -0.139   1.757   3.652                        (---------*--------)
5      -2.202  -0.307   1.589                (--------*---------)
6      -1.477   0.418   2.314                     (-------*---------)
                                   +---------+---------+---------+---------
                                 -4.0      -2.0       0.0       2.0


FACES = 4 subtracted from:

FACES   Lower   Center   Upper      +---------+---------+---------+---------
5      -3.959  -2.063  -0.168          (---------*--------)
6      -3.234  -1.338   0.557              (--------*---------)
                                   +---------+---------+---------+---------
                                 -4.0      -2.0       0.0       2.0


FACES = 5 subtracted from:

FACES   Lower   Center   Upper      +---------+---------+---------+---------
6      -1.171   0.725   2.621                     (---------*--------)
                                   +---------+---------+---------+---------
                                 -4.0      -2.0       0.0       2.0
```

A convenient summary of the results is below, where the treatment means are listed from highest to lowest with a solid line connecting those means that are not significantly different.

Means:	1.018	0.853	0.605	-0.320	-0.738	-1.045
Groups:	Happy	Angry	Disgusted	Neutral	Fearful	Sad

The mean dominance rating for the Sad face is significantly lower than the mean dominance ratings for the Happy and Angry faces. There are no other significant differences.

10.121 a.   Using MINITAB, the ANOVA results are:

ANOVA: Weight versus Diet, Size

```
Factor  Type   Levels  Values
Diet    fixed       2  1, 2
Size    fixed       2  1, 2


Analysis of Variance for Weight

Source      DF      SS      MS       F      P
Diet         1  0.0124  0.0124    0.22  0.645
Size         1  8.0679  8.0679  141.18  0.000
Diet*Size    1  0.0364  0.0364    0.64  0.432
Error       24  1.3715  0.0571
Total       27  9.4883
```

Source	df	SS	MS	F	P
Diet	1	0.0124	0.0124	0.22	0.645
Size	1	8.0679	8.0679	141.18	0.000
Interaction	1	0.0364	0.0364	0.64	0.432
Error	24	1.3715	0.0571		
Total	27	9.4883			

b.   To determine if Diet and Size interact, we test:

H_0: The factors Diet and Size do not interact
H_a: Diet and Size interact

The test statistic is $F = .64$ and p-value is $p = .432$. Since the p-value is so large, H_0 is not rejected. There is insufficient evidence to indicate Diet and Size interact for any reasonable value of α.

To determine if the mean weight differs for the two diets, we test:

H_0: $\mu_{Regular} = \mu_{Supplement}$
H_a: $\mu_{Regular} \neq \mu_{Supplement}$

The test statistic is $F = .22$ and p-value is $p = .645$. Since the p-value is so large, H_0 is not rejected. There is insufficient evidence to indicate the mean weight of kidneys differed for the two diets for any reasonable value of α.

To determine if the mean weight differs between the two sizes, we test:

H_0: $\mu_{Lean} = \mu_{Obese}$
H_a: $\mu_{Lean} \neq \mu_{Obese}$

The test statistic is $F = 141.18$ and the p-value is $p = .000$. Since the p-value is so small, H_0 is rejected. There is sufficient evidence to indicate the mean weight of kidneys differs for the two sizes for any reasonable value of α.

10.123 a.   df Photoperiod = $a - 1 = 2 - 1 = 1$;   df Gender = $b - 1 = 2 - 1 = 1$

df $P \times G = (a - 1)(b - 1) = (2 - 1)(2 - 1) = 1$

df Error = $n - ab = 124 - 2(2) = 120$;   df Total = $n - 1 = 124 - 1 = 123$

The ANOVA table is:

Source	df
Photoperiod	1
Gender	1
$P \times G$	1
Error	120
Total	123

b.   The test for interaction was not significant. This implies that the weight gain for each gender did not depend on the photoperiod.

c.   The *p*-value for the main effect photoperiod was less than .001. Since this *p*-value is so small, H_0 is rejected. There is sufficient evidence to indicate that the weight gain differs for the two photoperiods.

The *p*-value for the main effect gender was less than .001. Since this *p*-value is so small, H_0 is rejected. There is sufficient evidence to indicate that the weight gain differs for the two genders.

10.125 a.   This is a 2×2 factorial experiment.

b.   The two factors are the tent type (treated or untreated) and location (inside or outside). There are $2 \times 2 = 4$ treatments. The four treatments are (treated, inside), (treated, outside), (untreated, inside), and (untreated, outside).

c.   The response variable is the number of mosquito bites received in a 20 minute interval.

d.   There is sufficient evidence to indicate interaction is present. This indicates that the effect of the tent type on the number of mosquito bites depends on whether the person is inside or outside.

10.127 Using SAS, the ANOVA results are:

The SAS System

The ANOVA Procedure

Dependent Variable: CORROSION

Source	DF	Sum of Squares	Mean Square	F Value	Pr > F
Model	5	72.68833333	14.53766667	155.30	<.0001
Error	6	0.56166667	0.09361111		
Corrected Total	11	73.25000000			

R-Square	Coeff Var	Root MSE	CORROSION Mean
0.992332	3.170563	0.305959	9.650000

Source	DF	Anova SS	Mean Square	F Value	Pr > F
EXPOSURE	2	63.10500000	31.55250000	337.06	<.0001
SYSTEM	3	9.58333333	3.19444444	34.12	0.0004

The ANOVA Procedure

Tukey's Studentized Range (HSD) Test for CORROSION

NOTE: This test controls the Type I experimentwise error rate, but it generally has a higher Type II

error rate than REGWQ.

Alpha	0.05
Error Degrees of Freedom	6
Error Mean Square	0.093611
Critical Value of Studentized Range	4.89559
Minimum Significant Difference	0.8648

Means with the same letter are not significantly different.

Tukey Grouping		Mean	N	SYSTEM
	A	11.0667	3	3
	B	9.7333	3	2
	B			
C	B	9.0667	3	1
C				
C		8.7333	3	4

To determine if there are differences in the epoxy treatment means, we test:

H_0: $\mu_1 = \mu_2 = \mu_3 = \mu_4$

H_a: At least two treatment means differ

From the printout, the test statistic is $F = 34.12$ and the p-value is $p = 0.0004$.

Since the *p*-value is so small, H_0 is rejected. There is sufficient evidence to indicate a difference in the epoxy treatment means among the 4 groups for any reasonable value of α.

From the Tukey multiple comparison procedure, the mean corrosion rate for System 3 is significantly greater than the mean corrosion rate for any of the other 3 Systems. The mean corrosion rate for System 2 is significantly greater than the mean corrosion rate for System 4. No other significant differences exist. The System with the lowest corrosion rate is either System 1 or System 4. There is no significant difference in the mean corrosion rate between these two systems and they are both in the lowest group.

Simple Linear Regression

11.1   A deterministic model does not allow for random error or variation, whereas a probabilistic model does. An example where a deterministic model would be appropriate is:

Let y = cost of a 2×4 piece of lumber and x = length (in feet)

The model would be $y = \beta_1 x$. There should be no variation in price for the same length of wood.

An example where a probabilistic model would be appropriate is:

Let y = sales per month of a commodity and
   x = amount of money spent advertising

The model would be $y = \beta_0 + \beta_1 x + \varepsilon$. The sales per month will probably vary even if the amount of money spent on advertising remains the same.

11.3   The "line of means" is the deterministic component in a probabilistic model.

11.5

a.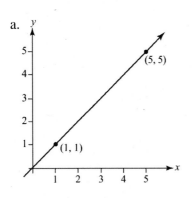

b.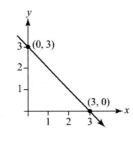

c.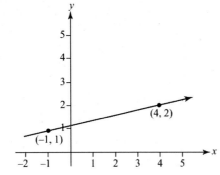

d.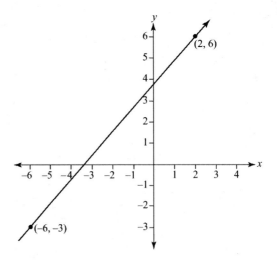

11.7   The two equations are:

$$4 = \beta_0 + \beta_1(-2) \text{ and } 6 = \beta_0 + \beta_1(4)$$

Subtracting the first equation from the second, we get

$$6 = \beta_0 + 4\beta_1$$
$$-\left(4 = \beta_0 - 2\beta_1\right)$$
$$\overline{}$$
$$2 = 6\beta_1 \Rightarrow \beta_1 = \frac{2}{6} = \frac{1}{3}$$

Substituting $\beta_1 = \frac{1}{3}$ into the first equation, we get:

$$4 = \beta_0 + \frac{1}{3}(-2) \Rightarrow \beta_0 = 4 + \frac{2}{3} = \frac{14}{3}$$

The equation for the line is $y = \frac{14}{3} + \frac{1}{3}x$

11.9   To graph a line, we need two points. Pick two values for x, and find the corresponding y values by substituting the values of x into the equation.

a.   Let $x = 0 \Rightarrow y = 4 + (0) = 4$
     and $x = 2 \Rightarrow y = 4 + (2) = 6$

b.   Let $x = 0 \Rightarrow y = 5 - 2(0) = 5$
     and $x = 2 \Rightarrow y = 5 - 2(2) = 1$

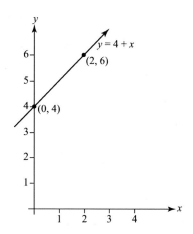

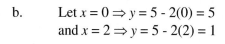

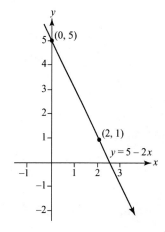

c.   Let $x = 0 \Rightarrow y = -4 + 3(0) = -4$
     and $x = 2 \Rightarrow y = -4 + 3(2) = 2$

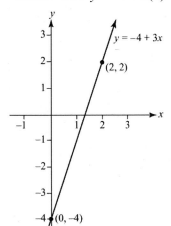

d.   Let $x = 0 \Rightarrow y = -2(0) = 0$
     and $x = 2 \Rightarrow y = -2(2) = -4$

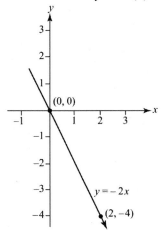

e.   Let $x = 0 \Rightarrow y = 0$
     and $x = 2 \Rightarrow y = 2$

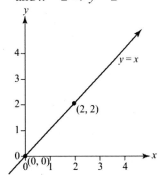

f.   Let $x = 0 \Rightarrow y = .5 + 1.5(0) = .5$
     and $x = 2 \Rightarrow y = .5 + 1.5(2) = 3.5$

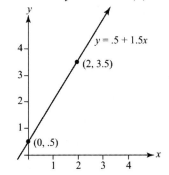

11.11   In regression, the error of prediction is the difference between the observed and predicted values of the dependent variable.

11.13   The statement "The estimates of β_0 and β_1 should be interpreted only within the sampled range of the independent variable, x." is true. We only have information about the relationship between y and x within the observed range of x.

11.15   From Exercise 11.14, $\hat{\beta}_0 = 7.10$ and $\hat{\beta}_1 = -.78$.

The fitted line is $\hat{y} = 7.10 - .78x$. To obtain values for $\hat{y}$, we substitute values of x into the equation and solve for $\hat{y}$.

a.

x	y	$\hat{y} = 7.10 - .78x$	$(y - \hat{y})$	$(y - \hat{y})^2$
7	2	1.64	.36	.1296
4	4	3.98	.02	.0004
6	2	2.42	−.42	.1764
2	5	5.54	−.54	.2916
1	7	6.32	.68	.4624
1	6	6.32	−.32	.1024
3	5	4.76	.24	.0576
			$\sum(y - \hat{y}) = 0.02$	$SSE = \sum(y - \hat{y})^2 = 1.2204$

b.

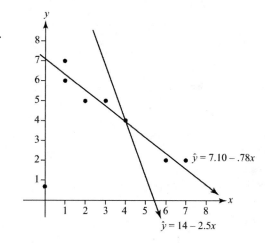

c.

x	y	$\hat{y} = 14 - 2.5x$	$(y - \hat{y})$	$(y - \hat{y})^2$
7	2	−3.5	5.5	30.25
4	4	4	0	0
6	2	−1	3	9
2	5	9	−4	16
1	7	11.5	−4.5	20.25
1	6	11.5	−5.5	30.25
3	5	6.5	−1.5	2.25
			$\sum(y - \hat{y}) = -7$	$SSE = 108.00$

11.17   a.

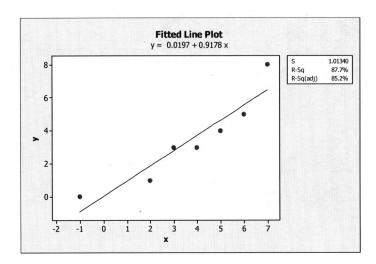

b.   As x increases, y tends to increase. Thus, there appears to be a positive, linear relationship between y and x.

c. $\hat{\beta}_1 = \dfrac{SS_{xy}}{SS_{xx}} = \dfrac{39.8571}{43.4286} = .9177616 \approx .918$

   $\hat{\beta}_0 = \bar{y} - \hat{\beta}_1\bar{x} = 3.4286 - .9177616(3.7143) = .0197581 \approx .020$

d.   The line appears to fit the data quite well.

e. $\hat{\beta}_0 = .020$ The estimated mean value of y when $x = 0$ is .020.

   $\hat{\beta}_1 = .918$ The estimated change in the mean value of y for each unit change in x is .918.

   These interpretations are valid only for values of x in the range from -1 to 7.

11.19   a.   The equation for the straight-line model is $y = \beta_0 + \beta_1 x + \varepsilon$.

b.   The least squares line is $\hat{y} = 19.393 - 8.036x$.

c. $\hat{\beta}_0 = 19.393$. The estimate of the mean wicking length when the concentration is 0 is 19.393 mm.

   $\hat{\beta}_1 = -8.036$. For each unit increase in antibody concentration, the mean wicking length is estimated to decrease by 8.036mm.

11.21   a.   **Radiata Pine**: For each unit increase in the natural logarithm of the number of blade cycles, the mean stress is estimated to decrease by 2.50.

   Hoop Pine: For each unit increase in the natural logarithm of the number of blade cycles, the mean stress is estimated to decrease by 2.36.

b. **Radiata Pine:** The mean stress is estimated to be 97.37 when the natural logarithm of the number of blade cycles is zero.

   Hoop Pine: The mean stress is estimated to be 122.03 when the natural logarithm of the number of blade cycles is zero.

c. Based on these results, it appears that the Hoop Pine is stronger and more fatigue resistant. When the natural logarithm of the number of blade cycles is zero, the mean stress for the Hoop Pine is greater than the mean stress for the Radiata Pine. As the natural logarithm of the number of blade cycles increases, the mean stress for the Radiata Pine decreases at a faster rate than does the Hoop Pine.

11.23 a. Using MINITAB, the least squares line is:

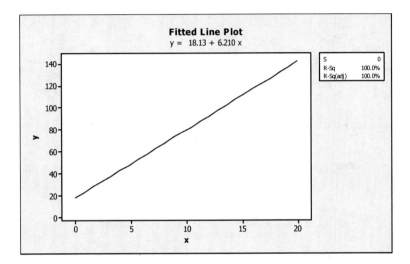

The slope of the line is positive – as x increases, y also increases.

b. $\hat{\beta}_0 = 18.13$. Since we have no information as to whether $x = 0$ is within the observed range, we will assume that it is. Then, the estimate of the mean magnitude when redshift level is 0 is 18.13.

c. $\hat{\beta}_1 = 6.21$. For each unit increase in redshift level, the mean value of magnitude is estimated to increase by 6.21.

11.25 a. The straight-line model is $y = \beta_0 + \beta_1 x + \varepsilon$.

b.   Using MINITAB, the results are:

Regression Analysis: Accuracy versus Distance

```
The regression equation is
Accuracy = 250 - 0.629 Distance


Predictor        Coef   SE Coef        T       P
Constant       250.14     14.23    17.58   0.000
Distance      -0.62944   0.04759  -13.23   0.000


S = 2.23639    R-Sq = 82.2%    R-Sq(adj) = 81.7%


Analysis of Variance

Source          DF       SS       MS        F       P
Regression       1   874.99   874.99   174.95   0.000
Residual Error  38   190.06     5.00
Total           39  1065.04
```

The least squares prediction equation is: $\hat{y} = 250.14 - .6294x$

c.   The estimated y-intercept is $\hat{\beta}_o = 250.14$. Since 0 is not in the observed range for values of driving distance, the y-intercept has no meaning.

d.   The estimated slope of the line is $\hat{\beta}_1 = -.6294$. For each additional yard of driving distance, the mean driving accuracy value will decrease by an estimated .6294.

e.   The slope will help determine if the golfer's concern is valid. If there is no significant linear relationship between driving accuracy and driving distance, then the golfer would have nothing to worry about. However, if there is a significant linear relationship between driving accuracy and driving distance and the relationship is negative, then as the driving distance increases, the mean driving accuracy will decrease.

11.27   a.   The hypothesized model would be $y = \beta_0 + \beta_1 x + \varepsilon$.

b.   The slope will be positive because the researcher's theory indicates a linearly increasing relationship.

c.   Some preliminary calculations are:

$$\sum x_i = 300 \qquad \sum y_i = 98,494 \qquad \sum x_i y_i = 1,473,555$$
$$\sum x_i^2 = 4900 \qquad \sum y_i^2 = 456,565,950$$

$$SS_{xy} = \sum x_i y_i - \frac{\sum x_i \sum y_i}{n} = 1,473,555 - \frac{300(98,494)}{24} = 242,380$$

$$SS_{xx} = \sum x_i^2 - \frac{\left(\sum x_i\right)^2}{n} = 4900 - \frac{300^2}{24} = 1150$$

$$\hat{\beta}_1 = \frac{SS_{xy}}{SS_{xx}} = \frac{242,380}{1150} = 210.7652174$$

For each additional resonance, the mean sound wave frequency is estimated to increase by 210.765.

$$\hat{\beta}_0 = \bar{y} - \hat{\beta}_1\bar{x} = \frac{98,494}{24} - (210.7652174)\left(\frac{300}{24}\right) = 1469.351449$$

Since 0 is not in the observed range of resonances, $\hat{\beta}_0$ has no interpretation other than being the y-intercept.

11.29   a.   Using MINITAB, the results of fitting a regression model are:

Regression Analysis: IdealHt versus Height

```
The regression equation is
IdealHt = 86.0 - 0.260 Height


Predictor        Coef   SE Coef      T      P
Constant       86.018     4.786  17.97  0.000
Height        -0.26046   0.07035  -3.70  0.000


S = 3.64336   R-Sq = 8.6%   R-Sq(adj) = 8.0%


Analysis of Variance

Source           DF        SS      MS      F      P
Regression        1    181.93  181.93  13.71  0.000
Residual Error  145   1924.74   13.27
Total           146   2106.67
```

The fitted regression line is $\hat{y} = 86.0 - .260x$.

Because the parameter estimate for Height is negative (-.26046), this confirms what the researchers found.

   b.   Yes, this is the correct interpretation.

c.   Using MINITAB, a plot of the data is:

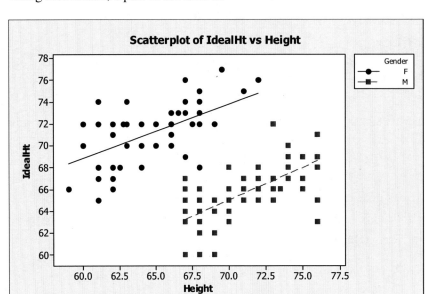

For the female students (round dots), there appears to be a positive relationship between the student's height and her ideal height. This same trend appears for the male students (squares).

d.   Using MINITAB, the results of fitting the female data are:

Regression Analysis: F-IdealHt versus F-Height

```
The regression equation is
F-IdealHt = 39.3 + 0.493 F-Height


Predictor     Coef   SE Coef      T       P
Constant     39.304     6.206   6.33   0.000
F-Height    0.49261   0.09602   5.13   0.000


S = 2.32153    R-Sq = 27.3%    R-Sq(adj) = 26.3%


Analysis of Variance

Source          DF       SS       MS       F       P
Regression       1   141.84   141.84   26.32   0.000
Residual Error  70   377.27     5.39
Total           71   519.11
```

The fitted regression line is $\hat{y} = 39.3 + .493x$.

The estimated slope of the line is .493. For each additional inch of height for females, the mean ideal partner's height is estimated to increase by .493 inches.

e.   Using MINITAB, the results of fitting the male data are:

Regression Analysis: M-IdealHt versus M-Height

```
The regression equation is
M-IdealHt = 23.3 + 0.596 M-Height


Predictor      Coef   SE Coef      T       P
Constant     23.274     6.333   3.67   0.000
M-Height    0.59630   0.08902   6.70   0.000


S = 2.05688    R-Sq = 38.1%    R-Sq(adj) = 37.2%


Analysis of Variance

Source           DF       SS       MS      F       P
Regression        1   189.82   189.82  44.87   0.000
Residual Error   73   308.84     4.23
Total            74   498.67
```

The fitted regression line is $\hat{y} = 23.3 + .596x$.

The estimated slope of the line is .596. For each additional inch of height for males, the mean ideal partner's height is estimated to increase by .596 inches.

f.   By fitting separate lines for the males and females, the results now match the theory developed by the anthropologists.

11.31   Using MINITAB, the results are:

Regression Analysis: Mass versus Time

```
The regression equation is
Mass = 5.22 - 0.114 Time


Predictor      Coef   SE Coef       T       P
Constant     5.2207    0.2960   17.64   0.000
Time        -0.11402  0.01032  -11.05   0.000


S = 0.857257   R-Sq = 85.3%    R-Sq(adj) = 84.6%


Analysis of Variance

Source           DF       SS       MS       F       P
Regression        1   89.794   89.794  122.19   0.000
Residual Error   21   15.433    0.735
Total            22  105.227
```

The fitted regression line is $\hat{y} = 5.2207 - .11402x$. Since the estimate of the slope is negative, the data indicate that the mass of the spill tends to diminish as time increases. For each additional minute, the mean mass is estimated to decrease by .11402 pounds.

11.33   See Figure 11.7 in the text.

11.35   a. $\text{SSE} = \text{SS}_{yy} - \hat{\beta}_1 \text{SS}_{xy} = 95 - .75(50) = 57.5$ $s^2 = \dfrac{\text{SSE}}{n-2} = \dfrac{57.5}{20-2} = 3.19444$

   b. $\text{SS}_{yy} = \sum y^2 - \dfrac{\left(\sum y\right)^2}{n} = 860 - \dfrac{50^2}{40} = 797.5$

   $\text{SSE} = \text{SS}_{yy} - \hat{\beta}_1 \text{SS}_{xy} = 797.5 - .2(2700) = 257.5$

   $s^2 = \dfrac{\text{SSE}}{n-2} = \dfrac{257.5}{40-2} = 6.776315789 \approx 6.7763$

   c. $\text{SS}_{yy} = \sum (y_i - \bar{y})^2 = 58$ $\hat{\beta}_1 = \dfrac{\text{SS}_{xy}}{\text{SS}_{xx}} = \dfrac{91}{170} = .535294117$

   $\text{SSE} = \text{SS}_{yy} - \hat{\beta}_1 \text{SS}_{xy} = 58 - .535294117(91) = 9.2882353 \approx 9.288$

   $s^2 = \dfrac{\text{SSE}}{n-2} = \dfrac{9.2882353}{10-2} = 1.161029413 \approx 1.1610$

11.37 $\text{SSE} = \text{SS}_{yy} - \hat{\beta}_1 \text{SS}_{xy}$

   where $\text{SS}_{yy} = \sum y_i^2 - \dfrac{\left(\sum y_i\right)^2}{n}$

For Exercise 11.14,
   $\sum y_i^2 = 159$ $\sum y_i = 31$

   $\text{SS}_{yy} = \sum y^2 - \dfrac{\left(\sum y\right)^2}{n} = 159 - \dfrac{31^2}{7} = 159 - 137.2857143 = 21.7142857$

   $\text{SS}_{xy} = -26.2857143$ $\hat{\beta}_1 = -.779661017$

Therefore, $\text{SSE} = 21.7142857 - (-.779661017)(-26.2857143) = 1.22033896 \approx 1.2203$

   $s^2 = \dfrac{\text{SSE}}{n-2} = \dfrac{1.22033896}{7-2} = .244067792$, $s = \sqrt{.244067792} = .4940$

We would expect most of the observations to fall within $2s = 2(.4940) = .988$ units of the least squares prediction line.

For Exercise 11.17,

   $\sum x_i = 26$ $\sum y_i = 24$ $\sum x_i y_i = 129$ $\sum x_i^2 = 140$ $\sum y_i^2 = 124$

$$SS_{xy} = \sum x_i y_i - \frac{\left(\sum x_i \sum y_i\right)}{n} = 129 - \frac{(26)(24)}{7} = 129 - 89.14285714 = 39.85714286$$

$$SS_{xx} = \sum x_i^2 - \frac{\left(\sum x_i\right)^2}{n} = 140 - \frac{(26)^2}{7} = 140 - 96.57142857 = 43.42857143$$

$$SS_{yy} = \sum y_i^2 - \frac{\left(\sum y_i\right)^2}{n} = 124 - \frac{(24)^2}{7} = 124 - 82.28571429 = 41.71428571$$

$$\hat{\beta}_1 = \frac{SS_{xy}}{SS_{xx}} = \frac{39.85714286}{43.42857143} = .917763157$$

$$SSE = SS_{yy} - \hat{\beta}_1 SS_{xy} = 41.71428571 - (.917763157)(39.85714286)$$
$$= 41.71428571 - 36.57941726 = 5.13486841 \approx 5.1349$$

$$s^2 = \frac{SSE}{n-2} = \frac{5.13486841}{7-2} = 1.026973682 \qquad s = \sqrt{1.026973682} = 1.0134$$

We would expect most of the observations to fall within $2s = 2(1.0134) = 2.0268$ units of the least squares prediction line.

11.39 a.   From the printout:

SSE = 22.268, $s^2 = MSE = 5.567$, and $s = \sqrt{5.567} = 2.3594$.

b.   We would expect most of the observations to fall within $2s = 2(2.3594) = 4.7188$ mm of the least squares prediction line.

11.41 a.   From the printout, SSE = 2760, $s^2 = 307.0$, and $s = 17.5111$.

b.   We would expect most of the observations to be within $2s = 2(17.5111) = 35.0222$ of the least squares line.

11.43 a.   From Exercise 11.26,

$$\sum y_i = 3,781.1 \qquad \sum y_i^2 = 651,612.45 \qquad SS_{xy} = -3,882.3686 \qquad \hat{\beta}_1 = -.305444503$$

$$SS_{yy} = \sum y_i^2 - \frac{\left(\sum y_i\right)^2}{n} = 651,612.45 - \frac{3,781.1^2}{22} = 1,761.6677$$

$$SSE = SS_{yy} - \hat{\beta}_1 SS_{xy} = 1,761.6677 - (-.305444503)(-3,882.3686) = 575.8195525$$

$$s^2 = \frac{SSE}{n-2} = \frac{575.8195525}{22-2} = 28.79097763$$

$$s = \sqrt{s^2} = \sqrt{28.79099763} = 5.36572247$$

We would expect most of the observations to be within $2s = 2(5.3657) = 10.7314$ of the least squares line.

b.　From Exercise 11.26,

$$\sum y_i = 3,764.2 \qquad \sum y_i^2 = 645,221.16 \qquad SS_{xy} = -3,442.16 \qquad \hat{\beta}_1 = -.270811187$$

$$SS_{yy} = \sum y_i^2 - \frac{\left(\sum y_i\right)^2}{n} = 645,221.16 - \frac{3,764.2^2}{22} = 1,166.54$$

$$SSE = SS_{yy} - \hat{\beta}_1 SS_{xy} = 1,166.54 - (-.270811187)(-3,442.16) = 234.3645646$$

$$s^2 = \frac{SSE}{n-2} = \frac{234.3645646}{22-2} = 11.71822823$$

$$s = \sqrt{s^2} = \sqrt{11.71822823} = 3.423189774$$

We would expect most of the observations to be within $2s = 2(3.4232) = 6.8464$ of the least squares line.

c.　Since the standard deviation ($s = 3.423$) for the reading scores is smaller than the standard deviation for the math scores ($s = 5.366$), the reading score can be more accurately predicted than the math score.

11.45　a.　Using MINITAB, the results of fitting the male data are:

Regression Analysis: M-IdealHt versus M-Height

```
The regression equation is
M-IdealHt = 23.3 + 0.596 M-Height


Predictor     Coef   SE Coef      T       P
Constant    23.274     6.333   3.67   0.000
M-Height   0.59630   0.08902   6.70   0.000


S = 2.05688   R-Sq = 38.1%   R-Sq(adj) = 37.2%


Analysis of Variance

Source          DF       SS       MS       F       P
Regression       1   189.82   189.82   44.87   0.000
Residual Error  73   308.84     4.23
Total           74   498.67
```

The fitted regression line is $\hat{y} = 23.3 + .596x$.

The estimate of σ is $s = 2.05688$.

b.   Using MINITAB, the results of fitting the female data are:

Regression Analysis: F-IdealHt versus F-Height

```
The regression equation is
F-IdealHt = 39.3 + 0.493 F-Height


Predictor     Coef  SE Coef     T      P
Constant    39.304    6.206  6.33  0.000
F-Height   0.49261  0.09602  5.13  0.000


S = 2.32153    R-Sq = 27.3%    R-Sq(adj) = 26.3%


Analysis of Variance

Source          DF       SS      MS      F      P
Regression       1   141.84  141.84  26.32  0.000
Residual Error  70   377.27    5.39
Total           71   519.11
```

The fitted regression line is $\hat{y} = 39.3 + .493x$.

The estimate of σ is $s = 2.32153$.

c.   The student's height is a more accurate predictor of ideal partner's height for the males than for the females because the estimate of the standard deviation for the males is smaller than the estimate for the females.

11.47   In the equation, $E(y) = \beta_0 + \beta_1 x$, the value of β_1 is 0 if x has no linear relationship with y.

11.49   If you are running a one-tailed test of β_1 in simple linear regression, you divide the p-value from the computer printout by 2 to obtain the correct p-value of the one-tailed test.

11.51   a.   For confidence coefficient .95, $\alpha = 1 - .95 = .05$ and $\alpha/2 = .05/2 = .025$. From Table VI, Appendix A, with df $= n - 2 = 12 - 2 = 10$, $t_{.025} = 2.228$.

The 95% confidence interval for β_1 is:

$$\hat{\beta}_1 \pm t_{.025} s_{\hat{\beta}_1} \Rightarrow 31 \pm 2.228(.5071) \Rightarrow 31 \pm 1.13 \Rightarrow (29.87, 32.13)$$

where $s_{\hat{\beta}_1} = \dfrac{s}{\sqrt{SS_{xx}}} = \dfrac{3}{\sqrt{35}} = .5071$

For confidence coefficient .90, $\alpha = 1 - .90 = .10$ and $\alpha/2 = .10/2 = .05$. From Table VI, Appendix A, with df $= 10$, $t_{.05} = 1.812$.

The 90% confidence interval for β_1 is:

$$\hat{\beta}_1 \pm t_{.05} s_{\hat{\beta}_1} \Rightarrow 31 \pm 1.812(.5071) \Rightarrow 31 \pm .92 \Rightarrow (30.08, 31.92)$$

b. $s^2 = \dfrac{SSE}{n-2} = \dfrac{1960}{18-2} = 122.5$, $s = \sqrt{s^2} = 11.0680$

For confidence coefficient, .95, $\alpha = 1 - .95 = .05$ and $\alpha/2 = .05/2 = .025$. From Table VI, Appendix A, with df $= n - 2 = 18 - 2 = 16$, $t_{.025} = 2.120$. The 95% confidence interval for β_1 is:

$$\hat{\beta}_1 \pm t_{.025} s_{\hat{\beta}_1} \Rightarrow 64 \pm 2.120(2.0207) \Rightarrow 64 \pm 4.28 \Rightarrow (59.72, 68.28)$$

where $s_{\hat{\beta}_1} = \dfrac{s}{\sqrt{SS_{xx}}} = \dfrac{11.0680}{\sqrt{30}} = 2.0207$

For confidence coefficient .90, $\alpha = 1 - .90 = .10$ and $\alpha/2 = .10/2 = .05$. From Table VI, Appendix A, with df $= 16$, $t_{.05} = 1.746$.

The 90% confidence interval for β_1 is:

$$\hat{\beta}_1 \pm t_{.05} s_{\hat{\beta}_1} \Rightarrow 64 \pm 1.746(2.0207) \Rightarrow 64 \pm 3.53 \Rightarrow (60.47, 67.53)$$

c. $s^2 = \dfrac{SSE}{n-2} = \dfrac{146}{24-2} = 6.6364$, $s = \sqrt{s^2} = 2.5761$

For confidence coefficient .95, $\alpha = 1 - .95 = .05$ and $\alpha/2 = .05/2 = .025$. From Table VI, Appendix A, with df $= n - 2 = 24 - 2 = 22$, $t_{.025} = 2.074$. The 95% confidence interval for β_1 is:

$$\hat{\beta}_1 \pm t_{.025} s_{\hat{\beta}_1} \Rightarrow -8.4 \pm 2.074(.322) \Rightarrow -8.4 \pm .67 \Rightarrow (-9.07, -7.73)$$

where $s_{\hat{\beta}_1} = \dfrac{s}{\sqrt{SS_{xx}}} = \dfrac{2.5761}{\sqrt{64}} = .3220$

For confidence coefficient .90, $\alpha = 1 - .90 = .10$ and $\alpha/2 = .10/2 = .05$. From Table VI, Appendix A, with df $= 22$, $t_{.05} = 1.717$.

The 90% confidence interval for β_1 is:

$$\hat{\beta}_1 \pm t_{.05} s_{\hat{\beta}_1} \Rightarrow -8.4 \pm 1.717(.322) \Rightarrow -8.4 \pm .55 \Rightarrow (-8.95, -7.85)$$

11.53　a & c.　Using MINITAB, a scatterplot of the data is:

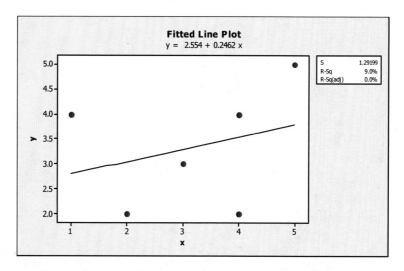

b.　Some preliminary calculations are:

$$\sum x_i = 19 \qquad \sum y_i = 20 \qquad \sum x_i y_i = 66$$

$$\sum x_i^2 = 71 \qquad \sum y_i^2 = 74$$

$$SS_{xy} = \sum x_i y_i - \frac{\sum x_i \sum y_i}{n} = 66 - \frac{19(20)}{6} = 2.666666667$$

$$SS_{xx} = \sum x_i^2 - \frac{\left(\sum x_i\right)^2}{n} = 71 - \frac{19^2}{6} = 10.833333333$$

$$\hat{\beta}_1 = \frac{SS_{xy}}{SS_{xx}} = \frac{2.66666667}{10.833333333} = .246153846$$

$$\hat{\beta}_0 = \bar{y} - \hat{\beta}_1\bar{x} = \frac{20}{6} - (.246153846)\left(\frac{19}{6}\right) = 2.553846155$$

The least squares prediction equation is $\hat{y} = 2.554 + .246x$

d.　Some preliminary calculations are:

$$SS_{yy} = \sum y_i^2 - \frac{\left(\sum y_i\right)^2}{n} = 74 - \frac{20^2}{6} = 7.3333333333$$

$$SSE = SS_{yy} - \hat{\beta}_1 SS_{xy} = 7.3333333333 - (.246153846)(2.666666667) = 6.6769$$

$$s^2 = \frac{\text{SSE}}{n-2} = \frac{6.6769}{6-2} = 1.6692 \qquad s = \sqrt{s^2} = \sqrt{1.6692} = 1.292$$

The test statistic is $t = \dfrac{\hat{\beta}_1 - 0}{s_{\hat{\beta}_1}} = \dfrac{.246}{\dfrac{1.292}{\sqrt{10.83333333}}} = .627$

e.   To determine if x and y are linearly related, we test:

H_0: $\beta_1 = 0$
H_a: $\beta_1 \neq 0$

The test statistic is $t = .627$.

The rejection region requires $\alpha / 2 = .01 / 2 = .005$ in each tail of the t distribution with df $= n - 2 = 6 - 2 = 4$. From Table VI, Appendix A, $t_{.005} = 4.604$. The rejection region is $t < -4.604$ or $t > 4.604$.

Since the observed value of the test statistic does not fall in the rejection region ($t = .627 \ngtr 4.604$), H_0 is not rejected. There is insufficient evidence to indicate that x and y are linearly related at $\alpha = .01$.

f.   For confidence coefficient .99, $\alpha = 1 - .99 = .01$ and $\alpha / 2 = .01 / 2 = .005$. From Table VI, Appendix A, with df $= n - 2 = 6 - 2 = 4$, $t_{.005} = 4.604$. The 99% confidence interval is:

$$\hat{\beta}_1 \pm t_{.005} s_{\hat{\beta}_1} \Rightarrow \hat{\beta}_1 \pm t_{.005} \frac{s}{\sqrt{\text{SS}_{xx}}} \Rightarrow .246 \pm 4.604 \frac{1.292}{\sqrt{10.8333333}}$$

$$\Rightarrow .246 \pm 1.807 \Rightarrow (-1.561, 2.053)$$

We are 99% confident that the true value of β_1 is between -1.561 and 2.053.

11.55   a.   Using MINITAB, a scatterplot of the data is:

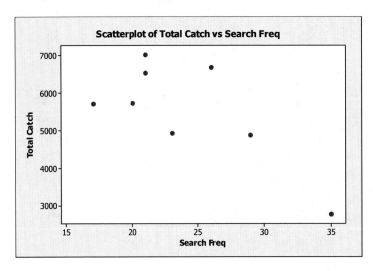

There appears to be a negative linear trend – as search frequency increases, the total catch tends to decrease.

b. From the printout, the least squares prediction equation is: $\hat{y} = 9{,}658.24 - 171.573x$. For each additional unit increase in search frequency, the mean total catch is estimated to decrease by 171.573 kilograms.

c. To determine if the total catch is negatively linearly related to search frequency, we test:

H_0: $\beta_1 = 0$
H_a: $\beta_1 < 0$

d. From the printout, the p-value is $p = .0402/2 = .0201$.

e. Since the p-value is less than α ($p = .0201 < .05$), H_0 is rejected. There is sufficient evidence to indicate that the total catch is negatively linearly related to search frequency at $\alpha = .05$.

11.57 a. From Exercises 11.26 and 11.43, $SS_{xx} = 12{,}710.55318$, $\hat{\beta}_1 = -.305$, and $s = 5.3657$,

To determine if math score and percentage of students below the poverty level are negatively linearly related, we test:

H_0: $\beta_1 = 0$
H_a: $\beta_1 < 0$

The test statistic is $t = \dfrac{\hat{\beta}_1 - 0}{s_{\hat{\beta}_1}} = \dfrac{-.305}{\dfrac{5.3657}{\sqrt{12{,}710.55318}}} = -6.41$

The rejection region requires $\alpha = .01$ in the lower tail of the t distribution with df $= n - 2 = 22 - 2 = 20$. From Table VI, Appendix A, $t_{.01} = 2.582$. The rejection region is $t < -2.582$.

Since the observed value of the test statistic falls in the rejection region ($t = -6.41 < -2.582$), H_0 is rejected. There is sufficient evidence to indicate the math score and percentage of students below the poverty level are negatively linearly related at $\alpha = .01$.

b. For confidence coefficient .99, $\alpha = 1 - .99 = .01$ and $\alpha/2 = .01/2 = .005$. From Table VI, Appendix A, with df $= n - 2 = 22 - 2 = 20$, $t_{.005} = 2.845$. The 99% confidence interval is:

$\hat{\beta}_1 \pm t_{.005}s_{\hat{\beta}_1} \Rightarrow \hat{\beta}_1 \pm t_{.005}\dfrac{s}{\sqrt{SS_{xx}}} \Rightarrow -.305 \pm 2.845\dfrac{5.3657}{\sqrt{12{,}710.55318}}$

$\Rightarrow -.305 \pm .135 \Rightarrow (-.440, -.170)$

We are 99% confident that the change in the mean value of math scores for each unit change in percentage of students below the poverty level is between $-.440$ and $-.170$.

11.59   Some preliminary calculations are:

$$SS_{yy} = \sum y_i^2 - \frac{\left(\sum y_i\right)^2}{n} = 769.72 - \frac{(135.8)^2}{24} = 1.3183333$$

$$SSE = SS_{yy} - \hat{\beta}_1 SS_{xy} = 1.3183333 - (-.002310625)(-130.441667) = 1.016931523$$

$$s^2 = \frac{SSE}{n-2} = \frac{1.016931523}{22} = .046224 \qquad s = \sqrt{.046224} = .215$$

$$s_{\hat{\beta}_1} = \sqrt{\frac{s^2}{SS_{xx}}} = \sqrt{\frac{.046224}{56,452.958}} = .000905$$

For confidence level .95, $\alpha = .05$ and $\alpha/2 = .05/2 = .025$. From Table VI, Appendix A with $df = n - 2 = 24 - 2 = 22$, $t_{.025} = 2.074$.

The confidence interval is:

$$\hat{\beta}_1 \pm t_{.025} s_{\hat{\beta}_1} \Rightarrow -.0023 \pm 2.074(.000905) \Rightarrow -.0023 \pm .0019 \Rightarrow (-.0042, -.0004)$$

We are 95% confident that the change in the mean sweetness index for each one unit change in the pectin is between $-.0042$ and $-.0004$.

11.61   a.   The model relating Anthropogenic Index, y, to Natural Origin Index, x, is:

$$y = \beta_0 + \beta_1 x + \varepsilon$$

b.   Some preliminary calculations are:

$$\sum x_i = 1,047.22 \qquad\qquad \sum y_i = 1,283.46 \qquad\qquad \sum x_i y_i = 32,048.2$$

$$\sum x_i^2 = 24,607.5 \qquad\qquad \sum y_i^2 = 54,184.8$$

$$SS_{xy} = \sum x_i y_i - \frac{\sum x_i \sum y_i}{n} = 32,048.2 - \frac{1,047.22(1,283.46)}{54} = 7,158.10776$$

$$SS_{xx} = \sum x_i^2 - \frac{\left(\sum x_i\right)^2}{n} = 24,607.5 - \frac{1,047.22^2}{54} = 4,298.80133$$

$$\hat{\beta}_1 = \frac{SS_{xy}}{SS_{xx}} = \frac{7,158.10776}{4,298.80133} = 1.665140399$$

$$\hat{\beta}_0 = \bar{y} - \hat{\beta}_1\bar{x} = \frac{1,283.46}{54} - (1.665140399)\left(\frac{1,047.22}{54}\right) = -8.524228315$$

The least squares prediction equation is $\hat{y} = -8.524 + 1.665x$

c. $\hat{\beta}_1 = 1.665$. For each unit increase in the Natural Origin Index, the mean Anthropogenic Index is estimated to increase by 1.665.

$\hat{\beta}_0 = 2.522$. Since $x = 0$ is not in the observed range of the Natural Origin Index, $\hat{\beta}_0$ has no interpretation other than the y-intercept.

d. Some preliminary calculations are:

$$SS_{yy} = \sum y_i^2 - \frac{\left(\sum y_i\right)^2}{n} = 54,184.8 - \frac{1,283.46^2}{54} = 23,679.80793$$

$$SSE = SS_{yy} - \hat{\beta}_1 SS_{xy} = 23,679.80793 - (1.665140399)(7,158.10776) = 11,760.55352$$

$$s^2 = \frac{SSE}{n-2} = \frac{11,760.55352}{54-2} = 226.1644908$$

$$s = \sqrt{s^2} = \sqrt{226.1644908} = 15.03876627$$

To determine if the natural origin index and anthropogenic index are positively linearly related, we test:

H_0: $\beta_1 = 0$
H_a: $\beta_1 > 0$

The test statistic is $t = \dfrac{\hat{\beta}_1 - 0}{s_{\hat{\beta}_1}} = \dfrac{1.665}{\dfrac{15.0388}{\sqrt{4,298.80133}}} = 7.259$

The rejection region requires $\alpha = .05$ in the upper tail of the t distribution with df $= n - 2$ $= 54 - 2 = 52$. From Table VI, Appendix A, $t_{.05} \approx 1.678$. The rejection region is $t > 1.678$.

Since the observed value of the test statistic falls in the rejection region ($t = 7.259 > 1.678$), H_0 is rejected. There is sufficient evidence to indicate the natural origin index and anthropogenic index are positively linearly related at $\alpha = .05$.

e. For confidence coefficient .95, $\alpha = 1 - .95 = .05$ and $\alpha / 2 = .05 / 2 = .025$. From Table VI, Appendix A, with df $= n - 2 = 54 - 2 = 52$, $t_{.025} \approx 2.011$. The 95% confidence interval is:

$$\hat{\beta}_1 \pm t_{.025} s_{\hat{\beta}_1} \Rightarrow \hat{\beta}_1 \pm t_{.025} \frac{s}{\sqrt{SS_{xx}}} \Rightarrow 1.665 \pm 2.011 \frac{15.0688}{\sqrt{4,298.80133}}$$

$$\Rightarrow 1.665 \pm .461 \Rightarrow (1.204, \quad 2.126)$$

We are 95% confident that the change in the anthropogenic index for each unit increase in the natural origin index is between 1.204 and 2.126.

11.63   a.   To determine if body-plus-head rotation and active head movement are positively linearly related, we test:

H_0: $\beta_1 = 0$

H_a: $\beta_1 > 0$

The test statistic is $t = \dfrac{\hat{\beta}_1 - 0}{s_{\hat{\beta}_1}} = \dfrac{.88 - 0}{.14} = 6.286$

The rejection region requires $\alpha = .05$ in the upper tail of the t distribution with df $= n - 2$ $= 39 - 2 = 37$. From Table VI, Appendix A, $t_{.05} \approx 1.687$. The rejection region is $t > 1.687$.

Since the observed value of the test statistic falls in the rejection region ($t = 6.286 >$ 1.687), H_0 is rejected. There is sufficient evidence to indicate that body-plus-head rotation and active head movement are positively linearly related at $\alpha = .05$.

b.   For confidence level .90, $\alpha = .10$ and $\alpha / 2 = .10 / 2 = .05$. From Table VI, Appendix A, with df $= n - 2 = 39 - 2 = 37$, $t_{.05} \approx 1.687$. The confidence interval is:

$$\hat{\beta}_1 \pm s_{\hat{\beta}_1} \Rightarrow .88 \pm 1.687(.14) \Rightarrow .88 \pm .2362 \Rightarrow (.6438, 1.1162)$$

We are 90% confident that the true value of β_1 is between .6438 and 1.1162.

c.   Because the interval in part b contains the value 1, there is no evidence that the true slope of the line differs from 1.

11.65   a.   From the printout, $\hat{\beta}_o = .515$, $\hat{\beta}_1 = .000021$, and $s = .0370$.

b.   To determine if there is a positive linear relationship between elevation and slugging percentage, we test:

H_0: $\beta_1 = 0$

H_a: $\beta_1 > 0$

From the printout, the test statistic is $t = 2.89$ and the p-value is $p = .008/2 = .004$.

Since the p-value is less than $\alpha = .01$ ($p = .004 < .01$), H_o is rejected. There is sufficient evidence to indicate a positive linear relationship between elevation and slugging percentage at $\alpha = .01$.

c. Using MINITAB, the scatterplot is:

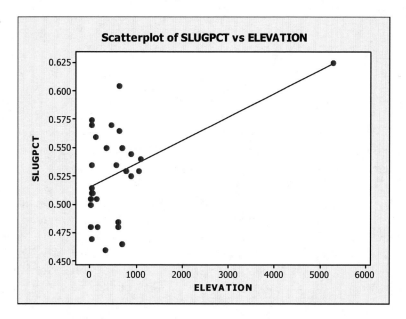

The data point corresponding to Denver is the data point furthest to the right. This point is very influential. If this point were deleted, there probably would not be a relationship between slugging percentage and elevation.

d. Using MINITAB with Denver removed, the results are:

Regression Analysis: SLUGPCT2 versus ELEVATION2

```
The regression equation is
SLUGPCT2 = 0.515 + 0.000020 ELEVATION2


Predictor          Coef     SE Coef       T       P
Constant        0.51537     0.01066   48.33   0.000
ELEVATION2   0.00002012  0.00002034    0.99   0.332


S = 0.0376839   R-Sq = 3.6%   R-Sq(adj) = 0.0%


Analysis of Variance

Source           DF         SS         MS      F      P
Regression        1   0.001389   0.001389   0.98  0.332
Residual Error   26   0.036922   0.001420
Total            27   0.038311
```

From the printout, $\hat{\beta}_o = .515$, $\hat{\beta}_1 = .000020$, and $s = .0377$.

To determine if there is a positive linear relationship between elevation and slugging percentage, we test:

H_0: $\beta_1 = 0$
H_a: $\beta_1 > 0$

From the printout, the test statistic is $t = 0.99$ and the p-value is $p = .332/2 = .166$.

Since the p-value is not less than $\alpha = .01$ ($p = .166 > .01$), H_0 is not rejected. There is insufficient evidence to indicate a positive linear relationship between elevation and slugging percentage when Denver is deleted using $\alpha = .01$.

The "thin air" theory appears to be valid. No other city has an elevation that is close to that of Denver's. If Denver is not included in the analysis, there is no relationship between the slugging percentage and the elevation. Without Denver, the elevations of the cities range from 10 feet to 1082 feet. The elevation of Denver is 5277. The slugging percentage at Denver stadium is also higher than any other city. Even if we changed the value of the elevation of Denver to 1500 feet, the linear relationship between slugging percentage and elevation is still statistically significant.

11.67   The statement "The correlation coefficient is a measure of the strength of the linear relationship between x and y." is a true statement.

11.69   a. $r = 1$ implies x and y have a perfect positive linear relationship.

   b. $r = -1$ implies x and y have a perfect negative linear relationship.

   c. $r = 0$ implies x and y are not linearly related.

   d. $r = .90$ implies x and y have a positive linear relationship. Since r is close to 1, the strength of the relationship is very high.

   e. $r = .10$ implies x and y have a positive linear relationship. Since r is close to 0, the relationship is very weak.

   f. $r = -.88$ implies x and y have a negative linear relationship. Since r is close to -1, the relationship is fairly strong.

11.71   a.   Using MINITAB, the scatterplot of the data is:

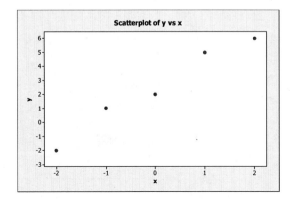

Some preliminary calculations are:

$$\sum x = 0 \qquad \sum x^2 = 10 \qquad \sum xy = 20 \qquad \sum y = 12 \qquad \sum y^2 = 70$$

$$SS_{xy} = \sum xy - \frac{\sum x \sum y}{n} = 20 - \frac{0(12)}{5} = 20$$

$$SS_{xx} = \sum x^2 - \frac{\left(\sum x\right)^2}{n} = 10 - \frac{0^2}{5} = 10$$

$$SS_{yy} = \sum y^2 - \frac{\left(\sum y\right)^2}{n} = 70 - \frac{12^2}{5} = 41.2$$

$$r = \frac{SS_{xy}}{\sqrt{SS_{xx}SS_{yy}}} = \frac{20}{\sqrt{10(41.2)}} = .9853 \qquad r^2 = .9853^2 = .9709$$

b. Using MINITAB, the scatterplot of the data is:

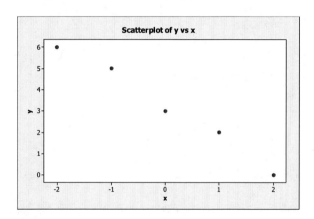

Some preliminary calculations are:

$$\sum x = 0 \qquad \sum x^2 = 10 \qquad \sum xy = -15 \qquad \sum y = 16 \qquad \sum y^2 = 74$$

$$SS_{xy} = \sum xy - \frac{\sum x \sum y}{n} = -15 - \frac{0(16)}{5} = -15$$

$$SS_{xx} = \sum x^2 - \frac{\left(\sum x\right)^2}{n} = 10 - \frac{0^2}{5} = 10$$

$$SS_{yy} = \sum y^2 - \frac{\left(\sum y\right)^2}{n} = 74 - \frac{16^2}{5} = 22.8$$

$$r = \frac{SS_{xy}}{\sqrt{SS_{xx}SS_{yy}}} = \frac{-15}{\sqrt{10(22.8)}} = -.9934 \qquad r^2 = (-.9934)^2 = .9868$$

c.   Using MINITAB, the scatterplot is:

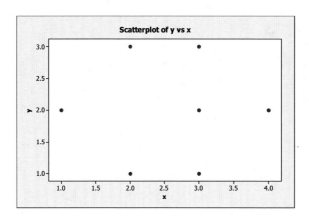

Some preliminary calculations are:

$$\sum x = 18 \qquad \sum x^2 = 52 \qquad \sum xy = 36 \qquad \sum y = 14 \qquad \sum y^2 = 32$$

$$SS_y = \sum xy - \frac{\sum x \sum y}{n} = 36 - \frac{18(14)}{7} = 0$$

$$SS_{xx} = \sum x^2 - \frac{\left(\sum x\right)^2}{n} = 52 - \frac{18^2}{7} = 5.71428571$$

$$SS_{yy} = \sum y^2 - \frac{\left(\sum y\right)^2}{n} = 32 - \frac{14^2}{7} = 4$$

$$r = \frac{SS_{xy}}{\sqrt{SS_{xx}SS_{yy}}} = \frac{0}{\sqrt{5.71428571(4)}} = 0 \qquad r^2 = 0^2 = 0$$

d.   Using MINITAB, the scatterplot of the data is:

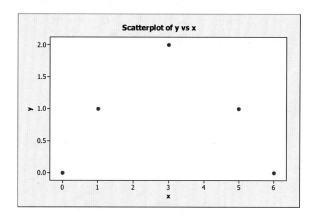

Some preliminary calculations are:

$$\sum x = 15 \qquad \sum x^2 = 71 \qquad \sum xy = 12 \qquad \sum y = 4 \qquad \sum y^2 = 6$$

$$SS_{xy} = \sum xy - \frac{\sum x \sum y}{n} = 12 - \frac{15(4)}{5} = 0$$

$$SS_{xx} = \sum x^2 - \frac{\left(\sum x\right)^2}{n} = 71 - \frac{15^2}{5} = 26$$

$$SS_{yy} = \sum y^2 - \frac{\left(\sum y\right)^2}{n} = 6 - \frac{4^2}{5} = 2.8$$

$$r = \frac{SS_{xy}}{\sqrt{SS_{xx}SS_{yy}}} = \frac{0}{\sqrt{26(2.8)}} = 0 \qquad r^2 = 0^2 = 0$$

11.73   From Exercises 11.17 and 11.37,

$$r^2 = 1 - \frac{SSE}{SS_{yy}} = 1 - \frac{5.13486841}{41.7142857} = .8769$$

87.69% of the total sample variability around $\bar{y}$ is explained by the linear relationship between y and x.

11.75   a. $r^2 = .18$. 18% of the sample variability around the sample mean number of points scored is explained by the linear relationship between the number of points scored and number of yards from the opposing goal line.

b. $r = -\sqrt{.18} = -.424$. The value of r is negative because the estimated slope of the regression line is negative.

11.77   a.   The correlation coefficient of $r = .50$ indicates that the strength of the positive linear relationship between baseline and follow-up physical activity among the obese adults is moderate. Since the p-value is not less than α $(p = .07 \not< .05)$, there is no evidence to indicate correlation coefficient differs from 0 at $\alpha = .05$.

b.   Using MINITAB, a possible scatterplot with 13 data points that would yield a value of $r = .50$ is:

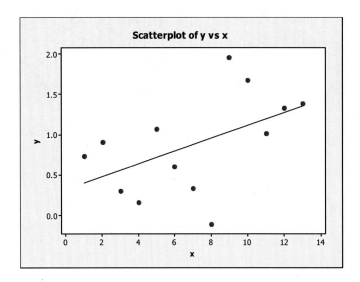

c.   Since the p-value is not small $(p = .66)$ there is no evidence to indicate a linear relationship between baseline and follow-up physical activity among the normal adults. The correlation coefficient of $r = -.12$ indicates that the strength of the negative linear relationship is extremely weak.

d.   Using MINITAB, a possible scatterplot with 15 data points that would yield a value of $r = -.12$ is:

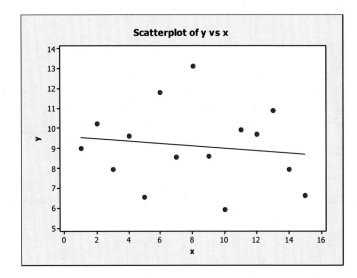

11.79   a. $r = .43$. Since the value is not particularly close to 1, there is a moderately weak positive linear relationship between time allotted to sports and audience ratings.

       b. $r^2 = .43^2 = .1849$. 18.49% of the total sample variability around the sample mean audience rating is explained by the linear relationship between audience rating and total time allotted to sports.

11.81   a. $r = -.726$. Since this value is somewhat close to -1, there is a moderately strong negative linear relationship between the number of online courses taken and the weekly quiz grade.

       b.   To determine if a negative correlation exists, we test:

$$H_0: \ \rho = 0$$
$$H_a: \ \rho < 0$$

The test statistic is $t = \dfrac{r\sqrt{n-2}}{\sqrt{1-r^2}} = \dfrac{-.726\sqrt{24-2}}{\sqrt{1-(-.726)^2}} = -4.95$.

The rejection region requires $\alpha = .05$ in the lower tail of the t distribution with df $= n - 2 = 24 - 2 = 22$. From Table VI, Appendix A, $t_{.05} = 1.717$. The rejection region is t < -1.717.

Since the observed value of the test statistic falls in the rejection region (t = -4.95 < -1.717), H_o is rejected. There is sufficient evidence to indicate a negative linear relationship between the number of online courses taken and the weekly quiz grade.

11.83   a.   **Piano**: $r = .447$
Because this value is near .5, there is s slight positive linear relationship between recognition exposure time and goodness of view for piano.

             Bench: $r = -.057$
Because this value is extremely close to 0, there is an extremely weak negative linear relationship between recognition exposure time and goodness of view for bench.

             Motorbike: $r = .619$
Because this value is near .5, there is a moderate positive linear relationship between recognition exposure time and goodness of view for motorbike.

             Armchair: $r = .294$
Because this value is fairly close to 0, there is a weak positive linear relationship between recognition exposure time and goodness of view for armchair.

             Teapot: $r = .949$
Because this value is very close to 1, there is a strong positive linear relationship between recognition exposure time and goodness of view for teapot.

b. **Piano**: $r^2 = (.447)^2 = .1998$
    19.98% of the total sample variability around the sample mean recognition exposure time is explained by the linear relationship between the recognition exposure time and the goodness of view for piano.

    Bench: $r^2 = (-.057)^2 = .0032$
    .32% of the total sample variability around the sample mean recognition exposure time is explained by the linear relationship between the recognition exposure time and the goodness of view for bench.

    Motorbike: $r^2 = (.619)^2 = .3832$
    38.32% of the total sample variability around the sample mean recognition exposure time is explained by the linear relationship between the recognition exposure time and the goodness of view for motorbike.

    Armchair: $r^2 = (.294)^2 = .0864$
    8.64% of the total sample variability around the sample mean recognition exposure time is explained by the linear relationship between the recognition exposure time and the goodness of view for armchair.

    Teapot: $r^2 = (.949)^2 = .9006$
    90.06% of the total sample variability around the sample mean recognition exposure time is explained by the linear relationship between the recognition exposure time and the goodness of view for teapot.

c. The test is:

$$H_0: \ \beta_1 = 0$$
$$H_a: \ \beta_1 \neq 0$$

The following are the values of α and $t_{\alpha/2}$ that correspond to df $= n - 2 = 25 - 2 = 23$ from Table VI, Appendix A.

α	.20	.10	.05	.02	.01	.002	.001
$t_{\alpha/2}$	1.319	1.714	2.069	2.500	2.807	3.485	3.767

Piano: $t = 2.40$
$2.069 < 2.40 < 2.500$. Thus, $p \approx .025$
For levels of significance greater than $\alpha = .025$, H_0 can be rejected. There is sufficient evidence to indicate that there is a linear relationship between goodness of view and recognition exposure time for piano for $\alpha > .025$.

Bench: $t = .27$
$.27 < 1.319$. Thus, $p > .2$
H_0 is not rejected. There is insufficient evidence to indicate that there is a linear relationship between goodness of view and recognition exposure time for bench for $\alpha \leq .2$.

Motorbike: $t = 3.78$
$3.78 > 3.767$. Thus, $p < .001$

H_0 can be rejected for $\alpha \geq .001$. There is sufficient evidence to indicate that there is a linear relationship between goodness of view and recognition exposure time for motorbike for $\alpha \geq .001$.

Armchair: $t = 1.47$
$1.319 < 1.47 < 1.714$. Thus, $p \approx .15$
H_0 cannot be rejected for levels of significance $\alpha < .15$. There is insufficient evidence to indicate that there is a linear relationship between goodness of view and recognition exposure time for armchair for $\alpha < .15$.

Teapot: $t = 14.50$
$14.50 > 3.767$. Thus, $p < .001$
H_0 can be rejected for $\alpha \geq .001$. There is sufficient evidence to indicate that there is a linear relationship between goodness of view and recognition exposure time for teapot for $\alpha \geq .001$.

11.85   a.   To determine whether average payoff and punishment use are negatively correlated, we test:

$$H_0:\ \beta_1 = 0$$
$$H_a:\ \beta_1 < 0$$

b.   Since the p-value is very small ($p = .001$), we would reject H_0 for any value of α greater than .001. There is sufficient evidence to indicate average payoff and punishment use are negatively correlated for $\alpha > .001$.

c.   No. Just because two variables are significantly correlated does not mean that one variable causes the other. Something else could be causing both of the variables to react in certain ways.

11.87   Using the values computed in Exercise 11.60:

$$r = \frac{SS_{xy}}{\sqrt{SS_{xx}SS_{yy}}} = \frac{48.125}{\sqrt{379.9375(18.75)}} = .5702$$

Because r is moderately small, there is a rather weak positive linear relationship between blood lactate concentration and perceived recovery.

$r^2 = .5702^2 = .3251$.

32.51% of the sample variance of blood lactate concentration around the sample mean is explained by the linear relationship between blood lactate concentration and perceived recovery.

11.89   For a given x, y is the actual value of the dependent variable and $E(y)$ is the mean value of y at the given value of x.

11.91   The statement "The greater the deviation between x and $\bar{x}$, the wider the prediction interval for y will be." is true. The further x gets from $\bar{x}$, the less precise the interval will be.

11.93   a. $\hat{\beta_1} = \dfrac{SS_{xy}}{SS_{xx}} = \dfrac{28}{32} = .875$

$\hat{\beta_0} = \bar{y} - \hat{\beta_1}\bar{x} = 4 - .875(3) = 1.375$

The least squares line is $\hat{y} = 1.375 + .875x$.

b.   The least squares line is:

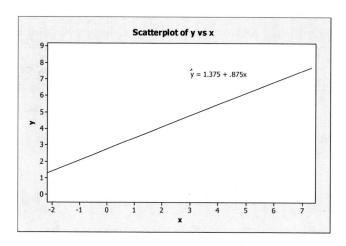

c. $SSE = SS_{yy} - \hat{\beta_1}SS_{xy} = 26 - .875(28) = 1.5$

d. $s^2 = \dfrac{SSE}{n-2} = \dfrac{1.5}{10-2} = .1875$

e. $s = \sqrt{.1875} = .4330$

The form of the confidence interval is $\hat{y} \pm t_{\alpha/2}\, s\sqrt{\dfrac{1}{n} + \dfrac{(x_p - \bar{x})^2}{SS_{xx}}}$

For $x_p = 2.5$, $\hat{y} = 1.375 + .875(2.5) = 3.5625$

For confidence coefficient .95, $\alpha = .05$ and $\alpha/2 = .05/2 = .025$. From Table VI, Appendix A, with df $= n - 2 = 10 - 2 = 8$, $t_{.025} = 2.306$. The confidence interval is:

$$3.5625 \pm 2.306(.4330)\sqrt{\dfrac{1}{10} + \dfrac{(2.5-3)^2}{32}} \Rightarrow 3.5625 \pm .3279 \Rightarrow (3.2346, 3.8904)$$

f.   The form of the prediction interval is $\hat{y} \pm t_{\alpha/2}\, s\sqrt{1 + \dfrac{1}{n} + \dfrac{(x_p - \bar{x})^2}{SS_{xx}}}$

For $x_p = 4$, $\hat{y} = 1.375 + .875(4) = 4.875$

For confidence coefficient .95, $\alpha = .05$ and $\alpha / 2 = .05 / 2 = .025$. From Table VI, Appendix A, with df $= n - 2 = 10 - 2 = 8$, $t_{.025} = 2.306$. The prediction interval is:

$$4.875 \pm 2.306(.4330)\sqrt{1 + \frac{1}{10} + \frac{(4-3)^2}{32}} \Rightarrow 4.875 \pm 1.062 \Rightarrow (3.813, 5.937)$$

11.95   a, b. The scattergram is:

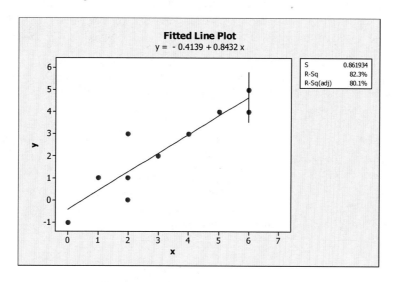

c. $\mathrm{SSE} = \mathrm{SS}_{yy} - \hat{\beta}_1 \mathrm{SS}_{xy} = 33.6 - .843(32.8) = 5.9496$

$s^2 = \dfrac{\mathrm{SSE}}{n-2} = \dfrac{5.9496}{10-2} = .7437$ $s = \sqrt{.7437} = .8623$

$\bar{x} = \dfrac{31}{10} = 3.1$

The form of the confidence interval is $\hat{y} \pm t_{\alpha/2}\, s\sqrt{\dfrac{1}{n} + \dfrac{(x_{\mathrm{p}} - \bar{x})^2}{\mathrm{SS}_{xx}}}$

For $x_{\mathrm{p}} = 6$, $\hat{y} = -.414 + .843(6) = 4.644$

For confidence coefficient .95, $\alpha = .05$ and $\alpha / 2 = .05 / 2 = .025$. From Table VI, Appendix A, with df $= n - 2 = 10 - 2 = 8$, $t_{.025} = 2.306$. The confidence interval is:

$$4.644 \pm 2.306(.8623)\sqrt{\frac{1}{10} + \frac{(6-3.1)^2}{38.9}} \Rightarrow 4.644 \pm 1.118 \Rightarrow (3.526, 5.762)$$

d.   For $x_{\mathrm{p}} = 3.2$, $\hat{y} = -.414 + .843(3.2) = 2.284$

The confidence interval is:

$$2.284 \pm 2.306(.8623)\sqrt{\frac{1}{10} + \frac{(3.2-3.1)^2}{38.9}} \Rightarrow 2.284 \pm .630 \Rightarrow (1.654, 2.914)$$

For $x_p = 0$, $\hat{y} = -.414 + .843(0) = -.414$

The confidence interval is:

$$-.414 \pm 2.306(.8623)\sqrt{\frac{1}{10} + \frac{(0-3.1)^2}{38.9}} \Rightarrow -.414 \pm 1.171 \Rightarrow (-1.585, .757)$$

e. The width of the confidence interval for the mean value of y depends on the distance x_p is from $\bar{x}$. The width of the interval for $x_p = 3.2$ is the smallest because 3.2 is the closest to $\bar{x} = 3.1$. The width of the interval for $x_p = 0$ is the widest because 0 is the farthest from $\bar{x} = 3.1$.

11.97   a.   To predict the average payoff for a single player who used punishment 10 times, you would form a prediction interval for the individual payoff when punishment is 10.

  b.   To predict the mean of the average payoff for all players who used punishment 10 times, you would form a confidence interval for the mean payoff when punishment is 10.

11.99   For observation 7, the 95% prediction interval for the actual annual rainfall when the maximum daily temperature is 11.4 is (92.298, 125.104). We are 95% confident that the actual rainfall in a site with a maximum daily temperature of 11.4 is between 92.298 and 125.104.

11.101   Answers may vary. One possible answer is:

The 90% confidence interval for $x = 220.00$ is (5.64898, 5.83848). We are 90% confident that the mean sweetness index of all orange juice samples will be between 5.64898 and 5.83848 parts per million when the pectin value is 220.00.

11.103   a.   From Exercise 11.29 d, $\hat{y} = 39.304 + .493x$ and $s = 2.32153$. From Exercise 11.58 b, $SS_{xx} = 584.6528$.

For $x = 66$, $\hat{y} = 39.304 + .493(66) = 71.842$.

For confidence coefficient .95, $\alpha = .05$ and $\alpha/2 = .05/2 = .025$. From Table VI, Appendix A, with $df = n - 2 = 72 - 2 = 70$, $t_{.025} \approx 2.00$. The 95% prediction interval is:

$$\hat{y} \pm t_{\alpha/2} s\sqrt{1 + \frac{1}{n} + \frac{(x_p - \bar{x})^2}{SS_{xx}}} \Rightarrow 71.842 \pm 2.00(2.32153)\sqrt{1 + \frac{1}{72} + \frac{(66-64.5694)^2}{584.6528}}$$
$$\Rightarrow 71.842 \pm 4.683 \Rightarrow (67.159, 76.525)$$

We are 95% confident that the actual ideal partner's height is between 67.159 and 76.525 inches when a female's height is 66 inches.

  b.   From Exercise 11.29 e, $\hat{y} = 23.274 + .596x$ and $s = 2.05688$. From Exercise 11.58 a, $SS_{xx} = 534.3467$.

For $x = 66$, $\hat{y} = 23.274 + .596(66) = 62.61$.

For confidence coefficient .95, $\alpha = .05$ and $\alpha / 2 = .05 / 2 = .025$. From Table VI, Appendix A, with $df = n - 2 = 75 - 2 = 73$, $t_{.025} \approx 2.00$. The 95% prediction interval is:

$$\hat{y} \pm t_{\alpha/2} s \sqrt{1 + \frac{1}{n} + \frac{\left(x_p - \bar{x}\right)^2}{SS_{xx}}} \Rightarrow 62.61 \pm 2.00(2.05688)\sqrt{1 + \frac{1}{75} + \frac{\left(66 - 71.0933\right)^2}{534.3467}}$$

$$\Rightarrow 62.61 \pm 4.239 \Rightarrow (58.371, 66.849)$$

We are 95% confident that the actual ideal partner's height is between 58.371 and 66.849 inches when a male's height is 66 inches.

c. The prediction for the ideal partner's height for males might be invalid because 66 is outside the range of observed values for male heights. The smallest observed value of male heights is 67 inches. We are not sure if the relationship between the two variables remains the same outside the observed range.

11.105 a. Using MINITAB, the output is:

```
Predicted Values for New Observations

New
Obs    Fit    SE Fit     99% CI           99% PI
 1   3.510   0.196   (2.955, 4.066)   (1.020, 6.000)


Values of Predictors for New Observations

New
Obs   Time
  1   15.0
```

From the printout, the 99% confidence interval for the mean mass of all spills with an elapsed time of 15 minutes is (2.955, 4.066). We are 99% confident that the mean mass of all spills with an elapsed time of 15 minutes is between 2.955 and 4.066.

b. From the printout, the 99% prediction interval for the actual mass of a spill with an elapsed time of 15 minutes is (1.020, 6.000). We are 99% confident that the actual mass of a spill with an elapsed time of 15 minutes is between 1.020 and 6.000.

c. The prediction interval in part b is larger than the confidence interval in part a. This will always be the case. The confidence interval gives a range of values for the mean of a distribution. The prediction interval gives a range of values for the actual value of a distribution. Once a mean is located the actual values can still vary around it. This is why the prediction interval is always wider than the confidence interval for the mean.

11.107 a. From Exercise 11.46, $SS_{xx} = 3000$ and $\bar{x} = 50$.

Also, for Brand A, $s = 1.211$; for Brand B, $s = .610$.

For Brand A, $\hat{y} = 6.62 - .0727(45) = 3.349$, while for Brand B, $\hat{y} = 9.31 - .1077(45) = 4.464$.

The degrees of freedom for both brands is $n - 2 = 15 - 2 = 13$. For confidence coefficient .90, (i.e., for all parts of this question), $\alpha = .10$ and $\alpha/2 = .10/2 = .05$. From Table VI, Appendix A, with df = 13, $t_{.05} = 1.771$.

The form of both confidence intervals is $\hat{y} \pm t_{\alpha/2}\, s\sqrt{\dfrac{1}{n} + \dfrac{(x_p - \bar{x})^2}{SS_{xx}}}$

For Brand A, we obtain:

$$3.349 \pm 1.771(1.211)\sqrt{\frac{1}{15} + \frac{(45 - 50)^2}{3000}} \Rightarrow 3.349 \pm .587 \Rightarrow (2.762, 3.936)$$

For Brand B, we obtain:

$$4.464 \pm 1.771(.610)\sqrt{\frac{1}{15} + \frac{(45 - 50)^2}{3000}} \Rightarrow 4.464 \pm .296 \Rightarrow (4.168, 4.760)$$

The interval for Brand A is wider, caused by the larger value of s.

b. The form of both prediction intervals is $\hat{y} \pm t_{\alpha/2}\, s\sqrt{1 + \dfrac{1}{n} + \dfrac{(x_p - \bar{x})^2}{SS_{xx}}}$

For Brand A, we obtain:

$$3.349 \pm 1.771(1.211)\sqrt{1 + \frac{1}{15} + \frac{(45 - 50)^2}{3000}} \Rightarrow 3.349 \pm 2.224 \Rightarrow (1.125, 5.573)$$

For Brand B, we obtain:

$$4.464 \pm 1.771(.610)\sqrt{1 + \frac{1}{15} + \frac{(45 + 50)^2}{3000}} \Rightarrow 4.464 \pm 1.120 \Rightarrow (3.344, 5.584)$$

Again, the interval is wider for Brand A, caused by the larger value of s. Each of these intervals is wider than its counterpart from part **a**, since, for the same x, a prediction interval for an individual y is always wider than a confidence interval for the mean of y. This is due to an individual observation having a greater variance than the variance of the mean of a set of observations.

c. To obtain a confidence interval for the life of a brand A cutting tool that is operated at 100 meters per minute, we use:

$$\hat{y} \pm t_{\alpha/1}s\sqrt{1 + \frac{1}{n} + \frac{(x_p - \bar{x})}{SS_{xx}}}$$

For $x = 100$, $\hat{y} = 6.62 - .0727(100) = -.65$.

The degrees of freedom are $n - 2 = 15 - 2 = 13$. For confidence coefficient .95, $\alpha = .05$ and $\alpha/2 = .05/2 = .025$. From Table VI, Appendix A, with df = 13, $t_{.025} = 2.160$.

Here, we obtain:

$$-.65 \pm 2.160(1.211)\sqrt{1 + \frac{1}{15} + \frac{(100-50)^2}{3000}} \Rightarrow -.65 \pm 3.606 \Rightarrow (-4.256, 2.956)$$

The additional assumption would be that the straight line model fits the data well for the x's actually observed all the way up to the value under consideration, 100. Clearly from the estimated value of $-.65$, this is not true (usually, negative "useful lives" are not found).

11.109 Using MINITAB, a scatterplot of the data is:

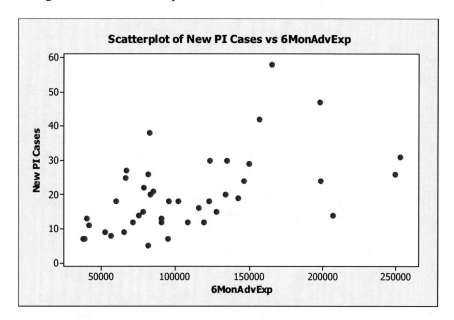

It appears that there is a positive linear relationship between the 6-month cumulative advertising expenditure and the number of new PI cases at the law firm.

The descriptive statistics for the two variables are:

Descriptive Statistics: New PI Cases, 6MonAdvExp

Variable	N	Mean	StDev	Minimum	Q1	Median	Q3	Maximum
New PI Cases	42	20.05	11.34	5.00	12.00	18.00	26.00	58.00
6MonAdvExp	42	108779	54104	37490	70293	93059	136611	253011

Using MINITAB, the results of fitting the regression line are:

Regression Analysis: New PI Cases versus 6MonAdvExp

```
The regression equation is
New PI Cases = 7.77 + 0.000113 6MonAdvExp


Predictor         Coef     SE Coef     T      P
Constant         7.767       3.385   2.29  0.027
6MonAdvExp  0.00011289  0.00002793   4.04  0.000


S = 9.67521    R-Sq = 29.0%    R-Sq(adj) = 27.2%


Analysis of Variance

Source          DF      SS      MS      F      P
Regression       1  1529.5  1529.5  16.34  0.000
Residual Error  40  3744.4    93.6
Total           41  5273.9
```

The fitted regression line is $\hat{y} = 7.77 + .000113x$.

To determine if there is a positive linear relationship between the 6-month cumulative advertising expenditures and the number of new PI cases, we test:

$$H_0: \ \beta_1 = 0$$
$$H_a: \ \beta_1 > 0$$

The test statistic is $t = 4.04$ and the p-value is $p = 0.000/2 = 0.000$. Since the p-value is so small, H_0 is rejected for any reasonable value of α. There is sufficient evidence to indicate a positive linear relationship between the 6-month cumulative advertising expenditure and the number of new PI cases. The more money spent on advertising, the greater the number of new PI cases.

From the printout, the value of r^2 is 29%. Thus, 29% of the total sample variability around the sample mean number of new PI cases is explained by the linear relationship between the number of new PI cases and the 6-month cumulative advertising expenditure. With only 29% of the variability explained, we know that there might be other factors not considered that can help explain the variability in the number of new PI cases.

11.111   The general form of the straight-line model for $E(y)$ is $E(y) = \beta_0 + \beta_1 x$.

11.113   The statement "In simple linear regression, about 95% of the y-values in the sample will fall within 2s of their respective predicted values." is a true statement.

11.115 a.

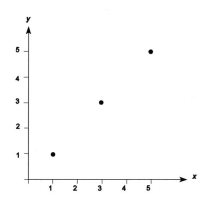

b.   One possible line is $\hat{y} = x$.

x	y	$\hat{y}$	$y - \hat{y}$
1	1	1	0
3	3	3	0
5	5	5	0
			0

For this example $\sum (y - \hat{y}) = 0$

A second possible line is $\hat{y} = 3$.

x	y	$\hat{y}$	$y - \hat{y}$
1	1	3	-2
3	3	3	0
5	5	3	2
			0

For this example $\sum (y - \hat{y}) = 0$

c.   Some preliminary calculations are:

$$\sum x_i = 9 \qquad \sum x_i^2 = 35 \qquad \sum x_i y_i = 35 \qquad \sum y_i = 9 \qquad \sum y_i^2 = 35$$

$$SS_{xy} = \sum x_i \sum y_i - \frac{\sum x_i \sum y_i}{n} = 35 - \frac{9(9)}{3} = 8$$

$$SS_{xx} = \sum x_i^2 - \frac{\left(\sum x_i\right)^2}{n} = 35 - \frac{9^2}{3} = 8$$

$$SS_{yy} = \sum y_i^2 - \frac{\left(\sum y_i\right)^2}{n} = 35 - \frac{9^2}{3} = 8$$

$$\hat{\beta}_1 = \frac{SS_{xy}}{SS_{xx}} = \frac{8}{8} = 1 \qquad\qquad \hat{\beta}_0 = \bar{y} - \hat{\beta}_1 \bar{x} = \frac{9}{3} - 1\left(\frac{9}{3}\right) = 0$$

The least squares line is $\hat{y} = 0 + 1x = x$.

d.   For $\hat{y} = x$, $\text{SSE} = \text{SS}_{yy} - \hat{\beta}_1 \text{SS}_{xy} = 8 - 1(8) = 0$

For $\hat{y} = 3$, $\text{SSE} = \sum (y_i - \hat{y}_i)^2 = (1-3)^2 + (3-3)^2 + (5-3)^2 = 8$

The least squares line has the smallest SSE of all possible lines.

11.117 a.   The straight-line model is $y = \beta_0 + \beta_1 x + \varepsilon$. Based on the theory, we would expect the metal level to decrease as the distance increases. Thus, the slope should be negative.

b.   Yes. As the distance from the plant increases, the concentration of calcium tends to decrease.

c.   No. As the distance from the plant increases, the concentration of arsenic tends to increase.

11.119 a.   We would expect the slope of the line from modeling the relationship between the mean number of games won and the team's batting average to be positive. As the team's batting average increases, we would expect that the number of games won should also increase.

b.   Using MINITAB, a scatterplot of the data is:

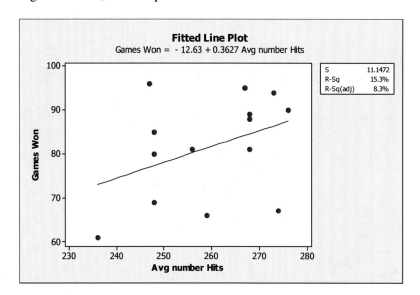

The graph indicates that batting average and games won have a positive linear relationship. As batting average increases, the number of games won tends to increase.

c.   From the printout, $\hat{\beta}_0 = -12.62556$ and $\hat{\beta}_1 = 0.36269$. The least squares line is:
$\hat{y} = -12.62556 + .36269x$.

d.   The least squares line is plotted on the graph in part **a**. The data appear to fit the data somewhat well.

e. $\hat{\beta}_0 = -12.62556$. Since there is no team with a batting average of 0, this has no meaning other than the y-intercept.

$\hat{\beta}_1 = .36269$. For each additional hit per 1000 at bats, the mean number of games won is estimated to increase by .36269.

f. To determine if the mean number of games won by a major league baseball team is positively linearly related to the team's batting average, we test:

H_0: $\beta_1 = 0$
H_a: $\beta_1 > 0$

The test statistic is $t = 1.47$ and the p-value is $p = 0.1660$. Since we are running only a one-tailed test, we must divide the p-value by 2. Thus, the p-value of this test is $p = 0.1660/2 = 0.0830$. Since the p-value is not less than $\alpha = .05$ ($p = 0.0830 > .05$), H_0 is not rejected. There is insufficient evidence to indicate the mean number of games won by a major league baseball team is positively linearly related to the team's batting average at $\alpha = .05$

g. From the printout, $r^2 =$ R-Square $= .1535$. 15.35% of the sample variability around the sample mean number of games won is explained by the linear relationship between the number of games won and the batting average of the team.

h. No. There is no evidence of a positive linear relationship between the number of hits and games won.

11.121 a. The equation for the straight-line model is $y = \beta_0 + \beta_1 x + \varepsilon$.

b. Some preliminary calculations are:

$$\bar{y} = \frac{\sum y}{n} = \frac{398}{9} = 44.2222$$

$$\bar{x} = \frac{\sum x}{n} = \frac{1,444}{9} = 160.4444$$

$$SS_{xy} = \sum xy - \frac{\sum x \sum y}{n} = 60,428 - \frac{1,444(398)}{9} = -3,428.88889$$

$$SS_{xx} = \sum x^2 - \frac{(\sum x)^2}{n} = 235,866 - \frac{1,444^2}{9} = 4,184.2222$$

$$\hat{\beta}_1 = \frac{SS_{xy}}{SS_{xx}} = \frac{-3,428.88889}{4,184.2222} = -.819480593$$

$$\hat{\beta}_0 = \bar{y} - \hat{\beta}_1\bar{x} = \frac{398}{9} - (-.819480593)\left(\frac{1,444}{9}\right) = 175.7033307$$

The fitted model is $\hat{y} = 175.7033 - .8195x$.

c. $\hat{\beta}_0 = 175.7033$. Since $x = 0$ (age of fish) is not in the observed range, $\hat{\beta}_0$ has no practical meaning.

d. $\hat{\beta}_1 = -.8195$. For each additional day of age, the mean number of strikes is estimated to decrease by .8195 strikes.

e.   Some additional calculations are:

$$\sum y_i = 398 \qquad \sum y_i^2 = 22,078$$

$$SS_{yy} = \sum y_i^2 - \frac{\left(\sum y_i\right)^2}{n} = 22,078 - \frac{398^2}{9} = 4,477.55556$$

$$SSE = SS_{yy} - \hat{\beta}_1 SS_{xy} = 4,477.55556 - (-.819480593)(-3,428.88889) = 1,667.647659$$

$$s^2 = \frac{SSE}{n-2} = \frac{1,667.647659}{9-2} = 238.2353799$$

$$s = \sqrt{s^2} = \sqrt{238.2353799} = 15.43487544$$

H_o: $\beta_1 = 0$
H_a: $\beta_1 < 0$

The test statistic is $t = \dfrac{\hat{\beta}_1 - 0}{s_{\hat{\beta}_1}} = \dfrac{-.8195 - 0}{15.4349 \Big/ \sqrt{4,184.2222}} = -3.434$

The rejection region requires $\alpha = .10$ in the lower tail of the t distribution with df $= n - 2$ $= 9 - 2 = 7$. From, Table VI, Appendix A, $t_{.10} = 1.415$. The rejection region is $t < -1.415$.

Since the observed value of the test statistic falls in the rejection region ($t = -3.434 < -1.415$), H_0 is rejected. There is sufficient evidence to indicate a negative linear relationship exists between the number of strikes and the age of the fish at $\alpha = .10$.

11.123 a.   The simple linear regression model relating y to x is $y = \beta_0 + \beta_1 x + \varepsilon$.

b. $r^2 = .92$. 92% of the total sample variability around the sample mean metal uptake is explained by the linear relationship between metal uptake and the final concentration of metal in the solution.

11.125 a. Using MINITAB, the results are:

Regression Analysis: Cost versus JIF

```
The regression equation is
Cost = 560 + 63 JIF


Predictor   Coef   SE Coef     T      P
Constant   560.1    168.4    3.33   0.003
JIF         63.3    113.2    0.56   0.581


S = 454.248    R-Sq = 1.2%    R-Sq(adj) = 0.0%


Analysis of Variance

Source          DF        SS       MS      F      P
Regression       1     64381    64381   0.31   0.581
Residual Error  26   5364863   206341
Total           27   5429244
```

The fitted regression equation is $\hat{y} = 560.1 + 63.3x$.

For each additional unit increase in the Journal Impact Factor (JIF), the mean cost of the journal is estimated to increase by \$63.30.

b. From the printout, $s = 454.248$. We would expect to predict cost within $2s = 2(454.248)$ = \$908.50.

c. For confidence level .95, $\alpha = .05$ and $\alpha / 2 = .05 / 2 = .025$. From Table VI, Appendix A with df $= n - 2 = 28 - 2 = 26$, $t_{.025} = 2.056$.

The confidence interval is:

$$\hat{\beta}_1 \pm t_{.025} s_{\hat{\beta}_1} \Rightarrow 63.3 \pm 2.056(113.2) \Rightarrow 63.3 \pm 232.7 \Rightarrow (-169.4,\ 296.0)$$

We are 95% confident that for each unit increase in JIF, the mean cost of the journal is estimated to change from -\$169.40 to \$296.00.

d. Using MINITAB, the results are:

Regression Analysis: Cost versus Cites

```
The regression equation is
Cost = 326 + 1.48 Cites


Predictor    Coef   SE Coef     T      P
Constant    326.5    109.1    2.99   0.006
Cites      1.4801   0.3953    3.74   0.001


S = 368.331    R-Sq = 35.0%    R-Sq(adj) = 32.5%
```

```
Analysis of Variance

Source             DF      SS         MS      F      P
Regression          1  1901880   1901880  14.02  0.001
Residual Error     26  3527364    135668
Total              27  5429244
```

The fitted regression equation is $\hat{y} = 326.5 + 1.48x$.

For each additional unit increase in the number of citations for the journal over the past 5 years, the mean cost of the journal is estimated to increase by $1.48.

From the printout, $s = 368.331$. We would expect to predict cost within $2s = 2(368.331) = \$736.66$.

For confidence level .95, $\alpha = .05$ and $\alpha / 2 = .05 / 2 = .025$. From Table VI, Appendix A with df $= n - 2 = 28 - 2 = 26$, $t_{.025} = 2.056$.

The confidence interval is:

$$\hat{\beta}_1 \pm t_{.025} s_{\hat{\beta}_1} \Rightarrow 1.48 \pm 2.056(.3953) \Rightarrow 1.48 \pm .81 \Rightarrow (.67, 2.29)$$

We are 95% confident that for each additional citation, the mean cost of the journal is estimated to increase from $.67 to $2.29.

e. Using MINITAB, the results are:

Regression Analysis: Cost versus RPI

```
The regression equation is
Cost = 339 + 197 RPI


27 cases used, 1 cases contain missing values


Predictor    Coef   SE Coef    T      P
Constant    338.9    156.6   2.16  0.040
RPI         197.21    86.24  2.29  0.031


S = 423.363   R-Sq = 17.3%   R-Sq(adj) = 14.0%


Analysis of Variance

Source             DF      SS         MS      F      P
Regression          1   937326    937326   5.23  0.031
Residual Error     25  4480916    179237
Total              26  5418242
```

The fitted regression equation is $\hat{y} = 338.9 + 197.21x$.

For each additional unit increase in the Relative Price Index (RPI), the mean cost of the journal is estimated to increase by $197.21.

From the printout, $s = 423.363$. We would expect to predict cost within $2s = 2(423.363)$ $= \$846.73$.

For confidence level .95, $\alpha = .05$ and $\alpha / 2 = .05 / 2 = .025$. From Table VI, Appendix A with df $= n - 2 = 28 - 2 = 26$, $t_{.025} = 2.056$.

The confidence interval is:

$$\hat{\beta}_1 \pm t_{.025} s_{\hat{\beta}_1} \Rightarrow 197.21 \pm 2.056(86.24) \Rightarrow 197.21 \pm 177.31 \Rightarrow (19.90, 374.52)$$

We are 95% confident that for each unit increase in RPI, the mean cost of the journal is estimated to increase from $19.90 to $374.52.

11.127 a. Using MINITAB, the scattergram of the data is:

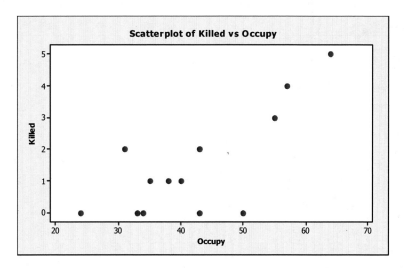

From the graph, it appears that as nest box tit occupancy increases, the number of flycatchers killed also increases.

   b. Some preliminary calculations are:

$$\sum x_i = 582 \qquad \sum y_i = 20 \qquad \sum x_i y_i = 1,009$$

$$\sum x_i^2 = 25,844 \qquad \sum y_i^2 = 62$$

$$SS_{xy} = \sum x_i y_i - \frac{\sum x_i \sum y_i}{n} = 1,009 - \frac{582(20)}{14} = 177.5714286$$

$$SS_{xx} = \sum x_i^2 - \frac{\left(\sum x_i\right)^2}{n} = 25,844 - \frac{582^2}{14} = 1,649.42857$$

$$\hat{\beta}_1 = \frac{SS_{xy}}{SS_{xx}} = \frac{177.5714286}{1,649.42857} = .107656331$$

For each unit increase in the nest box tit occupancy percent, the mean number of flycatchers killed is estimated to increase by .1077.

$$\hat{\beta}_0 = \bar{y} - \hat{\beta}_1 \bar{x} = \frac{20}{14} - (.107656331)\left(\frac{582}{14}\right) = -3.046856046$$

Since $x = 0$ is not in the observed range, $\hat{\beta}_0$ has no meaning other than the y-intercept.

c.   Some additional calculations are:

$$SS_{yy} = \sum y_i^2 - \frac{\left(\sum y_i\right)^2}{n} = 62 - \frac{20^2}{14} = 33.42857143$$

$$SSE = SS_{yy} - \hat{\beta}_1 SS_{xy} = 33.42857143 - (.107656331)(177.5714286) = 14.31188294$$

$$s^2 = \frac{SSE}{n-2} = \frac{14.31188294}{14-2} = 1.192656912 \quad s = \sqrt{s^2} = \sqrt{1.192656912} = 1.0921$$

To determine if the model is useful in predicting the number of flycatchers killed, we test:

H_0: $\beta_1 = 0$
H_a: $\beta_1 \neq 0$

The test statistic is $t = \dfrac{\hat{\beta}_1 - 0}{s_{\hat{\beta}_1}} = \dfrac{.1077 - 0}{1.0921 / \sqrt{1,649.42857}} = 4.005$

The rejection region requires $\alpha / 2 = .05 / 2 = .025$ in each tail of the t distribution. From Table VI, Appendix A, with df $= n - 2 = 14 - 2 = 12$, $t_{.025} = 2.179$. The rejection region is $t < -2.179$ or $t > 2.179$.

Since the observed value of the test statistic falls in the rejection region ($t = 4.005 > 2.179$), H_0 is rejected. There is sufficient evidence to indicate that the model is useful in predicting the number of flycatchers killed at $\alpha = .05$.

d. $r = \dfrac{SS_{xy}}{\sqrt{SS_{xx}SS_{yy}}} = \dfrac{177.5714286}{\sqrt{1,649.42857(33.42857143)}} = .7562$

The value of r is moderately close to 1. Thus, there is a moderately strong positive linear relationship between the number of flycatchers killed and the nest box tit occupancy.

$r^2 = .7562^2 = .572$. 57.2% of the sample variability around the sample mean number of flycatchers killed is explained by the linear relationship between the number of flycatchers killed and the nest box tit occupancy.

e.   From part c, $s = 1.0921$. We would expect most of the observed number of flycatchers killed to fall within $2s = 2(1.0921) = 2.1842$ units of their predicted values.

f.   Since there is a significant linear relationship between the number of flycatchers killed and the nest box tit occupancy, we would recommend using this model. In addition, the r^2 value is .572. Although this is not great, over half of the sample variation is explained by the model.

11.129 Using MINITAB, the output for fitting the least squares line is:

Regression Analysis: Impur versus Temp

```
The regression equation is
Impur = - 13.5 - 0.0528 Temp


Predictor        Coef    SE Coef       T       P
Constant      -13.490      2.074   -6.51   0.000
Temp        -0.052829   0.007728   -6.84   0.000


S = 0.133347   R-Sq = 85.4%   R-Sq(adj) = 83.6%


Analysis of Variance

Source           DF       SS       MS       F       P
Regression        1  0.83089  0.83089   46.73   0.000
Residual Error    8  0.14225  0.01778
Total             9  0.97315


Unusual Observations

Obs  Temp   Impur     Fit  SE Fit  Residual  St Resid
  2  -265  0.2020  0.5093  0.0492   -0.3073    -2.48R
  3  -256  0.2040  0.0339  0.1038    0.1701     2.03RX

R denotes an observation with a large standardized residual.
X denotes an observation whose X value gives it large influence.


Predicted Values for New Observations

New
Obs     Fit  SE Fit       95% CI             95% PI
  1  0.9320  0.0558  (0.8034, 1.0605)  (0.5987, 1.2652)


Values of Predictors for New Observations

New
Obs  Temp
  1  -273
```

a. $\hat{\beta}_0 = -13.490$. Since $x = 0$ is not in the observed range, $\hat{\beta}_0$ has no meaning since. It is the y-intercept.

$\hat{\beta}_1 = -.052829$. For each additional degree, the mean proportion of impurity passing through helium is estimated to decrease by .052829 units.

b.   For confidence coefficient .95, $\alpha = .05$ and $\alpha / 2 = .05 / 2 = .025$. From Table VI, Appendix A, with df $= n - 2 = 10 - 2 = 8$, $t_{.025} = 2.306$. The confidence interval is:

$$\hat{\beta}_1 \pm t_{\alpha/2} s_{\hat{\beta}_1} \Rightarrow -.0528 \pm 2.306(.007728) \Rightarrow -.0528 \pm .0178 \Rightarrow (-.0706, -.0350)$$

We are 95% confident that the change in mean proportion of impurity passing through helium for each additional degree is between $-.0706$ and $-.0350$. Since 0 is not in the interval, there is evidence to indicate that temperature contributes information about the proportion of impurity passing through helium.

c. $r^2 = R$-sq $= .854$. 85.4% of the total sample variation around the mean proportion of impurity is explained by the linear relationship between proportion of impurity and temperature.

d.   From the printout, the 95% prediction interval is (.5987, 1.2652). We are 95% confident that the actual percentage of impurity passing through solid helium at -273°C is between 59.87% and 126.52%. Since we cannot have more than 100% of the impurity passing through solid helium, the interval is from 59.87% to 100%.

e.   We cannot be sure that the relationship between the proportion of impurity passing through helium and temperature is the same outside the observed range.

11.131 a.   A scattergram of the data is:

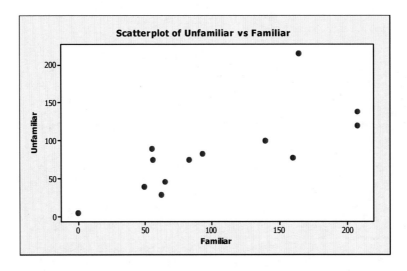

It appears that there is a positive linear relationship between total number of words with unfamiliar partner and with familiar partner.

b.   A straight line model would be $y = \beta_0 + \beta_1 x + \varepsilon$.

c. Some preliminary calculations are:

$$\sum x_i = 1,341 \qquad \sum y_i = 1,099 \qquad \sum x_i y_i = 145,607$$

$$\sum x_i^2 = 189,963 \qquad \sum y_i^2 = 127,397$$

$$SS_{xy} = \sum x_i y_i - \frac{\sum x_i \sum y_i}{n} = 145,607 - \frac{1,341(1,099)}{13} = 32,240.9231$$

$$SS_{xx} = \sum x_i^2 - \frac{\left(\sum x_i\right)^2}{n} = 189,963 - \frac{1,341^2}{13} = 51,633.6923$$

$$\hat{\beta}_1 = \frac{SS_{xy}}{SS_{xx}} = \frac{32,240.9231}{51,633.6923} = .624416377$$

$$\hat{\beta}_0 = \bar{y} - \hat{\beta}_1 \bar{x} = \frac{1,099}{13} - (.624416377)\left(\frac{1,341}{13}\right) = 20.12751061$$

The least squares prediction equation is $\hat{y} = 20.1275 + .6244x$

d. $\hat{\beta}_0 = 20.1275$. Since $x = 0$ is in the observed range, the mean number of words in conversation with an unfamiliar partner is estimated to be 20.1275 when the number of words in conversation with a familiar partner is 0.

$\hat{\beta}_1 = .6244$. For each additional word in conversation with a familiar partner, the mean number of words in conversation with an unfamiliar partner is estimated to increase by .6244.

11.133 Some preliminary calculations are:

$$\sum x = 4305 \qquad \sum x^2 = 1,652,025 \qquad \sum xy = 76,652,695$$

$$\sum y = 201,558 \qquad \sum y^2 = 3,571,211,200$$

a. $\hat{\beta}_1 = \dfrac{\sum xy}{\sum x^2} = \dfrac{76,652,695}{1,652,025} = 46.39923427 \approx 46.3992$

The least squares line is $\hat{y} = 46.3992x$.

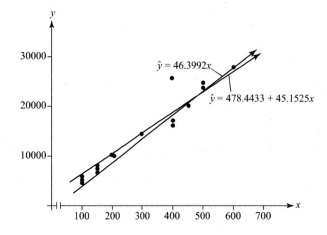

b. $SS_{yy} = \sum xy - \dfrac{\sum x \sum y}{n} = 76{,}652{,}695 - \dfrac{4305(201{,}558)}{15} = 18{,}805{,}549$

$SS_{xx} = \sum x^2 - \dfrac{(\sum x)^2}{n} = 1{,}652{,}025 - \dfrac{4305^2}{15} = 416{,}490$

$\hat{\beta}_1 = \dfrac{SS_{xy}}{SS_{xx}} = \dfrac{18{,}805{,}549}{416{,}490} = 45.15246224 \approx 45.1525$

$\hat{\beta}_0 = \bar{y} - \hat{\beta}_1 \bar{x} = \dfrac{201{,}558}{15} - 45.15246224\left(\dfrac{4305}{15}\right) = 478.4433$

The least squares line is $\hat{y} = 478.4433 + 45.1525x$.

c.   Because $x = 0$ is not in the observed range, we are trying to represent the data on the observed interval with the best fitting line. We are not concerned with whether the line goes through (0, 0) or not.

d.   Some preliminary calculations are:

$SS_{yy} = \sum y^2 - \dfrac{(\sum y)^2}{n} = 3{,}571{,}211{,}200 - \dfrac{201{,}558^2}{15} = 862{,}836{,}042$

$SSE = SS_{yy} - \hat{\beta}_1 SS_{xy} = 862{,}836{,}042 - 45.15246224(18{,}805{,}549) = 13{,}719{,}200.88$

$s^2 = \dfrac{SSE}{n-2} = \dfrac{13{,}719{,}200.88}{15-2} = 1{,}055{,}323.145 \qquad s = \sqrt{s^2} = 1027.2892$

H_0: $\beta_0 = 0$
H_a: $\beta_0 \neq 0$

The test statistic is $t = \dfrac{\hat{\beta}_0 - 0}{s\sqrt{\dfrac{1}{n} + \dfrac{\bar{x}^2}{SS_{xx}}}} = \dfrac{478.443}{1027.2892\sqrt{\dfrac{1}{15} + \dfrac{287^2}{416{,}490}}} = .906$

The rejection region requires $\alpha / 2 = .10 / 2 = .05$ in each tail of the t distribution with df $= n - 2 = 15 - 2 = 13$. From Table VI, Appendix A, $t_{.05} = 1.771$. The rejection region is $t < -1.771$ or $t > 1.771$.

Since the observed value of the test statistic does not fall in the rejection region ($t = .906$ $\not> 1.771$), H_0 is not rejected. There is insufficient evidence to indicate β_0 is different from 0 at $\alpha = .10$. Thus, β_0 should not be included in the model.

11.135 a. Some preliminary calculations are:

$$\sum x_i = 1,098 \qquad \sum y_i = 121.8 \qquad \sum x_i y_i = 4,294.6$$

$$\sum x_i^2 = 56,712 \qquad \sum y_i^2 = 597.52$$

$$SS_{xy} = \sum x_i y_i - \frac{\sum x_i \sum y_i}{n} = 4,294.6 - \frac{1,098(121.8)}{27} = -658.6$$

$$SS_{xx} = \sum x_i^2 - \frac{\left(\sum x_i\right)^2}{n} = 56,712 - \frac{1,098^2}{27} = 12,060$$

$$\hat{\beta}_1 = \frac{SS_{xy}}{SS_{xx}} = \frac{-658.6}{12,060} = -.054610281$$

$$\hat{\beta}_0 = \bar{y} - \hat{\beta}_1 \bar{x} = \frac{121.8}{27} - (-.054610281)\left(\frac{1,098}{27}\right) = 6.731929244$$

$$SS_{yy} = \sum y_i^2 - \frac{\left(\sum y_i\right)^2}{n} = 597.52 - \frac{121.8^2}{27} = 48.06666667$$

$$SSE = SS_{yy} - \hat{\beta}_1 SS_{xy} = 48.06666667 - (-.054610281)(-658.6) = 12.10033563$$

$$r^2 = 1 - \frac{SSE}{SS_{yy}} = 1 - \frac{12.10033563}{48.066666667} = .7483 \quad 74.83\% \text{ of the total sample variability around}$$

the sample mean fertility rate is explained by the linear relationship between the fertility rate and the contraceptive prevalence. This is not 90%.

b. From above, the fitted regression line is $\hat{y} = 6.732 - .0546x$

If the contraceptive use increases by 18 percent, then the mean fertility rate will decrease by an estimated $(.0546)(18) = .9828$. This is approximately 1. The researchers' statement is supported by the data.

Multiple Regression and Model Building

12.1   a. $E(y) = \beta_0 + \beta_1 x_1 + \beta_2 x_2$

b. $E(y) = \beta_0 + \beta_1 x_1 + \beta_2 x_2 + \beta_3 x_3 + \beta_4 x_4$

c. $E(y) = \beta_0 + \beta_1 x_1 + \beta_2 x_2 + \beta_3 x_3 + \beta_4 x_4 + \beta_5 x_5$

12.3   The 6 steps in a multiple regression analysis are:

1.   Hypothesize the deterministic component of the model. This model relates the mean, $E(y)$, to the independent variables $x_1, x_2, \ldots, x_k$. This involves the choice of the independent variables to be included in the model.

2.   Use the sample data to estimate the unknown model parameters β_0, β_1, β_2, ..., β_k in the model.

3.   Specify the probability distribution of the random error term, ε, and estimate the standard deviation of the distribution, σ.

4.   Check the assumptions on ε are satisfied, and make model modifications if necessary.

5.   Statistically evaluate the usefulness of the model.

6.   When satisfied that the model is useful, use it for prediction, estimation, and other purposes.

12.5   To test the overall adequacy of a multiple regression model, you would use a global test. The null hypothesis would be that all of the beta parameters (except β_0) are 0 and the alternative hypothesis would be that at least one beta parameter is not 0.

12.7   a.   We are given $\hat{\beta}_2 = 2.7$, $s_{\hat{\beta}_2} = 1.86$, and $n = 30$.

H_0: $\beta_2 = 0$
H_a: $\beta_2 \neq 0$

The test statistic is $t = \dfrac{\hat{\beta}_2 - 0}{s_{\hat{\beta}_2}} = \dfrac{2.7}{1.86} = 1.45$

The rejection region requires $\alpha / 2 = .05 / 2 = .025$ in each tail of the t distribution with df $= n - (k + 1) = 30 - (3 + 1) = 26$. From Table VI, Appendix A, $t_{.025} = 2.056$. The rejection region is $t < -2.056$ or $t > 2.056$.

Since the observed value of the test statistic does not fall in the rejection region ($t = 1.45 \ngtr 2.056$), H_0 is not rejected. There is insufficient evidence to indicate $\beta_2 \neq 0$ at $\alpha = .05$.

b. We are given $\hat{\beta}_3 = .93$, $s_{\hat{\beta}_3} = .29$, and $n = 30$.

Test H_0: $\beta_3 = 0$
H_a: $\beta_3 \neq 0$

The test statistic is $t = \dfrac{\hat{\beta}_3 - 0}{s_{\hat{\beta}_3}} = \dfrac{.93}{.29} = 3.21$

The rejection region is the same as part **a**, $t < -2.056$ or $t > 2.056$.

Since the observed value of the test statistic falls in the rejection region ($t = 3.21 > 2.056$), H_0 is rejected. There is sufficient evidence to indicate $\beta_3 \neq 0$ at $\alpha = .05$.

c. $\hat{\beta}_3$ has a smaller estimated standard error than $\hat{\beta}_2$. Therefore, the test statistic is larger for $\hat{\beta}_3$ even though $\hat{\beta}_3$ is smaller than $\hat{\beta}_2$.

12.9 The number of degrees of freedom available for estimating σ^2 is $n - (k + 1)$ where k is the number of variables in the regression model. Each additional independent variable placed in the model causes a corresponding decrease in the degrees of freedom.

12.11 a. Yes. Since $R^2 = .92$ is close to 1, this indicates the model provides a good fit. Also, SSE = .33 is fairly small which indicates the model provides a good fit.

b. H_0: $\beta_1 = \beta_2 = \cdots = \beta_5 = 0$
H_a: At least one $\beta_i \neq 0$

The test statistic is $F = \dfrac{R^2 / k}{(1 - R^2)/[n - (k+1)]} = \dfrac{.92/5}{(1 - .92)/[30 - (5+1)]} = 55.2$

The rejection region requires $\alpha = .05$ in the upper tail of the F distribution with $v_1 = k = 5$ and $v_2 = n - (k+1) = 30 - (5+1) = 24$. From Table IX, Appendix A, $F_{.05} = 2.62$. The rejection region is $F > 2.62$.

Since the observed value of the test statistic falls in the rejection region ($F = 55.2 > 2.62$), H_0 is rejected. There is sufficient evidence to indicate the model is useful in predicting y at $\alpha = .05$.

12.13 a. Two properties of the errors of prediction that result from using the method of least squares are the sum of the errors is 0 and the sum of the squares of the errors (SSE) is minimized.

b. $\hat{\beta}_4 = .42$. For each unit increase in the number of shortest paths between peers, the mean lead-user rating is estimated to increase by .42, holding all of the other variables constant.

c.   Since the *p*-value is less than α ($p = .002 < .05$), H_0 is rejected. There is sufficient evidence to indicate a linear relationship between the number of shortest paths between peers and lead-user rating, adjusted for the other variables in the model for $\alpha = .05$.

12.15 a.   The model would be: $E(y) = \beta_0 + \beta_1 x_1 + \beta_2 x_2$

b.   To determine if the length of an entangled whale increases linearly as the water depth increases for entanglements that are a fixed distance from land, we test:

$$H_0: \beta_2 = 0$$
$$H_a: \beta_2 > 0$$

c.   Since the *p*-value is less than α ($p = .013 < .05$), H_0 is rejected. There is sufficient evidence to indicate the length of an entangled whale increases linearly as the water depth increases for entanglements that are a fixed distance from land for $\alpha = .05$.

12.17 a.   From the output, the least squares prediction equation is
$\hat{y} = 3.70 + .34x_1 + .49x_2 + .72x_3 + 1.14x_4 + 1.51x_5 + .26x_6 - .14x_7 - .10x_8 - .10x_9$.

b. $\hat{\beta}_0 = 3.70$.   This value has no meaningful interpretation.

$\hat{\beta}_1 = .34$.   We estimate the mean number of runs scored by a team in a season to increase by .34 for each unit increase in the number of walks, when all the other independent variables are held fixed.

$\hat{\beta}_2 = .49$.   We estimate the mean number of runs scored by a team in a season to increase by .49 for each additional single, when all the other independent variables are held fixed.

$\hat{\beta}_3 = .72$.   We estimate the mean number of runs scored by a team in a season to increase by .72 for each additional double, when all the other independent variables are held fixed.

$\hat{\beta}_4 = 1.14$.   We estimate the mean number of runs scored by a team in a season to increase by 1.14 for each additional triple, when all the other independent variables are held fixed.

$\hat{\beta}_5 = 1.51$.   We estimate the mean number of runs scored by a team in a season to increase by 1.51 for each additional home run, when all the other independent variables are held fixed.

$\hat{\beta}_6 = .26$.   We estimate the mean number of runs scored by a team in a season to increase by .26 for each additional stolen base, when all the other independent variables are held fixed.

$\hat{\beta}_7 = -.14$.   We estimate the mean number of runs scored by a team in a season to decrease by .14 for each additional player caught stealing, when all the other independent variables are held fixed.

$\hat{\beta}_8 = -.10$.   We estimate the mean number of runs scored by a team in a season to decrease by .10 for each additional strikeout, when all the other independent variables are held fixed.

$\hat{\beta}_9 = -.10$.   We estimate the mean number of runs scored by a team in a season to decrease by .10 for each additional out, when all the other independent variables are held fixed.

c. H_0: $\beta_7 = 0$
H_a: $\beta_7 < 0$

The test statistic is $t = \dfrac{\hat{\beta}_7 - 0}{s_{\hat{\beta}_7}} = \dfrac{-.14 - 0}{.14} = -1$

The rejection region requires $\alpha = .05$ in the lower tail of the t distribution. From Table VI, Appendix A, with df $= n - (k + 1) = 234 - (9 + 1) = 224$, $t_{.05} \approx 1.645$. The rejection region is $t < -1.645$.

Since the observed value of the test statistic does not fall in the rejection region ($t = -1.00 \not< -1.645$), H_0 is not rejected. There is insufficient evidence to indicate that there is a negative linear relationship between the number of runs scored and the number of times caught stealing at $\alpha = .05$.

d.   For confidence coefficient .95, $\alpha = .05$ and $\alpha / 2 = .05 / 2 = .025$. From Table VI, Appendix A, with df $= n - (k + 1) = 234 - (9 + 1) = 224$, $t_{.025} \approx 1.96$. The 95% confidence interval is:

$\hat{\beta}_5 \pm t_{.025} s_{\hat{\beta}_5} \Rightarrow 1.51 \pm 1.96(.05) \Rightarrow 1.51 \pm .098 \Rightarrow (1.412, 1.608)$

We are 95% confident that for each additional home run, the mean number of runs scored will increase by anywhere from 1.412 to 1.608, holding all the other variables constant.

12.19   a.   The first order model is $E(y) = \beta_0 + \beta_1 x_1 + \beta_2 x_2 + \beta_3 x_3$

b. $R^2 = .08$. 8% of the sample variation in the frequency of marijuana use in the past 6 months is explained by the model containing severity of inattention, severity of impulsivity-hyperactivity, and level of oppositional-defiant and conduct disorder.

c.   The global F-test has the hypotheses:

H_0: $\beta_1 = \beta_2 = \beta_3 = 0$
H_a: At least one $\beta_i \neq 0$

The global *F*-test yielded a *p*-value less than .01. Since the *p*-value is so small, there is evidence to reject H_0 for $\alpha > .01$. There is sufficient evidence to indicate that at least one of the independent variables contributes to the prediction of the frequency of marijuana use in the past 6 months at $\alpha > .01$.

d.   The test for β_1:

H_0: $\beta_1 = 0$
H_a: $\beta_1 \neq 0$

The *p*-value for the test is less than .01. Since the *p*-value is so small, there is evidence to reject H_0 for $\alpha > .01$. There is sufficient evidence to indicate that severity of inattention contributes to the prediction of the frequency of marijuana use in the past 6 months, adjusting for severity of impulsivity-hyperactivity and level of oppositional-defiant and conduct disorder fixed at $\alpha > .01$.

e.   The test for β_2:

H_0: $\beta_2 = 0$
H_a: $\beta_2 \neq 0$

The *p*-value for the test is greater than .05. Since the *p*-value is not small, there is no evidence to reject H_0 for $\alpha \leq .05$. There is insufficient evidence to indicate that severity of impulsivity-hyperactivity contributes to the prediction of the frequency of marijuana use in the past 6 months, adjusting for severity of inattention and level of oppositional-defiant and conduct disorder fixed at $\alpha \leq .05$.

f.   The test for β_3:

H_0: $\beta_3 = 0$
H_a: $\beta_3 \neq 0$

The *p*-value for the test is greater than .05. Since the *p*-value is not small, there is no evidence to reject H_0 for $\alpha \leq .05$. There is insufficient evidence to indicate that level of oppositional-defiant and conduct disorder contributes to the prediction of the frequency of marijuana use in the past 6 months, adjusting for severity of inattention and severity of impulsivity-hyperactivity fixed at $\alpha \leq .05$.

12.21   a.   Let *y* = arsenic level, x_1 = latitude, x_2 = longitude, and x_3 = depth. The first-order model would be:

$$E(y) = \beta_o + \beta_1 x_1 + \beta_2 x_2 + \beta_3 x_3$$

b.   Using MINITAB, the results are:

Regression Analysis: ARSENIC versus LATITUDE, LONGITUDE, DEPTH-FT

```
The regression equation is
ARSENIC = - 86868 - 2219 LATITUDE + 1542 LONGITUDE - 0.350 DEPTH-FT


327 cases used, 1 cases contain missing values


Predictor      Coef  SE Coef      T      P
Constant     -86868    31224  -2.78  0.006
LATITUDE    -2218.8    526.8  -4.21  0.000
LONGITUDE    1542.2    373.1   4.13  0.000
DEPTH-FT    -0.3496   0.1566  -2.23  0.026


S = 103.301   R-Sq = 12.8%   R-Sq(adj) = 12.0%


Analysis of Variance

Source            DF        SS      MS      F      P
Regression         3    505770  168590  15.80  0.000
Residual Error   323   3446791   10671
Total            326   3952562


Source      DF   Seq SS
LATITUDE     1   132448
LONGITUDE    1   320144
DEPTH-FT     1    53179
```

The fitted regression model is:

$$\hat{y} = -86,868 - 2,218.8x_1 + 1,542.2x_2 - .3496x_3$$

c. $\hat{\beta}_0 = -86,868$. This value has no meaningful interpretation because $x_1 = 0$, $x_2 = 0$, and $x_3 = 0$ are not in the observed range.

$\hat{\beta}_1 = -2,218.8$. We estimate the mean arsenic level will decrease by 2,218.8 for each unit increase in latitude, holding longitude and depth constant.

$\hat{\beta}_2 = 1,542.2$. We estimate the mean arsenic level will increase by 1,542.2 for each unit increase in longitude, holding latitude and depth constant.

$\hat{\beta}_3 = -.3496$. We estimate the mean arsenic level will decrease by .3496 for each unit increase in depth, holding latitude and longitude constant.

d. $S = 103.301$. We would expect about 95% of the observed values of arsenic levels to fall within $2s = 2(103.301) = 206.602$ units of their least squares predicted values.

e. R^2 = R-Sq = .128. 12.8% of the sample variation in the arsenic level is explained by the model containing latitude, longitude and depth.

R_a^2 = R-Sq(adj) = .120. 12.0% of the sample variation in the arsenic level is explained by the model containing latitude, longitude and depth, adjusted for the sample size and the number of independent variables in the model.

f.   To determine if the overall model is useful for predicting y, we test:

H_0: $\beta_1 = \beta_2 = \beta_3 = 0$
H_a: At lease one $\beta_i \neq 0$

The test statistic is $F = 15.80$ and the p-value is $p = .000$.

Since the p-value is less than $\alpha = .05$ $(p = .000 < .05)$, H_0 is rejected. There is sufficient evidence to indicate that at least one of the independent variables is useful in predicting arsenic levels at $\alpha = .05$.

g.   Even though the independent variables are significant in predicting arsenic levels, the R^2 value and adjusted R^2 values are very low. Not very much of the sample variation in the arsenic levels can be explained by the model. In addition, the estimate of the standard deviation ($s = 103.301$) is very large compared with the actual values of the data. I would not recommend using this model.

12.23 a.   Using MINITAB, the results of fitting the first-order model are:

Regression Analysis: DESIRE versus GENDER, SELFESTM, BODYSAT, IMPREAL

```
The regression equation is
DESIRE = 14.0 - 2.19 GENDER - 0.0479 SELFESTM - 0.322 BODYSAT + 0.493 IMPREAL


Predictor        Coef  SE Coef      T      P
Constant      14.0107   0.7753  18.07  0.000
GENDER        -2.1865   0.6766  -3.23  0.001
SELFESTM     -0.04794  0.03669  -1.31  0.193
BODYSAT       -0.3223   0.1435  -2.25  0.026
IMPREAL        0.4931   0.1274   3.87  0.000


S = 2.25087   R-Sq = 49.8%   R-Sq(adj) = 48.5%


Analysis of Variance

Source           DF       SS      MS      F      P
Regression        4   827.83  206.96  40.85  0.000
Residual Error  165   835.95    5.07
Total           169  1663.79
```

```
Source     DF   Seq SS
GENDER      1   674.64
SELFESTM    1    57.66
BODYSAT     1    19.62
IMPREAL     1    75.91
```

The least squares prediction equation is:

$$\hat{y} = 14.0107 - 2.1865(\text{GENDER}) - .04794(\text{SELFESTM}) - .3223(\text{BODYSAT}) + .493(1\text{IMPREAL})$$

b. $\hat{\beta}_0 = 14.0107$.   This has no meaning other than the y-intercept.

$\hat{\beta}_1 = -2.1865$.   The mean value of desire to have cosmetic surgery is estimated to be 2.1865 units lower for males than females, holding all other variables constant.

$\hat{\beta}_2 = -0.04794$.   For each unit increase in self-esteem, the mean value of desire to have cosmetic surgery is estimated to decrease by 0.04794 units, holding all other variables constant.

$\hat{\beta}_3 = -0.3223$.   For each unit increase in body satisfaction, then mean value of desire to have cosmetic surgery is estimated to decrease by .3223 units, holding all other variables constant.

$\hat{\beta}_4 = 0.4931$.   For each unit increase in impression of reality TV, the mean value of desire to have cosmetic surgery is estimated to increase by 0.4931 units, holding all other variables constant.

c.   To determine if the overall model is useful for predicting desire to have cosmetic surgery, we test:

H_0: $\beta_1 = \beta_2 = \beta_3 = \beta_4 = 0$
H_a: At least one $\beta_i \neq 0$

From the printout, the test statistic is $F = 40.85$ and the p-value is $p = 0.000$.

Since the p-value is less than α $(p = 0.000 < .01)$, H_0 is rejected. There is sufficient evidence to indicate the overall model is useful for predicting desire to have cosmetic surgery at $\alpha = .01$.

d. R_a^2 is the preferred measure of model fit. From the printout, $R_a^2 = .485$. This indicates that 48.5% of the total sample variation in desire values is explained by the model containing gender, self-esteem, body satisfaction and impression of reality TV, adjusting for the sample size and the number of variables in the model.

e. To determine if the desire to have cosmetic surgery decreases linearly as level of body satisfaction increases, we test:

$H_0: \beta_3 = 0$
$H_a: \beta_3 < 0$

From the printout, the test statistic is $t = -2.25$ and the p-value is $p = .026/2 = .013$.

Since the p-value is less than α ($p = 0.013 < .05$), H_0 is rejected. There is sufficient evidence to indicate the desire to have cosmetic surgery decreases linearly as level of body satisfaction increases at $\alpha = .05$.

f. For confidence coefficient .95, $\alpha = 1 - .95 = .05$ and $\alpha/2 = .05/2 = .025$. From Table VI, Appendix A, with df $= n - (k+1) = 170 - (4+1) = 165$, $t_{.025} \approx 1.98$. The 95% confidence interval is

$$\hat{\beta}_4 \pm t_{\alpha/2} s_{\hat{\beta}_4} \Rightarrow 0.4931 \pm 1.98(.1274) \Rightarrow 0.4931 \pm 0.2523 \Rightarrow (0.2408, 0.7454)$$

We are 95% confident that the increase in mean desire for cosmetic surgery is between .2408 and .7454 for each unit increase in impression of reality TV, holding all other variables constant.

12.25 a. Let y = heat rate, x_1 = speed, x_2 = inlet temperature, x_3 = exhaust temperature, x_4 = cycle pressure ratio, and x_5 = air flow rate. The first-order model would be:

$$E(y) = \beta_o + \beta_1 x_1 + \beta_2 x_2 + \beta_3 x_3 + \beta_4 x_4 + \beta_5 x_5$$

b. Using MINITAB, the results are:

Regression Analysis: HEATRATE versus RPM, INLET-TEMP, ...

```
The regression equation is
HEATRATE = 13614 + 0.0888 RPM - 9.20 INLET-TEMP + 14.4 EXH-TEMP + 0.4
CPRATIO - 0.848 AIRFLOW


Predictor       Coef  SE Coef       T      P
Constant     13614.5    870.0   15.65  0.000
RPM          0.08879  0.01391    6.38  0.000
INLET-TEMP    -9.201    1.499   -6.14  0.000
EXH-TEMP      14.394    3.461    4.16  0.000
CPRATIO         0.35    29.56    0.01  0.991
AIRFLOW      -0.8480   0.4421   -1.92  0.060

S = 458.828    R-Sq = 92.4%    R-Sq(adj) = 91.7%


Analysis of Variance

Source           DF         SS        MS       F      P
Regression        5  155055273  31011055  147.30  0.000
Residual Error   61   12841935    210524
Total            66  167897208
```

```
Source        DF    Seq SS
RPM            1   119598530
INLET-TEMP     1    26893467
EXH-TEMP       1     7784225
CPRATIO        1        4623
AIRFLOW        1      774427
```

The fitted regression model is:

$$\hat{y} = 13,614.5 + .08879x_1 - 9.201x_2 + 14.394x_3 + .35x_4 - .848x_5$$

c. $\hat{\beta}_o = 13,614.5$. This value has no meaningful interpretation other than the y-intercept because $x_1 = 0$, $x_2 = 0$, $x_3 = 0$, $x_4 = 0$, and $x_5 = 0$, are not in the observed range.

$\hat{\beta}_1 = .08879$. We estimate the mean heat rate will increase by .08879 units for each unit increase in speed, holding inlet temperature, exhaust temperature, cycle pressure ratio, and air flow rate constant.

$\hat{\beta}_2 = -9.201$. We estimate the mean heat rate will decrease by 9.201 units for each unit increase in inlet temperature, holding speed, exhaust temperature, cycle pressure ratio, and air flow rate constant.

$\hat{\beta}_3 = 14.394$. We estimate the mean heat rate will increase by 14.394 units for each unit increase in exhaust temperature, holding speed, inlet temperature, cycle pressure ratio, and air flow rate constant.

$\hat{\beta}_4 = .35$. We estimate the mean heat rate will increase by .35 units for each unit increase in cycle pressure ratio, holding speed, exhaust temperature, inlet temperature, and air flow rate constant.

$\hat{\beta}_5 = -.8480$. We estimate the mean heat rate will decrease by .848 units for each unit increase in air flow rate constant, holding speed, inlet temperature, exhaust temperature, and cycle pressure ratio.

d. $s = 458.828$. We would expect about 95% of the observed values of heat rate to fall within $2s = 2(458.828) = 917.656$ units of their least squares predicted values.

e. $R_a^2 = $ R-Sq(adj) $= .917$ 91.7% of the sample variation in the heat rate is explained by the model containing speed, inlet temperature, exhaust temperature, cycle pressure ratio, and air flow rate, adjusted for the sample size and the number of independent variables in the model.

f. To determine if the overall model is useful for predicting y, we test:

H_0: $\beta_1 = \beta_1 = \beta_3 = \beta_4 = \beta_5 = 0$
H_a: At lease one $\beta_i \neq 0$

The test statistic is $F = 147.30$ and the p-value is $p = 0.000$.

Since the p-value is less than $\alpha = .01$ $(p = 0.000 < .01)$, H_0 is rejected. There is sufficient evidence to indicate that at least one of the independent variables is useful in predicting heat rate at $\alpha = .01$.

12.27   a.   The least squares prediction equation for the interstate highway model is:

$$\hat{y} = 1.81231 + .10875x_1 + .00017x_2$$

b. $\hat{\beta}_0 = 1.81231$. This value has no meaningful interpretation other than the y-intercept because $x_1 = 0$ and $x_2 = 0$ are not in the observed range.

$\hat{\beta}_1 = .10875$. We estimate the mean number of crashes per 3 years will increase by .10875 for each additional mile of roadway, holding AADT constant.

$\hat{\beta}_2 = .00017$. We estimate the mean number of crashes per 3 years will increase by .00017 for each additional vehicle per day, holding roadway mileage constant.

c.   For confidence coefficient .99, $\alpha = .01$ and $\alpha / 2 = .01 / 2 = .005$. From Table VI, Appendix A, with df $= n - (k + 1) = 100 - (2 + 1) = 97$, $t_{.005} \approx 2.62$. The 99% confidence interval is:

$$\hat{\beta}_1 \pm t_{.005}s_{\hat{\beta}_1} \Rightarrow .10875 \pm 2.62(.03166) \Rightarrow .10875 \pm .08295 \Rightarrow (.02580, \; .19170)$$

We are 99% confident that the increase in mean number of crashes for each additional mile of roadway is between .02580 and .19170, holding AADT constant.

d.   The 99% confident interval is:

$$\hat{\beta}_2 \pm t_{.005}s_{\hat{\beta}_2} \Rightarrow .00017 \pm 2.62(.00003) \Rightarrow .00017 \pm .00008 \Rightarrow (.00009, \; .00025)$$

We are 99% confident that the increase in mean number of crashes for each additional vehicle per day is between .00009 and .00025, holding roadway mileage constant.

e.   The least squares prediction equation for the non-interstate highway model is:

$$\hat{y} = 1.20785 + .06343x_1 + .00056x_2$$

$\hat{\beta}_o = 1.20785$. This value has no meaningful interpretation other than the y-intercept because $x_1 = 0$ and $x_2 = 0$ are not in the observed range.

$\hat{\beta}_1 = .06343$. We estimate the mean number of crashes per 3 years will increase by .06343 for each additional mile of roadway, holding AADT constant.

$\hat{\beta}_2 = .00056$. We estimate the mean number of crashes per 3 years will increase by .00056 for each additional vehicle per day, holding roadway mileage constant.

For confidence coefficient .99, $\alpha = .01$ and $\alpha/2 = .01/2 = .005$. From Table VI, Appendix A, with df $= n - (k+1) = 100 - (2+1) = 97$, $t_{.005} \approx 2.62$. The 99% confidence interval is:

$$\hat{\beta}_1 \pm t_{.005} s_{\hat{\beta}_1} \Rightarrow .06343 \pm 2.62(.01809) \Rightarrow .06343 \pm .04740 \Rightarrow (.01603, \ .11083)$$

We are 99% confident that the increase in mean number of crashes for each additional mile of roadway is between .01603 and .11083, holding AADT constant.

The 99% confident interval is:

$$\hat{\beta}_2 \pm t_{.005} s_{\hat{\beta}_2} \Rightarrow .00056 \pm 2.62(.00012) \Rightarrow .00056 \pm .00031 \Rightarrow (.00025, \ .00087)$$

We are 99% confident that the increase in mean number of crashes for each additional vehicle per day is between .00025 and .00087, holding roadway mileage constant.

12.29 The model we are fitting is $y = \beta_0 + \beta_1 x_1 + \beta_2 x_2 + ... + \beta_k x_k + \varepsilon$. The fitted model is $\hat{y} = \hat{\beta}_0 + \hat{\beta}_1 x_1 + \hat{\beta}_2 x_2 + ... + \hat{\beta}_k x_k$. Thus, we use the fitted model to predict the actual value of y. In addition, we could write the model as $E(y) = \beta_0 + \beta_1 x_1 + \beta_2 x_2 + ... + \beta_k x_k$. So, the fitted model will also estimate $E(y)$.

12.31 a. For $x_1 = 1$, $x_2 = 10$, $x_3 = 5$, and $x_4 = 2$, the predicted lead-user rating is:

$$\hat{y} = 3.58 + .01(1) - .06(10) - .01(5) + .42(2) = 3.78$$

b. For $x_1 = 0$, $x_2 = 8$, $x_3 = 10$, and $x_4 = 4$, the predicted lead-user rating is:

$$\hat{y} = 3.58 + .01(0) - .06(8) - .01(10) + .42(4) = 4.68$$

12.33 a. The confidence interval for student 1 is (13.4243, 14.3067). We are 95% confident that the true mean desire to have cosmetic surgery is between 13.4243 and 14.3067 for everyone who is female, has a self-esteem score of 24, has a body satisfaction score of 3, and has an impression of reality TV score of 4.

b. The confidence interval for student 4 is (8.7908, 10.8910). We are 95% confident that the true mean desire to have cosmetic surgery is between 8.7908 and 10.8910 for everyone who is male, has a self-esteem score of 22, has a body satisfaction score of 9, and has an impression of reality TV score of 4.

12.35 a. From the printout, the 95% prediction interval is (11,599.6, 13,665.5). We are 95% confident that the actual heat rate will be between 11,599.6 and 13,665.5 when the speed is 7,500 rpm, the inlet temperature is 1,000, the exhaust temperature is 525, the cycle pressure ratio is 13.5, and the air flow rate is 10.0.

b. From the printout, the 95% confidence interval for the mean is (12,157.9, 13,107.1). We are 95% confident that the mean heat rate will be between 12,157.9 and 13,107.1 when the speed is 7,500 rpm, the inlet temperature is 1,000, the exhaust temperature is 525, the cycle pressure ratio is 13.5, and the air flow rate is 10.0.

c.   The confidence interval for $E(y)$ will always be narrower than the prediction interval for y. The confidence interval for $E(y)$ is a confidence interval for the mean of a distribution. Once the mean is located, the actual values will vary around this mean. Thus, the width of the prediction interval includes the variation for locating the mean plus the variation of the actual values around the mean.

12.37   Using MINITAB, the results of the prediction are:

```
Predicted Values for New Observations

New
Obs    Fit   SE Fit        95% CI              95% PI
 1    11.75   16.66    (-21.19, 44.68)   (-183.76, 207.25)


Values of Predictors for New Observations

New
Obs   Miles   Length  Weight
 1     100     40.0     800
```

The 95% prediction interval is $(-183.76, 207.25)$. We are 95% confident that the actual DDT level of a fish caught 100 miles upstream with a length of 40 centimeters and a weight of 800 grams will be between -186.76 and 207.25 parts per million. Since DDT levels cannot be negative, the actual interval would be 0 to 207.25.

12.39   Yes, we agree. The estimated coefficients for latitude and depth are both negative. Thus, to maximize arsenic levels, we would want latitude and depth to be as low as possible. The estimated coefficient for longitude is positive. To maximize arsenic levels, we would want longitude to be as large as possible.

Using MINTAB to find the prediction interval:

```
Predicted Values for New Observations

New
Obs    Fit  SE Fit        95% CI              95% PI
 1   232.33   23.23   (186.63, 278.04)   (24.03, 440.64)X

X denotes a point that is an outlier in the predictors.


Values of Predictors for New Observations

New
Obs   LATITUDE   LONGITUDE   DEPTH-FT
 1     23.755      90.662       25.0
```

The 95% prediction interval is $(24.03, 440.64)$. We are 95% confident that the actual value of arsenic level is between 24.03 and 440.61 when latitude is minimized at 23.803, longitude is maximized at 90.662, and depth is minimized at 25.

12.41   a. $E(y) = \beta_0 + \beta_1 x_1 + \beta_2 x_2 + \beta_3 x_1 x_2$

b. $E(y) = \beta_0 + \beta_1 x_1 + \beta_2 x_2 + \beta_3 x_3 + \beta_4 x_1 x_2 + \beta_5 x_1 x_3 + \beta_6 x_2 x_3$

12.43   a. $R^2 = 1 - \dfrac{SSE}{SS_{yy}} = 1 - \dfrac{21}{479} = .956$

95.6% of the total sample variability of the y values is explained by this model.

b.   To test the utility of the model, we test:

H_0: $\beta_1 = \beta_2 = \beta_3 = 0$
H_a: At least one $\beta_i \neq 0$, $i = 1, 2, 3$

The test statistic is $F = \dfrac{R^2/k}{(1-R^2)/[n-(k+1)]} = \dfrac{.956/3}{(1-.956)/[32-(3+1)]} = 202.8$

The rejection region requires $\alpha = .05$ in the upper tail of the F distribution, with $v_1 = k = 3$ and $v_2 = n - (k+1) = 32 - (3+1) = 28$. From Table IX, Appendix A, $F_{.05} = 2.95$. The rejection region is $F > 2.95$.

Since the observed value of the test statistic falls in the rejection region ($F = 202.8 > 2.95$), H_0 is rejected. There is sufficient evidence that the model is adequate for predicting y at $\alpha = .05$.

c.

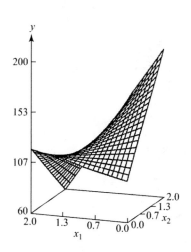

d.   To determine if x_1 and x_2 interact, we test:

H_0: $\beta_3 = 0$
H_a: $\beta_3 \neq 0$

The test statistic is $t = \dfrac{\hat{\beta}_3 - 0}{s_{\hat{\beta}_3}} = \dfrac{10}{4} = 2.5$.

The rejection region requires $\alpha/2 = .05/2 = .025$ in each tail of the t distribution with df $= n - (k+1) = 32 - (3+1) = 28$. From Table VI, Appendix A, $t_{.025} = 2.048$. The rejection region is $t < -2.048$ or $t > 2.048$.

Since the observed value of the test statistic falls in the rejection region ($t = 2.5 > 2.048$), H_0 is rejected. There is sufficient evidence to indicate that x_1 and x_2 interact at $\alpha = .05$.

12.45   a.   The equation for the interaction model is:

$$E(y) = \beta_0 + \beta_1 x_1 + \beta_2 x_2 + \beta_3 x_1 x_2$$

b.   When $x_2 = 10$, the model becomes:

$$E(y) = \beta_0 + \beta_1 x_1 + \beta_2(10) + \beta_3 x_1(10) + \varepsilon = \beta_0 + 10\beta_2 + (\beta_1 + 10\beta_3)x_1$$

Thus, the slope of the line is $\beta_1 + 10\beta_3$.

c.   When $x_2 = 25$, the model becomes:

$$E(y) = \beta_0 + \beta_1 x_1 + \beta_2(25) + \beta_3 x_1(25) + \varepsilon = \beta_0 + 25\beta_2 + (\beta_1 + 25\beta_3)x_1$$

Thus, the slope of the line is $\beta_1 + 25\beta_3$.

12.47   a.   A regression model containing the interaction between blade position and turntable speed would be:

$$E(y) = \beta_o + \beta_1 x_1 + \beta_2 x_2 + \beta_3 x_1 x_2$$

b.   "Blade position and turntable speed interact" means that the effect of turntable speed on the number of defects (y) depends on the blade position.

c.   This means that the interaction term in the model in part **a** would be significant. The slope of the line relating the number of defects and the turntable speed is different for different levels of blade position. Since the slope is steeper for lower values of cutting blade position, this implies that $\beta_3 < 0$.

12.49   a. $R^2 = .994$. This means that 99.4% of the total sample variation in the amplitude values is explained by the model containing spatial position of the probe, position of the cross, and the interaction of the two variables.

b.   If position of the probe and position of the cross interact, then the effect of position of the probe on the value of amplitude depends on the level of position of the cross.

c. A possible graph would look like the following:

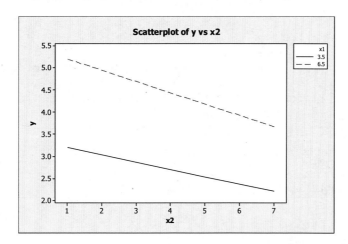

When $x_1 = 6.5$, $\hat{y} = .91 + .70(6.5) - .06x_2 - .03(6.5)x_2 = 5.46 - .255x_2$. The slope is -.255.
When $x_1 = 3.5$, $\hat{y} = .91 + .70(3.5) - .06x_2 - .03(3.5)x_2 = 3.36 - .165x_2$. The slope is -.165.
From the graph, you can see that the slope of the line corresponding to $x_1 = 6.5$ is steeper than the slope of the line corresponding to $x_1 = 3.5$.

12.51 a. $E(y) = \beta_0 + \beta_1 x_1 + \beta_2 x_2 + \beta_3 x_3 + \beta_4 x_4 + \beta_5 x_5 + \beta_6 x_1 x_2$

b. H_0: $\beta_4 = 0$

c. $t = 4.408$, p-value = .001

Since the p-value is so small, there is strong evidence to reject H_0. There is sufficient evidence to indicate that the strength of client-therapist relationship contributes information for the prediction of a client's reaction for any $\alpha > .001$.

d. Yes. For female therapists, the coefficient associated with age is β_1. For male therapists, the coefficient associated with age is $\beta_1 + \beta_6$. If β_6 is positive and highly significant, then the slope of the age line is more positive for males than for females.

e. $R^2 = .2946$. 29.46% of the variability in the client's reaction scores can be explained by this model.

12.53 a. The phrase "client credibility and linguistic delivery style interact" means that the affect of client credibility on likelihood depends on the level of linguistic delivery style.

b. To determine if the overall model is adequate for predicting likelihood, we test:

H_0: $\beta_1 = \beta_2 = \beta_3 = 0$
H_a: At least one $\beta_i \neq 0$, $i = 1, 2, 3$

c. From the table, the test statistic is $F = 55.35$ and the p-value is $p < 0.0005$. Since the p value is so small, H_0 is rejected. There is sufficient evidence to indicate the model is adequate in predicting likelihood for any reasonable level of α.

d.   To determine if client credibility and linguistic delivery style interact, we test:

H_0: $\beta_3 = 0$

H_a: $\beta_3 \neq 0$

e.   From the table, $t = 4.008$ and the p-value is $p < .005$. Since the p-value is so small, H_0 is rejected. There is sufficient evidence to indicate the that client credibility and linguistic delivery style interact for any reasonable level of α.

f.   The least squares prediction equation is $\hat{y} = 15.865 + .037x_1 - .678x_2 + .036x_1x_2$.

When client credibility is 22, the least squares line is:

$$\hat{y} = 15.865 + .037(22) - .678x_2 + .036(22)x_2 = 16.679 + .114x_2$$

When client credibility is 22, for each unit increase in linguistic delivery score, the mean likelihood score is estimated to increase by .114.

g.   When client credibility is 46, the least squares line is:

$$\hat{y} = 15.865 + .037(46) - .678x_2 + .036(46)x_2 = 17.567 + .978x_2$$

When client credibility is 46, for each unit increase in linguistic delivery score, the mean likelihood score is estimated to increase by .978.

12.55 a.   The model would be:

$$E(y) = \beta_o + \beta_1 x_1 + \beta_2 x_2 + \beta_3 x_3 + \beta_4 x_4 + \beta_5 x_5 + \beta_6 x_2 x_5 + \beta_7 x_3 x_5$$

b.   Using MINITAB, the results are:

Regression Analysis: HEATRATE versus RPM, INLET-TEMP

```
The regression equation is
HEATRATE = 13646 + 0.0460 RPM - 12.7 INLET-TEMP + 23.0 EXH-TEMP - 3.0
CPRATIO + 1.29 AIRFLOW + 0.0161 INxAIRF - 0.0414 EXxAIRF


Predictor        Coef   SE Coef       T       P
Constant        13646      1068   12.77   0.000
RPM           0.04599   0.01602    2.87   0.006
INLET-TEMP    -12.675     1.542   -8.22   0.000
EXH-TEMP       23.003     3.768    6.11   0.000
CPRATIO         -3.02     26.42   -0.11   0.909
AIRFLOW         1.288     3.563    0.36   0.719
INxAIRF      0.016149  0.003673    4.40   0.000
EXxAIRF      -0.04143   0.01098   -3.77   0.000


S = 404.693   R-Sq = 94.2%   R-Sq(adj) = 93.6%
```

```
Analysis of Variance

Source           DF        SS         MS        F       P
Regression        7   158234406   22604915   138.02   0.000
Residual Error   59     9662802     163776
Total            66   167897208


Source          DF      Seq SS
RPM              1   119598530
INLET-TEMP       1    26893467
EXH-TEMP         1     7784225
CPRATIO          1        4623
AIRFLOW          1      774427
INxAIRF          1      849347
EXxAIRF          1     2329786
```

The least squares prediction equation is:

$$\hat{y} = 13,646 + .04599x_1 - 12.675x_2 + 23.003x_3 - 3.02x_4 + 1.288x_5 + .016149x_2x_5 - .04143x_3x_5$$

c. To determine if inlet temperature and air flow rate interact to affect heat rate, we test:

$H_0: \beta_6 = 0$

$H_a: \beta_6 \neq 0$

From the printout, the test statistic is $t = 4.40$ and the p-value is $p = .000$.

Since the p-value is less than α $(p = .000 < .05)$, H_0 is rejected. There is sufficient evidence to indicate that inlet temperature and air flow rate interact to affect heat rate at $\alpha = .05$.

d. To determine if exhaust temperature and air flow rate interact to affect heat rate, we test:

$H_0: \beta_7 = 0$

$H_a: \beta_7 \neq 0$

From the printout, the test statistic is $t = -3.77$ and the p-value is $p = .000$.

Since the p-value is less than α $(p = .000 < .05)$, H_0 is rejected. There is sufficient evidence to indicate that exhaust temperature and air flow rate interact to affect heat rate at $\alpha = .05$.

e. Both of the interaction tests were significant. This means that the effect of inlet temperature on the heat rate is different at different levels of air flow and the effect of exhaust temperature on heat rate is different at different levels of air flow.

12.57 a. $E(y) = \beta_0 + \beta_1 x + \beta_2 x^2$

b. $E(y) = \beta_0 + \beta_1 x_1 + \beta_2 x_2 + \beta_3 x_1^2 + \beta_4 x_2^2 + \beta_5 x_1 x_2$

c. $E(y) = \beta_0 + \beta_1 x_1 + \beta_2 x_2 + \beta_3 x_3 + \beta_4 x_1^2 + \beta_5 x_2^2 + \beta_6 x_3^2 + \beta_7 x_1 x_2 + \beta_8 x_1 x_3 + \beta_9 x_2 x_3$

12.59   a. H_0: $\beta_2 = 0$

H_a: $\beta_2 \neq 0$

The test statistic is $t = \dfrac{\hat{\beta}_2 - 0}{s_{\hat{\beta}_2}} = \dfrac{.47 - 0}{.15} = 3.133$

The rejection region requires $\alpha / 2 = .05 / 2 = .025$ in each tail of the t distribution with df $= n - (k + 1) = 25 - (2 + 1) = 22$. From Table VI, Appendix A, $t_{.025} = 2.074$. The rejection region is $t < -2.074$ or $t > 2.074$.

Since the observed value of the test statistic falls in the rejection region ($t = 3.133 > 2.074$), H_0 is rejected. There is sufficient evidence to indicate the true relationship is given by the quadratic model at $\alpha = .05$.

b. H_0: $\beta_2 = 0$

H_a: $\beta_2 > 0$

The test statistic is the same as in part **a**, $t = 3.133$.

The rejection region requires $\alpha = .05$ in the upper tail of the t distribution with df $= 22$. From Table VI, Appendix A, $t_{.05} = 1.717$. The rejection region is $t > 1.717$.

Since the observed value of the test statistic falls in the rejection region ($t = 3.133 > 1.717$), H_0 is rejected. There is sufficient evidence to indicate the quadratic curve opens upward at $\alpha = .05$.

12.61   a.

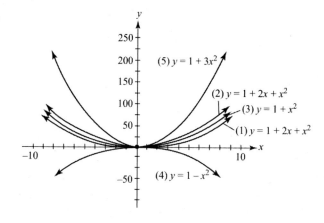

b.   It moves the graph to the right ($-2x$) or to the left ($+2x$) compared to the graph of $y = 1 + x^2$.

c.   It controls whether the graph opens up ($+x^2$) or down ($-x^2$). It also controls how steep the curvature is, i.e., the larger the absolute value of the coefficient of x^2, the narrower the curve is.

12.63   a. $\hat{\beta}_0 = 6.13$. This has no meaning other than the y-intercept, since $x = 0$ is not in the observed range.

$\hat{\beta}_1 = .141$. This has no meaning since the quadratic term is in the model. This is a shift parameter.

$\hat{\beta}_2 = -.0009$. Since the parameter estimate is negative, this indicates that the curve has a mound shape.

b. $R^2 = .226$. This means that 22.6% of the total sample variation in number of points scored is explained by the model containing number of yards from opposing goal line and number of yards from opposing goal line squared.

c.   No. If you add variables to the model, the value of R^2 will either increase or stay the same. We cannot tell whether the increase in the value of R^2 is significant or not by just looking at the 2 values.

d.   To determine if the quadratic model is a better fit than the linear model, we would test:

H_0: $\beta_2 = 0$
H_a: $\beta_2 \neq 0$

12.65   a.

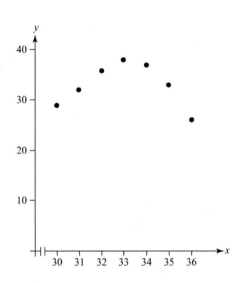

b.   If information was available only for $x = 30$, 31, 32, and 33, we would suggest a first-order model where $\beta_1 > 0$. If information was available only for $x = 33$, 34, 35, and 36, we would again suggest a first-order model where $\beta_1 < 0$. If all the information was available, we would suggest a second-order model.

12.67   a.   The complete second-order model is:

$$E(y) = \beta_o + \beta_1 x_1 + \beta_2 x_2 + \beta_3 x_1^2 + \beta_4 x_4^2 + \beta_5 x_1 x_2$$

b.   The terms that allow for a curvilinear relationship are $\beta_3 x_1^2$ and $\beta_4 x_4^2$.

12.69   a.   Using MINITAB, the graph is:

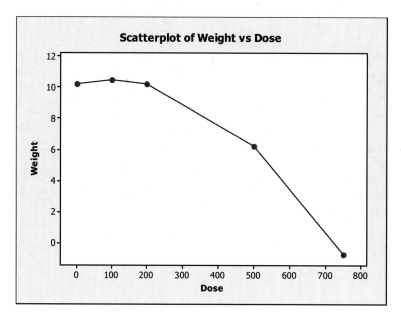

b.   For a dosage of 500 mg/kg, the estimated weight change would be:

$$\hat{y} = 10.25 + .0053(500) - .0000266(500^2) = 6.25$$

c.   For a dosage of 0 mg/kg, the estimated weight change would be:

$$\hat{y} = 10.25 + .0053(0) - .0000266(0^2) = 10.25$$

d.   For a dosage of 100 mg/kg, the estimated weight change would be:

$$\hat{y} = 10.25 + .0053(100) - .0000266(100^2) = 10.514$$

For a dosage of 200 mg/kg, the estimated weight change would be:

$$\hat{y} = 10.25 + .0053(200) - .0000266(200^2) = 10.246$$

For any dosage greater than 200 mg/kg, the weight change will be less than 10.246. Thus, a dosage of 200 mg/kg will be the largest dosage that yields an estimated weight change that is closest to, but below the estimated weight change for the control group (10.25).

12.71   a.   The complete 2nd-order model is: $E(y) = \beta_0 + \beta_1 x_1 + \beta_2 x_2 + \beta_3 x_1 x_2 + \beta_4 x_1^2 + \beta_5 x_2^2$

b. $R^2 = .402$. This means that 40.2% of the total sample variation of the satisfaction level values is explained by the complete 2nd-order model including needs and supplies.

c.   To determine if the overall model is adequate, we test:

H_0: $\beta_1 = \beta_2 = \beta_3 = \beta_4 = \beta_5 = 0$
H_a: At least one $\beta_i \neq 0$

The test statistic is $F = \dfrac{R^2/k}{(1-R^2)/[n-(k+1)]} = \dfrac{.402/5}{(1-.402)/[58-(5+1)]} = 6.99$.

The rejection region requires $\alpha = .10$ in the upper tail of the F-distribution with $v_1 = k = 5$ and $v_2 = n-(k+1) = 58-(5+1) = 52$. From Table VIII, Appendix A, $F_{.10} \approx 2.00$. The rejection region is $F > 2.00$.

Since the observed value of the test statistic falls in the rejection region ($F = 6.99 > 2.00$), H_0 is rejected. There is sufficient evidence to indicate at least one of the betas is different from 0 at $\alpha = .10$.

d.   To determine if needs is curvilinearly related to satisfaction, we test:

H_0: $\beta_4 = 0$
H_a: $\beta_4 \neq 0$

The test statistic is $t = \dfrac{\hat{\beta}_4 - 0}{s_{\hat{\beta}_4}} = \dfrac{-.181-0}{.302} = -.599$.

The rejection region requires $\alpha/2 = .10/2 = .05$ in each tail of the t-distribution with df $= n-(k+1) = 58-(5+1) = 52$. From Table VI, Appendix A, $t_{.05} \approx 1.684$. The rejection region is $t < -1.684$ or $t > 1.684$.

Since the observed value of the test statistic does not fall in the rejection region ($t = -.599 \not< -1.684$), H_0 is not rejected. There is insufficient evidence to indicate needs is curvilinearly related to satisfaction at $\alpha = .10$.

e.   To determine if supplies is curvilinearly related to satisfaction, we test:

H_0: $\beta_5 = 0$
H_a: $\beta_5 \neq 0$

The test statistic is $t = \dfrac{\hat{\beta}_5 - 0}{s_{\hat{\beta}_5}} = \dfrac{-.755-0}{.413} = -1.83$.

The rejection region requires $\alpha/2 = .10/2 = .05$ in each tail of the t-distribution with $df = n-(k+1) = 58-(5+1) = 52$. From Table VI, Appendix A, $t_{.05} \approx 1.684$. The rejection region is $t < -1.684$ or $t > 1.684$.

Since the observed value of the test statistic falls in the rejection region $(t = -1.83 < -1.684)$, H_0 is rejected. There is sufficient evidence to indicate supplies is curvilinearly related to satisfaction at $\alpha = .10$.

12.73   a.   Using MINITAB, the results are:

Regression Analysis: RATE versus EST, EST-SQ

```
The regression equation is
RATE = - 288 + 1.39 EST + 0.000035 EST-SQ


Predictor          Coef      SE Coef       T       P
Constant           -288         8049   -0.04   0.972
EST               1.395        3.651    0.38   0.706
EST-SQ       0.00003509   0.00009724    0.36   0.722


S = 31901.1    R-Sq = 45.9%    R-Sq(adj) = 40.8%


Analysis of Variance

Source             DF           SS          MS      F       P
Regression          2  18138955261  9069477631   8.91   0.002
Residual Error     21  21371254395  1017678781
Total              23  39510209656


Source  DF       Seq SS
EST      1  18006405335
EST-SQ   1    132549926
```

The fitted model is $\hat{y} = -288 + 1.398x + .0000351x^2$.

To determine if the incidence rate is curvilinearly related to estimated rate, we test:

H_0: $\beta_2 = 0$
H_a: $\beta_2 \neq 0$

From the printout, the test statistic is $t = .36$ and the p-value is $p = .722$.

Since the p-value is not less than α ($p = .722 \not< .05$), H_0 is not rejected. There is insufficient evidence that the incidence rate is curvilinearly related to estimated rate at $\alpha = .05$.

b.   Using MINITAB, the scatterplot is:

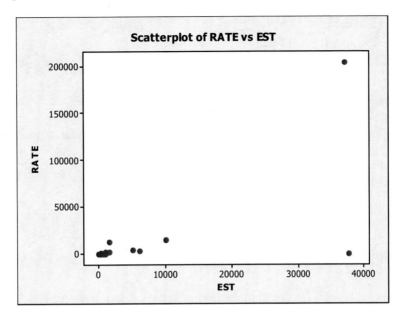

The point associated with Botulism is in the lower right corner of the graph. This point looks to be out of line with the other data points.

c.   Using MINITAB the new results are:

Regression Analysis: RATE versus EST, EST-SQ

```
The regression equation is
RATE = 735 - 0.081 EST + 0.000151 EST-SQ


Predictor          Coef       SE Coef        T      P
Constant          735.0         695.9     1.06   0.303
EST             -0.0810        0.3167    -0.26   0.801
EST-SQ       0.00015052    0.00000868    17.34   0.000


S = 2756.80    R-Sq = 99.6%    R-Sq(adj) = 99.6%


Analysis of Variance

Source          DF           SS           MS        F      P
Regression       2    39251490541   19625745270  2582.35  0.000
Residual Error  20      151998825      7599941
Total           22    39403489366


Source   DF      Seq SS
EST       1   36967483627
EST-SQ    1    2284006914
```

Now, if we test to see if the incidence rate is curvilinearly related to estimated rate:

H_0: $\beta_2 = 0$
H_a: $\beta_2 \neq 0$

From the printout, the test statistic is $t = 17.34$ and the p-value is $p = 0.000$.

Since the p-value is less than α ($p = .000 < .05$), H_0 is rejected. There is sufficient evidence that the incidence rate is curvilinearly related to estimated rate at $\alpha = .05$. The fit of the model has improved dramatically. The R^2 value for the new model is $R^2 = 99.6\%$ compared to the R^2 value for the first model, $R^2 = 45.9\%$.

12.75   Let $x = \begin{cases} 1 & \text{if qualitative variable assumes 2nd level} \\ 0 & \text{otherwise} \end{cases}$

The model is $E(y) = \beta_0 + \beta_1 x_1$.

β_0 = mean value of y when the qualitative variable assumes the first level.

β_1 = difference in the mean values of y between levels 2 and 1 of the qualitative variable.

12.77   a.   Level 1 implies $x_1 = x_2 = x_3 = 0$. $\hat{y} = 10.2$
Level 2 implies $x_1 = 1$ and $x_2 = x_3 = 0$. $\hat{y} = 10.2 - 4(1) = 6.2$
Level 3 implies $x_2 = 1$ and $x_1 = x_3 = 0$. $\hat{y} = 10.2 + 12(1) = 22.2$
Level 4 implies $x_3 = 1$ and $x_1 = x_2 = 0$. $\hat{y} = 10.2 + 2(1) = 12.2$

b.   The hypotheses are:

H_0: $\beta_1 = \beta_2 = \beta_3 = 0$
H_a: At least one $\beta_i \neq 0$, $i = 1, 2, 3$

12.79   a.   For gill nets, $x_1 = 0$ and $x_2 = 0$. Thus, $E(y) = \beta_0 + \beta_1(0) + \beta_2(0) = \beta_0$.

b. β_1 = difference in mean length between whales entangled in set nets and whales entangled in gill nets.

c.   To determine if the mean body lengths of entangled whales differ for the three types of fishing gear, we test:

H_0: $\beta_1 = \beta_2 = 0$
H_a: At least one $\beta_i \neq 0$

12.81   a.   Let $x_1 = \begin{cases} 1 & \text{if player is white} \\ 0 & \text{otherwise} \end{cases}$ $x_2 = \begin{cases} 1 & \text{if card availability is low} \\ 0 & \text{otherwise} \end{cases}$

$$x_3 = \begin{cases} 1 & \text{if player position is running back} \\ 0 & \text{otherwise} \end{cases} \qquad x_4 = \begin{cases} 1 & \text{if player position is wide receiver} \\ 0 & \text{otherwise} \end{cases}$$

$$x_5 = \begin{cases} 1 & \text{if player position is tight end} \\ 0 & \text{otherwise} \end{cases} \qquad x_6 = \begin{cases} 1 & \text{if player position is defensive lineman} \\ 0 & \text{otherwise} \end{cases}$$

$$x_7 = \begin{cases} 1 & \text{if player position is linebacker} \\ 0 & \text{otherwise} \end{cases} \qquad x_8 = \begin{cases} 1 & \text{if player position is defensive back} \\ 0 & \text{otherwise} \end{cases}$$

$$x_9 = \begin{cases} 1 & \text{if player position is offensive lineman} \\ 0 & \text{otherwise} \end{cases}$$

b.   The model for price (y) as a function of race is:

$$E(y) = \beta_0 + \beta_1 x_1 .$$

β_0 = mean price of card of black player
β_1 = difference in mean price of card between a white player and a black player

c.   The model for price (y) as a function of card availability is:

$$E(y) = \beta_0 + \beta_1 x_2 .$$

β_0 = mean price of card of high availability
β_1 = difference in mean price of card between a card of low availability and a card of high availability

d.   The model for price (y) as a function of position is:

$$E(y) = \beta_0 + \beta_1 x_3 + \beta_2 x_4 + \beta_3 x_5 + \beta_4 x_6 + \beta_5 x_7 + \beta_6 x_8 + \beta_7 x_9$$

β_0 = mean price of card of quarterback
β_1 = difference in mean price of card between a running back and a quarterback
β_2 = difference in mean price of card between a wide receiver and a quarterback
β_3 = difference in mean price of card between a tight end and a quarterback
β_4 = difference in mean price of card between a defensive lineman and a quarterback
β_5 = difference in mean price of card between a linebacker and a quarterback
β_6 = difference in mean price of card between a defensive back and a quarterback
β_7 = difference in mean price of card between a offensive lineman and a quarterback

12.83   a.   There would be a total of $2 \times 2 = 4$ different gene marker combinations at the two locations. They are:

AA   AB   BA   BB

b.   Let $x_1 = \begin{cases} 1 & \text{if gene marker is AA} \\ 0 & \text{otherwise} \end{cases}$ $x_2 = \begin{cases} 1 & \text{gene marker is AB} \\ 0 & \text{otherwise} \end{cases}$

$x_3 = \begin{cases} 1 & \text{gene marker is BA} \\ 0 & \text{otherwise} \end{cases}$

A model for $E(y)$ as a function of the gene marker combination is:

$$E(y) = \beta_0 + \beta_1 x_1 + \beta_2 x_2 + \beta_3 x_3$$

c. β_0 = mean extent of disease for gene marker combination BB

β_1 = difference in mean extent of disease between gene marker combinations AA and BB

β_2 = difference in mean extent of disease between gene marker combinations AB and BB

β_3 = difference in mean extent of disease between gene marker combinations BA and BB

d.   To determine if the overall model is useful for predicting extent of the disease, we test:

H_0: $\beta_1 = \beta_2 = \beta_3 = 0$
H_a: At least one $\beta_i \neq 0$

12.85   a.   Let $x_1 = \begin{cases} 1 \text{ if major depression only} \\ 0 \text{ if not} \end{cases}$ and $x_2 = \begin{cases} 1 \text{ if personality disorder only} \\ 0 \text{ if not} \end{cases}$

The model would be: $E(y) = \beta_0 + \beta_1 x_1 + \beta_2 x_2$.

b.   If there are no differences in the mean number of personality disorders for the three patient groups, then $\beta_1 = \beta_2 = 0$.

c.   To determine if the mean number of personality disorders for the major depression only patients is less than the corresponding mean for the patients with both major depression and personality disorder, we would test:

H_0: $\beta_1 = 0$
H_a: $\beta_1 < 0$

12.87   Let $x_1 = \begin{cases} 1 & \text{if Juror is female} \\ 0 & \text{otherwise} \end{cases}$ $x_2 = \begin{cases} 1 & \text{if no expert testimony is present} \\ 0 & \text{otherwise} \end{cases}$

To allow for the likelihood that female jurors would be more likely than male jurors to change a verdict from not guilty to guilty after deliberations if expert testimony is present, the model would have to include an interaction term between gender and expert testimony. The model would be:

$$E(y) = \beta_0 + \beta_1 x_1 + \beta_2 x_2 + \beta_3 x_1 x_2$$

A possible sketch of this relationship would be:

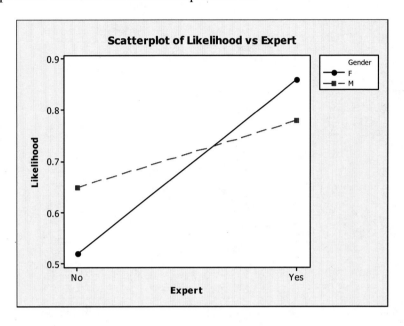

12.89 a.   Let y = mean body mass and $x = \begin{cases} 1 & \text{if flightless} \\ 0 & \text{otherwise} \end{cases}$

The model is $E(y) = \beta_0 + \beta_1 x$

b.   Let y = mean body mass and

$$x_1 = \begin{cases} 1 & \text{if vertebrates} \\ 0 & \text{otherwise} \end{cases} \qquad x_2 = \begin{cases} 1 & \text{if vegetables} \\ 0 & \text{otherwise} \end{cases} \qquad x_3 = \begin{cases} 1 & \text{if invertebrates} \\ 0 & \text{otherwise} \end{cases}$$

The model is $E(y) = \beta_o + \beta_1 x_1 + \beta_2 x_2 + \beta_3 x_3$

c.   Let y = mean egg length and

$$x_1 = \begin{cases} 1 & \text{if cavity within ground} \\ 0 & \text{otherwise} \end{cases} \qquad x_2 = \begin{cases} 1 & \text{if tree} \\ 0 & \text{otherwise} \end{cases} \qquad x_3 = \begin{cases} 1 & \text{if cavity above ground} \\ 0 & \text{otherwise} \end{cases}$$

The model is $E(y) = \beta_0 + \beta_1 x_1 + \beta_2 x_2 + \beta_3 x_3$

d.   Using MINTAB, the results are:

Regression Analysis: Body Mass versus Flt

```
The regression equation is
Body Mass = 641 + 30647 Flt


Predictor   Coef  SE Coef     T      P
Constant     641     2665  0.24  0.810
Flt        30647     5331  5.75  0.000


S = 26521.3   R-Sq = 20.3%   R-Sq(adj) = 19.7%


Analysis of Variance

Source           DF           SS           MS      F      P
Regression        1  23246186220  23246186220  33.05  0.000
Residual Error  130  91439576421    703381357
Total           131  1.14686E+11
```

The fitted model is $\hat{y} = 641 + 30,647x$

$\hat{\beta}_0 = 641$. The mean body mass for Volant birds is estimated to be 641.

$\hat{\beta}_1 = 30,647$. The difference in the mean body mass between Flightless birds and Volant birds is estimated to be 30,647.

e.   To determine if the model is useful in estimating body mass, we test:

H_0: $\beta_1 = 0$
H_a: $\beta_1 \neq 0$

The test statistic is $F = 33.05$ and the p-value is $p = .000$. Since the p-value is less than α ($p = .000 < .01$), H_0 is rejected. There is sufficient evidence to indicate a difference in mean body mass between the two types of flight capacity birds at $\alpha = .01$.

f.   Using MINITAB, the results are:

Regression Analysis: Body Mass versus DX1, DX2, DX3

```
The regression equation is
Body Mass = 903 + 2997 DX1 + 26206 DX2 - 660 DX3


Predictor   Coef  SE Coef      T      P
Constant     903     4171   0.22  0.829
DX1         2997    14298   0.21  0.834
DX2        26206     6090   4.30  0.000
DX3         -660     5772  -0.11  0.909


S = 27352.1   R-Sq = 16.5%   R-Sq(adj) = 14.5%
```

```
Analysis of Variance

Source            DF           SS           MS       F      P
Regression         3  18924196564   6308065521   8.43  0.000
Residual Error   128  95761566077    748137235
Total            131  1.14686E+11


Source  DF      Seq SS
X1       1    79962608
X2       1 18834456334
X3       1     9777622
```

The fitted model is $\hat{y} = 903 + 2{,}997x_1 + 26{,}206x_2 - 660x_3$

$\hat{\beta}_0 = 903$. The mean body mass for birds on a fish diet is estimated to be 903.

$\hat{\beta}_1 = 2{,}997$. The difference in the mean body mass between birds on a vertebrate diet and birds on a fish diet is estimated to be 2,997.

$\hat{\beta}_2 = 26{,}206$. The difference in the mean body mass between birds on a vegetable diet and birds on a fish diet is estimated to be 26,206.

$\hat{\beta}_3 = -660$. The difference in the mean body mass between birds on an invertebrate diet and birds on a fish diet is estimated to be -660.

g.   To determine if the model is useful in estimating body mass, we test:

H_0: $\beta_1 = \beta_2 = \beta_3 = 0$
H_a: At least one $\beta_i \neq 0$, $i = 1, 2, 3$

The test statistic is $F = 8.43$ and the p-value is $p = .000$. Since the p-value is less than α ($p = .000 < .01$), H_0 is rejected. There is sufficient evidence to indicate a difference in mean body mass among the four types of diets at $\alpha = .01$.

h.   Using MINITAB, the results are:

Regression Analysis: Egg Length versus NX1, NX2, NX3

```
The regression equation is
Egg Length = 73.7 - 9.13 NX1 - 45.0 NX2 - 39.5 NX3


130 cases used, 2 cases contain missing values


Predictor      Coef   SE Coef       T      P
Constant     73.732     4.907   15.03  0.000
NX1          -9.132     9.515   -0.96  0.339
NX2          -45.01     14.45   -3.12  0.002
NX3         -39.510     9.253   -4.27  0.000


S = 40.7622   R-Sq = 16.1%   R-Sq(adj) = 14.1%
```

```
Analysis of Variance

Source              DF        SS      MS      F       P
Regression           3     40230   13410   8.07   0.000
Residual Error     126    209356    1662
Total              129    249586


Source   DF   Seq SS
NX1       1      482
NX2       1     9455
NX3       1    30293
```

The fitted model is $\hat{y} = 73.732 - 9.132x_1 - 45.01x_2 - 39.51x_3$

$\hat{\beta}_0 = 73.732$. The mean egg length for nesting on the ground is estimated to be 73.732 mm.

$\hat{\beta}_1 = -9.132$. The difference in the mean egg length between birds nesting in a cavity within the ground and birds nesting on the ground is estimated to be -9.132.

$\hat{\beta}_2 = -45.01$. The difference in the mean egg length between birds nesting in a tree and birds nesting on the ground is estimated to be -45.01.

$\hat{\beta}_3 = -39.51$. The difference in the mean egg length between birds nesting in a cavity above ground and birds nesting on the ground is estimated to be -39.51.

i.   To determine if the model is useful in estimating body mass, we test:

H_0: $\beta_1 = \beta_2 = \beta_3 = 0$
H_a: At least one $\beta_i \neq 0$, $i = 1, 2, 3$

The test statistic is $F = 8.07$ and the p-value is $p = .000$. Since the p-value is less than α ($p = .000 < .01$), H_0 is rejected. There is sufficient evidence to indicate a difference in mean egg length among the four types of nesting sites at $\alpha = .01$.

12.91 a.   For months following the murder, $x_2 = 1$ and for Center, Texas, $x_1 = 0$. Thus, the mean violent crime rate would be:
$$E(y) = \beta_0 + \beta_1 x_1 + \beta_2 x_2 + \beta_3 x_1 x_2 = \beta_0 + \beta_1(0) + \beta_2(1) + \beta_3(0)(1) = \beta_0 + \beta_2$$

b.   For months following the murder, $x_2 = 1$ and for Jasper, Texas, $x_1 = 1$. Thus, the mean violent crime rate would be:
$$E(y) = \beta_0 + \beta_1 x_1 + \beta_2 x_2 + \beta_3 x_1 x_2 = \beta_0 + \beta_1(1) + \beta_2(1) + \beta_3(1)(1) = \beta_0 + \beta_1 + \beta_2 + \beta_3$$

c.   The difference between the mean violent crime rate for Jasper and Center following the murder would be: $\beta_0 + \beta_1 + \beta_2 + \beta_3 - (\beta_0 + \beta_2) = \beta_1 + \beta_3$

d.   For months before the murder, $x_2 = 0$ and for Center, Texas, $x_1 = 0$. Thus, the mean violent crime rate would be:
$$E(y) = \beta_0 + \beta_1 x_1 + \beta_2 x_2 + \beta_3 x_1 x_2 = \beta_0 + \beta_1(0) + \beta_2(0) + \beta_3(0)(0) = \beta_0$$

For months before the murder, $x_2 = 0$ and for Jasper, Texas, $x_1 = 1$. Thus, the mean violent crime rate would be:

$$E(y) = \beta_0 + \beta_1 x_1 + \beta_2 x_2 + \beta_3 x_1 x_2 = \beta_0 + \beta_1(1) + \beta_2(0) + \beta_3(1)(0) = \beta_0 + \beta_1$$

The difference between the mean violent crime rate for Jasper and Center before the murder would be: $\beta_0 + \beta_1 - \beta_0 = \beta_1$

e. If x_1 and x_2 do not interact, then the effect of x_1 on the mean crime rate would be the same for each level of x_2. Here, when $x_2 = 1$ (following the murder), the difference in the mean violent crime rate between Jasper and Center is $\beta_1 + \beta_3$. When $x_2 = 0$ (before the murder), the difference in the mean violent crime rate between Jasper and Center is β_1. These are not the same, so x_1 and x_2 interact.

f. Since the observed p-value is less than α, $(p < .001 < .01)$, H_0 is rejected. There is sufficient evidence to indicate β_3 is not 0 for $\alpha = .01$. This indicates that the 2 independent variables, time and place, interact to affect violent crime rate.

g. The mean violent crime rate in Center following the murder is $\beta_0 + \beta_2$. The mean violent crime rate in Center before the murder is β_0. The difference between following and before the murder is $\beta_0 + \beta_2 - \beta_0 = \beta_2$. The estimate of this is $\hat{\beta}_2 = -169$.

The mean violent crime rate in Jasper following the murder is $\beta_0 + \beta_1 + \beta_2 + \beta_3$. The mean violent crime rate in Jasper before the murder is $\beta_0 + \beta_1$. The difference between following and before the murder is $\beta_0 + \beta_1 + \beta_2 + \beta_3 - (\beta_0 + \beta_1) = \beta_2 + \beta_3$. The estimate of this is $\hat{\beta}_2 + \hat{\beta}_3 = -169 + 255 = 86$

12.93   a. The complete second-order model is $E(y) = \beta_0 + \beta_1 x_1 + \beta_2 x_1^2$

b. The new model is $E(y) = \beta_0 + \beta_1 x_1 + \beta_2 x_1^2 + \beta_3 x_2 + \beta_4 x_3$

where $x_2 = \begin{cases} 1 \text{ if level 2 of qualitative variable} \\ 0 \text{ otherwise} \end{cases}$ $x_3 = \begin{cases} 1 \text{ if level 3 of qualitative variable} \\ 0 \text{ otherwise} \end{cases}$

c. The model with the interaction terms is:

$$E(y) = \beta_0 + \beta_1 x_1 + \beta_2 x_1^2 + \beta_3 x_2 + \beta_4 x_3 + \beta_5 x_1 x_2 + \beta_6 x_1 x_3 + \beta_7 x_1^2 x_2 + \beta_8 x_1^2 x_3$$

d. The response curves will have the same shape if none of the interaction terms are present or if $\beta_5 = \beta_6 = \beta_7 = \beta_8 = 0$.

e. The response curves will be parallel lines if the interaction terms as well as the second-order terms are absent or if $\beta_2 = \beta_5 = \beta_6 = \beta_7 = \beta_8 = 0$.

f. The response curves will be identical if no terms involving the qualitative variable are present or $\beta_3 = \beta_4 = \beta_5 = \beta_6 = \beta_7 = \beta_8 = 0$.

12.95 a.   When $x_2 = x_3 = 0$, $E(y) = \beta_0 + \beta_1 x_1$

When $x_2 = 1$ and $x_3 = 0$, $E(y) = \beta_0 + \beta_2 + \beta_1 x_1$

When $x_2 = 0$ and $x_3 = 1$, $E(y) = \beta_0 + \beta + \beta_1 x_{13}$

b.   For level 1, $x_2 = x_3 = 0$. $\hat{y} = 44.8 + 2.2 x_1 + 9.4(0) + 15.6(0) = 44.8 + 2.2 x_1$

For level 2, $x_2 = 1$ and $x_3 = 0$.
$\hat{y} = 44.8 + 2.2 x_1 + 9.4(1) + 15.6(0) = 44.8 + 2.2 x_1 + 9.4 = 54.2 + 2.2 x_1$

For level 3, $x_2 = 0$ and $x_3 = 1$.
$\hat{y} = 44.8 + 2.2 x_1 + 9.4(0) + 15.6(1) = 44.8 + 2.2 x_1 + 15.6 = 60.4 + 2.2 x_1$

12.97 a.   The least squares prediction equation is $\hat{y} = 11.779 - 1.972 x_1 + .585 x_4 - .553 x_1 x_4$.

b.   For a male college student, $x_1 = 1$. The predictive level of desire is:
$\hat{y} = 11.779 - 1.972(1) + .585(5) - .553(1)(5) = 9.967$

c.   To determine if the model is adequate, we test:

H_0: $\beta_1 = \beta_2 = \beta_3 = 0$
H_a: At least one $\beta_i \neq 0$

From the printout, the test statistic is $F = 45.09$ and the p-value is $p < .0001$. Since the p-value is less than α $(p < .0001 < .10)$, H_0 is rejected. There is sufficient evidence to indicate the model is useful at $\alpha = .10$.

d. $R_a^2 = .4390$. This means that 43.9% of the total sample variation in level of desire scores is explained by the model containing gender, impression of reality TV and the interaction of gender and impression of reality TV, adjusting for the sample size and the number of parameters in the model.

e. $s = 2.35$. We would expect about 95% of the observed values of level of desire to fall within $2s = 2(2.35) = 4.70$ units of their least squares predicted values.

f.   To determine if gender and impression of reality TV interact, we test:

H_0: $\beta_3 = 0$
H_a: $\beta_3 \neq 0$

From the printout, the test statistic is $t = -2.00$ and the p-value is $p = .0467$. Since the p-value is less than α $(p = .0467 < .10)$, H_0 is rejected. There is sufficient evidence to indicate gender and impression of reality TV interact in the prediction of level of desire for cosmetic surgery at $\alpha = .10$.

g.   For female students, $x_1 = 0$. The predictive equation for females is:
$\hat{y} = 11.779 - 1.972(0) + .585 x_4 - .553(0) x_4 = 11.779 + .585 x_4$.

For every 1-point increase in impression of reality TV show, the level of desire is estimated to increase by .585 units for female students.

h. For male students, $x_1 = 1$. The predictive equation for males is:
$\hat{y} = 11.779 - 1.972(1) + .585x_4 - .553(1)x_4 = 9.807 + .032x_4$.

For every 1-point increase in impression of reality TV show, the level of desire is estimated to increase by .032 units for male students.

12.99 a. For obese smokers, $x_2 = 0$. The equation of the hypothesized line relating mean REE to time after smoking for obese smokers is:

$$E(y) = \beta_0 + \beta_1 x_1 + \beta_2(0) + \beta_3 x_1(0) = \beta_0 + \beta_1 x_1$$

The slope of the line is β_1.

b. For normal smokers, $x_2 = 1$. The equation of the hypothesized line relating mean REE to time after smoking for normal smokers is:

$$E(y) = \beta_0 + \beta_1 x_1 + \beta_2(1) + \beta_3 x_1(1) = (\beta_0 + \beta_2) + (\beta_1 + \beta_3)x_1$$

The slope of the line is $\beta_1 + \beta_3$.

c. The reported p-value is .044. Since the p-value is so small, there is evidence to indicate that interaction between time and weight is present for $\alpha > .044$.

12.101 a. Let x_1 = water depth of the entanglement, $x_2 = \begin{cases} 1 \text{ if set nets} \\ 0 \text{ if not} \end{cases}$, and $x_3 = \begin{cases} 1 \text{ if pots} \\ 0 \text{ if not} \end{cases}$.

The main-effects only model is: $E(y) = \beta_0 + \beta_1 x_1 + \beta_2 x_2 + \beta_3 x_3$

b. Using MINITAB, a possible graph of the main-effects only model might look like:

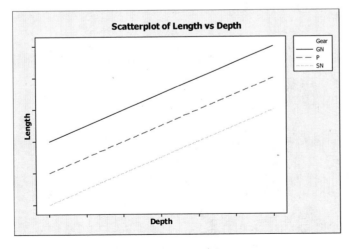

c. The interaction model would be: $E(y) = \beta_0 + \beta_1 x_1 + \beta_2 x_2 + \beta_3 x_3 + \beta_4 x_1 x_2 + \beta_5 x_1 x_3$

d.   Using MINITAB, the possible graph of an interaction model might look like:

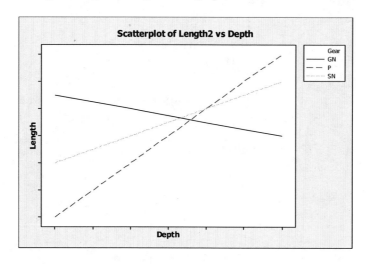

e.   For set nets, $x_2 = 1$ and $x_3 = 0$. The model for set nets would be:
$$E(y) = \beta_0 + \beta_1 x_1 + \beta_2(1) + \beta_3(0) + \beta_4 x_1(1) + \beta_5 x_1(0) = \beta_0 + \beta_2 + (\beta_1 + \beta_4)x_1$$
The rate of change of whale length with water depth for set nets is $\beta_1 + \beta_4$.

f.   For pots, $x_2 = 0$ and $x_3 = 1$. The model for set nets would be:
$$E(y) = \beta_0 + \beta_1 x_1 + \beta_2(0) + \beta_3(1) + \beta_4 x_1(0) + \beta_5 x_1(1) = \beta_0 + \beta_3 + (\beta_1 + \beta_5)x_1$$
The rate of change of whale length with water depth for set nets is $\beta_1 + \beta_5$.

g.   For gill nets, $x_2 = 0$ and $x_3 = 0$. The model for set nets would be:
$$E(y) = \beta_0 + \beta_1 x_1 + \beta_2(0) + \beta_3(0) + \beta_4 x_1(0) + \beta_5 x_1(0) = \beta_0 + \beta_1 x_1$$
The rate of change of whale length with water depth for set nets is β_1.

h.   To test whether the rate of change of whale length with water depth is the same for all three types of fishing gear, we test:

H_0: $\beta_4 = \beta_5 = 0$
H_a: At least one $\beta_i \neq 0$

12.103 a.   Let $x_2 = \begin{cases} 1 \text{ if low POS} \\ 0 \text{ if not} \end{cases}$ and $x_3 = \begin{cases} 1 \text{ if neutral POS} \\ 0 \text{ if not} \end{cases}$.

b.   The model with three parallel lines would be: $E(y) = \beta_0 + \beta_1 x_1 + \beta_2 x_2 + \beta_3 x_3$

c.   The model with three nonparallel lines would be:
$$E(y) = \beta_0 + \beta_1 x_1 + \beta_2 x_2 + \beta_3 x_3 + \beta_4 x_1 x_2 + \beta_5 x_1 x_3$$

d.   The model in part c would support the findings that the intention to leave was greater at the low level of POS than at the high level of POS.

12.105 a.   Let $x_1 = \begin{cases} 1 & \text{if water is from sub-surface flow} \\ 0 & \text{otherwise} \end{cases}$ $x_2 = \begin{cases} 1 & \text{if water is from overground flow} \\ 0 & \text{otherwise} \end{cases}$

and x_3 = silica concentration.

A model for $E(y)$ as a function of the water source and silica concentration is:

$$E(y) = \beta_0 + \beta_1 x_1 + \beta_2 x_2 + \beta_3 x_3$$

A possible graph would be:

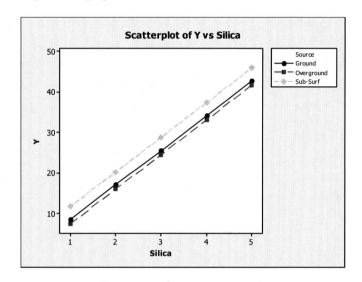

b.   A model allowing for the relationship between nitrate concentrations and silica concentrations to differ for the three water sources is:

$$E(y) = \beta_0 + \beta_1 x_1 + \beta_2 x_2 + \beta_3 x_3 + \beta_4 x_1 x_3 + \beta_5 x_2 x_3$$

A possible sketch of the relationships is:

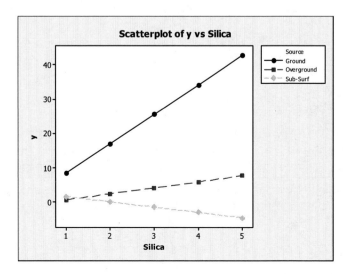

12.107 The models in parts **a** and **b** are nested:

The complete model is $E(y) = \beta_0 + \beta_1 x_1 + \beta_2 x_2$
The reduced model is $E(y) = \beta_0 + \beta_1 x_1$

The models in parts **a** and **d** are nested.

The complete model is $E(y) = \beta_0 + \beta_1 x_1 + \beta_2 x_2 + \beta_3 x_1 x_2$
The reduced model is $E(y) = \beta_0 + \beta_1 x_1 + \beta_2 x_2$

The models in parts **a** and **e** are nested.

The complete model is $E(y) = \beta_0 + \beta_1 x_1 + \beta_2 x_2 + \beta_3 x_1 x_2 + \beta_4 x_1^2 + \beta_5 x_2^2$
The reduced model is $E(y) = \beta_0 + \beta_1 x_1 + \beta_2 x_2$

The models in parts **b** and **c** are nested.

The complete model is $E(y) = \beta_0 + \beta_1 x_1 + \beta_2 x_1^2$
The reduced model is $E(y) = \beta_0 + \beta_1 x_1$

The models in parts **b** and **d** are nested.

The complete model is $E(y) = \beta_0 + \beta_1 x_1 + \beta_2 x_2 + \beta_3 x_1 x_2$
The reduced model is $E(y) = \beta_0 + \beta_1 x_1$

The models in parts **b** and **e** are nested.

The complete model is $E(y) = \beta_0 + \beta_1 x_1 + \beta_2 x_2 + \beta_3 x_1 x_2 + \beta_4 x_1^2 + \beta_5 x_2^2$
The reduced model is $E(y) = \beta_0 + \beta_1 x_1$

The models in parts **c** and **e** are nested.

The complete model is $E(y) = \beta_0 + \beta_1 x_1 + \beta_2 x_2 + \beta_3 x_1 x_2 + \beta_4 x_1^2 + \beta_5 x_2^2$
The reduced model is $E(y) = \beta_0 + \beta_1 x_1 + \beta_2 x_1^2$

The models in parts **d** and **e** are nested.

The complete model is $E(y) = \beta_0 + \beta_1 x_1 + \beta_2 x_2 + \beta_3 x_1 x_2 + \beta_4 x_1^2 + \beta_5 x_2^2$
The reduced model is $E(y) = \beta_0 + \beta_1 x_1 + \beta_2 x_2 + \beta_3 x_1 x_2$

12.109 A parsimonious model is a general linear model with a small number of β parameters. In situations where two competing models have essentially the same predictive power (as determined by an F-test), choose the model with the fewer number of β's.

12.111 a.   There are five β parameters in the complete model and three in the reduced model.

      b.   The hypotheses are:

H_0: $\beta_3 = \beta_4 = 0$

H_a: At least one $\beta_i \neq 0$, $i = 3, 4$

      c.   The test statistic is

$$F = \frac{(SSE_R - SSE_C)/(k - g)}{SSE_C /[n - (k+1)]} = \frac{(160.44 - 152.66)/(4-2)}{152.66/[20 - (4+1)]} = \frac{3.89}{10.1773} = .38$$

The rejection region requires $\alpha = .05$ in the upper tail of the F distribution with $v_1 = k - g = 4 - 2 = 2$ and $v_2 = n - (k+1) = 20 - (4+1) = 15$. From Table IX, Appendix A, $F_{.05} = 3.68$. The rejection region is $F > 3.68$.

Since the observed value of the test statistic does not fall in the rejection region ($F = .38$ $\ngtr$ 3.68), H_0 is not rejected. There is insufficient evidence to indicate the complete model is better than the reduced model at $\alpha = .05$.

12.113 a.   The two models are nested because all of the terms in the model in part **b** are contained in the model in part **c**. The complete model is

$$E(y) = \beta_0 + \beta_1 x_1 + \beta_2 x_2 + \beta_3 x_3 + \beta_4 x_1 x_2 + \beta_5 x_1 x_3$$

and the reduced model is $E(y) = \beta_0 + \beta_1 x_1 + \beta_2 x_2 + \beta_3 x_3$.

      b.   To compare the two models, the null hypothesis would be:

H_0: $\beta_4 = \beta_5 = 0$

      c.   If you reject H_0, then the complete model would be preferred. There is evidence that the interaction terms are significant.

      d.   If you fail to reject H_0, then the reduced model would be preferred. There is no evidence that the interaction terms are significant.

12.115 a.   Model 1: $R^2 = .101$. 10.1% of the total sample variation in supervisor-directed aggression score is explained by the model including the 4 independent variables.

Model 2: $R^2 = .555$. 55.5% of the total sample variation in supervisor-directed aggression score is explained by the model including the 8 independent variables.

b. To determine if the additional 4 variables are useful in the prediction of supervisor-directed aggression score, we test:

$$H_o: \beta_5 = \beta_6 = \beta_7 = \beta_8 = 0$$
$$H_a: \text{At least one } \beta_i \neq 0$$

c. Yes, the models are nested. All of the variables included in Model 1 are also included in model 2.

d. Since the p-value is very small ($p < .001$), there is evidence to reject H_0. There is sufficient evidence to indicate at least 1 of the additional 4 variables is useful in the prediction of supervisor-directed aggression score at $\alpha > .001$.

e. Model 3 would be:

$$E(y) = \beta_o + \beta_1(\text{Age}) + \beta_2(\text{Gender})$$
$$+ \beta_3(\text{Interactional injustice at 2nd job}) + \beta_4(\text{Abusive supervisor at 2nd job})$$
$$+ \beta_5(\text{Self-esteem}) + \beta_6(\text{History of agression})$$
$$+ \beta_7(\text{Interactional injustice at primary job})$$
$$+ \beta_8(\text{Abusive supervisor at primary job})$$
$$+ \beta_9(\text{Self-esteem})(\text{History of agression})$$
$$+ \beta_{10}(\text{Self-esteem})(\text{Interactional injustice at primary job})$$
$$+ \beta_{11}(\text{Self-esteem})(\text{Abusive supervisor at primary job})$$
$$+ \beta_{12}(\text{History of agression})(\text{Interactional injustice at primary job})$$
$$+ \beta_{13}(\text{History of agression})(\text{Abusive supervisor at primary job})$$
$$+ \beta_{14}(\text{Interactional injustice at primary job})(\text{Abusive supervisor at primary job})$$

f. Since the p-value is $p > .10$, there is no evidence to indicate that any of the interaction terms are useful in the prediction of supervisor-directed aggression score for $\alpha \leq .10$.

12.117 a. The hypothesized alternative model is

$$E(y) = \beta_0 + \beta_1 x_1 + \beta_2 x_2 + \beta_3 x_3 + \beta_4 x_4 + \beta_5 x_5 + \beta_6 x_6 + \beta_7 x_7 + \beta_8 x_8 + \beta_9 x_9 + \beta_{10} x_{10}$$

b. The null hypothesis would be $H_0: \beta_3 = \beta_4 = \beta_5 = \beta_6 = \beta_7 = \beta_8 = \beta_9 = \beta_{10} = 0$

c. The test was statistically significant. Thus, H_0 was rejected. There is sufficient evidence to indicate that at least one of the "control" variables contributes to the prediction of SAT-Math scores.

d. $R_a^2 = .79$. 79% of the sample variability of the SAT-Math scores around their means is explained by the proposed model relating SAT-Math scores to the 10 independent variables, adjusting for the sample size and the number of β parameters in the model.

e. For confidence coefficient .95, $\alpha = .05$ and $\alpha / 2 = .05 / 2 = .025$. From Table VI, Appendix A, with df $= n - (k + 1) = 3,492 - (10 + 1) = 3,481$, $t_{.025} = 1.96$. The 95% confidence interval is:

$$\hat{\beta}_2 \pm t_{.025} s_{\hat{\beta}_2} \Rightarrow 14 \pm 1.96(3) \Rightarrow 14 \pm 5.88 \Rightarrow (8.12, 19.88)$$

We are 95% confident that the difference in the mean SAT-Math scores between students who were coached and those who were not is between 8.12 and 19.88 points, holding all the other variables constant.

f. Yes. From Exercise 12.74, the confidence interval for β_2 was (13.12, 24.88). In part e, the confidence interval for β_2 was (8.12, 19.88). Even though coaching is significant in both models, the change in the mean SAT-Math scores is not as great if the control variables are added to the model.

g. The complete model would be:
$$E(y) = \beta_0 + \beta_1 x_1 + \beta_2 x_2 + \beta_3 x_3 + \beta_4 x_4 + \beta_5 x_5 + \beta_6 x_6 + \beta_7 x_7 + \beta_8 x_8 + \beta_9 x_9 + \beta_{10} x_{10} + \beta_{11} x_1 x_2$$
$$+ \beta_{12} x_3 x_2 + \beta_{13} x_4 x_2 + \beta_{14} x_5 x_2 + \beta_{15} x_6 x_2 + \beta_{16} x_7 x_2 + \beta_{17} x_8 x_2 + \beta_{18} x_9 x_2 + \beta_{19} x_{10} x_2$$

h. The null hypothesis would be $H_0: \beta_{11} = \beta_{12} = \beta_{13} = \beta_{14} = \beta_{15} = \beta_{16} = \beta_{17} = \beta_{18} = \beta_{19} = 0$. To perform this test, you would fit the complete model specified in part **g**. You would also fit the reduced model specified in part **a**. Then, you would perform the test comparing the complete and reduced models.

12.119 a. To determine if there is any difference in the mean lengths of entangled whales for the three gear types, we test:

$H_0: \beta_2 = \beta_3 = \beta_4 = \beta_5 = 0$
$H_a:$ At least one $\beta_i \neq 0$

b. The reduced model would be: $E(y) = \beta_0 + \beta_1 x_1$

c. If we reject H_0 above, we would conclude that the mean lengths of entangled whales for the three gear types differ.

d. To determine if the rate of change of whale length with water depth is the same for all three types of fishing gear, we test:

$H_0: \beta_4 = \beta_5 = 0$
$H_a:$ At least one $\beta_i \neq 0$

e. The reduced model would be: $E(y) = \beta_0 + \beta_1 x_1 + \beta_2 x_2 + \beta_3 x_3$

f. If we fail to reject H_0 above, we would conclude that the rate of change of whale length with water depth does not differ for the three types of fishing gear.

12.121 In a stepwise model, several steps are involved in selecting the model. First, all one-variable models are compared. The one variable model with the highest r^2 is selected. Next, all 2 variable models that include the previously selected variable are compared. The model with the highest R^2 is selected. This process is continued until no additional variables can be added to the model (no more contribute significantly to the prediction of y). In a standard regression model, the model is specified first and then all variables thought to be important are included in the model.

12.123 a.   The best one-variable predictor of y is the one whose t statistic has the largest absolute value. The t statistics for each of the variables are:

Independent Variable	$t = \dfrac{\hat{\beta}_i}{s_{\hat{\beta}_i}}$
x_1	$t = 1.6/.42 = 3.81$
x_2	$t = -.9/.01 = -90$
x_3	$t = 3.4/1.14 = 2.98$
x_4	$t = 2.5/2.06 = 1.21$
x_5	$t = -4.4/.73 = -6.03$
x_6	$t = .3/.35 = .86$

The variable x_2 is the best one-variable predictor of y. The absolute value of the corresponding t score is 90. This is larger than any of the others.

b.   Yes. In the stepwise procedure, the first variable entered is the one which has the largest absolute value of t, provided the absolute value of the t falls in the rejection region.

c.   Once x_2 is entered, the next variable that is entered is the one that, adjusted for x_2, has the largest absolute t value associated with it.

12.125 a.   The hypothesized equation of the final stepwise regression model is:

$E(y) = \beta_0 + \beta_1 x_1 + \beta_2 x_2$, where x_1 = leg length and x_2 = pressure.

b. $R^2 = .771$. This means that 77.1% of the sample variation in heel depth is explained by the model containing leg length and pressure.

c.   To determine if the overall model is useful in predicting heel depth, we test:

H_0: $\beta_1 = \beta_2 = 0$
H_a: At least one $\beta_i \neq 0$

Since the p-value is so small ($p < .001$), we would reject H_0 for any reasonable value of α. There is sufficient evidence to indicate the model is useful for predicting heel depth.

d.   In the first step of the stepwise regression, there were 6 individual t-tests run. In the second step, there are another $6 - 1 = 5$ individual t-tests run. In the third step, there are an additional $6 - 2 = 4$ individual t-tests run. Thus, at a minimum, there are a total of $6 + 5 + 4 = 15$ individual t-tests run.

e. Since there were so many individual *t*-tests run, the probability of making at least one Type I error during the stepwise analysis is very high. The probability of making at least one Type I error is 1 – probability of not making any Type I error in 15 tests. Thus, *P*(at least 1 Type I error) = 1 – *P*(no Type I errors) = $1 - (1 - \alpha)^{15}$.

12.127 a. In the first step of stepwise regression, 11 models are fit. These would be all the one-variable models.

b. In step two, all two variable models that include the variable selected in step one are fit. There will be a total of 10 models fit in this step.

c. In step 11, there is one model fit – the model containing all the independent variables.

d. The equation for *E*(*y*) is:

$$E(y) = \beta_0 + \beta_1 x_{11} + \beta_2 x_4 + \beta_3 x_2 + \beta_4 x_7 + \beta_5 x_{10} + \beta_6 x_1 + \beta_7 x_9 + \beta_8 x_3$$

e. $R^2 = .677$. 67.7% of the sample variation in the overall satisfaction with BRT scores is explained by the model containing the 8 independent variables.

f. In any stepwise regression, many *t*-tests are performed. This inflates the probability of committing a Type I error. In addition, there are certain combinations of variables that may never be reached because of the first few variables put in the model. Finally, you should consider including interaction and higher order terms in the model.

12.129 a. Since there were 11 potential independent variables, there were 11 *t*-tests performed in Step 1.

b. In Step 2, there were 11 – 1 = 10 *t*-tests performed.

c. The global test for testing whether at least one of the two variables is significant in predicting TME has a *p*-value of .001. Since this *p*-value is so small, there is evidence to indicate the model is useful in predicting TME. The value of R^2 is .988. This means that 98.8% of the sample variation in TME values is explained by the model containing both AMAP and NDF.

$\hat{\beta}_1 = 2.14$. For each unit increase in AMAP (amylopectin/amylase), the mean value of TME is estimated to increase by 2.14, holding NDF (neutral detergent fiber) constant.

$\hat{\beta}_2 = .16$. For each unit increase in NDF (neutral detergent fiber), the mean value of TME is estimated to increase by .16, holding AMAP (amylopectin/amylase) constant.

d. One should always be wary of using a model selected by stepwise regression. This stepwise regression model selected the variables AMAP and NDF. The best model for describing the dependent variable may not include either of these 2 variables. The stepwise procedure may prevent one from ever getting to the best model. Another reason to be wary is that a large number of *t*-tests have been conducted, leading to a high probability of making one or more Type I or Type II errors.

e.   The complete 2nd-order model for TME would be:

$$E(y) = \beta_0 + \beta_1 AMAP + \beta_2 NDF + \beta_3 AMAP \cdot NDF + \beta_4 AMAP^2 + \beta_5 NDF^2$$

f.   To determine if the terms in the model in part e that allow for curvature are statistically useful for predicting TME, we would test:

H_0: $\beta_4 = \beta_5 = 0$
H_a: At least one $\beta_i \neq 0$

12.131 A residual that is larger than $3s$ (in absolute value) is considered to be an outlier.

12.133 The statement "Regression models fit to time series data typically result in uncorrelated errors." is false. Time series data tend to be correlated.

12.135 Three indicators of a multicollinearity problem are:

1.   Significant correlations between pairs of independent variables.
2.   Nonsignificant t-tests for all (or nearly all) of the individual β parameters when the F-test for overall model adequacy is significant.
3.   Signs opposite from what is expected in the estimated β parameters.

12.137 Yes. x_2 and x_4 are highly correlated ($r_{24} = .93$), as well as x_4 and x_5 ($r_{45} = .86$). When highly correlated independent variables are present in a regression model, the results can be confusing. The researcher may want to include only one of the correlated variables.

12.139 If the highest correlation (in absolute value) for any pair of independent variables is $r = -.16$, the regression analysis would probably not exhibit multicollinearity problems because this shows that the independent variables are not highly correlated with each other.

12.141 a.   No, there is no evidence of extreme multicollinearity. The independent variables would be irritability, trait anger, and narcissism. The largest pairwise correlation (in absolute value) between independent variables is $r = .57$. This is not extremely large and would probably not lead to extreme multicollinearity.

b.   Yes, there would be some evidence of multicollinearity. The independent variables would be aggressive behavior, irritability, and trait anger. The largest pairwise correlation (in absolute value) between independent variables is $r = .80$. This is fairly large and would probably lead to a problem with multicollinearity.

12.143 The residual for the last observation is $y - \hat{y} = 87 - 29.63 = 57.37$. The standard deviation is $s = 24.68$. The number of standard deviations $\hat{y}$ is from y is found by dividing the residual by the value of the standard deviation or $57.37 / 24.68 = 2.32$. This point is more than 2 standard deviations from 0 but less than 3. Thus, this point is a possible outlier.

12.145 One possible reason for the sign of the estimated parameter multiplied by x_1 to be the opposite of what is expected is that company role of estimator and previous accuracy could be highly correlated to each other. If two independent variables are highly correlated to each other and appear in the regression equation together, multicollinearity could be present.

12.147 a.   Using MINITAB, the plots of the residuals versus each of the independent variables are:

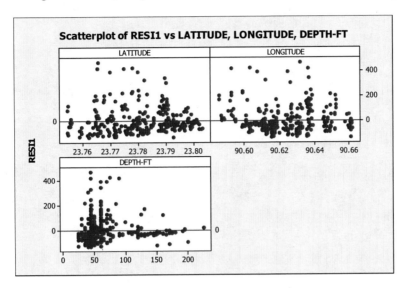

From each of the plots, there is no apparent U-shape to the residuals. This indicates that the model was probably not misspecified. Thus, the assumption of mean error = 0 is probably valid.

  b.   Using MINITAB, the residual plots are:

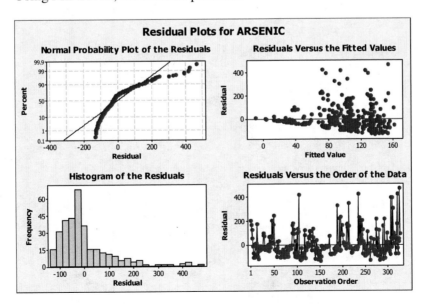

Using the plot of the residuals versus the fitted value, there appears to be an increase in the spread of the residuals as the fitted value increases. This implies that there is not constant error variance.

c. From the model, the estimate of the standard deviation is $s = 103.301$. Three standard deviations is $3(103.301) = 309.903$. There are 8 observations with residuals that are greater than 309.903. These are all potential outliers.

d. In part b, there is a histogram of the residuals and a normal probability plot. The histogram is skewed to the right, indicating that the error terms are not normally distributed. In addition, the normal probability plot is not a straight line. This also indicates that the error terms are not normally distributed.

e. Using MINITAB, the correlations of the variables are:

Correlations: ARSENIC, LATITUDE, LONGITUDE, DEPTH-FT

	ARSENIC	LATITUDE	LONGITUDE
LATITUDE	-0.186		
	0.001		
LONGITUDE	0.192	0.311	
	0.000	0.000	
DEPTH-FT	-0.245	0.151	-0.328
	0.000	0.006	0.000

Cell Contents: Pearson correlation
              P-Value

All of the pairs of independent variables have correlation coefficients that are less than .33 in absolute value. These correlations are very low. This indicates that there probably is not a problem with multicollinearity.

12.149 Using MINITAB, the plots of the residuals versus each of the independent variables are:

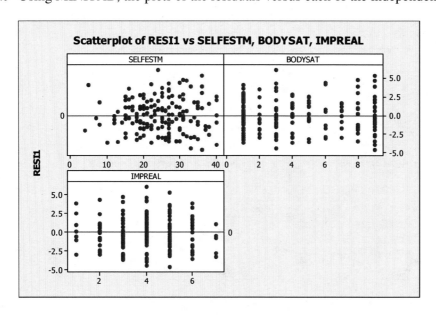

From the plot of the residuals versus self-esteem, there is no U-shaped pattern to the residuals. This implies that the relationship between desire and self-esteem has not been misspecified. Thus, the assumption of mean error = 0 is appropriate. From the analysis, the estimate of the standard deviation is $s = 2.25$. There are no points that are more than 3 standard deviations from the mean. Thus, there does not appear to be any outliers.

From the plot of the residuals versus body satisfaction, there is no U-shaped pattern to the residuals. This implies that the relationship between desire and body satisfaction has not been misspecified. Thus, the assumption of mean error = 0 is appropriate. In addition, there are no points that appear to be outliers.

From the plot of the residuals versus impression of reality TV, there is no U-shaped pattern to the residuals. This implies that the relationship between desire and impression of reality TV has not been misspecified. Thus, the assumption of mean error = 0 is appropriate. In addition, there are no points that appear to be outliers.

Using MINITAB, additional plots of the residuals are:

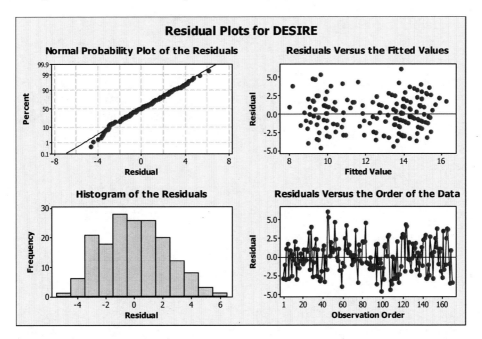

The plot of the residuals versus the fitted values does not appear to be funnel-shaped or football-shaped. Thus, the assumption of equal variance appears to be met.

The histogram of the residuals appears to be mound-shaped, which implies that the assumption of normal error terms is satisfied. The normal probability plot is very close to a straight line, also indicating that the error terms are normally distributed.

Using MINITAB, the correlations are:

Correlations: DESIRE, SELFESTM, BODYSAT, IMPREAL, GENDER

```
              DESIRE   SELFESTM   BODYSAT    IMPREAL
SELFESTM      -0.485
               0.000

BODYSAT       -0.644     0.757
               0.000     0.000

IMPREAL        0.132     0.167     0.143
               0.087     0.030     0.062

GENDER        -0.637     0.511     0.828     0.065
               0.000     0.000     0.000     0.398


Cell Contents: Pearson correlation
               P-Value
```

The correlation between body satisfaction and self-esteem is .757. This indicates a possible moderate problem with multicollinearity. Self-esteem and body satisfaction should probably not both appear in the model together. The correlations between impression of reality TV and both body satisfaction and self-esteem are below .2, indicating there is not a problem with multicollinearity.

12.151 $E(y) = \beta_0 + \beta_1 x_1 + \beta_2 x_2 + \beta_3 x_3$

where $x_1 = \begin{cases} 1, & \text{if level 2} \\ 0, & \text{otherwise} \end{cases}$ $x_2 = \begin{cases} 1, & \text{if level 3} \\ 0, & \text{otherwise} \end{cases}$ $x_3 = \begin{cases} 1, & \text{if level 4} \\ 0, & \text{otherwise} \end{cases}$

12.153 a. $E(y) = \beta_0 + \beta_1 x_1 + \beta_2 x_2 + \beta_3 x_3$

where $x_2 = \begin{cases} 1, & \text{if level 2} \\ 0, & \text{otherwise} \end{cases}$ $x_3 = \begin{cases} 1, & \text{if level 3} \\ 0, & \text{otherwise} \end{cases}$

b. $E(y) = \beta_0 + \beta_1 x_1 + \beta_2 x_1^2 + \beta_3 x_2 + \beta_4 x_3 + \beta_5 x_1 x_2 + \beta_6 x_1 x_3 + \beta_7 x_1^2 x_2 + \beta_8 x_1^2 x_3$

where x_1, x_2, and x_3 are as in part **a**.

12.155 a.   To determine if at least one of the β parameters is not zero, we test:

H_0: $\beta_1 = \beta_2 = \beta_3 = \beta_4 = 0$
H_a: At least one $\beta_i \neq 0$

The test statistic is $F = \dfrac{R^2 / k}{(1 - R^2)/[n - (k+1)]} = \dfrac{.83/4}{(1-.83)/[25-(4+1)]} = 24.41$

The rejection region requires $\alpha = .05$ in the upper tail of the F distribution with $v_1 = k = 4$ and $v_2 = n - (k = 1) = 25 - (4 + 1) = 20$. From Table IX, Appendix A, $F_{.05} = 2.87$. The rejection region is $F > 2.87$.

Since the observed value of the test statistic falls in the rejection region ($F = 24.41 > 2.87$), H_0 is rejected. There is sufficient evidence to indicate at least one of the β parameters is nonzero at $\alpha = .05$.

b. H_0: $\beta_1 = 0$
H_a: $\beta_1 < 0$

The test statistic is $t = \dfrac{\hat{\beta}_1 - 0}{s_{\hat{\beta}_1}} = \dfrac{-2.43 - 0}{1.21} = -2.01$

The rejection region requires $\alpha = .05$ in the lower tail of the t distribution with df = $n - (k + 1) = 25 - (4 + 1) = 20$. From Table VI, Appendix A, $t_{.05} = 1.725$. The rejection region is $t < -1.725$.

Since the observed value of the test statistic falls in the rejection region ($t = -2.01 < -1.725$), H_0 is rejected. There is sufficient evidence to indicate β_1 is less than 0 at $\alpha = .05$.

c. H_0: $\beta_2 = 0$
H_a: $\beta_2 > 0$

The test statistic is $t = \dfrac{\hat{\beta}_2 - 0}{s_{\hat{\beta}_2}} = \dfrac{.05 - 0}{.16} = .31$

The rejection region requires $\alpha = .05$ in the upper tail of the t distribution. From part **b** above, the rejection region is $t > 1.725$.

Since the observed value of the test statistic does not fall in the rejection region ($t = .31 \not> 1.725$), H_0 is not rejected. There is insufficient evidence to indicate β_2 is greater than 0 at $\alpha = .05$.

d. H_0: $\beta_3 = 0$
H_a: $\beta_3 \neq 0$

The test statistic is $t = \dfrac{\hat{\beta}_3 - 0}{s_{\hat{\beta}_3}} = \dfrac{.62 - 0}{.26} = 2.38$

The rejection region requires $\alpha / 2 = .05 / 2 = .025$ in each tail of the t distribution with df = 20. From Table VI, Appendix A, $t_{.025} = 2.086$. The rejection region is $t < -2.086$ or $t > 2.086$.

Since the observed value of the test statistic falls in the rejection region ($t = 2.38 > 2.086$), H_0 is rejected. There is sufficient evidence to indicate β_3 is different from 0 at $\alpha = .05$.

12.157 The error of prediction is smallest when the values of x_1, x_2, and x_3 are equal to their sample means. The further x_1, x_2, and x_3 are from their means, the larger the error. When $x_1 = 60$, $x_2 = .4$, and $x_3 = 900$, the observed values are outside the observed ranges of the x values. When $x_1 = 30$, $x_2 = .6$, and $x_3 = 1300$, the observed values are within the observed ranges and consequently the x values are closer to their means. Thus, when $x_1 = 30$, $x_2 = .6$, and $x_3 = 1300$, the error of prediction is smaller.

12.159 Even though SSE = 0, we cannot estimate σ^2 because there are no degrees of freedom corresponding to error. With three data points, there are only two degrees of freedom available. The degrees of freedom corresponding to the model is $k = 2$ and the degrees of freedom corresponding to error is $n - (k + 1) = 3 - (2 + 1) = 0$. Without an estimate for σ^2, no inferences can be made.

12.161 a. $R^2 = .31$. 31% of the sample variation in the ln(level of CO_2 emissions in current year) is explained by the model containing ln(foreign investments 16 years earlier), gross domestic investment 16 years earlier, trade exports 16 years earlier, ln(GNP 16 years earlier), agricultural production 16 years earlier, African country, and ln(level of CO_2 emissions 16 years earlier).

   b. H_0: $\beta_1 = \beta_2 = \cdots = \beta_7 = 0$
            H_a: At least one $\beta_i \neq 0$

   The test statistic is $F = \dfrac{R^2/k}{(1-R^2)/[n-(k+1)]} = \dfrac{.31/7}{(1-.31)/[66-(7+1)]} = 3.723$

   The rejection region requires $\alpha = .01$ in the upper tail of the F distribution. From Table XI, Appendix A, with $v_1 = k = 7$ and $v_2 = n - (k+1) = 66 - (7+1) = 58$, $F_{.01} \approx 2.95$.

   Since the observed value of the test statistic falls in the rejection region ($F = 3.723 > 2.95$), H_0 is rejected. There is sufficient evidence to indicate that at least one of the 7 independent variables contributes to the prediction of ln (level of CO_2 emissions in current year) at $\alpha = .01$.

   c.   We would not advise using individual t-tests on each of the independent variables to test for overall model adequacy. If we did, we would be very likely to make one or more errors in deciding which terms to retain in the model and which to exclude.

   d.   To determine if foreign investments 16 years earlier is a statistically useful predictor of CO_2 emissions in current year, we test:

        H_0: $\beta_1 = 0$
        H_a: $\beta_1 \neq 0$

   e.   The test statistic is $t = 2.52$ and the p-value is $p < .05$. Since the p-value is less than $\alpha = .05$, H_0 is rejected. There is sufficient evidence to indicate that foreign investment 16 years earlier is a statistically useful predictor of CO_2 emissions in current year.

f. The independent variables that are highly correlated are x_4 (ln(GNP 16 years earlier)) and x_5 (agricultural production 16 years earlier). The correlation coefficient is $r_{45} = -.84$. Several other variables have moderate correlation: x_1 and x_3 with $r_{13} = .57$, x_4 and x_6 with $r_{46} = -.53$, and x_5 and x_7 with $r_{57} = -.50$. If these highly correlated variables are included in the model at the same time, signs of the estimates of the beta parameters may be different than what is expected. Also, the global test could indicate that at least one of the variables contributes to the prediction of y, but none of the individual variable tests are significant.

12.163 a. The first order model for $E(y)$ as a function of the first five independent variables is:

$$E(y) = \beta_0 + \beta_1 x_1 + \beta_2 x_2 + \beta_3 x_3 + \beta_4 x_4 + \beta_5 x_5$$

b. To test the utility of the model, we test:

H_0: $\beta_1 = \beta_2 = \beta_3 = \beta_4 = \beta_5 = 0$
H_a: At least one $\beta_i \neq 0$, $i = 1, 2, 3, 4, 5$

The test statistic is $F = 34.47$ and the p-value is $p < .001$.

Since the p-value is so small, there is sufficient evidence to indicate the model is useful for predicting GSI at $\alpha > .001$.

$R^2 = .469$.   46.9% of the variability in the GSI scores is explained by the model including the first five independent variables.

c. The first order model for $E(y)$ as a function of the first seven independent variables is:

$$E(y) = \beta_0 + \beta_1 x_1 + \beta_2 x_2 + \beta_3 x_3 + \beta_4 x_4 + \beta_5 x_5 + \beta_6 x_6 + \beta_7 x_7$$

d. $R^2 = .603$ 60.3% of the variability in the GSI scores is explained by the model including the first seven independent variables.

e. Since the p-values associated with the variables DES and PDEQ-SR are both less than .001, there is evidence that both variables contribute to the prediction of GSI, adjusted for all the other variables already in the model for $\alpha > .001$.

12.165 a. The quadratic model would be:

$$E(y) = \beta_0 + \beta_1 x + \beta_2 x^2$$

b. From the plot, the β_2 would be positive because the points appear to form an upward curve.

c. This value of r^2 applies to the linear model, not the quadratic model. The linear model is:

$$E(y) = \beta_0 + \beta_1 x$$

12.167 a.   Let $x = \begin{cases} 1 & \text{if pond is enriched} \\ 0 & \text{if otherwise} \end{cases}$

The model is $E(y) = \beta_0 + \beta_1 x$

b. β_0 = mean mosquito larvae density in the natural pond

β_1 = difference in the mean mosquito larvae density between the enriched pond and the natural pond

c.   To determine if the mean larval density for the enriched pond exceeds the mean for the natural pond, we test:

H_0: $\beta_1 = 0$
H_a: $\beta_1 > 0$

d.   The p-value for the global F test is $p = .004$. Thus, the p-value for this test is $p = .004/2 = .002$. Since the p-value is so small, there is evidence to reject H_0. There is sufficient evidence to indicate that the mean larval densities differ for the enriched pond and the natural pond at $\alpha > .002$.

12.169 a. $R^2 = .362$. 36.2% of the variability in the active-caring scores can be explained by the model containing the variables self-esteem score, optimism score, and group cohesion score.

b.   To test the utility of the model, we test:

H_0: $\beta_1 = \beta_2 = \beta_3 = 0$
H_a: At least one $\beta_i \neq 0$, $i = 1, 2, 3$

The test statistic is:

$$F = \frac{R^2/k}{(1-R^2)[n-(k+1)]} = \frac{.362/3}{(1-.362)/[31-(3+1)]} = 5.11$$

The rejection region requires $\alpha = .05$ in the upper tail of the F distribution with $v_1 = k = 3$ and $v_2 = n-(k+1) = 31-(3+1) = 27$. From Table IX, Appendix A, $F_{.05} = 2.96$. The rejection region is $F > 2.96$.

Since the observed value of the test statistic falls in the rejection region ($F = 5.11 > 2.96$), H_0 is rejected. There is sufficient evidence that the model is useful in predicting AC score at $\alpha = .05$.

12.171 a.   **Regression Analysis: WeightChg versus Digest, Fiber**

```
The regression equation is
WeightChg = 12.2 - 0.0265 Digest - 0.458 Fiber

Predictor          Coef      SE Coef          T          P
Constant         12.180        4.402       2.77      0.009
Digest         -0.02654      0.05349      -0.50      0.623
Fiber           -0.4578       0.1283      -3.57      0.001
```

```
S = 3.519        R-Sq = 52.9%      R-Sq(adj) = 50.5%

Analysis of Variance

Source              DF          SS          MS          F          P
Regression           2       542.03      271.02      21.88      0.000
Residual Error      39       483.08       12.39
Total               41      1025.12
```

$$\hat{y} = 12.180 - .02654x_1 - .4578x_2$$

b. $\hat{\beta}_0 = 12.180$. This is the estimate of the y-intercept. It has no other interpretation.

$\hat{\beta}_1 = -.0265$. We estimate that the mean weight change will decrease by .0265% for each additional increase of 1% in digestion efficiency, with acid-detergent fibre held constant.

$\hat{\beta}_2 = -.458$. We estimate that the mean weight change will decrease by .458% for each additional increase of 1% in acid-detergent fibre, with digestion efficiency held constant.

c. To determine if digestion efficiency is a useful predictor of weight change, we test:

H_0: $\beta_1 = 0$
H_a: $\beta \neq 0$

The test statistic is $t = -.50$ and the p-value is $p = .623$. Since the p-value is greater than α ($p = .623 > .01$), H_0 is not rejected. There is insufficient evidence to indicate that digestion efficiency is a useful linear predictor of weight change at $\alpha = .01$.

d. For confidence coefficient .99, $\alpha = 1 - .99 = .01$ and $\alpha / 2 = .01 / 2 = .005$. From Table VI, Appendix A, with df $= n - (k + 1) = 42 - (2 + 1) = 39$, $t_{.005} \approx 2.704$. The 99% confidence interval is:

$$\hat{\beta}_2 \pm t_{.005} s_{\hat{\beta}_2} \Rightarrow -.4578 \pm 2.704(.1283) \Rightarrow -.4578 \pm .3469 \Rightarrow (-.8047, -.1109)$$

We are 99% confident that the change in mean weight change for each unit change in acid-detergent fiber, holding digestion efficiency constant is between $-.8047$ and $-.1109$.

e. $R^2 = \text{R-Sq} = .529$. 52.9% of the total sample variation in the weight change scores is explained by the model containing the independent variables digestion efficiency and acid-detergent fibre. $R_a^2 = \text{R-Sq(adj)} = .505$. 50.5% of the total sample variation in the weight change scores is explained by the model containing the independent variables digestion efficiency and acid-detergent fibre, adjusting for the sample size and the number of independent variables in the model. R_a^2 is the preferred measure of model fit because as more variables are added to the model, R_a^2 can actually decrease if the variables do not improve the model, where R^2 will either stay the same or increase as variables are added to the model.

f.   To determine if the overall model is useful in the prediction of weight change, we test:

H_0: $\beta_1 = \beta_2 = 0$
H_a: At least one $\beta_i \neq 0$, $i = 1, 2$

The test statistic is $F = 21.88$ and the p-value is $p = .000$. Since the p-value is less than α ($p = .000 < .05$), H_0 is rejected. There is sufficient evidence to indicate at least one of the variables digestion efficiency and acid-detergent fibre is useful in the prediction weight change at $\alpha = .05$.

g.   The first-order model that allows for different slopes is $E(y) = \beta_0 + \beta_1 x_1 + \beta_2 x_2 + \beta_3 x_1 x_2$

where x_1 = digestive efficiency and $x_2 = \begin{cases} 1 & \text{if plants} \\ 0 & \text{otherwise} \end{cases}$

h.   Using MINITAB, the results are:

Regression Analysis: WeightChg versus X1, X2, X1X2

```
The regression equation is
WeightChg = 8.14 - 0.016 X1 - 10.4 X2 + 0.095 X1X2


Predictor      Coef   SE Coef       T       P
Constant      8.144     8.446    0.96   0.341
X1          -0.0165    0.1329   -0.12   0.902
X2          -10.354     8.538   -1.21   0.233
X1X2         0.0948    0.1418    0.67   0.508


S = 3.88175   R-Sq = 44.1%   R-Sq(adj) = 39.7%


Analysis of Variance

Source           DF       SS      MS      F      P
Regression        3   452.54  150.85  10.01  0.000
Residual Error   38   572.58   15.07
Total            41  1025.12


Source  DF  Seq SS
X1       1  384.24
X2       1   61.57
X1X2     1    6.73
```

The least squares prediction equation is $\hat{y} = 8.14 - .016 x_1 - 10.4 x_2 + .095 x_1 x_2$.

i.   For diet = plants, $x_2 = 1$. The prediction equation is

$\hat{y} = 8.14 - .016 x_1 - 10.4(1) + .095 x_1(1) = -2.26 + .079 x_1$. The slope of the line is .079.

For each unit increase is diet efficiency, the mean gosling weight change is estimated to increase by .079 for goslings fed plants.

j. For diet = duck chow, $x_2 = 0$. The prediction equation is

$$\hat{y} = 8.14 - .016x_1 - 10.4(0) + .095x_1(0) = 8.14 - .016x_1.$$ The slope of the line is $-.016$.

For each unit increase is diet efficiency, the mean gosling weight change is estimated to decrease by .016 for goslings fed duck chow.

k. To determine if the slopes associated with the two diets are different, we test:

H_0: $\beta_3 = 0$
H_a: $\beta_3 \neq 0$

From the printout, the test statistic is $t = .67$ and the p-value is $p = .508$. Since the p-value is greater than α ($p = .508 > .05$), H_0 is not rejected. There is insufficient evidence to indicate the slopes associated with the two diets differ at $\alpha = .05$.

12.173 a. Variables that are moderately or highly correlated have correlation coefficients of .5 or higher in absolute value. There are only two pairs of variables which are moderately or highly correlated. They are "year GRE taken" and "years in graduate program" ($r = -.602$), and "race" and "foreign status" ($r = -.515$).

b. When independent variables that are highly correlated with each other are included in a regression model, the results may be confusing. Highly correlated independent variables contribute overlapping information in the prediction of the dependent variable. The overall global test can indicate that the model is useful in predicting the dependent variable, while the individual t-tests on the independent variables can indicate that none of the independent variables are significant. This happens because the individual t-tests tests for the significance of an independent variable are adjusted for the other independent variables in the model. Usually, only one of the independent variables that are highly correlated with each other are included in the regression model.

12.175 a. Since in the proposed model the funniness of the joke should increase and then decrease, the value of β_2 is expected to be negative.

b. To determine if the quadratic model relating pain to funniness rating is useful, we test:

H_0: $\beta_1 = \beta_2 = 0$
H_a: At least one $\beta_i = 0$, $i = 1, 2$

The test statistic is $F = 1.60$.

The rejection region requires $\alpha = .05$ in the upper tail of the F distribution with $v_1 = k = 2$ and $v_2 = n - (k+1) = 32 - (2+1) = 29$. From Table IX, Appendix A, $F_{.05} = 3.33$. The rejection region is $F > 3.33$.

Since the observed value of the test statistic does not fall in the rejection region ($F = 1.60 \not> 3.33$), H_0 is not rejected. There is insufficient evidence to indicate that the quadratic model relating pain to funniness rating is useful at $\alpha = .05$.

c. To determine if the quadratic model relating aggression/hostility to funniness rating is useful, we test:

H_0: $\beta_1 = \beta_2 = 0$
H_a: At least one $\beta_i = 0$, $i = 1, 2$

The test statistic is $F = 1.61$.

The rejection region requires $\alpha = .05$ in the upper tail of the F distribution with $v_1 = k = 2$ and $v_2 = n - (k + 1) = 32 - (2 + 1) = 29$. From Table IX, Appendix A, $F_{.05} = 3.33$. The rejection region is $F > 3.33$.

Since the observed value of the test statistic does not fall in the rejection region ($F = 1.61 \not> 3.33$), H_0 is not rejected. There is insufficient evidence to indicate that the quadratic model relating aggression/hostility to funniness rating is useful at $\alpha = .05$.

12.177 a. Using MINITAB, the scatterplot of the data is:

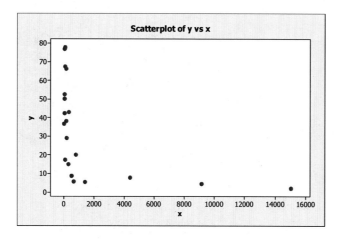

There is a curvilinear trend.

b. From MINITAB, the output is:

Regression Analysis: y versus x, x-sq

```
The regression equation is
y = 42.2 - 0.0114 x + 0.000001 x-sq


Predictor          Coef      SE Coef      T      P
Constant         42.247        5.712   7.40  0.000
x             -0.011404     0.005053  -2.26  0.037
x-sq         0.00000061   0.00000037   1.66  0.115


S = 21.8086   R-Sq = 34.9%   R-Sq(adj) = 27.2%
Analysis of Variance

Source           DF        SS       MS      F      P
Regression        2    4325.4   2162.7   4.55  0.026
Residual Error   17    8085.5    475.6
Total            19   12410.9


Source   DF   Seq SS
x         1   3013.3
x-sq      1   1312.1


Unusual Observations

Obs       x       y      Fit   SE Fit   Residual   St Resid
 16    9150    4.60   -11.21    16.24      15.81       1.09 X
 17   15022    2.20     8.09    21.40      -5.89      -1.41 X

X denotes an observation whose X value gives it large influence.
```

The fitted model is $\hat{y} = 42.2 + .0114x + .00000061x^2$

c. To determine if a curvilinear relationship exists, we test:

H_0: $\beta_2 = 0$
H_a: $\beta_2 \neq 0$

From MINITAB, the test statistic is $t = 1.66$ with p-value is $p = .115$. Since the p-value is greater than $\alpha = .05$, H_0 is not rejected. There is insufficient evidence to indicate that a curvilinear relationship exists between dissolved phosphorus percentage and soil loss at $\alpha = .05$.

12.179 a. $E(y) = \beta_0 + \beta_1 x_1 + \beta_2 x_6 + \beta_3 x_7$

where $x_6 = \begin{cases} 1 & \text{if condition is good} \\ 0 & \text{otherwise} \end{cases}$ $\qquad x_7 = \begin{cases} 1 & \text{if condition is fair} \\ 0 & \text{otherwise} \end{cases}$

b.   Using MINITAB, the plot is:

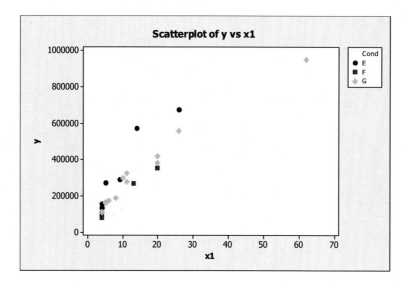

The model specified in part **a** seems appropriate. The points for E, F, and G cluster around three parallel lines.

c.   Using MINITAB, the output is

Regression Analysis: y versus x1, x6, x7

```
The regression equation is
y = 188875 + 15617 x1 - 103046 x6 - 152487 x7


Predictor     Coef   SE Coef       T      P
Constant     188875     28588    6.61  0.000
x1            15617      1066   14.66  0.000
x6          -103046     31784   -3.24  0.004
x7          -152487     39157   -3.89  0.001


S = 64623.6   R-Sq = 91.8%   R-Sq(adj) = 90.7%


Analysis of Variance

Source          DF          SS           MS      F       P
Regression       3  9.86170E+11  3.28723E+11  78.71  0.000
Residual Error  21  87700442851   4176211564
Total           24  1.07387E+12


Source  DF       Seq SS
x1       1  9.15776E+11
x6       1   7061463149
x7       1  63332198206
```

The fitted model is $\hat{y} = 188,875 + 15,617x_1 - 103,046x_6 - 152,487x_7$

For excellent condition, $x_6 = 0$ and $x_7 = 0$:
$$\hat{y} = 188,875 + 15,617x_1 - 103,046(0) - 152,487(0) = 188,875 + 15,617x_1$$

For good condition, $x_6 = 1$ and $x_7 = 0$:
$$\hat{y} = 188,875 + 15,617x_1 - 103,046(1) - 152,487(0) = 85,829 + 15,617x_1$$

For fair condition, $x_6 = 0$ and $x_7 = 1$:
$$\hat{y} = 188,875 + 15,617x_1 - 103,046(0) - 152,487(1) = 36,388 + 15,617x_1$$

d. Using MINITAB the plot is:

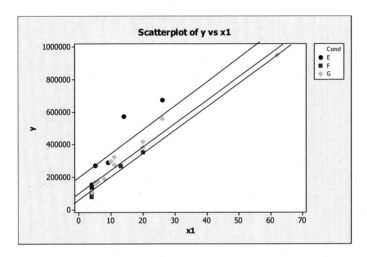

e. We must first fit a reduced model with just x_1, number of apartments. Using MINITAB, the output is:

Regression Analysis: y versus x1

```
The regression equation is
y = 101786 + 15525 x1


Predictor    Coef   SE Coef      T      P
Constant   101786     23291   4.37  0.000
x1          15525      1345  11.54  0.000


S = 82907.5   R-Sq = 85.3%   R-Sq(adj) = 84.6%


Analysis of Variance

Source          DF           SS          MS       F      P
Regression       1  9.15776E+11  9.15776E+11  133.23  0.000
Residual Error  23  1.58094E+11  6873656705
Total           24  1.07387E+12
```

To determine if the relationship between sale price and number of units differs depending on the physical condition of the apartments, we test:

H_0: $\beta_6 = \beta_7 = 0$
H_a: At least one $\beta_i \neq 0$, $i = 6, 7$

The test statistic is:

$$F = \frac{(\text{SSE}_R - \text{SSE}_C)/(k-g)}{\text{SSE}_C/[n-(k+1)]} = \frac{(1.58094 \times 10^{11} - 87700442851)/2}{4176211564} = 8.43$$

The rejection region requires $\alpha = .05$ in the upper tail of the F distribution with $v_1 = k - g = 3 - 1 = 2$ and $v_2 = n - (k+1) = 25 - (3+1) = 21$. From Table IX, Appendix A, $F_{.05} = 3.47$. The rejection region is $F > 3.47$.

Since the observed value of the test statistic falls in the rejection region ($F = 8.43 > 3.47$), H_0 is rejected. There is sufficient evidence to indicate that the relationship between sale price and number of units differs depending on the physical condition of the apartments at $\alpha = .05$.

f.   The pairwise correlations are:

```
        x1       x2       x3      x4       x5       x6
x2    -0.014
x3     0.800   -0.188
x4     0.224   -0.363   0.166
x5     0.878    0.027   0.673   0.089
x6     0.175   -0.447   0.271   0.112   0.020
x7    -0.128    0.392  -0.118   0.050  -0.238  -0.564
```

When highly correlated independent variables are present in a regression model, the results are confusing. The researchers may only want to include one of the variables. This may be the case for the variables: x_1 and x_3 $(r_{13} = .800)$, x_1 and x_5 $(r_{15} = .878)$, and x_3 and x_5 $(r_{35} = .673)$.

g.   Use the following plots to check the assumptions on ε.

standardized residuals vs x_1
standardized residuals vs x_2
standardized residuals vs x_3
standardized residuals vs x_4
standardized residuals vs x_5
standardized resisduals vs predicted values
histogram of the standardized residuals
normal probability plot

From the plots of the residuals, there do not appear to be any outliers no standardized residuals are larger than 2.38 in magnitude. In all the plots of the residuals vs x_i, there is no trend that would indicate non-constant variance (no funnel shape). In addition, there is no U or upside-down U shape that would indicate that any of the variables should be squared.   In the histogram of the residuals, the plot is fairly mound-shaped, that would indicate the residuals are approximately normally distributed. The normal probability plot is a fairly straight line. This indicates that the assumption of normality is valid. All of the assumptions appear to be met.

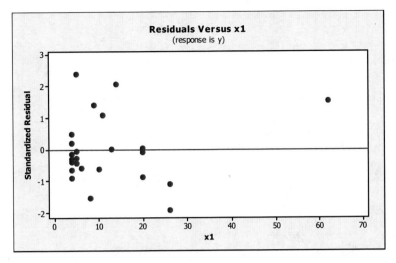

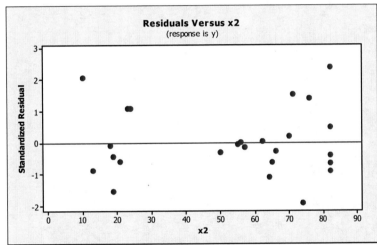

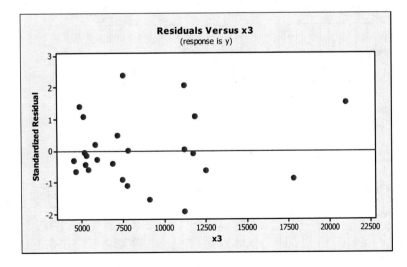

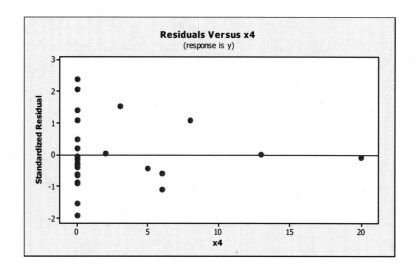

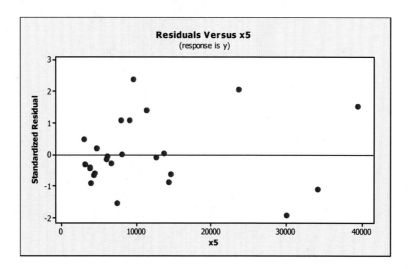

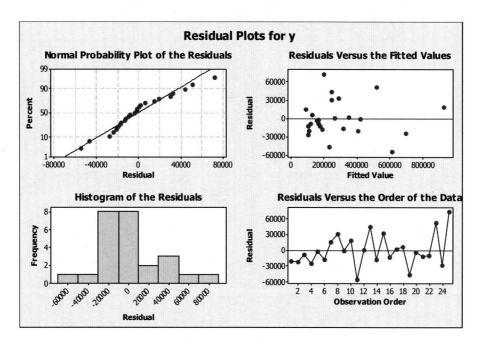

12.181 a.   In Step 1, all one-variable models are fit to the data. These models are of the form:

$$E(y) = \beta_0 + \beta_1 x_i$$

Since there are 7 independent variables, 7 models are fit. (Note: There are actually only 6 independent variables. One of the qualitative variables has three levels and thus two dummy variables. Some statistical packages will allow one to bunch these two variables together so that they are either both in or both out. In this answer, we are assuming that each x_i stands by itself.

b.   In Step 2, all two-varirable models are fit to the data, where the variable selected in Step 1 is one of the variables. These models are of the form:

$$E(y) = \beta_0 + \beta_1 x_1 + \beta_2 x_i$$

Since there are 6 independent variables remaining, 6 models are fit.

c.   In Step 3, all three-variable models are fit to the data, where the variables selected in Step 2 are two of the variables. These models are of the form:

$$E(y) = \beta_0 + \beta_1 x_1 + \beta_2 x_2 + \beta_3 x_i$$

Since there are 5 independent variables remaining, 5 models are fit.

d.   The procedure stops adding independent variables when none of the remaining variables, when added to the model, have a p-value less than some predetermined value. This predetermined value is usually $\alpha = .05$.

e.   Two major drawbacks to using the final stepwise model as the "best" model are:

(1)   An extremely large number of single β parameter t-tests have been conducted. Thus, the probability is very high that one or more errors have been made in including or excluding variables.

(2)   Often the variables selected to be included in a stepwise regression do not include the high-order terms. Consequently, we may have initially omitted several important terms from the model.

12.183 a.   (1)   Transect location is qualitative.
　　　　　(2)   Land use is qualitative
　　　　　(3)   Total number of trees in transect is quantitative.

b. $E(y) = \beta_0 + \beta_1 x_1$ where x_1 = total number of trees in a transect

c. $E(y) = \beta_0 + \beta_1 x_1 + \beta_2 x_2 + \beta_3 x_3$

where $x_2 = \begin{cases} 1 & \text{if small pasture field} \\ 0 & \text{otherwise} \end{cases}$ $x_3 = \begin{cases} 1 & \text{if small arable field} \\ 0 & \text{otherwise} \end{cases}$

A sketch of the response curve might be:

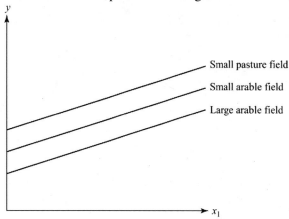

d. $E(y) = \beta_0 + \beta_1 x_1 + \beta_2 x_2 + \beta_3 x_3 + \beta_4 x_4$

where $x_4 = \begin{cases} 1 & \text{if land use is pasture} \\ 0 & \text{if land use is arable} \end{cases}$

The slopes are the same for any combination of transect location and land use. The slopes are all β_1.

e. $E(y) = \beta_0 + \beta_1 x_1 + \beta_2 x_2 + \beta_3 x_3 + \beta_4 x_4 + \beta_5 x_2 x_4 + \beta_6 x_3 x_4$

No, the slope is not affected because all the values of β_2, β_3, β_4, β_5 and β_6 are multiplied by values of 0 or 1 only.

f. $E(y) = \beta_0 + \beta_1 x_1 + \beta_2 x_2 + \beta_3 x_3 + \beta_4 x_4 + \beta_5 x_2 x_4 + \beta_6 x_3 x_4 + \beta_7 x_1 x_2 + \beta_8 x_1 x_3 + \beta_9 x_1 x_4$
$\qquad + \beta_{10} x_1 x_2 x_4 + \beta_{11} x_1 x_3 x_4$

Large arable field and land use arable: $x_2 = x_3 = x_4 = 0$.

$$E(y) = \beta_0 + \beta_1 x_1 + \beta_2(0) + \beta_3(0) + \beta_4(0) + \beta_5(0)(0) + \beta_6(0)(0) + \beta_7 x_1(0)$$
$$\qquad + \beta_8 x_1(0) + \beta_9 x_1(0) + \beta_{10} x_1(0)(0) + \beta_{11} x_1(0)(0)$$
$$= \beta_0 + \beta_1 x_1$$

The slope is β_1.

Large arable field and land use pasture: $x_2 = x_3 = 0, x_4 = 1$

$$E(y) = \beta_0 + \beta_1 x_1 + \beta_2(0) + \beta_3(0) + \beta_4(1) + \beta_5(0)(1) + \beta_6(0)(1) + \beta_7 x_1(0)$$
$$+ \beta_8 x_1(0) + \beta_9 x_1(1) + \beta_{10} x_1(0)(1) + \beta_{11} x_1(0)(1)$$
$$= (\beta_0 + \beta_4) + (\beta_1 + \beta_9) x_1$$

The slope is $(\beta_1 + \beta_9)$.

Small pasture field and land use arable: $x_2 = 1, x_3 = 0, x_4 = 0$

$$E(y) = \beta_0 + \beta_1 x_1 + \beta_2(1) + \beta_3(0) + \beta_4(0) + \beta_5(1)(0) + \beta_6(0)(0) + \beta_7 x_1(1)$$
$$+ \beta_8 x_1(0) + \beta_9 x_1(0) + \beta_{10} x_1(1)(0) + \beta_{11} x_1(0)(0)$$
$$= (\beta_0 + \beta_2) + (\beta_1 + \beta_7) x_1$$

The slope is $(\beta_1 + \beta_7)$.

Small pasture field and land use pasture: $x_2 = 1, x_3 = 0, x_4 = 1$

$$E(y) = \beta_0 + \beta_1 x_1 + \beta_2(1) + \beta_3(0) + \beta_4(1) + \beta_5(1)(1) + \beta_6(0)(1) + \beta_7 x_1(1)$$
$$+ \beta_8 x_1(0) + \beta_9 x_1(1) + \beta_{10} x_1(1)(1) + \beta_{11} x_1(0)(1)$$
$$= (\beta_0 + \beta_2 + \beta_4) + (\beta_1 + \beta_7 + \beta_9 + \beta_{10}) x_1$$

The slope is $(\beta_1 + \beta_7 + \beta_9 + \beta_{10})$.

Small arable field and land use arable: $x_2 = 0, x_3 = 1, x_4 = 0$

$$E(y) = \beta_0 + \beta_1 x_1 + \beta_2(0) + \beta_3(1) + \beta_4(0) + \beta_5(0)(0) + \beta_6(1)(0) + \beta_7 x_1(0)$$
$$+ \beta_8 x_1(1) + \beta_9 x_1(0) + \beta_{10} x_1(0)(0) + \beta_{11} x_1(1)(0)$$
$$= (\beta_0 + \beta_3) + (\beta_1 + \beta_8) x_1$$

The slope is $(\beta_1 + \beta_8)$.

Small arable field and land use pasture: $x_2 = 0, x_3 = 1, x_4 = 1$

$$E(y) = \beta_0 + \beta_1 x_1 + \beta_2(0) + \beta_3(1) + \beta_4(1) + \beta_5(0)(1) + \beta_6(1)(1) + \beta_7 x_1(0)$$
$$+ \beta_8 x_1(1) + \beta_9 x_1(1) + \beta_{10} x_1(0)(1) + \beta_{11} x_1(1)(1)$$
$$= (\beta_0 + \beta_3 + \beta_4 + \beta_6) + (\beta_1 + \beta_8 + \beta_9 + \beta_{11}) x_1$$

The slope is $(\beta_1 + \beta_8 + \beta_9 + \beta_{11})$.

12.185   All of the independent variables are continuous except District. Four dummy variables were created for District before using the stepwise regression. Also, note that several observations had missing values. Thus, only 217 observations were used in the analysis. Since Ratio was computed from the ratio of Price and DOT estimate, it was not used as a predictor variable. Using SAS and $\alpha = .10$ for keeping variables in the model, the Stepwise regression results are:

```
                      The REG Procedure
                        Model: MODEL1
                  Dependent Variable: PRICE

                  Stepwise Selection: Step 1

        Variable DOTEST Entered: R-Square = 0.9763 and C(p) = 14.2985


                      Analysis of Variance

                           Sum of           Mean
Source               DF    Squares         Square     F Value    Pr > F

Model                 1   687136244      687136244    8856.79    <.0001
Error               215    16680334          77583
Corrected Total     216   703816578


                  Parameter    Standard
        Variable   Estimate      Error    Type II SS   F Value   Pr > F

        Intercept  33.24768    22.19263      174129      2.24    0.1356
        DOTEST      0.90637     0.00963   687136244   8856.79    <.0001

               Bounds on condition number: 1, 1

----------------------------------------------------------------------------

                  Stepwise Selection: Step 2


        Variable STATUS Entered: R-Square = 0.9774 and C(p) = 5.9631


                      Analysis of Variance

                           Sum of           Mean
Source               DF    Squares         Square     F Value    Pr > F

Model                 2   687894714      343947357    4622.87    <.0001
Error               214    15921865          74401
Corrected Total     216   703816578

                      The REG Procedure
                        Model: MODEL1
                  Dependent Variable: PRICE

                  Stepwise Selection: Step 2

                  Parameter    Standard
        Variable   Estimate      Error    Type II SS   F Value   Pr > F

        Intercept -12.92808    26.10499       18247      0.25    0.6209
        DOTEST      0.91233     0.00961   669924194   9004.21    <.0001
        STATUS    132.16624    41.39440      758469     10.19    0.0016

            Bounds on condition number: 1.0392, 4.1569
```

```
--------------------------------------------------------------------------

                    Stepwise Selection: Step 3


           Variable DAYS Entered: R-Square = 0.9779 and C(p) = 3.4115


                           Analysis of Variance

                              Sum of        Mean
      Source          DF      Squares       Square    F Value   Pr > F

      Model            3     688228730    229409577   3134.76   <.0001
      Error          213      15587848        73182
      Corrected Total 216     703816578


                    Parameter   Standard
           Variable  Estimate     Error   Type II SS  F Value  Pr > F

           Intercept  -59.79987  33.93608     227239     3.11  0.0795
           DOTEST       0.88616   0.01552  238455655  3258.37  <.0001
           STATUS     139.40952  41.19370     838166    11.45  0.0008
           DAYS         0.35761   0.16739     334016     4.56  0.0338

             Bounds on condition number: 2.7653, 19.698
--------------------------------------------------------------------------


    All variables left in the model are significant at the 0.1000 level.


                          The REG Procedure
                           Model: MODEL1
                      Dependent Variable: PRICE

                    Stepwise Selection: Step 3

No other variable met the 0.1500 significance level for entry into the model.


                   Summary of Stepwise Selection

        Variable Variable  Number  Partial   Model
   Step Entered  Removed   Vars In R-Square R-Square  C(p)    F Value Pr > F

    1   DOTEST              1       0.9763   0.9763  14.2985  8856.79 <.0001
    2   STATUS              2       0.0011   0.9774   5.9631    10.19 0.0016
    3   DAYS                3       0.0005   0.9779   3.4115     4.56 0.0338
```

From the results, only three independent variables are selected to predict price using the stepwise regression and $\alpha = .05$. These variables are DOT estimate, Status, and Days.

A new model was fit with just the variables DOT estimate, Status, and Days using MINITAB. The results are:

Regression Analysis: Price versus Dotest, Status, Days

```
The regression equation is
  Price = - 59.8 + 0.886 Dotest + 139 Status + 0.358 Days


  Predictor      Coef   SE Coef      T       P
  Constant     -59.80     33.94   -1.76   0.079
  Dotest      0.88616   0.01552   57.08   0.000
  Status       139.41     41.19    3.38   0.001
  Days         0.3576    0.1674    2.14   0.034


  S = 270.522    R-Sq = 97.8%    R-Sq(adj) = 97.8%


  Analysis of Variance

  Source          DF          SS         MS        F       P
  Regression       3   688228730  229409577  3134.76   0.000
  Residual Error 213    15587848      73182
  Total          216   703816578
```

We see that this is a fairly good model. The *R*-square value is .9779. 97.79% of the variation in the PRICE values is explained by the model containing DODEST, STATUS, and DAYS.

We note that the parameter estimate for the parameter associated Bid Status is 139.40952. Since Bid Status is either 1 (fixed) or 0 (competitive), the PRICE for the fixed bid is estimated to be 139.40952 units higher than the competitive bid, all other variables held constant.

To check the model assumptions, we will look at the plots. The plots of the residuals versus the independent variables are:

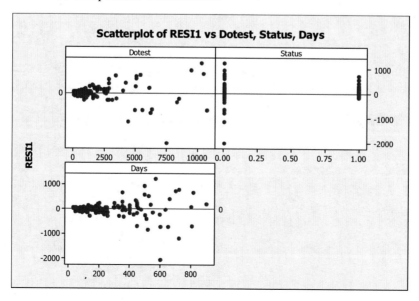

There are no 'U' or 'upside-down U' shapes to these plots. Thus, the relationship between PRICE and the independent variables appear to be linear. There are a total of 6 observations that are more than 3 standard deviations from the mean. These 6 points could be outliers.

Additional residual plots are:

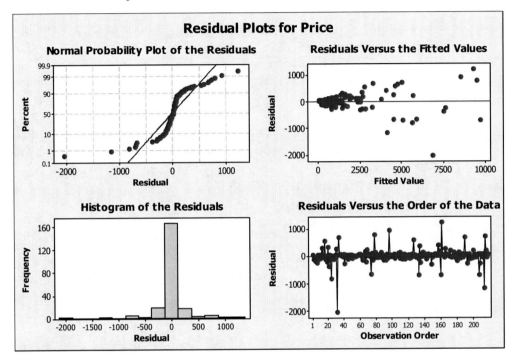

The histogram of the residuals does not look to be mound-shaped, so the assumption that the data are normal appears to be violated. In addition, the normal probability plot does not look like a straight line. This also indicates that the data are not normal.

Using the plot of the residuals against the fitted values, the spread of the data points appears to increase as the predicted value of y increases. This indicates that there is non-constant variance.

Thus, even though the fit of this model appears to be fairly good ($R^2 = .9779$), some of the model assumptions are not met. The data do not appear to be normally distributed. There appears that there are several (6) outliers and that the variance is not constant.

Additional models could be tried. One could look at interactions of the 3 variables included so far. When this is tried, the interaction term of DOTEST*STATUS is significant. However, the R-square value does not increase and the problems with the assumptions still exist.

Chapter

13

Categorical Data Analysis

13.1   The characteristics of the multinomial experiment are:

1.   The experiment consists of n identical trials.
2.   There are k possible outcomes to each trial.
3.   The probabilities of the k outcomes, denoted $p_1, p_2, \ldots, p_k$, remain the same from trial to trial, where $p_1 + p_2 + \cdots + p_k = 1$.
4.   The trials are independent.
5.   The random variables of interest are the counts $n_1, n_2, \ldots, n_k$ in each of the k cells.

The characteristics of the binomial are the same as those for the multinomial with $k = 2$.

13.3   a.   With df = 10, $\chi^2_{.05} = 18.3070$

b.   With df = 50, $\chi^2_{.990} = 29.7067$

c.   With df = 16, $\chi^2_{.10} = 23.5418$

d.   With df = 50, $\chi^2_{.005} = 79.4900$

13.5   a.   The rejection region requires $\alpha = .05$ in the upper tail of the χ^2 distribution with df $= k - 1 = 3 - 1 = 2$. From Table VII, Appendix A, $\chi^2_{.05} = 5.99147$. The rejection region is $\chi^2 > 5.99147$.

b.   The rejection region requires $\alpha = .10$ in the upper tail of the χ^2 distribution with df $= k - 1 = 5 - 1 = 4$. From Table VII, Appendix A, $\chi^2_{.10} = 7.77944$. The rejection region is $\chi^2 > 7.77944$.

c.   The rejection region requires $\alpha = .01$ in the upper tail of the χ^2 distribution with df $= k - 1 = 4 - 1 = 3$. From Table VII, Appendix A, $\chi^2_{.01} = 11.3449$. The rejection region is $\chi^2 > 11.3449$.

13.7   Some preliminary calculations are:

If the probabilities are the same, $p_{1,0} = p_{2,0} = p_{3,0} = p_{4,0} = .25$

$$E_1 = np_{1,0} = 205(.25) = 51.25$$
$$E_2 = E_3 = E_4 = 205(.25) = 51.25$$

a.   To determine if the multinomial probabilities differ, we test:

H_0: $p_1 = p_2 = p_3 = p_4 = .25$
H_a: At least one of the probabilities differs from .25

The test statistic is

$$\chi^2 = \sum \frac{[n_i - E_i]^2}{E_i} = \frac{(43-51.25)^2}{51.25} + \frac{(56-51.25)^2}{51.25} + \frac{(59-51.25)^2}{51.25} + \frac{(47-51.25)^2}{51.25} = 3.293$$

The rejection region requires $\alpha = .05$ in the upper tail of the χ^2 distribution with df = $k-1$ = 4 − 1 = 3. From Table VII, Appendix A, $\chi^2_{.05} = 7.81473$. The rejection region is $\chi^2 > 7.81473$.

Since the observed value of the test statistic does not fall in the rejection region $(\chi^2 = 3.293 \not> 7.81473)$, H_0 is not rejected. There is insufficient evidence to indicate the multinomial probabilities differ at $\alpha = .05$.

b. The Type I error is concluding the multinomial probabilities differ when, in fact, they do not.

The Type II error is concluding the multinomial probabilities are equal, when, in fact, they are not.

c. For confidence coefficient .95, $\alpha = .05$ and $\alpha/2 = .05/2 = .025$. From Table IV, Appendix A, $z_{.025} = 1.96$.

$\hat{p}_3 = 59/205 = .288$

The confidence interval is:

$$\hat{p} \pm z_{.025}\sqrt{\frac{\hat{p}\hat{q}}{n}} \Rightarrow .288 \pm 1.96\sqrt{\frac{.288(.712)}{205}} \Rightarrow .288 \pm .062 \Rightarrow (.226, .350)$$

13.9 a. The qualitative variable of interest is nonfunctional jaw habits. It has 4 levels: bruxism, teeth clenching, both bruxism and clenching, and neither.

b. A one-way table of the data is:

	Bruxism	Clench	Bruxism & Clench	Neither
n_i	3	11	30	16

c. To determine whether the percentages associated with the admitted habits differ, we test:

H_0: $p_1 = p_2 = p_3 = p_4 = .25$
H_a: At least one p_i differs from .25, $i = 1, 2, 3, 4$

d. $E_1 = E_2 = E_3 = E_4 = np_{i,0} = 60(.25) = 15$.

e.   The test statistic is

$$\chi^2 = \sum \frac{[n_i - E_i]^2}{E_i} = \frac{[3-15]^2}{15} + \frac{[11-15]^2}{15} + \frac{[30-15]^2}{15} + \frac{[16-15]^2}{15} = 25.733$$

f.   The rejection region requires $\alpha = .05$ in the upper tail of the χ^2 distribution with df = $k - 1 = 4 - 1 = 3$. From Table VII, Appendix A, $\chi^2_{.05} = 7.81473$. The rejection region is $\chi^2 > 7.81473$.

g.   Since the observed value of the test statistic falls in the rejection region $(\chi^2 = 25.733 > 7.81473)$, H_0 is rejected. There is sufficient evidence to indicate the percentages associated with the admitted habits differ at $\alpha = .05$.

h. $\hat{p}_3 = \frac{30}{60} = .5$

For confidence coefficient .95, $\alpha = .05$ and $\alpha / 2 = .05 / 2 = .025$. From Table IV, Appendix A, $z_{.025} = 1.96$. The 95% confidence interval is:

$$\hat{p}_i \pm z_{.025} \sqrt{\frac{\hat{p}_i \hat{q}_i}{n}} \Rightarrow .5 \pm 1.96 \sqrt{\frac{.5(.5)}{60}} \Rightarrow .5 \pm .127 \Rightarrow (.373, .627)$$

We are 95% confident that the true proportion of dental patients who admit to both habits is between .373 and .627.

13.11   a.   The qualitative variable of interest in this problem is the type of pottery found. There are 4 levels of the variable: burnished, monochrome, painted, and other.

b.   If all 4 types of pottery occur with equal probabilities, the values of p_1, p_2, p_3, and p_4 are all .25.

c.   To determine if one type of pottery is more likely to occur at the site than any other, we test:

H_0: $p_1 = p_2 = p_3 = p_4 = .25$
H_a: At least one $p_i \neq .25$ for $i = 1, 2, 3, 4$

d.   Some preliminary calculations are:

$E_1 = E_2 = E_3 = E_4 = np_{i,0} = 837(.25) = 209.25$.

$$\chi^2 = \sum \frac{[n_i - E_i]^2}{E_i} = \frac{[133 - 209.25]^2}{209.25} + \frac{[460 - 209.25]^2}{209.25} + \frac{[183 - 209.25]^2}{209.25} + \frac{[61 - 209.25]^2}{209.25}$$
$$= 436.591$$

e. The *p*-value is $p = P(\chi^2 \geq 436.591)$. Using Table VII, Appendix A, with df = $k - 1$ = $4 - 1 = 3$, $p = P(\chi^2 \geq 436.591) < .005$.

Since the *p*-value is less than α ($p < .005 < .10$), H_0 is rejected. There is sufficient evidence to indicate at least one type of pottery is more likely to occur at the site than another at $\alpha = .10$.

13.13 a. If it is just as likely to have a boy as a girl, then for any child, $P(G) = \frac{1}{2}$ and $P(B) = \frac{1}{2}$. Since the gender of one child is independent of the next, $P(GG) = \frac{1}{2}(\frac{1}{2}) = \frac{1}{4} = .25$, $P(GB) = \frac{1}{2}(\frac{1}{2}) = \frac{1}{4} = .25$, $P(BG) = \frac{1}{2}(\frac{1}{2}) = \frac{1}{4} = .25$, and $P(BB) = \frac{1}{2}(\frac{1}{2}) = \frac{1}{4} = .25$.

b. Since each probability is .25, the expected number of families of each combination is $np_i = 42,888(.25) = 10,722$.

c. The test statistics is

$$\chi^2 = \sum \frac{[n_i - E_i]^2}{E_i} = \frac{[9,523 - 10,722]^2}{10,722} + \frac{[11,118 - 10,722]^2}{10,722} + \frac{[10,913 - 10,722]^2}{10,722}$$
$$+ \frac{[11,334 - 10,722]^2}{10,722} = 187.04$$

d. The rejection region requires $\alpha = .10$ in the upper tail of the χ^2 distribution with *df* = $k - 1 = 4 - 1 = 3$. From Table VII, Appendix A, $\chi^2_{.10} = 6.25139$. The rejection region is $\chi^2 > 6.25139$.

Since the observed value of the test statistic falls in the rejection region ($\chi^2 = 187.04 > 6.25139$), H_0 is rejected. There is sufficient evidence to indicate that at least one probability of gender configuration for a two-child family is not .25 at $\alpha = .10$.

e. The expected numbers of families for each gender configuration are:

GG: $np_1 = 42,888(.23795) = 10,205.2$
GB: $np_2 = 42,888(.24985) = 10,715.6$
BG: $np_3 = 42,888(.24985) = 10,715.6$
BB: $np_4 = 42,888(.26235) = 11,251.7$

The test statistics is

$$\chi^2 = \sum \frac{[n_i - E_i]^2}{E_i} = \frac{[9,523 - 10,205.2]^2}{10,205.2} + \frac{[11,118 - 10,715.6]^2}{10,715.6} + \frac{[10,913 - 10,715.6]^2}{10,715.6}$$
$$+ \frac{[11,334 - 11,251.7]^2}{11,251.7} = 64.95$$

Since the observed value of the test statistic falls in the rejection region $(\chi^2 = 64.95 > 6.25139)$, H_0 is rejected. There is sufficient evidence to indicate that at least one probability of gender configuration for a two-child family is not the proposed probability at $\alpha = .10$.

13.15 a. There were 1,470 responses that were missing. In addition, 14 responses were 8 = Don't know and 7 responses were 9 = Missing. Those responding with 8 or 9 were not included. The frequency table is:

Response	Frequency
1	450
2	627
3	219
4	23
Totals	**1319**

b. To determine if the true proportions in each category are equal, we test:

H_0: $p_1 = p_2 = p_3 = p_4 = .25$
H_a: At least one p_i differs

c. $E_1 = E_2 = E_3 = E_4 = np_{i,0} = 1,319(.25) = 329.75$

d. The test statistic is

$$\chi^2 = \sum \frac{[n_i - E_i]^2}{E_i}$$

$$= \frac{[450 - 329.75]^2}{329.75} + \frac{[627 - 329.75]^2}{329.75} + \frac{[219 - 329.75]^2}{329.75} + \frac{[23 - 329.75]^2}{329.75} = 634.36$$

e. The rejection region requires $\alpha = .10$ in the upper tail of the χ^2 distribution with df = $k - 1 = 4 - 1 = 3$. From Table VII, Appendix A, $\chi^2_{.10} = 6.25139$. The rejection region is $\chi^2 > 6.25139$.

Since the observed value of the test statistic falls in the rejection region $(\chi^2 = 634.36 > 6.25139)$, H_0 is rejected. There is sufficient evidence to indicate at least one proportion differs at $\alpha = .10$.

f. To determine if the true proportions follow the proportions given, we test:

H_0: $p_1 = .30, p_2 = .50, p_3 = .15, p_4 = .05$
H_a: At least one p_i differs from its hypothesized value

$E_1 = np_{1,0} = 1,319(.30) = 395.7$ $E_2 = np_{2,0} = 1,319(.50) = 659.5$
$E_3 = np_{3,0} = 1,319(.15) = 197.85$ $E_4 = np_{4,0} = 1,319(.05) = 65.95$

The test statistic is

$$\chi^2 = \sum \frac{[n_i - E_i]^2}{E_i}$$

$$= \frac{[450 - 395.7]^2}{395.7} + \frac{[627 - 659.5]^2}{659.5} + \frac{[219 - 197.85]^2}{197.85} + \frac{[23 - 65.95]^2}{65.95} = 39.29$$

The rejection region requires $\alpha = .10$ in the upper tail of the χ^2 distribution with df $= k - 1 = 4 - 1 = 3$. From Table VII, Appendix A, $\chi^2_{.10} = 6.25139$. The rejection region is $\chi^2 > 6.25139$.

Since the observed value of the test statistic falls in the rejection region $(\chi^2 = 39.29 > 6.25139)$, H_0 is rejected. There is sufficient evidence to indicate at least one proportion differs from its hypothesized value at $\alpha = .10$.

13.17   Some preliminary calculations are:

$E_1 = np_{1,0} = 2,097(.02) = 41.94$
$E_2 = np_{2,0} = 2,097(.25) = 524.25$
$E_3 = np_{3,0} = 2,097(.73) = 1,530.81$

To determine if the distribution of the E4/E4 genotypes for the population of young adults differs from the norm, we test:

H_0: $p_1 = .02$; $p_2 = .25$; $p_3 = .73$
H_a: At least one p_i differs from its hypothesized value

The test statistic is

$$\chi^2 = \sum \frac{[n_i - E_i]^2}{E_i} = \frac{[56 - 41.94]^2}{41.94} + \frac{[517 - 524.25]^2}{524.25} + \frac{[1524 - 1530.81]^2}{1530.81} = 4.84$$

The rejection region requires $\alpha = .05$ in the upper tail of the χ^2 distribution with $df = k - 1 = 3 - 1 = 2$. From Table VII, Appendix A, $\chi^2_{.05} = 5.99146$. The rejection region is $\chi^2 > 5.99146$.

Since the observed value of the test statistic does not fall in the rejection region $(\chi^2 = 4.84 \not> 5.99146)$, H_0 is not rejected. There is insufficient evidence to indicate the distribution of the E4/E4 genotypes for the population of young adults differs from the norm at $\alpha = .05$.

13.19   a.   Some preliminary calculations are:

$$E_A = np_{A,0} = 700(.09) = 63 \qquad E_B = np_{B,0} = 700(.02) = 14$$
$$E_C = np_{C,0} = 700(.02) = 14 \qquad E_D = np_{D,0} = 700(.04) = 28$$
$$E_E = np_{E,0} = 700(.12) = 84 \qquad E_F = np_{F,0} = 700(.02) = 14$$
$$E_G = np_{G,0} = 700(.03) = 21 \qquad E_H = np_{H,0} = 700(.02) = 14$$
$$E_I = np_{I,0} = 700(.09) = 63 \qquad E_J = np_{J,0} = 700(.01) = 7$$
$$E_K = np_{K,0} = 700(.01) = 7 \qquad E_L = np_{L,0} = 700(.04) = 28$$
$$E_M = np_{M,0} = 700(.02) = 14 \qquad E_N = np_{N,0} = 700(.06) = 42$$
$$E_O = np_{O,0} = 700(.08) = 56 \qquad E_P = np_{P,0} = 700(.02) = 14$$
$$E_Q = np_{Q,0} = 700(.01) = 7 \qquad E_R = np_{R,0} = 700(.06) = 42$$
$$E_S = np_{S,0} = 700(.04) = 28 \qquad E_T = np_{T,0} = 700(.06) = 42$$
$$E_U = np_{U,0} = 700(.04) = 28 \qquad E_V = np_{V,0} = 700(.02) = 14$$
$$E_W = np_{W,0} = 700(.02) = 14 \qquad E_X = np_{X,0} = 700(.01) = 7$$
$$E_Y = np_{Y,0} = 700(.02) = 14 \qquad E_Z = np_{Z,0} = 700(.01) = 7$$
$$E_{BLANK} = np_{BL,0} = 700(.02) = 14$$

To determine if the ScrabbleExpress™ presents the player with "unfair word selection opportunities" that are different from the Scrabble™ board game, we test:

H_0: $p_A = .09, p_B = .02, p_C = .02, p_D = .04, p_E = .12, p_F = .02, p_G = .03, p_H = .02, p_I = .09,$
$p_J = .01, p_K = .01, p_L = .04, p_M = .02, p_N = .06, p_O = .08, p_P = .02, p_Q = .01, p_R = .06,$
$p_S = .04, p_T = .06, p_U = .04, p_V = .02, p_W = .02, p_X = .01, p_Y = .02, p_Z = .01,$
$p_{BLANK} = .02$

H_a: At least one p_i differs from its hypothesized value, i = A, B, …, BLANK

The test statistic is

$$\chi^2 = \sum \frac{[n_i - E_i]^2}{E_i} = \frac{[39-63]^2}{63} + \frac{[18-14]^2}{14} + \frac{[30-14]^2}{14} + \cdots + \frac{[34-14]^2}{14} = 360.48$$

The rejection region requires $\alpha = .05$ in the upper tail of the χ^2 distribution with df = $k - 1 = 27 - 1 = 26$. From Table VII, Appendix A, $\chi^2_{.05} = 38.8852$. The rejection region is $\chi^2 > 38.8852$.

Since the observed value of the test statistic falls in the rejection region ($\chi^2 = 360.48 > 38.8852$), H_0 is rejected. There is sufficient evidence to indicate the ScrabbleExpress™ presents the player with "unfair word selection opportunities" that are different from the Scrabble™ board game at $\alpha = .05$.

b.   The true proportion of letters drawn from the board game is

$$p_A + p_E + p_I + p_O + p_U = .09 + .12 + .09 + .08 + .04 = .42$$

The number of vowels drawn by the electronic game is 39 + 31 + 25 + 20 + 21 = 136. The proportion of vowels drawn is 136 / 700 = .194.

For confidence coefficient .95, $\alpha = .05$ and $\alpha/2 = .05/2 = .025$. From Table IV, Appendix A, $z_{.025} = 1.96$. The 95% confidence interval is:

$$\hat{p}_i \pm z_{.025}\sqrt{\frac{\hat{p}_i\hat{q}_i}{n}} \Rightarrow .194 \pm 1.96\sqrt{\frac{.194(.806)}{700}} \Rightarrow .194 \pm .029 \Rightarrow (.165, \ .223)$$

We are 95% confident that the true proportion of vowels drawn by the electronic game is between .165 and .223. The actual proportion of vowels drawn in the board game is .42. Since this value is not contained in the 95% confidence interval, one can conclude that the ScrabbleExpress^{TM} game produces too few vowels in the 7-letter draws.

13.21   A contingency table with fixed marginals is a table where the column totals for one qualitative variable are known in advance. For example, we could ask a sample of 150 males and a sample of 150 females their preference to a particular brand. The column totals for gender are known in advance.

13.23   The conditions that are required for a valid chi-square test of data from a contingency table are:

1. The n observed counts are a random sample from the population of interest.
2. The sample size, n, will be large enough so that, for every cell, the expected count, E_{ij}, will be equal to 5 or more.

13.25   a. H_0:   The row and column classifications are independent
            H_a: The row and column classifications are dependent

b.   The test statistic is $\chi^2 = \sum\sum\frac{[n_{ij} - \hat{E}_{ij}]^2}{\hat{E}_{ij}}$

The rejection region requires $\alpha = .01$ in the upper tail of the χ^2 distribution with df $= (r-1)(c-1) = (2-1)(3-1) = 2$. From Table VII, Appendix A, $\chi^2_{.01} = 9.21034$. The rejection region is $\chi^2 > 9.21034$.

c.   The expected cell counts are:

$$\hat{E}_{11} = \frac{R_1C_1}{n} = \frac{96(25)}{167} = 14.37 \qquad \hat{E}_{21} = \frac{R_2C_1}{n} = \frac{71(25)}{167} = 10.63$$

$$\hat{E}_{12} = \frac{R_1C_2}{n} = \frac{96(64)}{167} = 36.79 \qquad \hat{E}_{22} = \frac{R_2C_2}{n} = \frac{71(64)}{167} = 27.21$$

$$\hat{E}_{13} = \frac{R_1C_3}{n} = \frac{96(78)}{167} = 44.84 \qquad \hat{E}_{23} = \frac{R_2C_3}{n} = \frac{71(78)}{167} = 33.16$$

d.   The test statistic is

$$\chi^2 = \sum\sum\frac{[n_{ij} - \hat{E}_{ij}]^2}{\hat{E}_{ij}} = \frac{(9-14.37)^2}{14.37} + \frac{(34-36.79)^2}{36.79} + \frac{(53-44.84)^2}{44.84} + \frac{(16-10.63)^2}{10.63}$$

$$+ \frac{(30-27.21)^2}{27.21} + \frac{(25-33.16)^2}{33.16} = 8.71$$

Since the observed value of the test statistic does not fall in the rejection region $(\chi^2 = 8.71 \not> 9.21034)$, H_0 is not rejected. There is insufficient evidence to indicate the row and column classifications are dependent at $\alpha = .01$.

13.27   Some preliminary calculations are:

$$\hat{E}_{11} = \frac{R_1 C_1}{n} = \frac{154(134)}{439} = 47.007 \qquad \hat{E}_{21} = \frac{186(134)}{439} = 56.774$$

$$\hat{E}_{12} = \frac{154(163)}{439} = 57.180 \qquad \hat{E}_{22} = \frac{186(163)}{439} = 69.062$$

$$\hat{E}_{13} = \frac{154(142)}{439} = 49.813 \qquad \hat{E}_{23} = \frac{186(142)}{439} = 60.164$$

$$\hat{E}_{31} = \frac{99(134)}{439} = 30.219 \qquad \hat{E}_{33} = \frac{99(142)}{439} = 32.023$$

$$\hat{E}_{32} = \frac{99(163)}{439} = 36.759$$

To determine if the row and column classifications are dependent, we test:

H_0: The row and column classifications are independent
H_a: The row and column classifications are dependent

The test statistic is

$$\chi^2 = \sum\sum \frac{[n_{ij} - \hat{E}_{ij}]^2}{\hat{E}_{ij}} = \frac{(40-47.007)^2}{47.007} + \frac{(72-57.180)^2}{57.180} + \frac{(42-49.813)^2}{49.813} + \frac{(63-56.774)^2}{56.774}$$

$$+ \frac{(53-69.062)^2}{69.062} + \frac{(70-60.164)^2}{60.164} + \frac{(31-32.023)^2}{32.023} + \frac{(38-36.759)^2}{36.759} + \frac{(30-32.023)^2}{32.023}$$

$$= 12.33$$

The rejection region requires $\alpha = .05$ in the upper tail of the χ^2 distribution with df = $(r-1)(c-1) = (3-1)(3-1) = 4$. From Table VII, Appendix A, $\chi^2_{.05} = 9.48773$. The rejection region is $\chi^2 > 9.48773$.

Since the observed value of the test statistic falls in the rejection region $(\chi^2 = 12.33 > 9.48773)$, H_0 is rejected. There is sufficient evidence to indicate the row and column classifications are dependent at $\alpha = .05$.

13.29.   a. $\hat{p}_1 = \dfrac{6}{77} = .078$

b. $\hat{p}_2 = \dfrac{11}{70} = .157$

c.   The proportion of girls drawing a dog is almost 2 times the proportion of boys drawing a dog. It is possible that the likelihood of drawing a dog depends on gender.

d.   To determine if the likelihood of drawing a dog depends on gender, we test:

H_0: Presence of a dog and gender are independent
H_a: Presence of a dog and gender are dependent

e.   From the printout, the test statistic is $\chi^2 = 2.250$ and the p-value is $p = 0.134$.

Since the p-value is not less than $\alpha = .05$ $(p = .134 \not< .05)$, H_0 is not rejected. There is insufficient evidence to indicate the likelihood of drawing a dog depends on gender at $\alpha = .05$.

f.   Using MINITAB, the results are:

Tabulated statistics: TV, GENDER

```
Using frequencies in NUMBER


Rows: TV    Columns: GENDER

           Boy    Girl      All

No          66      61      127
         66.52   60.48   127.00

Yes         11       9       20
         10.48    9.52    20.00

All         77      70      147
         77.00   70.00   147.00

Cell Contents:       Count
                     Expected count


Pearson Chi-Square = 0.064, DF = 1, P-Value = 0.801
Likelihood Ratio Chi-Square = 0.064, DF = 1, P-Value = 0.801
```

To determine if the likelihood of drawing a TV in the bedroom depends on gender, we test:

H_0: Presence of TV and gender are independent
H_a: Presence of TV and gender are dependent

From the printout, the test statistic is $\chi^2 = .064$ and the p-value is $p = 0.801$.

Since the p-value is not less than $\alpha = .05$ $(p = .801 \not< .05)$, H_0 is not rejected. There is insufficient evidence to indicate the likelihood of drawing TV depends on gender at $\alpha = .05$.

13.31   a. $\hat{p}_1 = \dfrac{111}{170} = .653$

b. $\hat{p}_2 = \dfrac{61}{110} = .555$

c. $\hat{p}_3 = \dfrac{11}{30} = .355$

d.   The sample proportions range from .367 to .653. It appears that the proportion of news stories that are deceptive depends on story tone.

e.   To determine whether the authenticity of a news story depends on tone, we test:

H_0: Authenticity of news story and tone are independent
H_a: Authenticity of news story and tone are dependent

f.   From the printout, the test statistic is $\chi^2 = 10.427$ and the p-value is $p = .005$. Since the p-value is less than α $(p = .005 < .05)$, H_0 is rejected. There is sufficient evidence to indicate that the authenticity of a news story depends on tone at $\alpha = .05$.

13.33   a.   The two categorical variables are Risk of masculinity and type of event. Risk of masculinity has two levels: High and Low. Type of event has two levels: Violent event and Avoided-Violent event.

b.   The experimental units are 1,507 newly incarcerated men.

c. $E_{11} = \dfrac{R_1 C_1}{n} = \dfrac{379(1,037)}{1,507} = 260.798$

d. $E_{12} = \dfrac{R_1 C_2}{n} = \dfrac{379(470)}{1,507} = 118.202$ $E_{21} = \dfrac{R_2 C_1}{n} = \dfrac{1,128(1,037)}{1,507} = 776.202$

$E_{22} = \dfrac{R_2 C_2}{n} = \dfrac{1,128(470)}{1,507} = 351.798$

e.

$$\chi^2 = \sum\sum \dfrac{[n_{ij} - \hat{E}_{ij}]^2}{\hat{E}_{ij}} = \dfrac{[236 - 260.798]^2}{260.798} + \dfrac{[143 - 118.202]^2}{118.202}$$

$$+ \dfrac{[801 - 776.202]^2}{776.202} + \dfrac{[327 - 351.798]^2}{351.798} = 10.101$$

f.   To determine whether event type depends on high/low risk masculinity, we test:

H_0: Event Type and High/Low Risk Masculinity are independent
H_a: Event Type and High/Low Risk Masculinity are dependent

The test statistic is $\chi^2 = 10.101$.

The rejection region requires $\alpha = .05$ in the upper tail of the χ^2 distribution with df $= (r - 1)(c - 1) = (2 - 1)(2 - 1) = 1$. From Table VII, Appendix A, $\chi^2_{.05} = 3.84146$. The rejection region is $\chi^2 > 3.84146$.

Since the observed value of the test statistic falls in the rejection region $(\chi^2 = 10.101 > 3.84146)$, H_0 is rejected. There is sufficient evidence that event type depends on high/low risk masculinity at $\alpha = .05$.

13.35. Some preliminary calculations are:

$$\hat{E}_{11} = \frac{R_1 C_1}{n} = \frac{192(115)}{526} = 41.98 \qquad \hat{E}_{12} = \frac{R_1 C_2}{n} = \frac{192(411)}{526} = 150.02$$

$$\hat{E}_{21} = \frac{R_2 C_1}{n} = \frac{157(115)}{526} = 34.33 \qquad \hat{E}_{22} = \frac{R_2 C_2}{n} = \frac{157(411)}{526} = 122.67$$

$$\hat{E}_{31} = \frac{R_3 C_1}{n} = \frac{177(115)}{526} = 38.70 \qquad \hat{E}_{32} = \frac{R_3 C_2}{n} = \frac{177(411)}{526} = 138.30$$

To determine if the proportion of students diagnosed with MR depends on the IQ test/retest method, we test:

H_0: MR diagnosis and Test/Retest are independent
H_a: MR diagnosis and Test/Retest are dependent

The test statistic is

$$\chi^2 = \sum\sum \frac{[n_{ij} - \hat{E}_{ij}]^2}{E_{ij}} = \frac{[25-41.98]^2}{41.98} + \frac{[54-34.33]^2}{34.33} + \frac{[36-38.70]^2}{38.70} + \frac{[167-150.02]^2}{150.02}$$
$$+ \frac{[103-122.67]^2}{122.67} + \frac{[141-138.30]^2}{138.30} = 23.46$$

The rejection region requires $\alpha = .01$ in the upper tail of the χ^2 distribution with df $= (r-1)(c-1) = (3-1)(2-1) = 2$. From Table VII, Appendix A, $\chi^2_{.01} = 9.21034$. The rejection region is $\chi^2 > 9.21034$.

Since the observed value of the test statistic falls in the rejection region $(\chi^2 = 23.46 > 9.21034)$, H_0 is rejected. There is sufficient evidence to indicate the proportion of students diagnosed with MR depends on the IQ test/retest method at $\alpha = .01$.

13.37 Some preliminary calculations are:

$$\hat{E}_{11} = \frac{R_1 C_1}{n} = \frac{2097(149)}{6560} = 47.630 \qquad \hat{E}_{12} = \frac{R_1 C_2}{n} = \frac{2097(1647)}{6560} = 526.488$$

$$\hat{E}_{13} = \frac{R_1 C_3}{n} = \frac{2097(4764)}{6560} = 1522.882 \qquad \hat{E}_{21} = \frac{R_2 C_1}{n} = \frac{2182(149)}{6560} = 49.561$$

$$\hat{E}_{22} = \frac{R_2 C_2}{n} = \frac{2182(1647)}{6560} = 547.828 \qquad \hat{E}_{23} = \frac{R_2 C_3}{n} = \frac{2182(4764)}{6560} = 1584.611$$

$$\hat{E}_{31} = \frac{R_3C_1}{n} = \frac{2281(149)}{6560} = 51.809 \qquad \hat{E}_{32} = \frac{R_3C_2}{n} = \frac{2281(1647)}{6560} = 572.684$$

$$\hat{E}_{33} = \frac{R_3C_3}{n} = \frac{2281(4764)}{6560} = 1656.507$$

To determine if there are significant genotype differences across the three age groups, we test:

H_0: Genotype and age group are independent

H_a: Genotype and age group are dependent

The test statistic is

$$\chi^2 = \sum \frac{\left[n_{ij} - \hat{E}_{ij}\right]^2}{\hat{E}_{ij}} = \frac{[56-47.630]^2}{47.630} + \frac{[517-526.488]^2}{526.488} + \frac{[1524-1522.882]^2}{1522.882}$$
$$+ \frac{[45-49.561]^2}{49.561} + \frac{[566-547.828]^2}{547.828} + \frac{[1571-1584.611]^2}{1584.611}$$
$$+ \frac{[48-51.809]^2}{51.809} + \frac{[564-572.684]^2}{572.684} + \frac{[1669-1656.507]^2}{1656.507} = 3.288$$

The rejection region requires $\alpha = .05$ in the upper tail of the χ^2 distribution with $df = (r-1)(c-1) = (3-1)(3-1) = 4$. From Table VII, Appendix A, $\chi^2_{.05} = 9.48773$. The rejection region is $\chi^2 > 9.48773$.

Since the observed value of the test statistics does not fall in the rejection region $(\chi^2 = 3.288 \not> 9.48773)$, H_0 is not rejected. There is insufficient evidence to indicate there are significant genotype differences across the three age groups at $\alpha = .05$.

13.39   Using MINITAB, the results of the analyses are:

Results for: AIRTHR~1.MTP

Tabulated statistics: Instruction, Strategy

```
Rows: Instruction   Columns: Strategy

              Guess    Other    TTBC     All

Cue              5        6       13      24
             7.000    8.500    8.500  24.000

Pattern          9       11        4      24
             7.000    8.500    8.500  24.000

All             14       17       17      48
            14.000   17.000   17.000  48.000

Cell Contents:         Count
                       Expected count
```

```
Pearson Chi-Square = 7.378, DF = 2, P-Value = 0.025
Likelihood Ratio Chi-Square = 7.668, DF = 2, P-Value = 0.022
```

To determine if the choice of heuristic strategy depends on type of instruction provided, we test:

H_0: Choice of heuristic strategy and instruction provided are independent
H_0: Choice of heuristic strategy and instruction provided are dependent

From the printout, the test statistic is $\chi^2 = 7.378$ and the p-value is $p = .025$.

Since the p-value is less than $\alpha = .05$ $(p = .025 < .05)$, H_0 is rejected. There is sufficient evidence to indicate the choice of heuristic strategy depends on type of instruction provided at $\alpha = .05$.

If $\alpha = .01$, then the p-value is not less than $\alpha (p = .025 \not< .01)$. H_0 would not be rejected at $\alpha = .01$.

13.41   a.   Using MINITAB, the results of the analyses are:

Tabulated statistics: SHSS, CAHS

```
Using frequencies in Fr

Rows: SHSS   Columns: CAHS

             L       M       H      V      All

L           32      14       2      0       48
          18.09   16.25   11.45   2.22   48.00

M           11      14       6      0       31
          11.68   10.49    7.39   1.43   31.00

H            6      14      19      3       42
          15.83   14.22   10.02   1.94   42.00

V            0       2       4      3        9
           3.39    3.05    2.15   0.42    9.00

All         49      44      31      6      130
          49.00   44.00   31.00   6.00  130.00

Cell Contents:        Count
                      Expected count


Pearson Chi-Square = 60.103, DF = 9
Likelihood Ratio Chi-Square = 59.638, DF = 9

* WARNING * 1 cells with expected counts less than 1
* WARNING * Chi-Square approximation probably invalid

* NOTE * 7 cells with expected counts less than 5
```

From the printout, 7 cells have expected counts less than 5. In order for the test to be valid, all of the cells should have expected counts greater than 5. Thus, we should not proceed with the analysis.

b.   Combining the High and Very High categories, the new table is:

		CAHS LEVEL		
SHSS: C LEVEL		**Low**	**Medium**	**High/Very High**
	Low	32	14	2
	Medium	11	14	6
	High/Very High	6	16	29

c.   The expected cell counts are:

$$\hat{E}_{11} = \frac{R_1 C_1}{n} = \frac{48(49)}{130} = 18.0923 \qquad \hat{E}_{21} = \frac{R_2 C_1}{n} = \frac{31(49)}{130} = 11.6846$$

$$\hat{E}_{12} = \frac{R_1 C_2}{n} = \frac{48(44)}{130} = 16.2462 \qquad \hat{E}_{22} = \frac{R_2 C_2}{n} = \frac{31(44)}{130} = 10.4923$$

$$\hat{E}_{13} = \frac{R_1 C_3}{n} = \frac{48(37)}{130} = 13.6615 \qquad \hat{E}_{23} = \frac{R_2 C_3}{n} = \frac{31(37)}{130} = 8.8231$$

$$\hat{E}_{31} = \frac{R_3 C_1}{n} = \frac{51(49)}{130} = 19.2231 \qquad \hat{E}_{32} = \frac{R_3 C_2}{n} = \frac{51(44)}{130} = 17.2615$$

$$\hat{E}_{33} = \frac{R_3 C_3}{n} = \frac{51(37)}{130} = 14.5154$$

Since all of the expected cell counts are now 5 or greater, the assumption is met.

d.   To determine whether CAHS Levels and SHSS:C Levels are dependent, we test:

H_0: CAHS Levels and SHSS:C Levels are independent
H_a: CAHS Levels and SHSS:C Levels are dependent

The test statistic is

$$\chi^2 = \sum\sum \frac{[n_{ij} - \hat{E}_{ij}]^2}{\hat{E}_{ij}} = \frac{(32-18.0923)^2}{18.0923} + \frac{(14-16.2462)^2}{16.2462} + \frac{(2-13.6615)^2}{13.6615} + \frac{(11-11.6846)^2}{11.6846}$$

$$+ \frac{(14-10.4923)^2}{10.4923} + \frac{(6-8.8231)^2}{8.8231} + \frac{(6-19.2231)^2}{19.2231} + \frac{(16-17.2615)^2}{17.2615} + \frac{(29-14.5154)^2}{14.5154} = 46.71$$

The rejection region requires $\alpha = .05$ in the upper tail of the χ^2 distribution with df = $(r-1)(c-1) = (3-1)(3-1) = 4$. From Table VII, Appendix A, $\chi^2_{.05} = 9.48773$. The rejection region is $\chi^2 > 9.48773$.

Since the observed value of the test statistic falls in the rejection region $(\chi^2 = 46.70 > 9.48773)$, H_0 is rejected. There is sufficient evidence to indicate that CAHS Levels and SHSS:C Levels are dependent at $\alpha = .05$.

13.43   a.   Some preliminary calculations are:

$$\hat{E}_{11} = \frac{R_1 C_1}{n} = \frac{31(24)}{38} = 19.58 \qquad \hat{E}_{12} = \frac{R_1 C_2}{n} = \frac{31(14)}{38} = 11.42$$

$$\hat{E}_{21} = \frac{R_2 C_1}{n} = \frac{7(24)}{38} = 4.42 \qquad \hat{E}_{22} = \frac{R_2 C_2}{n} = \frac{7(14)}{38} = 2.58$$

To determine whether the vaccine is effective in treating the MN strain of HIV, we test:

   H_0:   Vaccine and MN strain are independent
   H_a: Vaccine and MN strain are dependent

The test statistic is

$$\chi^2 = \sum\sum \frac{[n_{ij} - \hat{E}_{ij}]^2}{\hat{E}_{ij}} = \frac{(22-19.58)^2}{19.58} + \frac{(9-11.42)^2}{11.42} + \frac{(2-4.42)^2}{4.42} + \frac{(5-2.58)^2}{2.58} = 4.407$$

The rejection region requires $\alpha = .05$ in the upper tail of the χ^2 distribution with df = $(r-1)(c-1) = (2-1)(2-1) = 1$. From Table VII, Appendix A, $\chi^2_{.05} = 3.84146$. The rejection region is $\chi^2 > 3.84146$.

Since the observed value of the test statistic falls in the rejection region $(\chi^2 = 4.407 > 3.84146)$, H_0 is rejected. There is sufficient evidence to indicate that the vaccine is effective in treating the MN strain of HIV at $\alpha = .05$.

   b.   The necessary assumptions are:

   1. The n observed counts are a random sample from the population of interest.

   2. The sample size, n, will be large enough so that, for every cell, the expected count, $E(n_{ij})$, will be equal to 5 or more.

   For this example, the second assumption is violated. Two of the expected cell counts are less than 5. What we have computed for the test statistic may not have a χ^2 distribution.

   c.   For this contingency table:

$$p = \frac{\binom{7}{2}\binom{31}{22}}{\binom{38}{24}} = \frac{\dfrac{7!}{2!5!}\dfrac{31!}{22!9!}}{\dfrac{38!}{24!14!}} = .0438$$

   d.   If vaccine and MN strain are independent, then the proportion of positive results should be relatively the same for both patient groups. In the two tables presented, the proportion of positive results for the vaccinated group is smaller than the proportion for the original table.

For the first table,

$$p = \frac{\binom{7}{1}\binom{31}{23}}{\binom{38}{24}} = \frac{\frac{7!}{1!6!}\frac{31!}{23!8!}}{\frac{38!}{24!14!}} = .0057$$

For the second table:

$$p = \frac{\binom{7}{0}\binom{31}{24}}{\binom{38}{24}} = \frac{\frac{7!}{0!7!}\frac{31!}{24!7!}}{\frac{38!}{24!14!}} = .0003$$

e.   The p-value for Fisher's exact test is $p = .0438 + .0057 + .0003 = .0498$. Since the p-value is small, there is evidence to reject H_0. There is sufficient evidence to indicate that the vaccine is effective in treating the MN strain of HIV at $\alpha > .0498$.

13.45   The statement "Rejecting the null hypothesis in a chi-square test for independence implies that a causal relationship between the two categorical variables exists." is false. If the variables are dependent (reject H_0), it simply implies that they are related. One cannot infer causal relationships from contingency table analysis.

13.47   a.   Some preliminary calculations are:

$$\hat{E}_{11} = \frac{R_1 C_1}{n} = \frac{50(50)}{250} = 10 \qquad \hat{E}_{21} = \frac{R_2 C_1}{n} = \frac{100(50)}{250} = 20$$

$$\hat{E}_{12} = \frac{R_1 C_2}{n} = \frac{50(90)}{250} = 18 \qquad \hat{E}_{22} = \frac{R_2 C_2}{n} = \frac{100(90)}{250} = 36$$

$$\hat{E}_{13} = \frac{R_1 C_3}{n} = \frac{50(110)}{250} = 22 \qquad \hat{E}_{23} = \frac{R_2 C_3}{n} = \frac{100(100)}{250} = 44$$

$$\hat{E}_{31} = \frac{R_3 C_1}{n} = \frac{100(50)}{250} = 20 \qquad \hat{E}_{32} = \frac{R_3 C_2}{n} = \frac{100(90)}{250} = 36$$

$$\hat{E}_{33} = \frac{R_3 C_3}{n} = \frac{100(110)}{250} = 44$$

To determine if the rows and columns are dependent, we test:

H_0: Rows and columns are independent
H_a: Rows and columns are dependent

The test statistic is $\chi^2 = \sum\sum \frac{[n_{ij} - E_{ij}]^2}{\hat{E}_{ij}} = \frac{(20-10)^2}{10} + \cdots + \frac{(30-44)^2}{44} = 54.14$

The rejection region requires $\alpha = .05$ in the upper tail of the χ^2 distribution with df = $(r-1)(c-1) = (3-1)(3-1) = 4$. From Table VII, Appendix A, $\chi^2_{.05} = 9.48773 = 9.48773$. The rejection region is $\chi^2 > 9.48773$.

Since the observed value of the test statistic falls in the rejection region $(\chi^2 = 54.14 > 9.48773)$, H_0 is rejected. There is sufficient evidence to indicate a dependence between rows and columns at $\alpha = .05$.

b. No, the analysis remains identical.

c. Yes, the assumptions on the sampling differ.

d. The percentages are in the table below.

		Column		
	1	**2**	**3**	**Totals**
Row 1	$\dfrac{20}{50} \times 100\% = 40\%$	$\dfrac{20}{90} \times 100\% = 22.2\%$	$\dfrac{10}{110} \times 100\% = 9.1\%$	$\dfrac{50}{250} \times 100\% = 20\%$
2	$\dfrac{10}{50} \times 100\% = 20\%$	$\dfrac{20}{90} \times 100\% = 22.2\%$	$\dfrac{70}{110} \times 100\% = 63.6\%$	$\dfrac{100}{250} \times 100\% = 40\%$
3	$\dfrac{20}{50} \times 100\% = 40\%$	$\dfrac{50}{90} \times 100\% = 55.6\%$	$\dfrac{30}{110} \times 100\% = 37.3\%$	$\dfrac{100}{250} \times 100\% = 40\%$

e. The graph is:

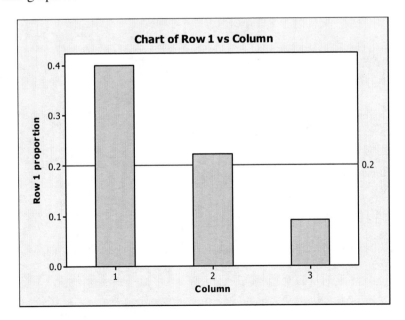

If the rows and columns are independent, then the proportion of observations for each column should be relatively the same. In this graph, the percentages are quite different, supporting the results of the test in part **a**.

f.   The graph for row 2 is:

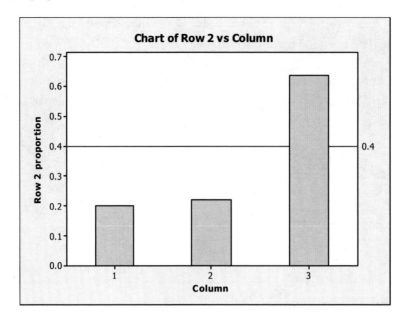

The percentages are quite different, supporting the results of the test in part **a**.

g.   The graph for row 3 is:

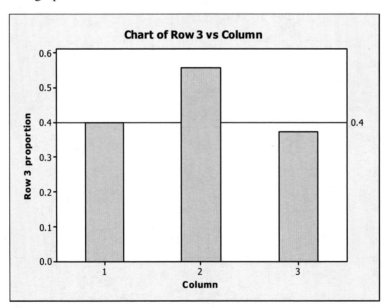

The percentages are quite different, supporting the results of the test in part **a**.

13.49 a.   The qualitative variable of interest in the study is location of major sports franchises. There are three levels of the variable – downtown, central city, and suburban areas.

b.   Let p_1 = proportion of facilities in downtown, p_2 = proportion if facilities in central cities, and p_3 = proportion of facilities in suburban areas. To determine whether the proportions of major sports facilities in downtown, central city, and suburban areas are the same in 1997 as they were in 1985, we test:

H_0: $p_1 = .40, p_2 = .30, p_3 = .30$
H_a: At least one of the probabilities differs from the hypothesized value

c. If H_0 is true, we would expect the following number of sports facilities in the 3 locations:

$E_1 = n\,p_{1,0} = 113\,(.40) = 45.2$

$E_2 = n\,p_{2,0} = 113\,(.30) = 33.9$

$E_3 = n\,p_{3,0} = 113\,(.30) = 33.9$

d. The test statistic is $\chi^2 = \sum \dfrac{[n_i - E_i]^2}{E_i} = \dfrac{(58 - 45.2)^2}{45.2} + \dfrac{(26 - 33.9)^2}{33.9} + \dfrac{(29 - 33.9)^2}{33.9} = 6.17$

e. Using Table VII, Appendix A, with df $= k - 1 = 3 - 1 = 2$, $P(\chi^2 > 5.99147) = .05$ and $P(\chi^2 > 7.37776) = .025$. Thus, $.025 < P(\chi^2 > 6.17) < .05$.

Since the p-value is less than α ($p < .05$), H_0 is rejected. There is sufficient evidence to indicate the proportions of major sports facilities in downtown, central city, and suburban areas are different in 1997 from what they were in 1985 at $\alpha = .05$.

13.51 a. The 2×2 contingency table is:

LLD

Genetic Trait	Yes	No	Total
Yes	21	15	36
No	150	306	456
Total	171	321	492

b. Some preliminary calculations are:

$\hat{E}_{11} = \dfrac{R_1 C_1}{n} = \dfrac{36(171)}{492} = 12.51$ $\qquad$ $\hat{E}_{12} = \dfrac{R_1 C_2}{n} = \dfrac{36(321)}{492} = 23.49$

$\hat{E}_{21} = \dfrac{R_2 C_1}{n} = \dfrac{456(171)}{492} = 158.49$ $\qquad$ $\hat{E}_{22} = \dfrac{R_2 C_2}{n} = \dfrac{456(321)}{492} = 297.51$

To determine if the genetic trait occurs at a higher rate in LDD patients than in the controls, we test:

H_0: Genetic trait and LDD group are independent
H_a: Genetic trait and LDD group are dependent

The test statistic is

$\chi^2 = \sum \sum \dfrac{[n_{ij} - \hat{E}_{ij}]^2}{\hat{E}_{ij}} = \dfrac{(21 - 12.51)^2}{12.51} + \dfrac{(15 - 23.49)}{23.49} + \dfrac{(150 - 158.49)^2}{158.49}$

$+ \dfrac{(306 - 297.51)^2}{297.51} = 9.52$

The rejection region requires $\alpha = .01$ in the upper tail of the χ^2 distribution with df = $(r-1)(c-1) = (2-1)(2-1) = 1$. From Table VII, Appendix A, $\chi^2_{.01} = 6.63490$. The rejection region is $\chi^2 > 6.63490$.

Since the observed value of the test statistic falls in the rejection region $(\chi^2 = 9.52 > 6.63490)$, H_0 is rejected. There is sufficient evidence to indicate that the genetic trait occurs at a higher rate in LDD patients than in the controls at $\alpha = .01$.

c. To construct a bar graph, we will first compute the proportion of LDD/Control patients that have the genetic trait. Of the LLD patients, 21 / 171 = .123 have the trait. For the Controls, 15 / 321 = .043. For the row totals, 36 / 492 = .073. The bar graph is:

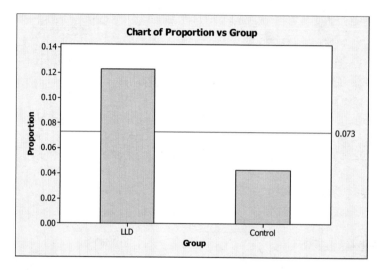

Because the bars are not close to being the same height, it indicates that the genetic trait appears at a higher rate among LLD patients than among the controls. This supports the conclusion of the test in part **b**.

13.53   Some preliminary calculations are:

$$\hat{E}_{11} = \frac{R_1C_1}{n} = \frac{62(106)}{275} = 23.90 \qquad \hat{E}_{12} = \frac{R_1C_2}{n} = \frac{62(112)}{275} = 25.25$$

$$\hat{E}_{13} = \frac{R_1C_3}{n} = \frac{62(57)}{275} = 12.85 \qquad \hat{E}_{21} = \frac{R_2C_1}{n} = \frac{59(106)}{275} = 22.74$$

$$\hat{E}_{22} = \frac{R_2C_2}{n} = \frac{59(112)}{275} = 24.03 \qquad \hat{E}_{23} = \frac{R_2C_3}{n} = \frac{59(57)}{275} = 12.23$$

$$\hat{E}_{31} = \frac{R_3C_1}{n} = \frac{59(106)}{275} = 22.74 \qquad \hat{E}_{32} = \frac{R_3C_2}{n} = \frac{59(112)}{275} = 24.03$$

$$\hat{E}_{33} = \frac{R_3C_3}{n} = \frac{59(57)}{275} = 12.23 \qquad \hat{E}_{41} = \frac{R_4C_1}{n} = \frac{95(106)}{275} = 36.62$$

$$\hat{E}_{42} = \frac{R_4C_2}{n} = \frac{95(112)}{275} = 38.69 \qquad \hat{E}_{43} = \frac{R_4C_3}{n} = \frac{95(57)}{275} = 19.69$$

To determine if political strategy of ethnic groups depends on world region, we test:

H_0: Political strategy and world region are independent
H_a: Political strategy and world region are dependent

The test statistic is

$$\chi^2 = \sum\sum \frac{[n_{ij} - \hat{E}_{ij}]^2}{\hat{E}_{ij}} = \frac{(24 - 23.90)^2}{23.90} + \frac{(31 - 25.25)^2}{25.25} + \frac{(7 - 12.85)^2}{12.85} + \frac{(32 - 22.74)^2}{22.74}$$

$$+ \frac{(23 - 24.03)^2}{24.03} + \frac{(4 - 12.23)^2}{12.23} + \frac{(11 - 22.74)^2}{22.74} + \frac{(22 - 24.03)^2}{24.03}$$

$$+ \frac{(26 - 12.23)^2}{12.23} + \frac{(39 - 36.62)^2}{36.62} + \frac{(36 - 38.69)^2}{38.69} + \frac{(20 - 19.69)^2}{19.69} = 35.409$$

The rejection region requires $\alpha = .10$ in the upper tail of the χ^2 distribution with df = $(r-1)(c-1) = (4-1)(3-1) = 6$. From Table VII, Appendix A, $\chi^2_{.10} = 10.6446$. The rejection region is $\chi^2 > 10.6446$.

Since the observed value of the test statistic falls in the rejection region ($\chi^2 = 35.409 > 10.6446$), H_0 is rejected. There is sufficient evidence to indicate that the political strategy of the ethnic groups depends on world region at $\alpha = .10$.

To graph the data, we will first compute the percent of observations in each category of Political Strategy for each World Region. To do this, we divide each cell frequency by the row total and then multiply by 100%. Using MINITAB, a graph of the data is:

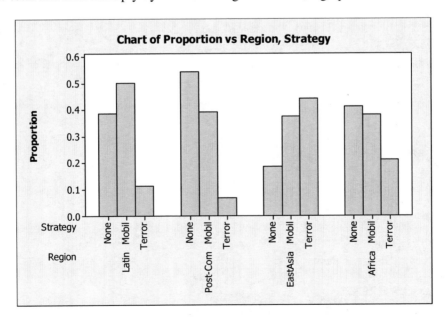

The graph supports the results of the test above. The percents in each level of Political Strategy differ for the different values of World Region.

13.55   a.   Some preliminary calculations:

$$E_1 = np_{1,0} = 28(.25) = 7$$
$$E_2 = np_{2,0} = 28(.25) = 7$$
$$E_3 = np_{3,0} = 28(.25) = 7$$
$$E_4 = np_{4,0} = 28(.25) = 7$$

To determine if the true proportions of intellectually disabled elderly patients in each of the hearing loss categories differ, we test:

H_0: $p_1 = p_2 = p_3 = p_4 = .25$
H_a: At least one p_i differs from .25, $i = 1, 2, 3, 4$

The test statistic is $\chi^2 = \sum \dfrac{[n_i - E_i]^2}{E_i} = \dfrac{(7-7)^2}{7} + \dfrac{(7-7)^2}{7} + \dfrac{(9-7)^2}{7} + \dfrac{(5-7)^2}{7} = 1.143$

The rejection region requires $\alpha = .05$ in the upper tail of the χ^2 distribution with df = $k - 1 = 4 - 1 = 3$. From Table VII, Appendix A, $\chi^2_{.05} = 7.81473 = 7.81473$. The rejection region is $\chi^2 > 7.81473$.

Since the observed value of the test statistic does not fall in the rejection region $(\chi^2 = 1.143 \not> 7.81473)$, H_0 is not rejected. There is insufficient evidence to indicate that at least one of the true proportions of intellectually disabled elderly patients in each of the hearing loss categories differs at $\alpha = .05$.

b. $\hat{p} = \dfrac{x}{n} = \dfrac{5}{28} = .179$

For confidence coefficient .90; $\alpha = .10$ and $\alpha / 2 = .10 / 2 = .05$. From Table IV, Appendix A, $z_{.05} = 1.645$. The confidence interval is

$$\hat{p} \pm z_{.05}\sqrt{\dfrac{\hat{p}\hat{q}}{n}} \Rightarrow .179 \pm 1.645\sqrt{\dfrac{(.179)(1-.179)}{28}} \Rightarrow .179 \pm .119 \Rightarrow (.060, .298)$$

We are 90% confident that the true proportion of disabled elderly patients with severe hearing loss is between .060 and .298.

13.57   First, the values of the variables will be defined.

1.   The variable WAR has values 1, 2, and 9. 1 = Support, 2 = Oppose, and 9 = Don't know or refused to answer.

2.   The variable INTERNET has values 1, 2, and 9. 1 = Yes, 2 = No, and 9 = Don't know or refused to answer.

3.   The variable PARTY has values 1, 2, 3, 4, 5, and 9. 1 = Republican, 2 = Democrat, 3 = Independent, 4 = No preference, 5 = Other, and 9 = Don't know or refused to answer.

4.   The variable VET has values 1, 2, 3, 4, and 9. 1 = Yes, I have, 2 = Yes, other household member, 3 = Yes, both, 4 = No, and 9 = Don't know or refused to answer.

5.   The variable IDEO has values 1, 2, 3, 4, 5, and 9. 1 = Very conservative, 2 = Conservative, 3 = Moderate, 4 = Liberal, 5 = Very liberal, and 9 = Don't know or refused to answer.

6.   The variable RACE has values 1, 2, 3, 4, 5, and 9. 1 = White, 2 = Black or African American, 3 = Asian or Pacific Islander, 4 = Mixed, 5 = Native American, 6 = Other and 9 = Don't know or refused to answer.

7.   The variable INCRANGE has values 1, 2, 3, 4, 5, 6, 7, 8, and 9. 1 = <$10,000, 2 = $10,000-under $20,000, 3 = $20,000-under $30,000, 4 = $30,000-under $40,000, 5 = $40,000-under $50,000, 6 = $50,000-under $75,000, 7 = $75,000 to under $100,000, 8 = Over $100,000, and 9 = Don't know or refused to answer.

8.   The variable CUMMUNIT has values 1, 2, 3, and 9. 1 = Urban, 2 = Suburban, 3 = Rural, and 9 = Don't know or refused to answer.

To do the analysis, all values of 9 were converted to missing values and are not included in the analysis. For all tests, $\alpha = .05$ was used.

WAR versus INTERNET. Using MINITAB, the results of the analyses are:

Tabulated statistics: War, Internet

```
Rows: War   Columns: Internet

              1        2  Missing      All

1           703      399        4     1102
          77.94    76.29        *    77.33

2           199      124        1      323
          22.06    23.71        *    22.67

Missing      27       38        0        *
              *        *        *        *

All         902      523        *     1425
         100.00   100.00        *   100.00

Cell Contents:       Count
                     % of Column


Pearson Chi-Square = 0.512, DF = 1, P-Value = 0.474
Likelihood Ratio Chi-Square = 0.510, DF = 1, P-Value = 0.475
```

To determine if Support of Iraq War and access of internet are dependent, we test:

H_0: Support of the Iraq War and access of internet are independent
H_a: Support of the Iraq War and access of internet are dependent

The test statistic is $\chi^2 = .512$ and the p-value is $p = .474$. Since the p-value is not less than $\alpha = .05$, H_0 is not rejected. There is insufficient evidence to indicate that Support of Iraq war and access of internet are dependent at $\alpha = .05$.

WAR versus PARTY. Since only 3 respondents selected "Other" I combined the "Other" group with the "No preference" group" so that the expected counts in all cells are at least 5. Using MINITAB, the results of the analyses are:

Tabulated statistics: War, Party

```
Rows: War    Columns: Party

               1        2        3        4   Missing      All

1            451      262      315       53       25      1081
           95.55    60.23    73.77    80.30        *     77.21

2             21      173      112       13        5       319
            4.45    39.77    26.23    19.70        *     22.79

Missing        8       27       21        5        4         *
               *        *        *        *        *         *

All          472      435      427       66        *      1400
          100.00   100.00   100.00   100.00        *    100.00

Cell Contents:      Count
                    % of Column
```

```
Pearson Chi-Square = 164.761, DF = 3, P-Value = 0.000
Likelihood Ratio Chi-Square = 189.289, DF = 3, P-Value = 0.000
```

To determine if Support of Iraq War and Party are dependent, we test:

H_0: Support of the Iraq War and Party are independent
H_a: Support of the Iraq War and Party are dependent

The test statistic is $\chi^2 = 164.761$ and the p-value is $p = 0.000$. Since the p-value is less than $\alpha = .05$, H_0 is rejected. There is sufficient evidence to indicate that Support of Iraq war and Party are dependent at $\alpha = .05$.

If you look at the column percents in the above table, we see that 95.55% of the Republicans support the Iraq War, while only 60.23% of Democrats support the Iraq War. In addition, 73.33% of the Independents support the war and 80.3% of those with no preference or other support the war.

WAR versus VET. Using MINITAB, the results of the analyses are:

Tabulated statistics: War, vet

```
Rows: War    Columns: vet

                1         2        3         4  Missing       All

1             209       237       27       631        2      1104
            84.27     76.45    77.14     75.66        *     77.37

2              39        73        8       203        1       323
            15.73     23.55    22.86     24.34        *     22.63

Missing         6        15        0        44        0         *
                *         *        *         *        *         *

All           248       310       35       834        *      1427
           100.00    100.00   100.00    100.00        *    100.00

Cell Contents:        Count
                      % of Column


Pearson Chi-Square = 8.295, DF = 3, P-Value = 0.040
Likelihood Ratio Chi-Square = 8.858, DF = 3, P-Value = 0.031
```

To determine if Support of Iraq War and Veteran status are dependent, we test:

H_0: Support of the Iraq War and Veteran status are independent
H_a: Support of the Iraq War and Veteran status are dependent

The test statistic is $\chi^2 = 8.295$ and the p-value is $p = 0.040$. Since the p-value is less than $\alpha = .05$, H_0 is rejected. There is sufficient evidence to indicate that Support of Iraq War and Veteran status are dependent at $\alpha = .05$.

If you look at the column percents in the above table, we see that 84.27% of the veterans support the Iraq War, while 76.45% of those who have a family member who is a veteran support the Iraq War. In addition, 77.14% of those who are vets and have a family member who is a vet support the war and 75.66% of those who are not veterans support the war.

WAR versus IDEO. Using MINITAB, the results of the analyses are:

Tabulated statistics: War, ideo

```
Rows: War    Columns: ideo

                1        2        3        4        5  Missing      All

1              52      453      415      102       21       63     1043
            75.36    89.88    78.75    52.85    34.43        *    77.03

2              17       51      112       91       40       13      311
            24.64    10.12    21.25    47.15    65.57        *    22.97

Missing         2       17       19       11        2       14        *
                *        *        *        *        *        *        *

All            69      504      527      193       61        *     1354
           100.00   100.00   100.00   100.00   100.00        *   100.00

Cell Contents:        Count
                      % of Column


Pearson Chi-Square = 174.386, DF = 4, P-Value = 0.000
Likelihood Ratio Chi-Square = 161.296, DF = 4, P-Value = 0.000
```

To determine if Support of Iraq War and Political views are dependent, we test:

H_0: Support of the Iraq War and Political views are independent
H_a: Support of the Iraq War and Political views are dependent

The test statistic is $\chi^2 = 174.386$ and the p-value is $p = 0.000$. Since the p-value is less than $\alpha = .05$, H_0 is rejected. There is sufficient evidence to indicate that Support of Iraq War and Political views are dependent at $\alpha = .05$.

If you look at the column percents in the above table, we see that 75.36% of the Very Conservatives support the Iraq War, while 89.88% of the Conservatives support the Iraq War. In addition, 78.75% of Moderates support the war, 52.85% of Liberals support the war, and 34.43% of the Very Liberals support the war.

WAR versus RACE. Because there were less than 24 respondents for the race categories Asian/Pacific Islander, Mixed, Native American, and Other, these 4 categories were combined. Using MINITAB, the results of the analyses are:

Tabulated statistics: War, Race

```
Rows: War    Columns: Race

                1        2        3   Missing      All

1             989       51       54        12     1094
            81.00    46.79    72.97         *    77.92

2             232       58       20        14      310
            19.00    53.21    27.03         *    22.08

Missing        50        9        4         2        *
               *        *        *         *        *

All          1221      109       74         *     1404
           100.00   100.00   100.00         *   100.00


Cell Contents:       Count
                     % of Column
```

```
Pearson Chi-Square = 69.181, DF = 2, P-Value = 0.000
Likelihood Ratio Chi-Square = 57.984, DF = 2, P-Value = 0.000
```

To determine if Support of Iraq War and Race are dependent, we test:

H_0: Support of the Iraq War and Race are independent
H_a: Support of the Iraq War and Race are dependent

The test statistic is $\chi^2 = 69.181$ and the p-value is $p = 0.000$. Since the p-value is less than $\alpha = .05$, H_0 is rejected. There is sufficient evidence to indicate that Support of Iraq War and Race are dependent at $\alpha = .05$.

If you look at the column percents in the above table, we see that 81.00% of the Whites support the Iraq War, while 46.79% of the Blacks support the Iraq War. In addition, 72.96% of the Others support the war.

WAR versus INCRANGE Using MINITAB, the results of the analyses are:

Tabulated statistics: War, incrange

```
Rows: War    Columns: incrange

                 1         2         3         4         5         6         7         8

1               37        79       122       112       103       196       132       134
             61.67     73.15     77.71     77.78     83.06     82.35     77.19     74.03

2               23        29        35        32        21        42        39        47
             38.33     26.85     22.29     22.22     16.94     17.65     22.81     25.97

Missing         12        10         4         9         2         5         2         3
                 *         *         *         *         *         *         *         *

All             60       108       157       144       124       238       171       181
            100.00    100.00    100.00    100.00    100.00    100.00    100.00    100.00

            Missing       All

1               191       915
                 *      77.35

2                56       268
                 *      22.65

Missing          18         *
                 *         *

All               *      1183
                 *    100.00

Cell Contents:        Count
                      % of Column


Pearson Chi-Square = 16.387, DF = 7, P-Value = 0.022
Likelihood Ratio Chi-Square = 15.684, DF = 7, P-Value = 0.028
```

To determine if Support of Iraq War and Income range are dependent, we test:

H_0: Support of the Iraq War and Income range are independent
H_a: Support of the Iraq War and Income range are dependent

The test statistic is $\chi^2 = 16.387$ and the p-value is $p = 0.022$. Since the p-value is less than $\alpha = .05$, H_0 is rejected. There is sufficient evidence to indicate that Support of Iraq War and Income range are dependent at $\alpha = .05$.

If you look at the column percents in the above table, we see that those in the income ranges of \$40,000 to under \$50,000 and from \$50,000 to under \$75,000 support the Iraq War the most. Those in the lowest income range and the highest income range tend to not support the Iraq War as much as those in the middle income brackets.

WAR versus COMMUNIT Using MINITAB, the results of the analyses are:

Tabulated statistics: War, communit

```
Rows: War    Columns: communit

              1        2        3      All

1           235      605      266     1106
          69.73    78.27    83.13    77.34

2           102      168       54      324
          30.27    21.73    16.88    22.66

Missing      15       35       15        *
             *        *        *        *

All         337      773      320     1430
         100.00   100.00   100.00   100.00

Cell Contents:      Count
                    % of Column
```

```
Pearson Chi-Square = 17.618, DF = 2, P-Value = 0.000
Likelihood Ratio Chi-Square = 17.309, DF = 2, P-Value = 0.000
```

To determine if Support of Iraq War and Community are dependent, we test:

H_0: Support of the Iraq War and Community are independent
H_a: Support of the Iraq War and Community are dependent

The test statistic is $\chi^2 = 17.618$ and the p-value is $p = 0.000$. Since the p-value is less than $\alpha = .05$, H_0 is rejected. There is sufficient evidence to indicate that Support of Iraq War and Community are dependent at $\alpha = .05$.

If you look at the column percents in the above table, we see that 69.73% of those who live in the urban areas support the Iraq War, while 78.27% of those who live in suburban areas support the Iraq War. In addition, 83.13% of those who live in Rural areas support the war.

13.59　a.　Some preliminary calculations are:

$E_1 = np_{1,0} = 85(.26) = 22.1$
$E_2 = np_{2,0} = 85(.30) = 25.5$
$E_3 = np_{3,0} = 85(.11) = 9.35$
$E_4 = np_{4,0} = 85(.14) = 11.9$
$E_5 = np_{5,0} = 85(.19) = 16.15$

To determine of probabilities differ from the hypothesized values, we test:

H_0: $p_1 = .26, p_2 = .30, p_3 = .11, p_4 = .14, p_5 = .19$
H_a: At least one of the probabilities differs from its hypothesized value.

The test statistic is

$$\chi^2 = \sum \frac{[n_i - E_i]^2}{E_i} = \frac{(32-22.1)^2}{22.1} + \frac{(26-25.5)^2}{25.5} + \frac{(15-9.35)^2}{9.35} + \frac{(6-11.9)^2}{11.9} + \frac{(6-16.15)^2}{16.15} = 17.16$$

The rejection region requires $\alpha = .05$ in the upper tail of the χ^2 distribution with $df = k - 1 = 5 - 1 = 4$. From Table VII, Appendix A, $\chi^2_{.05} = 9.48773$. The rejection region is $\chi^2 > 9.48773$.

Since the observed value of the test statistic falls in the rejection region $(\chi^2 = 17.16 > 9.48773)$, reject H_0. There is sufficient evidence to indicate the probabilities differ from their hypothesized values at $\alpha = .05$.

b. $\hat{p}_1 = \frac{32}{85} = .376$

For confidence coefficient .95, $\alpha = .05$ and $\alpha / 2 = .05 / 2 = .025$. From Table IV, Appendix A, $z_{.025} = 1.96$. The 95% confidence interval is:

$$\hat{p}_i \pm z_{.025}\sqrt{\frac{\hat{p}_i \hat{q}_i}{n}} \Rightarrow .376 \pm 1.96\sqrt{\frac{.376(.624)}{85}} \Rightarrow .376 \pm .103 \Rightarrow (.273, \ .479)$$

We are 95% confident that the true proportion of Avonex MS patients who are exacerbation-free during a 2-year period is between .273 and .479.

c.   From previous studies, it is known that 26% of the MS patients on a placebo experienced no exacerbations in a 2-year period. Since .26 does not fall in the 95% confidence interval, there is evidence that Avonex patients are more likely to have no exacerbations than the placebo patients at $\alpha = .05$.

13.61   Some preliminary calculations are:

$$\hat{E}_{11} = \frac{R_1 C_1}{n} = \frac{12(10)}{24} = 5 \qquad\qquad \hat{E}_{12} = \frac{R_1 C_2}{n} = \frac{12(14)}{24} = 7$$

$$\hat{E}_{21} = \frac{R_2 C_1}{n} = \frac{12(10)}{24} = 5 \qquad\qquad \hat{E}_{22} = \frac{R_2 C_2}{n} = \frac{12(14)}{24} = 7$$

To determine if a relationship exists between food choice and whether or not chickadees fed on gypsy moth eggs, we test:

H_0: Food choice and whether or not chickadees fed on gypsy moth eggs are independent
H_a: Food choice and whether or not chickadees fed on gypsy moth eggs are dependent

The test statistic is $\chi^2 = \sum\sum \frac{[n_{ij} - \hat{E}_{ij}]^2}{\hat{E}_{ij}} = \frac{(2-5)^2}{5} + \frac{(10-7)^2}{7} + \frac{(8-5)^2}{5} + \frac{(4-7)^2}{7} = 6.171$

The rejection region requires $\alpha = .10$ in the upper tail of the χ^2 distribution with df $= (r-1)(c-1) = (2-1)(2-1) = 1$. From Table VII, Appendix A, $\chi^2_{.10} = 2.70554$. The rejection region is $\chi^2 > 2.70554$.

Since the observed value of the test statistic falls in the rejection region $(\chi^2 = 6.171 > 2.70554)$, H_0 is rejected. There is sufficient evidence to indicate that a relationship exists between food choice and whether or not chickadees fed on gypsy moth eggs at $\alpha = .10$.

13.63   a.   Some preliminary calculations are:

$E_1 = np_{1,0} = 1,000(.50) = 500$ $E_2 = np_{2,0} = 1,000(.22) = 220$
$E_3 = np_{3,0} = 1,000(.11) = 110$ $E_4 = np_{4,0} = 1,000(.17) = 170$

To determine if the data disagree with the percentages reported by Nielson/NetRatings, we test:

H_0: $p_1 = .50$, $p_2 = .22$, $p_3 = .11$, and $p_4 = .17$
H_a: At least one p_i differs from it hypothesized value

The test statistic is

$$\chi^2 = \sum \frac{[n_i - E_i]^2}{E_i}$$

$$= \frac{[487-500]^2}{500} + \frac{[245-220]^2}{220} + \frac{[121-110]^2}{110} + \frac{[147-170]^2}{170} = 7.39$$

The rejection region requires $\alpha = .05$ in the upper tail of the χ^2 distribution with df $= k - 1 = 4 - 1 = 3$. From Table VII, Appendix A, $\chi^2_{.05} = 7.81473$. The rejection region is $\chi^2 > 7.81473$.

Since the observed value of the test statistic does not fall in the rejection region $(\chi^2 = 7.39 \not> 7.81473)$, H_0 is not rejected. There is insufficient evidence to indicate the data disagree with the percentages reported by Nielson/NetRatings at $\alpha = .05$.

b. $\hat{p}_1 = \dfrac{487}{1000} = .487$

For confidence coefficient .95, $\alpha = .05$ and $\alpha / 2 = .05 / 2 = .025$. From Table IV, Appendix A, $z_{.025} = 1.96$. The 95% confidence interval is:

$$\hat{p}_1 \pm z_{.025} \sqrt{\frac{\hat{p}_1 \hat{q}_1}{n}} \Rightarrow .487 \pm 1.96 \sqrt{\frac{.487(.513)}{1000}} \Rightarrow .487 \pm .031 \Rightarrow (.456, .518)$$

We are 95% confident that the true proportion of all Internet searches that use the Google Search engine is between .456 and .518.

13.65   a.   The observed frequencies for the 3 groups are:

Mail-only – 262
Internet-only – 43
Both – 135

Some preliminary calculations are:

$E_1 = E_2 = E_3 = np_{1,0} = 440(1/3) = 146.67$

To determine if the proportions of mail-only, internet-only and both users are different, we test:

H_0: $p_1 = p_2 = p_3 = 1/3$
H_a: At least one p_i differs from it hypothesized value

The test statistic is

$$\chi^2 = \sum \frac{[n_i - E_i]^2}{E_i} = \frac{[262 - 146.67]^2}{146.67} + \frac{[43 - 146.67]^2}{146.67} + \frac{[135 - 146.67]^2}{146.67} = 164.89$$

The rejection region requires $\alpha = .05$ in the upper tail of the χ^2 distribution with df = $k - 1 = 3 - 1 = 2$. From Table VII, Appendix A, $\chi^2_{.05} = 5.99147$. The rejection region is $\chi^2 > 5.99147$.

Since the observed value of the test statistic falls in the rejection region $(\chi^2 = 164.89 > 5.99147)$, H_0 is rejected. There is sufficient evidence to indicate the proportions of mail-only, internet-only and both users are different at $\alpha = .05$.

Using MINITAB, a graph of the categories is:

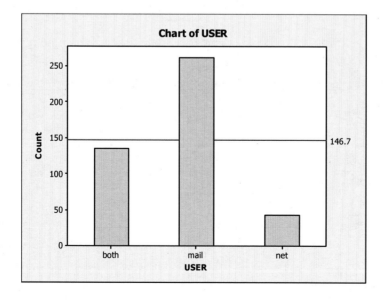

b.   **Type of User vs Gender**:

Using MINITAB, the contingency table is:

Tabulated statistics: USER, GENDER

```
Rows: USER    Columns: GENDER

          Female     Male      All

both         104       31      135
           77.04    22.96   100.00
           32.70    25.41    30.68

mail         178       84      262
           67.94    32.06   100.00
           55.97    68.85    59.55

net           36        7       43
           83.72    16.28   100.00
           11.32     5.74     9.77

All          318      122      440
           72.27    27.73   100.00
          100.00   100.00   100.00

Cell Contents:      Count
                    % of Row
                    % of Column


Pearson Chi-Square = 6.797, DF = 2, P-Value = 0.033
Likelihood Ratio Chi-Square = 7.105, DF = 2, P-Value = 0.029
```

To determine if coupon user type is related to gender, we test:

H_0: Type of coupon user and gender are independent
H_0: Type of coupon user and gender are dependent

From the printout, the test statistic is $\chi^2 = 6.797$ and the p-value is $p = .003$.

Since the observed p-value is less than $\alpha = .05$ ($p = .003 < .05$), H_0 is rejected. There is sufficient evidence to indicate the type of coupon user is related to gender at $\alpha = .05$.

Using MINITAB, a graph of the results is:

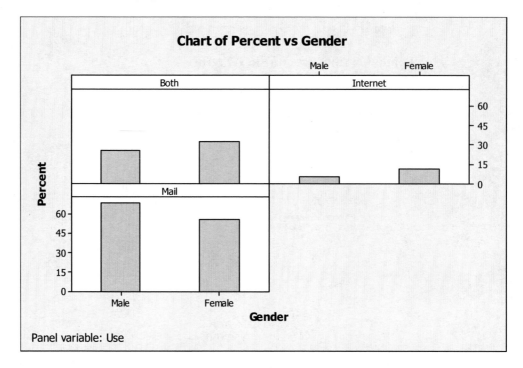

The majority of male and female users use coupons distributed through the mail. However, the proportion of male users who use coupons distributed through the mail is larger than the proportion of females. Very few users use coupons distributed via the Internet, but a higher proportion of females use coupons distributed via the Internet than males.

Type of User vs Education:

Using MINITAB, the contingency table is:

Tabulated statistics: USER, EDUC

```
Rows: USER   Columns: EDUC

            COLL     GRAD      HS     PROF      All

both          62       38      19       16      135
            45.93    28.15   14.07    11.85   100.00
            34.83    28.79   31.67    22.86    30.68

mail          96       85      34       47      262
            36.64    32.44   12.98    17.94   100.00
            53.93    64.39   56.67    67.14    59.55

net           20        9       7        7       43
            46.51    20.93   16.28    16.28   100.00
            11.24     6.82   11.67    10.00     9.77

All          178      132      60       70      440
            40.45    30.00   13.64    15.91   100.00
           100.00   100.00  100.00   100.00   100.00

Cell Contents:       Count
                     % of Row
                     % of Column


Pearson Chi-Square = 6.587, DF = 6, P-Value = 0.361
Likelihood Ratio Chi-Square = 6.786, DF = 6, P-Value = 0.341
```

To determine if coupon user type is related to education, we test:

H_0: Type of coupon user and education are independent
H_0: Type of coupon user and education are dependent

From the printout, the test statistic is $\chi^2 = 6.587$ and the p-value is $p = .361$.

Since the observed p-value is not less than $\alpha = .05$ ($p = .361 \not< .05$), H_0 is not rejected. There is insufficient evidence to indicate the type of coupon user is related to education at $\alpha = .05$.

Using MINITAB, a graph of the results is:

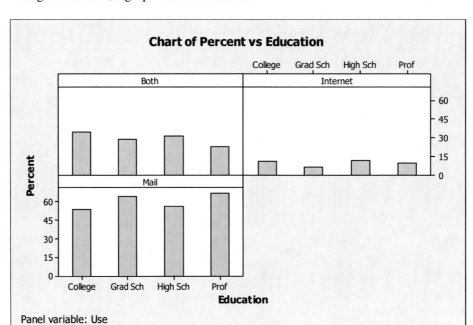

For each category of use, the proportions of each education level are about the same. This corresponds to the conclusion above that coupon use is not related to education level.

Type of User vs Work Status:

Using MINITAB, the contingency table is:

Tabulated statistics: USER, WORK

```
Rows: USER    Columns: WORK

                 1        2        3        4      All

both            90       13       17       15      135
             66.67     9.63    12.59    11.11   100.00
             33.71    25.00    33.33    21.43    30.68

mail           148       31       31       52      262
             56.49    11.83    11.83    19.85   100.00
             55.43    59.62    60.78    74.29    59.55

net             29        8        3        3       43
             67.44    18.60     6.98     6.98   100.00
             10.86    15.38     5.88     4.29     9.77

All            267       52       51       70      440
             60.68    11.82    11.59    15.91   100.00
            100.00   100.00   100.00   100.00   100.00

Cell Contents:        Count
                      % of Row
                      % of Column
```

```
Pearson Chi-Square = 11.687, DF = 6, P-Value = 0.069
Likelihood Ratio Chi-Square = 12.208, DF = 6, P-Value = 0.057

* NOTE * 1 cells with expected counts less than 5
```

To determine if coupon user type is related to work status, we test:

H_0: Type of coupon user and work status are independent
H_a: Type of coupon user and work status are dependent

From the printout, the test statistic is $\chi^2 = 11.687$ and the p-value is $p = .069$.

Since the observed p-value is not less than $\alpha = .05$ ($p = .069 \not< .05$), H_0 is not rejected. There is insufficient evidence to indicate the type of coupon user is related to work status at $\alpha = .05$.

Using MINITAB, a graph of the results is:

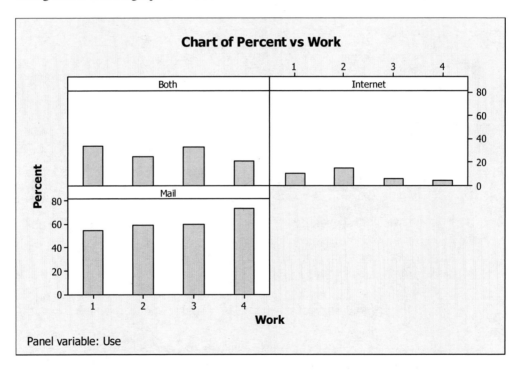

Again, for each category of use, the proportions of each work level are about the same. This corresponds to the conclusion above that coupon use is not related to work level.

Type of User vs Coupon satisfaction:

Using MINITAB, the contingency table is:

Tabulated statistics: USER, SATISF

```
Rows: USER    Columns: SATISF

              No      Some      Yes      All

both           3         9       123      135
            2.22      6.67     91.11   100.00
            8.57     11.25     37.85    30.68

mail          28        62       172      262
           10.69     23.66     65.65   100.00
           80.00     77.50     52.92    59.55

net            4         9        30       43
            9.30     20.93     69.77   100.00
           11.43     11.25      9.23     9.77

All           35        80       325      440
            7.95     18.18     73.86   100.00
          100.00    100.00    100.00   100.00

Cell Contents:        Count
                      % of Row
                      % of Column


Pearson Chi-Square = 30.418, DF = 4, P-Value = 0.000
Likelihood Ratio Chi-Square = 34.934, DF = 4, P-Value = 0.000

* NOTE * 1 cells with expected counts less than 5
```

To determine if coupon user type is related to coupon satisfaction, we test:

H_0: Type of coupon user and coupon satisfaction are independent
H_0: Type of coupon user and coupon satisfaction are dependent

From the printout, the test statistic is $\chi^2 = 30.418$ and the *p*-value is *p* = .000.

Since the observed *p*-value is less than $\alpha = .05$ (*p* = .000 < .05), H_0 is rejected. There is sufficient evidence to indicate the type of coupon user is related to coupon satisfaction at $\alpha = .05$.

Using MINITAB, a graph of the results is:

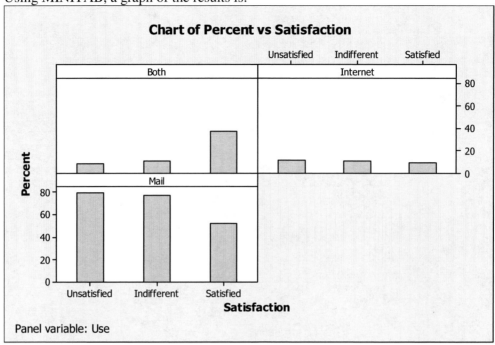

Again, very few users use coupons distributed via the Internet. The proportions of the different levels of satisfaction are about the same for this category of coupon use. For coupons distributed through the mail, the proportions of users who were unsatisfied or indifferent are larger than the proportion satisfied. However, for users of coupons distributed through both the mail and the Internet, the proportion of users who were satisfied is greater then the proportions of users unsatisfied or indifferent.

13.67   a.

$$\chi^2 = \sum \frac{[n_i - E_i]^2}{E_i} = \frac{(26-23)^2}{23} + \frac{(146-136)^2}{136} + \frac{(361-341)^2}{341} + \frac{(143-136)^2}{136} + \frac{(13-23)^2}{23}$$
$$= 9.647$$

   b.   From Table VII, Appendix A, with df = 5, $\chi^2_{.05} = 11.0705$

   c.   No. Since the observed value of the test statistic does not fall in the rejection region ($\chi^2 = 9.647 \not> 11.0705$), H_0 is not rejected. There is insufficient evidence to indicate the salary distribution is nonnormal for $\alpha = .05$.

   d.   The p-value = $P(\chi^2 \geq 9.647)$.

       Using Table VII, Appendix A, with df = 5, $.05 < P(\chi^2 \geq 9.647) < .10$.

13.69   Using MINITAB, the results of the analyses are:

Tabulated statistics: Candidate, Time

```
Using frequencies in Fr


Rows: Candidate    Columns: Time

               1        2        3        4        5        6      All

Coppin        55       51      109       98       88      104      505
           13.89    13.56    13.37    13.98    14.08    14.04    13.82
            54.7     52.0    112.6     96.9     86.4    102.4    505.0

Montes       133      117      255      211      186      227     1129
           33.59    31.12    31.29    30.10    29.76    30.63    30.90
           122.4    116.2    251.8    216.6    193.1    229.0   1129.0

Smith        208      208      451      392      351      410     2020
           52.53    55.32    55.34    55.92    56.16    55.33    55.28
           218.9    207.9    450.5    387.5    345.5    409.6   2020.0

All          396      376      815      701      625      741     3654
          100.00   100.00   100.00   100.00   100.00   100.00   100.00
           396.0    376.0    815.0    701.0    625.0    741.0   3654.0

Cell Contents:      Count
                    % of Column
                    Expected count

Pearson Chi-Square = 2.284, DF = 10, P-Value = 0.994
Likelihood Ratio Chi-Square = 2.272, DF = 10, P-Value = 0.994
```

To determine if time period and votes received by the candidates are independent, we test:

H_0: Votes received by candidates and time period are independent
H_a: Votes received by candidates and time period are dependent

From the printout, the test statistic is $\chi^2 = 2.284$ and the p-value is $p = .994$.

Since the p-value is so large, H_0 would not be rejected for any reasonable value of α. There is insufficient evidence to indicate that the votes received by the candidates are dependent on the time period. In other words, there is no evidence that the percentages of votes received by the candidates change over time. If one looks at the column percentages (the second number in each cell above), the values are very similar for each time period.

A graph of the column percents is:

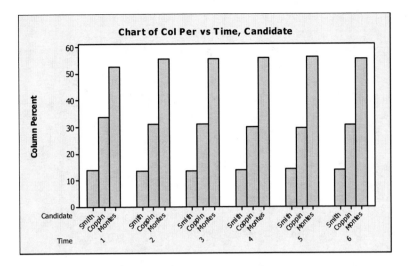

As one can see from the graph, the percent of votes received by the candidates are essentially the same for each time period.

All of the above indicates that the probability of a candidate receiving votes is independent of the time period. However, this does not necessarily imply a rigged election. If each time period is considered a random sample of voters, then the percentage of votes received by each candidate should be similar to the actual percentage of voters who favor each candidate.

Nonparametric Statistics

14.1   The sign test is preferred to the *t*-test when the population from which the sample is selected is not normal.

14.3   a. $P(x \geq 6) = 1 \leq P(x-5) = 1 - .937 = .063$

      b. $P(x \geq 5) = 1 \leq P(x-4) = 1 - .500 = .500$

      c. $P(x \geq 8) = 1 \leq P(x-7) = 1 - .996 = .004$

      d. $P(x \geq 10) = 1 \leq P(x-9) = 1 - .849 = .151$

$$\mu = np = 15(.5) = 7.5 \text{ and } \sigma = \sqrt{npq} = \sqrt{15(.5)(.5)} = 1.9365$$

$$P(x \geq 10) \approx P\left(z \geq \frac{(10-.5)-7.5}{1.9365} \right) = P(z \geq 1.03) = .5 - .3485 = .1515$$

      e. $P(x \geq 15) = 1 \leq P(x \leq 14) = 1 - .788 = .212$

$$\mu = np = 25(.5) = 12.5 \text{ and } \sigma = \sqrt{npq} = \sqrt{25(.5)(.5)} = 2.5$$

$$P(x \geq 15) \approx P\left(z \geq \frac{(15-.5)-12.5}{2.5} \right) = P(z \geq .80) = .5 - .2881 = .2119$$

14.5   To determine if the median is greater than 80, we test:

H_0: $\eta = 80$
H_a: $\eta > 80$

The test statistic is S = number of measurements greater than $80 = 16$.

The *p*-value = $P(x \geq 16)$ where x is a binomial random variable with $n = 25$ and $p = .5$. From Table II,

$$p\text{-value} = P(x \geq 16) = 1 - P(x \leq 15) = 1 - .885 = .115$$

Since the *p*-value = $.115 > \alpha = .10$, H_0 is not rejected. There is insufficient evidence to indicate the median is greater than 80 at $\alpha = .10$.

We must assume the sample was randomly selected from a continuous probability distribution.

Note: Since $n \geq 10$, we could use the large-sample approximation.

14.7   a.   To determine if the median molar depth of all cheek teeth differs from 15, we test:

$$H_0: \eta = 15$$
$$H_a: \eta \neq 15$$

b.   The sign test is appropriate because the data do not appear to be normally distributed and the sample size is only $n = 18$.

c.   The test statistic is $S = 14$.

d.   From the printout, the p-value is $p = .0309$. Since the p-value is less than α ($p = .0309 < .05$), H_0 is rejected. There is sufficient evidence to indicate the median molar depth of all cheek teeth differs from 15 at $\alpha = .05$.

14.9   a.   To determine if female college students have a median emotional scale score higher than 2.8, we test:

$$H_0: \eta = 2.8$$
$$H_a: \eta > 2.8$$

b.   The test statistics is $S = \{$number of observations greater than 2.8$\} = 21$. Because $n = 30$, we will use the large sample approximation:

$$z = \frac{(S - .5) - .5n}{.5\sqrt{n}} = \frac{(21 - .5) - .5(30)}{.5\sqrt{30}} = 2.01$$

c.   The p-value $= P(z \geq 2.01) = .5 - .4778 = .0222$ using Table II, Appendix A.

d.   Since the p-value is not less than α ($p = .0222 \not< .01$), H_0 is not rejected. There is insufficient evidence to indicate that female college students have a median emotional scale score higher than 2.8 at $\alpha = .01$.

14.11   a.   If the data do not come from a normal distribution, then the test performed in Exercise 8.35 is not valid.

b.   If the data are not normal, then we could use the sign test for the median with

$$H_0: \eta = 95$$
$$H_a: \eta \neq 95$$

c. $S_1 = \{$Number of observations less than 95$\} = 5$ and
$S_2 = \{$Number of observations greater than 95$\} = 2$.

The test statistic is $S = $ larger of S_1 and $S_2 = 5$.

d.   The p-value $= 2P(x \geq 5)$ where x is a binomial random variable with $n = 7$ and $p = .5$. From Table II,

$$p\text{-value} = 2P(x \geq 5) = 2(1 - P(x \leq 4)) = 2(1 - .773) = .454$$

e.   In Exercise 8.65, we used $\alpha = .10$. Since the p-value is not less than α ($p = .454 \nless .10$), H_0 is not rejected. There is insufficient evidence to indicate the median trap spacing measurement differs from 95 at $\alpha = .10$.

14.13   To determine if 50% of the ingots have a freckle index of 10 or higher, we test:

H_o: $\eta = 10$
H_a: $\eta > 10$

S = number of measurements greater than 10 = 10.

The p-value = $P(x \geq 10)$ where x is a binomial random variable with $n = 17$ and $p = .5$. (One observation is eliminated because it is equal to 10.) Using MINITAB, the p-value = $P(x \geq 10) = 1 - P(x \leq 9) = 1 - .6855 = .3145$.

Since the p-value is greater than $\alpha = .01$ ($p = .3145 > .01$), H_0 is not rejected. There is insufficient evidence to indicate that 50% of the ingots have a freckle index of 10 or higher at $\alpha = .01$.

14.15   To determine if the population median skidding distance is more than 400 meters, we test:

H_0: $\eta = 400$
H_a: $\eta > 400$

The test statistic is S = number of measurements greater than 400 = 8.

The p-value = $P(x \geq 8)$ where x is a binomial random variable with $n = 19$ and $p = .5$. (One observation is eliminated because it is equal to 400.) Using MINITAB,

$$P(x \geq 8) = 1 - P(x \leq 7) = 1 - .180 = .820$$

Since the p-value is greater than $\alpha = .10$ ($p = .820 > .10$), H_0 is not rejected. There is insufficient evidence to indicate the population median skidding distance is more than 400 meters at $\alpha = .10$.

14.17   The statement "If the rank sum for sample 1 is much larger than the rank sum for sample 2 when $n_1 = n_2$, then the distribution of population 1 is likely to be shifted to the right of the distribution of population 2." is true.

14.19   a.   The test statistic is T_1, the rank sum of population A (because $n_1 < n_2$).

The rejection region is $T_1 \leq 41$ or $T_1 \geq 71$, from Table XII, Appendix A, with $n_1 = 7$, $n_2 = 8$, and $\alpha = .10$.

b.   The test statistic is T_1, the rank sum of population A (because $n_1 = n_2$).

The rejection region is $T_1 \geq 50$, from Table XII, Appendix A, with $n_1 = 6$, $n_2 = 6$, and $\alpha = .05$.

c.   The test statistic is T_1, the rank sum of population A (because $n_1 < n_2$).

The rejection region is $T_1 \leq 43$, from Table XII, Appendix A, with $n_1 = 7$, $n_2 = 10$, and $\alpha = .025$.

d.   Since $n_1 = n_2 = 20$, the test statistic is $z = \dfrac{T_1 - \dfrac{n_1(n_1 + n_2 + 1)}{2}}{\sqrt{\dfrac{n_1 n_2 (n_1 + n_2 + 1)}{12}}}$

The rejection region is $z < -z_{\alpha/2}$ or $z > z_{\alpha/2}$. For $\alpha = .05$ and $\alpha / 2 = .05 / 2 = .025$, $z_{.025} = 1.96$ from Table IV, Appendix A. The rejection region is $z < -1.96$ or $z > 1.96$.

14.21   To determine if distribution A is shifted to the left of distribution B, we test:

H_0: The probability distributions of treatments A and B are identical
H_a: The probability distribution of treatment A lies to the left of that for treatment B

The test statistic is

$$z = \dfrac{T_1 - \dfrac{n_2(n_1 + n_2 + 1)}{2}}{\sqrt{\dfrac{n_1 n_2 (n_1 + n_2 + 1)}{12}}} = \dfrac{173 - \dfrac{15(15 + 15 + 1)}{2}}{\sqrt{\dfrac{15(15)(15 + 15 + 1)}{12}}} = \dfrac{-59.5}{24.1091} = -2.47$$

The rejection region requires $\alpha = .05$ in the lower tail of the z distribution. From Table IV, Appendix A, $z_{.05} = 1.645$. The rejection region is $z < -1.645$.

Since the observed value of the test statistic falls in the rejection region ($-2.47 < -1.645$), H_0 is rejected. There is sufficient evidence to conclude that distribution A is shifted to the left of distribution B for $\alpha = .05$.

14.23

Sample from Population 1 (A)	Rank	Sample from Population 2 (B)	Rank
15	13	5	2.5
10	8.5	12	10.5
12	10.5	9	6.5
16	14	9	6.5
13	12	8	4.5
8	4.5	4	1
	$T_1 = 62.5$	5	2.5
		10	8.5
			$T_2 = 42.5$

a.   To determine if there is a shift in location for the two distributions, we test:

H_0: The two sampled populations have identical probability distributions
H_a: The probability distribution for population A is shifted to the left or to the right of that for B

The test statistic is $T_1 = 62.5$ since sample A has the smallest number of measurements.

The null hypothesis will be rejected if $T_1 \leq T_L$ or $T_1 \geq T_U$ where T_L and T_U correspond to $\alpha = .05$ (two-tailed), $n_1 = 6$ and $n_2 = 8$. From Table XII, Appendix A, $T_L = 29$ and $T_U = 61$.

   Reject H_0 if $T_1 \leq 29$ or $T_1 \geq 61$.

Since the observed value of the test statistic falls in the rejection region ($T_1 = 62.5 \geq 61$), H_0 is rejected. There is sufficient evidence to indicate population A is shifted to the left or right of population B at $\alpha = .05$.

b. H_0: The two sampled populations have identical probability distributions
H_a: The probability distribution for population A is shifted to the right of population B

The test statistic remains $T_1 = 62.5$.

The null hypothesis will be rejected if $T_1 \geq T_U$ where T_U corresponds to $\alpha = .05$ (one-tailed), $n_1 = 6$ and $n_2 = 8$. From Table XII, Appendix A, $T_U = 58$.

   Reject H_0 if $T_1 \geq 58$.

Since the observed value of the test statistic falls in the rejection region ($T_1 = 62.5 \geq 58$), H_0 is rejected. There is sufficient evidence to indicate population A is shifted to the right of population B at $\alpha = .05$.

14.25   a.   The appropriate test would be the Wilcoxon Rank Sum Test.

b.   To determine if the low-handicapped golfers tend to have higher X-factors than high-handicapped golfers, we test:

H_0: The distributions of the X-factors for the two groups of golfers are identical

H_a: The distribution of the X-factors for the low-handicapped golfers is shifted to the right of that for the high-handicapped golfers

c.   Let T_1 = rank sum of low-handicapped golfers and T_2 = rank sum of high-handicapped golfers. Since $n_1 = 8$ and $n_2 = 7$, the test statistic is T_2. The rejection region is $T_2 \leq T_L$ where $\alpha = .05$ (one-tailed). From Table XII, Appendix A, $T_L = 41$. The rejection region is $T_2 \leq 41$.

d.   Since the p-value is not less than α ($p = .487 \nless .05$), H_0 is not rejected. There is insufficient evidence to indicate the low-handicapped golfers tend to have higher X-factors than high-handicapped golfers at $\alpha = .05$.

14.27   a.   To determine if the distribution of FNE scores for bulimic students is shifted above the corresponding distribution for female students with normal eating habits, we test:

H_0: The distribution of FNE scores for bulimic students is the same as the corresponding distribution for female students with normal eating habits

H_a: The distribution of FNE scores for bulimic students is shifted above the corresponding distribution for female students with normal eating habits

    b.   The data ranked are:

Bulimic Students	Rank	Normal Students	Rank
21	21.5	13	8.5
13	8.5	6	1
10	4.5	16	13.5
20	19.5	13	8.5
25	25	8	3
19	17	19	17
16	13.5	23	23
21	21.5	18	15
24	24	11	6
13	8.5	19	17
14	11	7	2
		10	4.5
	$T_1 = 174.5$	15	12
		20	19.5
			$T_2 = 150.5$

    c.   The sum of the ranks of the 11 FNE scores for bulimic students is $T_1 = 174.5$.

    d.   The sum of the ranks of the 14 FNE scores for normal students is $T_2 = 150.5$.

    e.   The test statistic is $z = \dfrac{T_1 - \dfrac{n_1(n_1 + n_2 + 1)}{2}}{\sqrt{\dfrac{n_1 n_2 (n_1 + n_2 + 1)}{12}}} = \dfrac{174.5_1 - \dfrac{11(11 + 14 + 1)}{2}}{\sqrt{\dfrac{11(14)(11 + 14 + 1)}{12}}} = 1.72$

The rejection region requires $\alpha = .10$ in the upper tail of the z distribution. From Table IV, Appendix A, $z_{.10} = 1.28$. The rejection region is $z > 1.28$.

    f.   Since the observed value of the test statistic falls in the rejection region ($z = 1.72 > 1.28$), H_0 is rejected. There is sufficient evidence to indicate that the distribution of FNE scores for bulimic students is shifted above the corresponding distribution for female students with normal eating habits at $\alpha = .10$.

14.29   First, we rank the data:

	Honey Dosage			DM Dosage			
Honey	Rank	Honey	Rank	DM	Rank	DM	Rank
12	52	12	52	4	5	6	11.5
11	44	8	21	6	11.5	8	21
15	65	12	52	9	28	12	52
11	44	9	28	4	5	12	52
10	37	11	44	7	16	4	5
13	59.5	15	65	7	16	12	52
10	37	10	37	7	16	13	59.5
4	5	15	65	9	28	7	16
15	65	9	28	12	52	10	37
16	68	13	59.5	10	37	13	59.5
9	28	8	21	11	44	9	28
14	62	12	52	6	11.5	4	5
10	37	10	37	3	1	4	5
6	11.5	8	21	4	5	10	37
10	37	9	28	9	28	15	65
8	21	5	9	12	52	9	28
11	44	12	52	7	16		
12	52						
			$T_1 = 1440.5$				$T_2 = 905.5$

To determine if honey is a preferable treatment for the cough and sleep difficulty, we test:

H_0: The distributions of the cough improvement scores for the two groups of children
are identical

H_a: The distribution of the cough improvement scores for the honey dosage children is
shifted to the right of that for the DM dosage children

The test statistic is

$$z = \frac{T_1 - \frac{n_1(n_1 + n_2 + 1)}{2}}{\sqrt{\frac{n_1 n_2 (n_1 + n_2 + 1)}{12}}} = \frac{1440.5 - \frac{35(35 + 33 + 1)}{2}}{\sqrt{\frac{35(33)(35 + 33 + 1)}{12}}} = \frac{233}{81.4939} = 2.86$$

The rejection region requires $\alpha = .05$ in the upper tail of the z distribution. From Table IV, Appendix A, $z_{.05} = 1.645$. The rejection region is $z > 1.645$.

Since the observed value of the test statistic falls in the rejection region ($z = 2.86 > 1.645$), H_0 is rejected. There is sufficient evidence to indicate that honey is a preferable treatment for the cough and sleep difficulty over DM dosage at $\alpha = .05$.

14.31   a.   Since the data are probably not from a normal distribution the Wilcoxon rank sum test should be used.

b.   To determine if the relational intimacy scores for the participants in the CMC group are lower than the relational intimacy scores for the participants in the FTF group, we test:

H_0: The probability distributions of those in the CMC group and those in the FTF group are identical

H_a: The probability distribution of the CMC group is shifted to the left of that for the FTF group

c.   The rejection region requires $\alpha = .10$ in the lower tail of the z distribution. From Table IV, Appendix A, $z_{.10} = 1.28$. The rejection region is $z < -1.28$.

d.   First, we rank the data:

CMC Group	Rank	FTF Group	Rank
4	34.5	5	
3	13.5	4	34.5
3	13.5	4	34.5
4	34.5	4	34.5
3	13.5	3	13.5
3	13.5	3	13.5
3	13.5	3	13.5
3	13.5	4	34.5
4	34.5	3	13.5
4	34.5	3	13.5
3	13.5	3	13.5
4	34.5	3	13.5
3	13.5	4	34.5
3	13.5	4	34.5
2	2	4	34.5
4	34.5	4	34.5
2	2	4	34.5
4	34.5	3	13.5
5	47	3	13.5
4	34.5	3	13.5
4	34.5	4	34.5
4	34.5	4	34.5
5	47	2	2
3	13.5	4	34.5
	$T_1 = 578$		$T_2 = 598$

The test statistic is

$$z = \frac{T_1 - \dfrac{n_2(n_1 + n_2 + 1)}{2}}{\sqrt{\dfrac{n_1 n_2 (n_1 + n_2 + 1)}{12}}} = \frac{578 - \dfrac{24(24 + 24 + 1)}{2}}{\sqrt{\dfrac{24(24)(24 + 24 + 1)}{12}}} = \frac{-10}{48.4974} = -.21$$

Since the observed value of the test statistic does not fall in the rejection region ($z = -.21 \not< -1.28$), H_0 is not rejected. There is insufficient evidence to indicate the relational intimacy scores for the participants in the CMC group are lower than the relational intimacy scores for the participants in the FTF group at $\alpha = .10$.

14.33   a.   Since the data can take on a very limited number of values, the data are probably not from a normal distribution.

b.   Using SAS, the output for the Wilcoxon test is:

```
                        The NPAR1WAY Procedure

              Wilcoxon Scores (Rank Sums) for Variable GSHWS
                        Classified by Variable COND

                        Sum of      Expected      Std Dev          Mean
        COND      N     Scores      Under H0      Under H0         Score
        -------------------------------------------------------------------
        ATIPS    98     8039.0      11123.0      476.284142     82.030612
        TIPS    128    17612.0      14528.0      476.284142    137.593750

                   Average scores were used for ties.


                       Wilcoxon Two-Sample Test

            Statistic                 8039.0000

            Normal Approximation
            Z                           -6.4741
            One-Sided Pr <  Z           <.0001
            Two-Sided Pr > |Z|          <.0001

            t Approximation
            One-Sided Pr <  Z           <.0001
            Two-Sided Pr > |Z|          <.0001

        Z includes a continuity correction of 0.5.


                        Kruskal-Wallis Test

            Chi-Square                 41.9273
            DF                               1
            Pr > Chi-Square             <.0001
```

To determine if there was a difference in the level of involvement in science homework assignments between TIPS and ATIPS students, we test:

H_0: The distributions of the involvement scores for science homework for the two groups have the same location

H_a: The distributions of the involvement scores for science homework for the TIPS students is shifted to the right or left of that for the ATIPS students

From the printout, the test statistic is $z = -6.4741$ and the p-value is $p < .0001$. Since the p-value is less than α ($p < .0001 < .05$), H_0 is rejected. There is sufficient evidence to indicate that there is a difference in the level of involvement in science homework assignments between TIPS and ATIPS students at $\alpha = .05$.

c. Using SAS, the output for the Wilcoxon test is:

```
                   The NPAR1WAY Procedure

         Wilcoxon Scores (Rank Sums) for Variable MTHHWS
                    Classified by Variable COND

                     Sum of      Expected       Std Dev        Mean
    COND      N      Scores      Under H0       Under H0       Score
    -------------------------------------------------------------------
    ATIPS    98      10936.0      11123.0      472.485707    111.591837
    TIPS    128      14715.0      14528.0      472.485707    114.960938

                Average scores were used for ties.


                    Wilcoxon Two-Sample Test

          Statistic                    10936.0000

          Normal Approximation
          Z                               -0.3947
          One-Sided Pr <  Z                0.3465
          Two-Sided Pr > |Z|               0.6930

          t Approximation
          One-Sided Pr <  Z                0.3467
          Two-Sided Pr > |Z|               0.6934

       Z includes a continuity correction of 0.5.


                      Kruskal-Wallis Test

          Chi-Square                       0.1566
          DF                                    1
          Pr > Chi-Square                  0.6923
```

To determine if there was a difference in the level of involvement in mathematics homework assignments between TIPS and ATIPS students, we test:

H_0: The distributions of the involvement scores for mathematics homework for the two groups have the same location

H_a: The distributions of the involvement scores for mathematics homework for the TIPS students is shifted to the right or left of that for the ATIPS students

From the printout, the test statistic is $z = -.3947$ and the p-value is $p = .6930$. Since the p-value is not less than α ($p = .6930 > .05$), H_0 is not rejected. There is insufficient evidence to indicate that there is a difference in the level of involvement in mathematics homework assignments between TIPS and ATIPS students at $\alpha = .05$.

d.   Using SAS, the output for the Wilcoxon test is:

```
                    The NPAR1WAY Procedure

            Wilcoxon Scores (Rank Sums) for Variable LAHWS
                      Classified by Variable COND

                      Sum of      Expected      Std Dev        Mean
          COND    N    Scores      Under H0      Under H0      Score
          -------------------------------------------------------------
          ATIPS   98   10496.0     11123.0       464.844149    107.102041
          TIPS   128   15155.0     14528.0       464.844149    118.398438

                 Average scores were used for ties.


                      Wilcoxon Two-Sample Test

              Statistic              10496.0000

              Normal Approximation
              Z                         -1.3478
              One-Sided Pr <  Z          0.0889
              Two-Sided Pr > |Z|         0.1777

              t Approximation
              One-Sided Pr <  Z          0.0895
              Two-Sided Pr > |Z|         0.1791

          Z includes a continuity correction of 0.5.


                       Kruskal-Wallis Test

              Chi-Square                 1.8194
              DF                              1
              Pr > Chi-Square            0.1774
```

To determine if there was a difference in the level of involvement in language arts homework assignments between TIPS and ATIPS students, we test:

H_0: The distributions of the involvement scores for language arts homework for the two groups have the same location

H_a: The distributions of the involvement scores for language arts homework for the TIPS students is shifted to the right or left of that for the ATIPS students

From the printout, the test statistic is $z = -1.3478$ and the p-value is $p = .1777$. Since the p-value is not less than α ($p = .1777 > .05$), H_0 is not rejected. There is insufficient evidence to indicate that there is a difference in the level of involvement in language arts homework assignments between TIPS and ATIPS students at $\alpha = .05$.

14.35   We assume that the probability distribution of differences is continuous so that the absolute differences will have unique ranks. Although tied (absolute) differences can be assigned average ranks, the number of ties should be small relative to the number of observations to assure validity.

14.37   a.   The hypotheses are:

H_0:   The two sampled populations have identical probability distributions

H_a: The probability distributions for population A is shifted to the right of that for population B

b.   Some preliminary calculations are:

Treatment A	B	Difference A – B	Rank of Absolute Difference
54	45	9	5
60	45	15	10
98	87	11	7
43	31	12	9
82	71	11	7
77	75	2	2.5
74	63	11	7
29	30	−1	1
63	59	4	4
80	82	−2	2.5
			$T_- = 3.5$

The test statistic is $T_- = 3.5$

The rejection region is $T_- \le 8$, from Table XIII, Appendix A, with $n = 10$ and $\alpha = .025$.

Since the observed value of the test statistic falls in the rejection region ($T_- = 3.5 \le 8$), H_0 is rejected. There is sufficient evidence to indicate the responses for A tend to be larger than those for B at $\alpha = .025$.

14.39 a. H_0: The two sampled populations have identical probability distributions
H_a: The probability distribution for population A is located to the right of that for population B

b.   The test statistic is:

$$z = \frac{T_+ - \frac{n(n+1)}{4}}{\sqrt{\frac{n(n+1)(2n+1)}{24}}} = \frac{354 - \frac{30(30+1)}{4}}{\sqrt{\frac{30(30+1)(60+1)}{24}}} = \frac{121.5}{48.6184} = 2.499$$

The rejection region requires $\alpha = .05$ in the upper tail of the z distribution. From Table IV, Appendix A, $z = 1.645$. The rejection region is $z > 1.645$.

Since the observed value of the test statistic falls in the rejection region ($z = 2.499 > 1.645$), H_0 is rejected. There is sufficient evidence to indicate population A is located to the right of that for population B at $\alpha = .05$.

c.   The p-value $= P(z \ge 2.499) = .5 - .4938 = .0062$ (using Table IV, Appendix A).

d.   The necessary assumptions are:

1.   The sample of differences is randomly selected from the population of differences.
2.   The probability distribution from which the sample of paired differences is drawn is continuous.

14.41   a.   Since two measurements were obtained from each patient (before and after), the measurements are not independent of each other. Each "before" measurement is paired with an "after" measurement.

      b.   The distributions for both before and after treatment are skewed to the right and the spread of the distribution before treatment is much larger than the spread after treatment. The distributions are not normal.

      c.   Since the p-value is so small ($p < .0001$), H_0 would be rejected. There is sufficient evidence to indicate the ichthyotherapy was effective in treating psoriasis for any reasonable value of α.

14.43   a.   From the printout, the test statistic is $z = -4.638$.

      b.   To determine if handling a museum object has a positive impact on a sick patient's well being, we test:

            H_0: The distributions of patient's health status are identical before and after handling museum objects
            H_a: The distribution of patient's health status after handling museum objects is shifted to the right of that before

        The test statistic is $z = -4.638$ and the p-value is $p = .000/2 = .000$.

        Since the p-value is less than α ($p < .000 < .01$), H_0 is rejected. There is sufficient evidence to indicate handling a museum object has a positive impact on a sick patient's well-being at $\alpha = .01$.

14.45   a.   To determine if the distributions of the FSI scores for good and average readers differ in location, we test:

            H_0: The distribution of the FSI scores of good readers is identical to the distribution of the FSI scores of average readers

            H_a: The distribution of the of the FSI scores of good readers is shifted to the right or left of the distribution of the FSI scores of average readers

b, c.

STRATEGY	FSI Scores		Difference	Rank of Absolute Difference
	Good Readers	Average Readers		
Word meaning	.38	.32	.06	2
Words in context	.29	.25	.04	1
Literal comprehension	.42	.25	.17	5
Draw inference from single string	.60	.26	.34	8
Draw inference from multiple string	.45	.31	.14	4
Interpretation of metaphor	.32	.14	.18	6.5
Find salient or main idea	.21	.03	.18	6.5
Form judgment	.73	.80	−.07	3

    d.   Positive rank sum $T_+ = 33$. Negative rank sum $T_- = 3$. The test statistic is the smaller of of T_+ and T_- and is $T_- = 3$.

    e.   The rejection region is $T_- \le 4$, from Table XIII, Appendix A, with $n = 8$ and $\alpha = .05$ (two-tails).

    f.   Since the observed value of the test statistic falls in the rejection region ($T_- = 3 \le 4$), H_0 is rejected. There is sufficient evidence to indicate the distributions of the FSI scores for good and average readers differ in location at $\alpha = .05$.

14.47   Some preliminary calculations:

Teacher	Pre-test	Post-test	Difference *Pre-Post*	Rank of Absolute Differences
1	53	74	-21	9
2	73	80	-7	8
3	70	94	-24	10
4	72	78	-6	6.5
5	77	78	-1	1.5
6	81	84	-3	5
7	73	71	2	3.5
8	87	88	-1	1.5
9	61	63	-2	3.5
10	76	83	-6	6.5
			Positive rank sum $T_+ = 3.5$	

To determine if the program was effective, we test:

    H_0: The distribution of racial tolerance before the program has the same location as the distribution of racial tolerance after program

    H_a: The distribution of racial tolerance before the program is shifted to the left of that for racial tolerance after the program

The test statistic is $T_+ = 3.5$.

Reject H_0 if $T_+ \leq T_0$ where T_0 is based on $\alpha = .01$ and $n = 10$ (one-tailed):

Reject H_0 if $T_+ \leq 5$ (from Table XIII, Appendix A)

Since the observed value of the test statistic falls in the rejection region ($T_+ = 3.5 \leq 5$), H_0 is rejected. There is sufficient evidence to indicate that the program was effective at $\alpha = .01$.

14.49   Some preliminary calculations are:

| | | Change in Transverse Strain | | | |
Day	Change in Temp.	Field Meas.	3D Model	Diff.	Rank of Absolute Difference
Oct. 24	-6.3	-58	-52	-6	3
Dec. 3	13.2	69	59	10	5
Dec. 15	3.3	35	32	3	2
Feb. 2	-14.8	-32	-24	-8	4
Mar. 25	1.7	-40	-39	-1	1
May 24	-.2	-83	-71	-12	6
					$T_+ = 7$

To determine if there is a shift in the change in transverse strain distributions between field measurements and the 3D model, we test:

H_0: The change in transverse strain distribution for field measurements is identical to the change in transverse strain distribution for the 3D model

H_a: The change in transverse strain distribution for field measurements is shifted to the right or left of the change in transverse strain distribution for the 3D model

The test statistic is T = smaller of T_- and T_+ which is $T_+ = 7$.

The rejection region is $T_+ \leq 1$, from Table XIII, Appendix A, with $n = 6$ and $\alpha = .05$ (two-tails).

Since the observed value of the test statistic does not fall in the rejection region ($T_+ = 7 \not\leq 1$), H_0 is not rejected. There is insufficient evidence to indicate there is a shift in the change in transverse strain distributions between field measurements and the 3D model at $\alpha = .05$.

14.51   Some preliminary calculations are:

Patient	Before Thickness	After Thickness	Difference	Rank of Absolute difference
1	11.0	11.5	−0.5	6
2	4.0	6.4	−2.4	20
3	6.3	6.1	0.2	3
4	12.0	10.0	2.0	15.5
5	18.2	14.7	3.5	24
6	9.2	7.3	1.9	14
7	7.5	6.1	1.4	12
8	7.1	6.4	0.7	8.5
9	7.2	5.7	1.5	13
10	6.7	6.5	0.2	3
11	14.2	13.2	1.0	10
12	7.3	7.5	−0.2	3
13	9.7	7.4	2.3	18.5
14	9.5	7.2	2.3	18.5
15	5.6	6.3	−0.7	8.5
16	8.7	6.0	2.7	21
17	6.7	7.3	−0.6	7
18	10.2	7.0	3.2	23
19	6.6	5.3	1.3	11
20	11.2	9.0	2.2	17
21	8.6	6.6	2.0	15.5
22	6.1	6.3	−0.2	3
23	10.3	7.2	3.1	22
24	7.0	7.2	−0.2	3
25	12.0	8.0	4.0	25

Negative rank sum $T_- = 50.5$
Positive rank sum $T_+ = 274.5$

To determine if the treatment for tendonitis tends to reduce the thickness of tendons, we test:

H_0: The distribution of the thickness before treatment has the same location as the distribution of the thickness after the treatment

H_a: The distribution of the thickness before the treatment is shifted to the right of that for the thickness after the treatment

The test statistics is T = smaller of T_- and T_+ which is $T_- = 50.5$.

Reject H_0 if $T_- \leq T_0$ where T_0 is based on $\alpha = .10$ and $n = 25$ (one-tailed).

Reject H_0 if $T_- \leq 101$ (From Table XIII, Appendix A)

(Note: There is no value for $\alpha = .10$ for a one-tailed test. However, if we reject H_0 for $\alpha = .05$, we would also reject H_0 for $\alpha = .10$.)

Since the observed value of the test statistic falls in the rejection region $(T_- = 50.5 \leq 101)$, H_0 is rejected. There is sufficient evidence to indicate the treatment for tendonitis tends to reduce the thickness of tendons at $\alpha = .10$.

14.53 The distribution of H is approximately a χ^2 if the null hypothesis is true and if the sample sizes, n_j, for each of the k distributions is more than 5.

14.55   a.   A completely randomized design was used.

   b.   The hypotheses are:

   H_0: The three probability distributions are identical
   H_a: At least two of the three probability distributions differ in location

   c.   The rejection region requires $\alpha = .01$ in the upper tail of the χ^2 distribution with df = $k - 1 = 3 - 1 = 2$. From Table VII, Appendix A, $\chi^2_{.01} = 9.21034$. The rejection region is $H > 9.21034$.

   d.   Some preliminary calculations are:

I		II		III	
Observation	**Rank**	**Observation**	**Rank**	**Observation**	**Rank**
34	5.5	24	2	72	14
56	10	18	1	101	20
65	12	27	3	91	19
59	11	41	7	76	16
82	18	34	5.5	80	17
70	13	42	8	75	15
45	9	33	4		
	$R_A = 78.5$		$R_B = 30.5$		$R_C = 101$

$$\bar{R}_A = \frac{R_A}{n_A} = \frac{78.5}{7} = 11.214 \qquad \bar{R}_B = \frac{R_B}{n_B} = \frac{30.5}{7} = 4.357$$

$$\bar{R}_C = \frac{R_C}{n_C} = \frac{101}{6} = 16.833 \qquad \bar{R} = \frac{1}{2}(n+1) = \frac{1}{2}(20+1) = 10.5$$

The test statistic is

$$H = \frac{12}{n(n+1)} \sum n_j (\bar{R}_j - \bar{R})^2$$

$$= \frac{12}{20(20+1)} \left[7(11.214 - 10.5)^2 + 7(4.357 - 10.5)^2 + 6(16.833 - 10.5)^2 \right] = 14.525$$

Since the observed value of the test statistic falls in the rejection region ($H = 14.5252 > 9.21034$), H_0 is rejected. There is sufficient evidence to indicate at least two of the three probability distributions differ in location at $\alpha = .01$.

14.57   a.   To determine if the probability distributions of interaction lengths differ among the three groups, we test:

   H_0: The probability distributions of interaction lengths for the three groups are identical
   H_a: At least two of the three probability distributions differ in location

b. The rejection region requires $\alpha = .05$ in the upper tail of the χ^2 distribution with $df = k-1$ $= 3 - 1 = 2$. From Table VII, Appendix A, $\chi^2_{.05} = 5.99147$. The rejection region is $H > 5.99147$.

c. The test statistic is $H = 1.1$ and the p-value is $p = .60$. Since the observed value of the test statistic does not fall in the rejection region ($H = 1.1 \not> 5.99147$), H_0 is not rejected. There is insufficient evidence to indicate the probability distributions of interaction lengths differ among the three groups at $\alpha = .05$.

14.59   a – d. The ranks and sums of ranks appear in the table:

Scopolamine	Rank	Placebo	Rank	No Drug	Rank
5	2.5	8	13	8	13
8	13	10	20.5	9	17
8	13	12	26.5	11	23.5
6	6.5	10	20.5	12	26.5
6	6.5	9	17	11	23.5
6	6.5	7	10	10	20.5
6	6.5	9	17	12	26.5
8	13	10	20.5	12	26.5
6	6.5				
4	1				
5	2.5				
6	6.5				
	$R_1 = 84$		$R_2 = 145$		$R_3 = 177$

e. $\bar{R}_1 = \dfrac{R_1}{n_1} = \dfrac{84}{12} = 7$ $\qquad \bar{R}_2 = \dfrac{R_2}{n_2} = \dfrac{145}{8} = 18.125$ $\qquad \bar{R}_3 = \dfrac{R_3}{n_3} = \dfrac{177}{8} = 22.125$

$$\bar{R} = \frac{1}{2}(n+1) = \frac{1}{2}(28+1) = 14.5$$

$$H = \frac{12}{n(n+1)} \sum n_j (\bar{R}_j - \bar{R})^2$$

$$= \frac{12}{28(28+1)} \left[12(7-14.5)^2 + 8(18.125 - 14.5)^2 + 8(22.125 - 14.5)^2 \right] = 18.403$$

f. To determine if the distributions of the number of word pairs recalled for the three groups have different locations, we test:

H_0: The three probability distributions are identical
H_a: At least two of the three probability distributions differ in location

The test statistic is $H = 18.403$.

The rejection region requires $\alpha = .05$ in the upper tail of the χ^2 distribution with df = $k - 1 = 3 - 1 = 2$. From Table VII, Appendix A, $\chi^2_{.05} = 5.99147$. The rejection region is $H > 5.99147$.

Since the observed value of the test statistic falls in the rejection region ($H = 18.403 > 5.99147$), H_0 is rejected. There is sufficient evidence to indicate the distributions of the number of word pairs recalled for the three groups have different locations at $\alpha = .05$.

g.   Re-ranking the data for only groups 1 and 2, we get:

Scopolamine	Rank	Placebo	Rank
5	2.5	8	12.5
8	12.5	10	18
8	12.5	12	20
6	6.5	10	18
6	6.5	9	15.5
6	6.5	7	10
6	6.5	9	15.5
8	12.5	10	18
6	6.5		
4	1		
5	2.5		
6	6.5		
	$T_1 = 82.5$		$T_2 = 127.5$

To determine if the distributions of the number of word pairs recalled for the Scopolamine group is shifted to the left of that for the Placebo group, we test:

H_0: The two probability distributions are identical
H_a: The probability distribution of the Scopolamine group is shifted to the left of that for the Placebo group

The test statistic is the T_2 because $n_2 < n_1$.

There is no table value for this problem since $n_1 = 12$ and the largest n in Table XII is 10. We will use the normal approximation even though both sample sizes are not greater than or equal to 10 ($n_2 = 8$).

The test statistic is $z = \dfrac{T_1 - \dfrac{n_1(n_1 + n_2 + 1)}{2}}{\sqrt{\dfrac{n_1 n_2 (n_1 + n_2 + 1)}{12}}} = \dfrac{82.5 - \dfrac{12(12 + 8 + 1)}{2}}{\sqrt{\dfrac{12(8)(12 + 8 + 1)}{12}}} = -3.36$

The rejection region requires $\alpha = .05$ in the lower tail of the z distribution. From Table IV, Appendix A, $z_{.05} = 1.645$. The rejection region is $z < -1.645$.

Since the observed value of the test statistic falls in the rejection region ($z = -3.36 < -1.645$), H_0 is rejected. There is sufficient evidence to indicate the distributions of the number of word pairs recalled for the Scopolamine group is shifted to the left of that for the Placebo group at $\alpha = .05$.

14.61 a. Some preliminary calculations are:

VHR	Rank	VRD	Rank	Control	Rank
-20	4	-12	6	51	20
-56	1	63	21	21	17
-34	3	12	15	8	14
0	11	-7	8.5	0	11
16	16	29	18	4	13
0	11				
-14	5				
-7	8.5				
-44	2				
43	19				
-11	7				
	$R_1 = 87.5$		$R_2 = 68.5$		$R_3 = 75$

$$\bar{R}_1 = \frac{R_1}{n_1} = \frac{87.5}{11} = 7.955, \quad \bar{R}_2 = \frac{R_2}{n_2} = \frac{68.5}{5} = 13.7, \quad \bar{R}_3 = \frac{R_3}{n_3} = \frac{75}{5} = 15,$$

$$\bar{R} = \frac{n+1}{2} = \frac{21+1}{2} = 11$$

To determine if the probability distributions of differences in pain intensity levels differ in location among the three treatments, we test:

H_0: The probability distributions of differences in pain intensity levels for the three groups are identical

H_a: At least two of the three probability distributions differ in location

The test statistic is

$$H = \frac{12}{n(n+1)} \sum n_j (\bar{R}_j - \bar{R})^2$$

$$= \frac{12}{21(21+1)} \left(11(7.955 - 11)^2 + 5(13.7 - 11)^2 + 5(15 - 11)^2 \right) = 5.674$$

The rejection region requires $\alpha = .05$ in the upper tail of the χ^2 distribution with $df = k - 1 = 3 - 1 = 2$. From Table VII, Appendix A, $\chi^2_{.05} = 5.99147$. The rejection region is $H > 5.99147$.

Since the observed value of the test statistic does not fall in the rejection region ($H = 5.674 \not> 5.99147$), H_0 is not rejected. There is insufficient evidence to indicate the probability distributions of differences in pain intensity levels differ in location among the three treatments at $\alpha = .05$.

b.   Some preliminary calculations are:

VHR	Rank	VRD+ Control	Rank
-20	4	-12	6
-56	1	63	21
-34	3	12	15
0	11	-7	8.5
16	16	29	18
0	11	51	20
-14	5	21	17
-7	8.5	8	14
-44	2	0	11
43	19	4	13
-11	7		
	$T_1 = 87.5$		$T_2 = 143.5$

To determine if the probability distributions of differences in pain intensity levels differ in location between the two treatments, we test:

H_0:   The probability distributions of differences in pain intensity levels for the two groups are identical

H_a: The probability distribution of differences in pain intensity levels for the VRH group is shifted to the right or left of that for the VRD and Control groups combined

The test statistic is

$$z = \frac{T_1 - \frac{n_1(n_1 + n_2 + 1)}{2}}{\sqrt{\frac{n_1 n_2(n_1 + n_2 + 1)}{12}}} = \frac{87.5 - \frac{11(11+10+1)}{2}}{\sqrt{\frac{11(10)(11+10+1)}{12}}} = \frac{-33.5}{14.2009} = -2.36$$

The rejection region requires $\alpha / 2 = .05 / 2 = .025$ in each tail of the z distribution. From Table IV, Appendix A, $z_{.025} = 1.96$. The rejection region is $z < -1.96$ or $z > 1.96$.

Since the observed value of the test statistic falls in the rejection region ($z = -2.36 < -1.96$), H_0 is rejected. There is sufficient evidence to indicate the probability distributions of differences in pain intensity levels differ in location between the VHR group and the combination of the VRD and control group at $\alpha = .05$.

14.63 Some preliminary calculations are:

AR	Rank	AC	Rank	A	Rank	P	Rank
.51	38	.50	36	.16	11.5	.58	41.5
.58	41.5	.30	21	.10	7.5	.12	9
.52	39	.47	33.5	.20	15	.62	44
.47	33.5	.36	25	.29	20	.43	29.5
.61	43	.39	26	-.14	3	.26	19
.00	4	.22	18	.18	13	.50	36
.32	23	.20	15	-.35	1	.44	31
.53	40	.21	17	.31	22	.20	15
.50	36	.15	10	.16	11.5	.42	28
.46	32	.10	7.5	.04	6	.43	29.5
.34	<u>24</u>	.02	<u>5</u>	-.25	<u>2</u>	.40	<u>27</u>
	$R_1 = 354$		$R_2 = 214$		$R_3 = 112.5$		$R_4 = 309.5$

$$\bar{R}_1 = \frac{R_1}{n_1} = \frac{354}{11} = 32.182 \qquad \bar{R}_2 = \frac{R_2}{n_2} = \frac{214}{11} = 19.455 \qquad \bar{R}_3 = \frac{R_3}{n_3} = \frac{112.5}{11} = 10.227$$

$$\bar{R}_4 = \frac{R_4}{n_4} = \frac{309.5}{11} = 28.136 \qquad \bar{R} = \frac{1}{2}(n+1) = \frac{1}{2}(44+1) = 22.5$$

To determine if the distributions of the task scores differ in location among the 4 groups, we test:

H_0: The distributions of the task scores for the 4 groups are identical
H_a: At least two of the four distributions differ in location

The test statistic is

$$H = \frac{12}{n(n+1)} \sum n_j (\bar{R}_j - \bar{R})^2$$

$$= \frac{12}{44(44+1)} [11(32.182 - 22.5)^2 + 11(19.455 - 22.5)^2 + 11(10.227 - 22.5)^2$$

$$+ 11(28.136 - 22.5)^2] = 19.027$$

The rejection region requires $\alpha = .05$ in the upper tail of the χ^2 distribution with df $= k - 1 = 4 - 1 = 3$. From Table VII, Appendix A, $\chi^2_{.05} = 7.81473$. The rejection region is $H > 7.81473$.

Since the observed value of the test statistic falls in the rejection region ($H = 19.027 > 7.81473$), H_0 is rejected. There is sufficient evidence to indicate the distributions of the task scores differ in location among the 4 groups at $\alpha = .05$. We can infer that the task scores are affected by the amount of alcohol and reward presented.

14.65   a.   Using MINITAB, the histograms of the three data sets are:

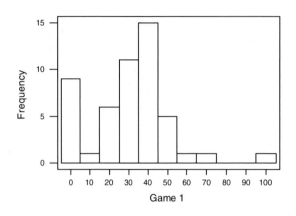

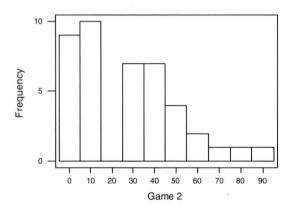

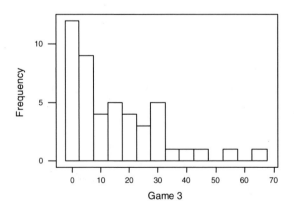

None of these three graphs look approximately normal. Thus, the assumption of normality necessary for ANOVA is likely violated.

b. Using MINITAB, the results of the Kruskal-Wallis analysis are:

```
Kruskal-Wallis Test on Percent

Game        N    Median    Ave Rank         Z
1          50    32.000        83.1      2.88
2          42    26.000        73.0      0.59
3          47     9.000        53.3     -3.49
Overall   139                  70.0

H = 13.61   DF = 2   P = 0.001
H = 13.66   DF = 2   P = 0.001  (adjusted for ties)
```

To determine if the percentages of names recalled are different for the three retrieval methods, we test:

H_0: The distributions of the three retrieval methods are identical
H_a: At least two of the three retrieval methods have probability distributions that differ in location

From the printout, the test statistic is $H = 13.66$ and the p-value is $p = .001$. Since the p-value is less than α ($p = .001 < .05$), H_0 is rejected. There is sufficient evidence to indicate that the percentages of names recalled are different for the three retrieval methods at $\alpha = .05$.

14.67 To correctly rank the data in a randomized block design, we would use b - for each block, rank the data across the treatments from smallest to largest.

14.69 a. The number of blocks, b, is 6.

b. H_0: The probability distributions for the four treatments are identical
H_a: At least two of the probability distributions differ in location

c. $\bar{R}_A = \dfrac{R_A}{b} = \dfrac{11}{6} = 1.833 \qquad \bar{R}_B = \dfrac{R_B}{b} = \dfrac{21}{6} = 3.5 \quad \bar{R}_C = \dfrac{R_C}{b} = \dfrac{21}{6} = 3.5$

$\bar{R}_D = \dfrac{R_D}{b} = \dfrac{7}{6} = 1.167 \qquad \bar{R} = \dfrac{k+1}{2} = \dfrac{4+1}{2} = 2.5$

The test statistic is

$$F_r = \frac{12b}{k(k+1)} \sum (\bar{R}_j - \bar{R})^2$$

$$= \frac{12(6)}{4(4+1)}\left((1.833-2.5)^2 + (3.5-2.5)^2 + (3.5-2.5)^2 + (1.167-2.5)^2\right) = 15.198$$

The rejection region requires $\alpha = .10$ in the upper tail of the χ^2 distribution with df = $k - 1 = 4 - 1 = 3$. From Table VII, Appendix A, $\chi^2_{.10} = 6.25139$. The rejection region is $F_r > 6.25139$.

Since the observed value of the test statistic falls in the rejection region (F_r = 15.198 > 6.25139), reject H_0. There is sufficient evidence to indicate a difference in the location for at least two of the four treatments at α = .10.

d.   The p-value = $P(F_r \geq 15.2) = P(\chi^2 \geq 15.198)$. With 3 degrees of freedom, F_r = 15.198 falls above $\chi^2_{.005}$; therefore, p-value < .005.

14.71 $R_1 = 16$ $R_2 = 7$ $R_3 = 23$ $R_4 = 14$

$$\bar{R}_1 = \frac{R_1}{b} = \frac{16}{6} = 2.667 \quad \bar{R}_2 = \frac{R_2}{b} = \frac{7}{6} = 1.167 \quad \bar{R}_3 = \frac{R_3}{b} = \frac{23}{6} = 3.833$$

$$\bar{R}_4 = \frac{R_4}{b} = \frac{14}{6} = 2.333 \quad \bar{R} = \frac{1}{2}(k+1) = \frac{1}{2}(4+1) = 2.5$$

To determine if at least two of the treatment probability distributions differ in location, we test:

H_0: The probability distributions of the four treatments are identical
H_a: At least two of the probability distributions differ in location

The test statistic is

$$F_r = \frac{12b}{k(k+1)}\sum(\bar{R}_j - \bar{R})^2$$

$$= \frac{12(6)}{(4)(4+1)}[(2.667-2.5)^2 + (1.167-2.5)^2 + (3.833-2.5)^2 + (2.333-2.5)^2] = 12.994$$

The rejection region requires α = .05 in the upper tail of the χ^2 distribution with df = $k-1$ = 4 − 1 = 3. From Table VII, Appendix A, $\chi^2_{.05} = 7.84173$. The rejection region is $F_r > 7.81473$.

Since the observed value of the test statistic falls in the rejection region (F_r = 12.994 > 7.81473), reject H_0. There is sufficient evidence to indicate a difference in the location for at least two of the probability distributions at α = .05.

14.73   a.   From the printout, the rank sums are R_1 = 25.5, R_2 = 11.0, R_3 = 18.5, and R_4 = 25.0.

b.   Some preliminary calculations are:

$$\bar{R}_1 = \frac{R_1}{b} = \frac{25.5}{8} = 3.1875, \quad \bar{R}_2 = \frac{R_2}{b} = \frac{11.0}{8} = 1.375, \quad \bar{R}_3 = \frac{R_3}{b} = \frac{18.5}{8} = 2.3125,$$

$$\bar{R}_4 = \frac{R_4}{b} = \frac{25}{8} = 3.125, \quad \bar{R} = \frac{k+1}{2} = \frac{4+1}{2} = 2.5$$

$$F_r = \frac{12b}{k(k+1)} \sum (\bar{R}_j - \bar{R})^2$$

$$= \frac{12(8)}{4(4+1)} \left((3.1875 - 2.5)^2 + (1.375 - 2.5)^2 + (2.3125 - 2.5)^2 + (3.125 - 2.5)^2 \right)$$

$$= 10.3875$$

From the printout, the value of the test statistic is $S = 10.39$. This agrees with the value computed above.

c. From the printout, the p-value is $p = .016$.

d. Since the p-value is less than α ($p = .016 < .05$), H_0 is rejected. There is sufficient evidence to indicate at least two of the distributions of heart rates differ in location among the 4 pre-slaughter phases at $\alpha = .05$.

14.75 a. The rejection region requires $\alpha = .01$ in the upper tail of the χ^2 distribution with df = $k - 1 = 3 - 1 = 2$. From Table VII, Appendix A, $\chi^2_{.01} = 9.21034$. The rejection region is $F_r > 9.21034$.

b. Since the observed value of the test statistic falls in the rejection region ($F_r = 19.16 > 9.21034$), H_0 is rejected. There is sufficient evidence to indicate the distributions of proportions of eye fixations on the interior mirror differ in location among the 3 treatments at $\alpha = .01$.

c. Since the observed value of the test statistic does not fall in the rejection region ($F_r = 7.80 \not> 9.21034$), H_0 is not rejected. There is insufficient evidence to indicate the distributions of proportions of eye fixations on the side mirror differ in location among the 3 treatments at $\alpha = .01$.

d. Since the observed value of the test statistic falls in the rejection region ($F_r = 20.67 > 9.21034$), H_0 is rejected. There is sufficient evidence to indicate the distributions of proportions of eye fixations on the speedometer differ in location among the 3 treatments at $\alpha = .01$.

14.77 Some preliminary calculations are:

Boxer	M1	Rank	R1	Rank	M5	Rank	R5	Rank
1	1243	1	1244	2	1291	4	1262	3
2	1147	2	1053	1	1169	3	1177	4
3	1247	1	1375	4	1309	2	1321	3
4	1274	2	1235	1	1290	4	1285	3
5	1177	2	1139	1	1233	3	1238	4
6	1336	2	1313	1	1366	4	1362	3
7	1238	1	1279	4	1275	3	1261	2
8	1261	2	1152	1	1289	4	1266	3
		$R_1 = 13$		$R_2 = 15$		$R_3 = 27$		$R_4 = 25$

$$\bar{R}_1 = \frac{R_1}{b} = \frac{13}{8} = 1.625, \qquad \bar{R}_2 = \frac{R_2}{b} = \frac{15}{8} = 1.875, \qquad \bar{R}_3 = \frac{R_3}{b} = \frac{27}{8} = 3.375,$$

$$\bar{R}_4 = \frac{R_4}{b} = \frac{25}{8} = 3.125, \qquad \bar{R} = \frac{k+1}{2} = \frac{4+1}{2} = 2.5$$

To compare the punching power means of the four interventions, we test:

H_0: The four probability distributions of punching power are the same
H_a: At least two of the probability distributions of punching power differ in location

The test statistic is

$$F_r = \frac{12b}{k(k+1)}\sum(\bar{R}_j - \bar{R})^2$$

$$= \frac{12(8)}{4(4+1)}\left((1.625-2.5)^2 + (1.875-2.5)^2 + (3.375-2.5)^2 + (3.125-2.5)^2\right) = 11.1$$

Since no α was given, we will use $\alpha = .05$. The rejection region requires $\alpha = .05$ in the upper tail of the χ^2 distribution with $df = k - 1 = 4 - 1 = 3$. From Table VII, Appendix A, $\chi^2_{.05} = 7.81473$. The rejection region is $F_r > 7.81473$.

Since the observed value of the test statistic falls in the rejection region ($F_r = 11.1 > 7.81473$), H_0 is rejected. There is sufficient evidence to indicate that the punching power means of the four interventions differ at $\alpha = .05$.

This is the same result that we obtained in Exercise 10.72.

14.79   The ranks and sums of ranks within each treatment are:

Week	Monday	Tuesday	Wednesday	Thursday	Friday
1	5	1	4	2	3
2	5	4	3	1	2
3	2.5	2.5	5	1	4
4	2	1	3.5	5	3.5
5	5	1	2	3	4
6	4	2	3	1	5
7	5	3.5	1.5	3.5	1.5
8	4	2	1	3	5
9	1	2	5	3	4
	$R_1 = 33.5$	$R_2 = 19$	$R_3 = 28$	$R_4 = 22.5$	$R_5 = 32$

$$\bar{R}_1 = \frac{R_1}{b} = \frac{33.5}{9} = 3.722, \qquad \bar{R}_2 = \frac{R_2}{b} = \frac{19}{9} = 2.111, \qquad \bar{R}_3 = \frac{R_3}{b} = \frac{28}{9} = 3.111,$$

$$\bar{R}_4 = \frac{R_4}{b} = \frac{22.5}{9} = 2.5, \qquad \bar{R}_5 = \frac{R_5}{b} = \frac{32}{9} = 3.555, \qquad \bar{R} = \frac{k+1}{2} = \frac{5+1}{2} = 3$$

To determine if the distributions of absentee rates differ in location fro the five days of the work week, we test:

H_0: The probabilities of the five days of the week are identical
H_a: The distributions of at least two days of the week differ in location

The test statistic is

$$F_r = \frac{12b}{k(k+1)}\sum(\bar{R}_j - \bar{R})^2$$

$$= \frac{12(9)}{5(5+1)}\left((3.722-3)^2 + (2.111-3)^2 + (3.111-3)^2 + (2.5-3)^2 + (3.555-3)^2\right) = 6.775$$

Since no α level was given, we will use $\alpha = .05$. The rejection region requires $\alpha = .05$ in the upper tail of the χ^2 distribution with df $= k - 1 = 5 - 1 = 4$. From Table VII, Appendix A, $\chi^2_{.05} = 9.48773$. The rejection region is $\chi^2 > 9.48773$.

Since the observed value of the test statistic does not fall in the rejection region $(\chi^2 = 6.775 \not> 9.48773)$, H_0 is not rejected. There is insufficient evidence to indicate the distributions of absentee rates differ in location fro the five days of the work week at $\alpha = .05$.

14.81   The conditions required for a valid Spearman's test are:

1.   The sample of experimental units on which the two variables are measured is randomly selected.
2.   The probability distributions of the two variables are continuous.

14.83   a.   From Table XIV with $n = 10$, $r_{s,\alpha} = r_{s,.025} = .648$. The rejection region is $r_s > .648$ or $r_s < -.648$.

   b.   From Table XIV with $n = 20$, $r_{s,\alpha} = r_{s,.025} = .450$. The rejection region is $r_s > .450$.

   c.   From Table XIV with $n = 30$, $r_{s,\alpha} = r_{s,.01} = .432$. The rejection region is $r_s < -.432$.

14.85   a. H_0: $\rho_s = 0$
      H_a: $\rho_s \neq 0$

   b.   The test statistic is $r_s = \dfrac{SS_{uv}}{\sqrt{SS_{uu}SS_{vv}}}$

x	Rank, u	y	Rank, v	u^2	v^2	uv
0	3	0	1.5	9	2.25	4.5
3	5.5	2	5	30.25	25	27.5
0	3	2	5	9	25	15
−4	1	0	1.5	1	2.25	1.5
3	5.5	3	7	30.25	49	38.5
0	3	1	3	9	9	9
4	7	2	5	49	25	35
	$\sum u = 28$		$\sum v = 28$	$\sum u^2 = 137.5$	$\sum v^2 = 137.5$	$\sum uv = 131$

$$SS_{uv} = \sum uv - \frac{(\sum u)(\sum v)}{n} = 131 - \frac{28(28)}{7} = 19$$

$$SS_{uu} = \sum u^2 - \frac{(\sum u)^2}{n} = 137.5 - \frac{(28)^2}{7} = 25.5$$

$$SS_{vv} = \sum v^2 - \frac{(\sum v)^2}{n} = 137.5 - \frac{(28)^2}{7} = 25.5$$

$$r_s = \frac{19}{\sqrt{25.5(25.5)}} = \frac{SS_{uv}}{\sqrt{SS_{uu}SS_{vv}}} = \frac{19}{\sqrt{25.5(25.5)}} = .745$$

Reject H_0 if $r_s < -r_{s,\alpha/2}$ or $r_s > r_{s,\alpha/2}$ where $\alpha/2 = .025$ and $n = 7$:

Reject H_0 if $r_s < -.786$ or $r_s > .786$ (from Table XIV, Appendix A).

Since the observed value of the test statistic does not fall in the rejection region, ($r_s = .745 \not> .786$), H_0 is not rejected. There is insufficient evidence to indicate x and y are correlated at $\alpha = .05$.

c. The p-value is $P(r_s \geq .745) + P(r_s \leq -.745)$. For $n = 7$, $r_s = .745$ is above $r_{s,.025}$ where $\alpha/2 = .025$ and below $r_{s,.05}$ where $\alpha/2 = .05$. Therefore, $2(.025) = .05 < p$-value $< 2(.05) = .10$.

d. The assumptions of the test are that the samples are randomly selected and the probability distributions of the two variables are continuous.

14.87 a, b. The ranks and Spearman correlation coefficient are:

Porosity	Rank, u	Diameter	Rank, v	u^2	v^2	uv
18.8	5	12.0	5	25	25	25
18.3	4	9.7	3	16	9	12
16.3	2	7.3	2	4	4	4
6.9	1	5.3	1	1	1	1
17.1	3	10.9	4	9	16	12
20.4	6	16.8	6	36	36	36
	$\sum u = 21$		$\sum v = 21$	$\sum u^2 = 91$	$\sum v^2 = 91$	$\sum uv = 90$

$$SS_{uv} = \sum uv - \frac{\left(\sum u\right)\left(\sum v\right)}{n} = 90 - \frac{21(21)}{6} = 16.5$$

$$SS_{uu} = \sum u^2 - \frac{\left(\sum u\right)^2}{n} = 91 - \frac{21^2}{6} = 17.5$$

$$SS_{vv} = \sum v^2 - \frac{\left(\sum v\right)^2}{n} = 91 - \frac{21^2}{6} = 17.5$$

$$r_s = \frac{SS_{uv}}{\sqrt{SS_{uu}SS_{vv}}} = \frac{16.5}{\sqrt{17.5(17.5)}} = .9429$$

Since r_s is very close to 1, apparent porosity and mean pore diameter are strongly, positively related.

c.   To determine if apparent porosity and mean pore diameter have a positive rank correlation, we test:

H_o: $\rho_s = 0$
H_a: $\rho_s > 0$

The test statistic is $r_s = .9429$.

Reject H_0 if $r_s > r_{s,\alpha}$ where $\alpha = .01$ and $n = 6$.

Reject H_0 if $r_s > .943$ (From Table XIV, Appendix A)

Since the observed value of r_s does not fall in the rejection region ($r_s = .9429 \not> .943$), H_0 is not rejected. There is insufficient evidence of a positive rank correlation between apparent porosity and mean pore diameter at $\alpha = .01$.

14.89   a.   Some preliminary calculations are:

Blood Lactate Level	Rank u	Perceived Recovery	Rank v	u^2	v^2	uv
3.8	3	7	1.5	9	2.25	4.5
4.2	5.5	7	1.5	30.25	2.25	8.25
4.8	7	11	3	49	9	21
4.1	4	12	5	16	25	20
5.0	8	12	5	64	25	40
5.3	9.5	12	5	90.25	25	47.5
4.2	5.5	13	7	30.25	49	38.5
2.4	1	17	9	1	81	9
3.7	2	17	9	4	81	18
5.3	9.5	17	9	90.25	81	85.5
5.8	12	18	11.5	144	132.25	138
6.0	14	18	11.5	196	132.25	161
5.9	13	21	14.5	169	210.25	188.5
6.3	15	21	14.5	225	210.25	217.5
5.5	12	20	13	121	169	143
6.5	16	24	16	256	256	256
	$\sum u = 136$		$\sum v = 136$	$\sum u^2 = 1495$	$\sum v^2 = 1490.5$	$\sum uv = 1396.25$

$$SS_{uv} = \sum uv - \frac{\left(\sum u\right)\left(\sum v\right)}{n} = 1396.25 - \frac{136(136)}{16} = 240.25$$

$$SS_{uu} = \sum u^2 - \frac{\left(\sum u\right)^2}{n} = 1495 - \frac{136^2}{16} = 339$$

$$SS_{vv} = \sum v^2 - \frac{\left(\sum v\right)^2}{n} = 1490.5 - \frac{136^2}{16} = 334.5$$

a.   The ranks of the blood lactate levels are stored in the "u" column above.

b.   The ranks of the perceived recovery values are stored in the "v" column above.

c. $r_s = \dfrac{SS_{uv}}{\sqrt{SS_{uu}SS_{vv}}} = \dfrac{240.25}{\sqrt{339(334.5)}} = .713$

Since r_s is relatively close to 1, blood lactate level and perceived recovery have a fairly strong, positive relationship.

d.   Reject H_0 if $r_s < -r_{s,\alpha/2}$ or if $r_s > r_{s,\alpha/2}$ where $\alpha/2 = .05$ and $n = 16$;

Reject H_0 if $r_s < -.425$ or $r_s > .425$ (from Table XIV, Appendix A.)

e. To determine if blood lactate level and perceived recovery are rank correlated, we test:

H_0: Blood lactate level and Perceived recovery are not rank correlated
H_a: Blood lactate level and Perceived recovery are rank correlated
The test statistic is $r_s = .713$.

Since the observed value of the test statistic falls in the rejection region, ($r_s = .713 >$ .425), H_0 is rejected. There is sufficient evidence to indicate that blood lactate level and perceived recovery are rank correlated at $\alpha = .10$.

14.91 a. Some preliminary calculations are:

Pair	Rank-Child	Rank-Parent	d	d^2
1	1	1	0	0
2	2	2	0	0
3	3	3	0	0
4	4	4	0	0
5	5	5	0	0
6	6	6	0	0
7	7	7	0	0
8	8	8	0	0
9	9	9	0	0
10	10	10	0	0
				$\Sigma d^2 = 0$

$$r_s = 1 - \frac{6\sum d^2}{n(n^2-1)} = 1 - \frac{6(0)}{10(100-1)} = 1$$

b. There is a perfect positive rank correlation between the BMI of children and the BMI of their parents.

c. The correlation found in part a is the rank correlation. This indicates that the ranks have a perfect linear relationship. However, the relationship between the actual child BMI values and parent BMI values could be curvilinear.

14.93   a.   Some preliminary calculations are:

Change KSADS-MRS	Rank, u	Improve CGI-BP	Rank, v	u^2	v^2	uv
80	18	6	18	324	324	324
65	17	5	17	289	289	289
20	15.5	4	15	240.25	225	232.5
−15	13	4	15	169	225	195
−50	9	4	15	81	225	135
20	15.5	3	12	240.25	144	186
−30	11	3	12	121	144	132
−70	5.5	3	12	30.25	144	66
−10	14	2	6	196	36	84
−25	12	2	6	144	36	72
−35	10	2	6	100	36	60
−65	7.5	2	6	56.25	36	45
−65	7.5	2	6	26.25	36	45
−70	5.5	2	6	30.25	36	33
−80	4	2	6	16	36	24
−90	2.5	2	6	6.25	36	15
−95	1	2	6	1	36	6
−90	2.5	1	1	6.25	1	2.5
$\sum u = 171$		$\sum v = 171$		$\sum u^2 = 2107$	$\sum v^2 = 2045$	$\sum uv = 1946$

$$SS_{uv} = \sum uv - \frac{\sum u \sum v}{n} = 1946 - \frac{171(171)}{18} = 321.5$$

$$SS_{uu} = \sum u^2 - \frac{\left(\sum u\right)^2}{n} = 2107 - \frac{171^2}{18} = 482.5$$

$$SS_{vv} = \sum v^2 - \frac{\left(\sum v\right)^2}{n} = 2045 - \frac{171^2}{18} = 420.5$$

$$r_s = \frac{SS_{uv}}{\sqrt{SS_{uu}SS_{vv}}} = \frac{321.5}{\sqrt{482.5(420.5)}} = .714$$

b.   To determine if there is a positive rank correlation between the two test score changes in the population of all pediatric patients with manic symptoms, we test:

H_0: $\rho_s = 0$

H_a: $\rho_s > 0$

The test statistic is $r_s = .714$.

Reject H_0 if $r_s > r_{s, .05}$ with $n = 18$

Reject H_0 if $r_s > .399$ (From Table XIV, Appendix A)

Since the observed value of the test statistic falls in the rejection region ($r_s = .714 > .399$), H_0 is rejected. There is sufficient evidence to indicate there is a positive rank correlation between the two test score changes in the population of all pediatric patients with manic symptoms at $\alpha = .05$.

14.95   a.   Some preliminary calculations are:

FCAT-Math	Rank, u	% Below Poverty	Rank, v	u^2	v^2	uv
166.4	7	91.7	22	49	484	154
159.6	4	90.2	21	16	441	84
159.1	3	86.0	20	9	400	60
155.5	1	83.9	19	1	361	19
164.3	5	80.4	18	25	324	90
169.8	9	76.5	17	81	289	153
155.7	2	76.0	16	4	256	32
165.2	6	75.8	15	36	225	90
175.4	13	75.6	14	169	196	182
178.1	16	75.0	13	256	169	208
167.1	8	74.7	12	64	144	96
177.0	15	63.2	11	225	121	165
174.2	11	52.9	10	121	100	110
175.6	14	48.5	9	196	81	126
170.8	10	39.1	8	100	64	80
175.1	12	38.4	7	144	49	84
182.8	20.5	34.3	6	420.25	36	123
180.3	18	30.3	4.5	324	20.25	81
178.8	17	30.3	4.5	289	20.25	76.5
181.4	19	29.6	3	361	9	57
182.8	20.5	26.5	2	420.25	4	41
186.1	22	13.8	1	484	1	22
$\sum u = 253$			$\sum v = 253$	$\sum u^2 = 3794.5$	$\sum v^2 = 3794.5$	$\sum uv = 2133.5$

$$SS_{uv} = \sum uv - \frac{\sum u \sum v}{n} = 2133.5 - \frac{253(253)}{22} = -776$$

$$SS_{uu} = \sum u^2 - \frac{\left(\sum u\right)^2}{n} = 3794.5 - \frac{253^2}{22} = 885$$

$$SS_{vv} = \sum v^2 - \frac{\left(\sum v\right)^2}{n} = 3794.5 - \frac{253^2}{22} = 885$$

$$r_s = \frac{SS_{uv}}{\sqrt{SS_{uu}SS_{vv}}} = \frac{-776}{\sqrt{885(885)}} = -.877$$

Since r_s is very close to -1, there is a fairly strong, negative relationship between FCAT math scores and the percentage of students below the poverty level.

b.   Some preliminary calculations are:

FCAT-Read	Rank, u	% Below Poverty	Rank, v	u^2	v^2	uv
165.0	7.5	91.7	22	56.25	484	165
157.2	1	90.2	21	1	441	21
164.4	5	86.0	20	25	400	100
162.4	3	83.9	19	9	361	57
162.5	4	80.4	18	16	324	72
164.9	6	76.5	17	36	289	102
162.0	2	76.0	16	4	256	32
165.0	7.5	75.8	15	56.25	225	112.5
173.7	14	75.6	14	196	196	196
171.0	11	75.0	13	121	169	143
169.4	9	74.7	12	81	144	108
172.9	13	63.2	11	169	121	143
172.7	12	52.9	10	144	100	120
174.9	16	48.5	9	256	81	144
174.8	15	39.1	8	225	64	120
170.1	10	38.4	7	100	49	70
181.4	20	34.3	6	400	36	120
180.6	19	30.3	4.5	361	20.25	85.5
178.0	18	30.3	4.5	324	20.25	81
175.9	17	29.6	3	289	9	51
181.6	21	26.5	2	441	4	42
183.8	22	13.8	1	484	1	22
$\sum u = 253$		$\sum v = 253$		$\sum u^2 = 3794.5$	$\sum v^2 = 3794.5$	$\sum uv = 2107$

$$SS_{uv} = \sum uv - \frac{\sum u \sum v}{n} = 2107 - \frac{253(253)}{22} = -802.5$$

$$SS_{uu} = \sum u^2 - \frac{\left(\sum u\right)^2}{n} = 3794.5 - \frac{253^2}{22} = 885$$

$$SS_{vv} = \sum v^2 - \frac{\left(\sum v\right)^2}{n} = 3794.5 - \frac{253^2}{22} = 885$$

$$r_s = \frac{SS_{uv}}{\sqrt{SS_{uu}SS_{vv}}} = \frac{-802.5}{\sqrt{885(885)}} = -.907$$

Since r_s is very close to -1, there is a fairly strong, negative relationship between FCAT reading scores and the percentage of students below the poverty level.

c. To determine whether the FCAT math scores and the percentage of students below the poverty level are negatively rank correlated, we test:

H_0: $\rho_s = 0$
H_a: $\rho_s < 0$

The test statistic is $r_s = -.877$.

Reject H_0 if $r_s < -r_{s,.01}$ with $n = 22$

Reject H_0 if $r_s < -.508$ (From Table XIV, Appendix A)

Since the observed value of the test statistic falls in the rejection region ($r_s = -.877 < -.508$), H_0 is rejected. There is sufficient evidence to indicate FCAT math scores and the percentage of students below the poverty level are negatively rank correlated at $\alpha = .01$.

d. To determine whether the FCAT reading scores and the percentage of students below the poverty level are negatively rank correlated, we test:

H_0: $\rho_s = 0$
H_a: $\rho_s < 0$

The test statistic is $r_s = -.907$.

Reject H_0 if $r_s < -r_{s,.01}$ with $n = 22$

Reject H_0 if $r_s < -.508$ (From Table XIV, Appendix A)

Since the observed value of the test statistic falls in the rejection region ($r_s = -.907 < -.508$), H_0 is rejected. There is sufficient evidence to indicate FCAT reading scores and the percentage of students below the poverty level are negatively rank correlated at $\alpha = .01$.

14.97   a.   Using MINITAB, histograms of the two data sets are:

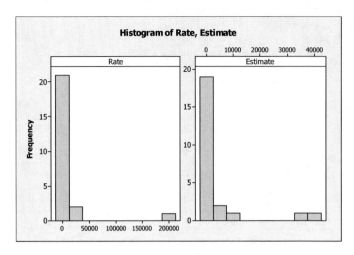

From the graphs, both data sets are skewed to the right and are not normal.

b.   The analysis will not be valid because the data are not normal.

c.   Some preliminary calculations are:

Incidence rate	Rank, u	Estimate	Rank, v	u^2	v^2	uv
.25	1	300	9	1	81	9
1	2	1000	16.5	4	272.25	33
1.75	3	691	14	9	196	42
2	4	20	6	16	36	24
3	5	17.5	2	25	4	10
5	6	.8	1	36	1	6
9	7	1000	16.5	49	272.25	115.5
10	8	150	4	64	16	32
22	10	326.5	10	100	100	100
23	11	146.5	3	121	9	33
39	12	370	11	144	121	132
98	13	400	12.5	169	156.25	162.5
119	14	225	8	196	64	112
152	15	200	6	225	36	90
179	16	200	6	256	36	96
936	17	400	12.5	289	156.25	212.5
1514	18	1500	19.5	324	380.25	351
1627	19	1000	16.5	361	272.25	313.5
2926	20	6000	22	400	484	440
4019	21	5000	21	441	441	441
12619	22	1500	19.5	484	380.25	429
14889	23	1000	16.5	529	272.25	379.5
203864	24	37000	23	576	529	552
15	9	37500	24	81	576	216
	$\sum u = 300$		$\sum v = 300$	$\sum u^2 = 4900$	$\sum v^2 = 4892$	$\sum uv = 4331.5$

$$SS_{uv} = \sum uv - \frac{\left(\sum u\right)\left(\sum v\right)}{n} = 4331.5 - \frac{300(300)}{24} = 581.5$$

$$SS_{uu} = \sum u^2 - \frac{\left(\sum u\right)^2}{n} = 4900 - \frac{300^2}{24} = 1150$$

$$SS_{vv} = \sum v^2 - \frac{\left(\sum v\right)^2}{n} = 4892 - \frac{300^2}{24} = 1142$$

$$r_s = \frac{SS_{uv}}{\sqrt{SS_{uu}SS_{vv}}} = \frac{581.5}{\sqrt{1150(1142)}} = .5074$$

Since the value of r_s is about half way between 0 and 1, there is a moderate positive relationship between incidence rate and estimate.

d.   To determine if there is a positive association between actual incidence and estimated incidence, we test:

H_0: $\rho_s = 0$
H_a: $\rho_s > 0$

The test statistic is $r_s = .5074$.

From Table XIV, with $n = 24$ and $\alpha = .01$, the rejection region is $r_s > .485$.

Since the observed value of r_s falls in the rejection region ($r_s = .5074 > .485$), H_0 is rejected. There is sufficient evidence to indicate a positive association between incidence rate and estimated rate at $\alpha = .01$.

14.99   a.   To compare two populations with independent samples, one would use the Wilcoxon Rank Sum test.

b.   To make an inference about a population median, one would use the sign test.

c.   To compare three or more populations with independent samples, one would use the Kruskal-Wallis H test.

d.   To make inferences about a rank correlation, one would use the Spearman's rank correlation coefficient test.

e.   To compare two populations with matched pairs, one would use the Wilcoxon Signed Rank Test.

f.   To compare three or more populations with a block design, one would use the Friedman's F_r test.

14.101 a. Some preliminary calculations are:

Pair	X	Rank u_i	Y	Rank v_i	u_i^2	v_i^2	$u_i v_i$
1	19	5	12	5	25	25	25
2	27	7	19	8	49	64	56
3	15	2	7	1	4	1	2
4	35	9	25	9	81	81	81
5	13	1	11	4	1	16	4
6	29	8	10	2.5	64	6.25	20
7	16	3.5	16	6	12.25	36	21
8	22	6	10	2.5	36	6.25	15
9	16	3.5	18	7	12.25	49	24.5
		$\sum u_i = 45$		$\sum v_i = 45$	$\sum u_i^2 = 284.5$	$\sum v_i^2 = 284.5$	$\sum u_i v_i = 248.5$

$$SS_{uv} = \sum u_i v_i - \frac{\sum u_i v_i}{n} = 248.5 - \frac{45(45)}{9} = 23.5$$

$$SS_{uu} = \sum u_i^2 - \frac{\left(\sum u_i\right)^2}{n} = 284.5 - \frac{45^2}{9} = 59.5$$

$$SS_{vv} = \sum v_i^2 - \frac{\left(\sum v_i\right)^2}{n} = 284.5 - \frac{45^2}{9} = 59.5$$

To determine if the Spearman rank correlation differs from 0, we test:

H_0: $\rho_s = 0$

H_a: $\rho_s \neq 0$

The test statistic is $r_s = \dfrac{SS_{uv}}{\sqrt{SS_{uu} SS_{vv}}} = \dfrac{23.5}{\sqrt{59.5(59.5)}} = .395$

Reject H_0 if $r_s < -r_{s, \alpha/2}$ or if $r_s > r_{s, \alpha/2}$ where $\alpha/2 = .025$ and $n = 9$:

Reject H_0 if $r_s < -.683$ or if $r_s > .683$ (from Table XIV, Appendix A)

Since the observed value of the test statistic does not fall in the rejection region ($r_s = .395 \not> .683$), H_0 is not rejected. There is insufficient evidence to indicate that Spearman's rank correlation between x and y is significantly different from 0 at $\alpha = .05$.

b.   Use the Wilcoxon signed rank test. Some preliminary calculations are:

Pair	X	Y	Difference	Rank of Absolute Difference
1	19	12	7	3
2	27	19	8	4.5
3	15	7	8	4.5
4	35	25	10	6
5	13	11	2	1.5
6	29	10	19	8
7	16	16	0	(eliminated)
8	22	10	12	7
9	16	18	−2	1.5
				$T_- = 1.5$

To determine if the probability distribution of x is shifted to the right of that for y, we test:

H_0:   The probability distributions are identical for the two variables
H_a: The probability distribution of x is shifted to the right of the probability distribution of y

The test statistic is $T = T_- = 1.5$

Reject H_0 if $T \leq T_0$ where T_0 is based on $\alpha = .05$ and $n = 8$ (one-tailed):

Reject H_0 if $T \leq 6$ (from Table XIII, Appendix A).

Since the observed value of the test statistic falls in the rejection region ($T = 1.5 \leq 6$), H_0 is rejected. There is sufficient evidence to conclude that the probability distribution of x is shifted to the right of that for y at $\alpha = .05$.

14.103   Some preliminary calculations are:

Block	1	Rank	2	Rank	3	Rank	4	Rank	5	Rank
1	75	4	65	1	74	3	80	5	69	2
2	77	3	69	1	78	4	80	5	72	2
3	70	4	63	1.5	69	3	75	5	63	1.5
4	80	3.5	69	1	80	3.5	86	5	77	2
		$R_1 = 14.5$		$R_2 = 4.5$		$R_3 = 13.5$		$R_4 = 20$		$R_5 = 7.5$

$$\bar{R}_1 = \frac{R_1}{b} = \frac{14.5}{4} = 3.625 \quad \bar{R}_2 = \frac{R_2}{b} = \frac{4.5}{4} = 1.125 \quad \bar{R}_3 = \frac{R_3}{b} = \frac{13.5}{4} = 3.375$$

$$\bar{R}_4 = \frac{R_4}{b} = \frac{20}{4} = 5 \quad \bar{R}_5 = \frac{R_5}{b} = \frac{7.5}{4} = 1.875 \quad \bar{R} = \frac{k+1}{2} = \frac{5+1}{2} = 3$$

To determine whether at least two of the treatment probability distributions differ in location, use Friedman F_r test.

H_0:   The five treatments have identical probability distributions
H_a: At least two of the populations have probability distributions that differ in location

The test statistic is

$$F_r = \frac{12b}{k(k+1)}\sum(\bar{R}_j - \bar{R})^2$$
$$= \frac{12(4)}{5(5+1)}\left((3.625-3)^2 + (1.125-3)^2 + (3.375-3)^2 + (5-3)^2 + (1.875-3)^2\right) = 14.9$$

The rejection region requires $\alpha = .05$ in the upper tail of the χ^2 distribution with df $= k - 1$ $= 5 - 1 = 4$. From Table VII, Appendix A, $= 9.48773$. The rejection region is $F_r > 9.48773$.

Since the observed value of the test statistic falls in the rejection region ($F_r = 14.9 >$ 9.48773), H_0 is rejected. There is sufficient evidence to indicate that at least two of the treatment means differ in location at $\alpha = .05$.

14.105   To determine if the median cesium amount of lichen differs from .003, we test:

H_0: $\eta = .003$
H_a: $\eta \neq .003$

The test statistic is $S = \{$number of observations greater than .003$\} = 8$ (from printout). The p-value is $p = .0391$. Since the p-value is less than $\alpha = .10$, H_0 is rejected. There is sufficient evidence to indicate the median cesium amount of lichen differs from .003 at $\alpha = .10$.

This result agrees with the t-test in Exercise 8.71.

14.107   Some preliminary calculations:

Theme	High School Teachers	Geography Alumni	Difference $T - A$	Rank of Absolute Differences
Tourism	10	2	8	11
Physical	2	1	1	1.5
Transportation	7	3	4	8
People	1	6	−5	9
History	2	5	−3	6
Climate	6	4	2	3.5
Forestry	5	8	−3	6
Agriculture	7	10	−3	6
Fishing	9	7	2	3.5
Energy	2	8	−6	10
Mining	10	11	−1	1.5
Manufacturing	12	12	0	(eliminated)
			Positive rank sum $T_+ = 27.5$	

To determine if the distributions of theme rankings for the two groups differ, we test:

H_0: The probability distributions for the two populations are identical
H_a: The probability distribution of the high school teachers is shifted to the right or left of the probability distribution of the geography alumni

The test statistic is $T_+ = 27.5$.

Reject H_0 if $T_+ \le T_0$ where T_0 is based on $\alpha = .05$ and $n = 11$ (two-tailed):

Reject H_0 if $T_+ \le 11$ (from Table XIII, Appendix A)

Since the observed value of the test statistic does not fall in the rejection region ($T_+ = 27.5 \not\le$ 11), H_0 is not rejected. There is insufficient evidence to indicate that the distributions of these rankings for the two groups differ at $\alpha = .05$. Practically, this means that the thematic content of a new atlas could be based on the views of either educators or geography alumni.

14.109 a.   Some preliminary calculations are:

Week	Number of Strikes	Rank u_i	Age of fish	Rank v_i	$d_i = u_i - v_i$	d_i^2
1	85	9	120	1	−8	64
2	63	8	136	2	−6	36
3	34	3	150	3	0	0
4	39	5	155	4	1	1
5	58	7	162	5	2	4
6	35	4	169	6	−2	4
7	57	6	178	7	−1	1
8	12	1	184	8	−7	49
9	15	2	190	9	−7	49

$$\sum d_i^2 = 208$$

$$r_s = 1 - \frac{6\sum d_i^2}{n(n^2-1)} = 1 - \frac{6(208)}{9(9^2-1)} = 1 - 1.733 = -.733$$

b.   To determine whether the number of strikes and age are negatively correlated, we test:

H_0: $\rho_s = 0$
H_a: $\rho_s < 0$

The test statistic is $r_s = -.733$.

Reject H_0 if $r_s < -r_{s,\alpha}$ where $\alpha = .01$ and $n = 9$.

Reject H_0 if $r_s < -.783$ (From Table XIV, Appendix A)

Since the observed value of the test statistic does not fall in the rejection region ($r_s = -.733 \not< -.783$), H_0 is not rejected. There is insufficient evidence to indicate that the number of strikes and age are negatively correlated at $\alpha = .01$.

14.111 a.   To determine if the median level of TCDD in the fat tissue of Vietnam vets exceeds 3 ppt, we test:

H_0: $\eta = 3$
H_a: $\eta > 3$

The test statistic is S = number of measurements greater than 3 = 14.

The p-value = $P(x \geq 14)$ where x is a binomial random variable with $n = 20$ and $p = .5$. From Table II, Appendix A,

$P(x \geq 14) = 1 - P(x \leq 13) = 1 - .942 = .058$

Since the p-value is greater than $\alpha = .05$ ($p = .058 > .05$), H_0 is not rejected. There is insufficient evidence to indicate the median level of TCDD in the fat tissue of Vietnam vets exceeds 3 ppt at $\alpha = .05$.

   b.   To determine if the median level of TCDD in the plasma of Vietnam vets exceeds 3 ppt, we test:

H_0: $\eta = 3$
H_a: $\eta > 3$

The test statistic is S = number of measurements greater than 3 = 12.

The p-value = $P(x \geq 12)$ where x is a binomial random variable with $n = 19$ and $p = .5$. (One observation is eliminated because it is equal to 3.0.)   Using MINITAB,

$P(x \geq 12) = 1 - P(x \leq 11) = 1 - .820 = .180$

Since the p-value is greater than $\alpha = .05$ ($p = .180 > .05$), H_0 is not rejected. There is insufficient evidence to indicate the median level of TCDD in the plasma of Vietnam vets exceeds 3 ppt at $\alpha = .05$.

c.   Some preliminary calculations are:

Fat	Plasma	Difference	Rank of Absolute Difference
4.9	2.5	2.4	11.5
6.9	3.5	3.4	16
10.0	6.8	3.2	15
4.4	4.7	-0.3	4
4.6	4.6	0	(eliminated)
1.1	1.8	-0.7	8
2.3	2.5	-0.2	2.5
5.9	3.1	2.8	14
7.0	3.1	3.9	17
5.5	3.0	2.5	13
7.0	6.9	0.1	1
1.4	1.6	-0.2	2.5
11.0	20.0	-9.0	19
2.5	4.1	-1.6	9
4.4	2.1	2.3	10
4.2	1.8	2.4	11.5
41.0	36.0	5.0	18
2.9	3.3	-0.4	5
7.7	7.2	0.5	6.5
2.5	2.0	0.5	6.5

$$T_- = 50$$
$$T_+ = 140$$

To determine if the distribution of TCDD levels in fat is shifted above or below the distribution of TCDD levels in plasma, we test:

H_0: The distribution of TCDD levels in fat is identical the distribution of TCDD levels in plasma

H_a: The distribution of TCDD levels in fat is shifted above or below the distribution of TCDD levels in plasma

The test statistic is $T =$ smaller of T_- and T_+ which is $T_- = 50$.

The rejection region is $T_- \leq 46$, from Table XIII, Appendix A, with $n = 19$ and $\alpha = .05$ (two-tails).

Since the observed value of the test statistic does not fall in the rejection region ($T_- = 50 \not\leq 46$), H_0 is not rejected. There is insufficient evidence to indicate the distribution of TCDD levels in fat is shifted above or below the distribution of TCDD levels in plasma at $\alpha = .05$.

d.   Some preliminary calculations are:

Fat	Rank, u	Plasma	Rank, v	u^2	v^2	uv
4.9	11	2.5	6.5	121	42.25	71.5
6.9	14	3.5	12	196	144	168
10.0	18	6.8	16	324	256	288
4.4	8.5	4.7	15	72.25	225	127.5
4.6	10	4.6	14	100	196	140
1.1	1	1.8	2.5	1	6.25	2.5
2.3	3	2.5	6.5	9	42.25	19.5
5.9	13	3.1	9.5	169	90.25	123.5
7.0	15.5	3.1	9.5	240.25	90.25	147.25
5.5	12	3.0	8	144	64	96
7.0	15.5	6.9	17	240.25	289	263.5
1.4	2	1.6	1	4	1	2
11.0	19	20.0	19	361	361	361
2.5	4.5	4.1	13	20.25	169	58.5
4.4	8.5	2.1	5	72.25	25	42.5
4.2	7	1.8	2.5	49	6.25	17.5
41.0	20	36.0	20	400	400	400
2.9	6	3.3	11	36	121	66
7.7	17	7.2	18	289	324	306
2.5	4.5	2.0	4	20.25	16	18
	$\sum u = 210$		$\sum v = 210$	$\sum u^2 = 2868.5$	$\sum v^2 = 2868.5$	$\sum uv = 2718.75$

$$SS_{uv} = \sum uv - \frac{\left(\sum u\right)\left(\sum v\right)}{n} = 2718.75 - \frac{210(210)}{20} = 513.75$$

$$SS_{uu} = \sum u^2 - \frac{\left(\sum u\right)^2}{n} = 2868.5 - \frac{210^2}{20} = 663.5$$

$$SS_{vv} = \sum v^2 - \frac{\left(\sum v\right)^2}{n} = 2868.5 - \frac{210^2}{20} = 663.5$$

$$r_s = \frac{SS_{uv}}{\sqrt{SS_{uu}SS_{vv}}} = \frac{513.75}{\sqrt{663.5(663.5)}} = .774$$

To determine if there is a positive association between the two TCDD measures, we test:

H_0: $\rho_s = 0$

H_a: $\rho_s > 0$

The test statistic is $r_s = .774$

From Table XIV, Appendix A, with $n = 20$ and $\alpha = .05$, the rejection region is $r_s > .377$.

Since the observed value of the test statistic falls in the rejection region ($r_s = .774 > .377$), H_0 is rejected. There is sufficient evidence to indicate that there is a positive association between the two TCDD measures at $\alpha = .05$.

14.113   a.   The Students' t procedure requires that the populations sampled from be normal. The distributions of the zinc measurements in the 3 locations may not be normal.

   b.   The ranks of the observations and the sums for the two groups are:

Text line	Rank	Intersection	Rank
.335	4	.393	7
.374	6	.353	5
.440	8	.285	1
		.295	2
		.319	3
	$T_1 = 18$		$T_2 = 18$

To determine if the distributions of zinc measurements for the text line and the intersection have different centers of location, we test:

H_0: The distributions of zinc measurements for the text line and the intersection are identical

H_a: The distribution of zinc measurements for the text line is shifted to the right or left of that for the intersection

The test statistic is $T_1 = 18$.

The null hypothesis will be rejected if $T_1 \leq T_L$ or $T_1 \geq T_U$ where $\alpha = .05$ (two-tailed), $n_1 = 3$ and $n_2 = 5$. From Table XII, Appendix A, $T_L = 6$ and $T_U = 21$. The rejection region is $T_1 \leq 6$ or $T_1 \geq 21$.

Since the observed value of the test statistic does not fall in the rejection region ($T_1 = 18 \nleq 6$ and $T_1 = 18 \ngeq 21$), H_0 is not rejected. There is insufficient evidence to indicate the distributions of zinc measurements for the text line and the intersection have different centers of location at $\alpha = .05$.

c.   The ranks of the observations and the sums for the two groups are:

Witness line	Rank	Intersection	Rank
.210	2	.393	9
.262	3	.353	8
.188	1	.285	4
.329	7	.295	5
.439	11	.319	6
.397	10		
	$T_1 = 34$		$T_2 = 32$

To determine if the distributions of zinc measurements for the witness line and the intersection have different centers of location, we test:

H_0: The distributions of zinc measurements for the witness line and the intersection are identical

H_a: The distribution of zinc measurements for the witness line is shifted to the right or left of that for the intersection

The test statistic is $T_2 = 32$.

The null hypothesis will be rejected if $T_2 \le T_L$ or $T_2 \ge T_U$ where $\alpha = .05$ (two-tailed), $n_1 = 6$ and $n_2 = 5$. From Table XII, Appendix A, $T_L = 19$ and $T_U = 41$. The rejection region is $T_2 \le 19$ or $T_2 \ge 41$.

Since the observed value of the test statistic does not fall in the rejection region ($T_2 = 32 \not\le 19$ and $T_2 = 32 \not\ge 41$), H_0 is not rejected. There is insufficient evidence to indicate the distributions of zinc measurements for the witness line and the intersection have different centers of location at $\alpha = .05$.

d.   Some preliminary calculations are:

Text line	Rank	Witness line	Rank	Inter-section	Rank
.335	8	.210	2	.393	11
.374	10	.262	3	.353	9
.440	14	.188	1	.285	4
		.329	7	.295	5
		.439	13	.319	6
		.397	12		
$R_1 = 32$		$R_2 = 38$		$R_3 = 35$	

$$\bar{R}_1 = \frac{R_1}{n_1} = \frac{32}{3} = 10.667, \qquad \bar{R}_2 = \frac{R_2}{n_2} = \frac{38}{6} = 6.333, \qquad \bar{R}_3 = \frac{R_3}{n_3} = \frac{35}{5} = 7,$$

$$\bar{R} = \frac{n+1}{2} = \frac{14+1}{2} = 7.5$$

To determine if the probability distributions of zinc measurements differ in location among the notebook locations, we test:

> H_0:   The probability distributions of zinc measurements for the three notebook locations are identical
>
> H_a: At least two of the three probability distributions differ in location

The test statistic is

$$H = \frac{12}{n(n+1)} \sum n_j (\bar{R}_j - \bar{R})^2$$

$$= \frac{12}{14(14+1)} \left(3(10.667 - 7.5)^2 + 6(6.333 - 7.5)^2 + 5(7 - 7.5)^2 \right) = 2.258$$

The rejection region requires $\alpha = .05$ in the upper tail of the χ^2 distribution with $df = k - 1$ $= 3 - 1 = 2$. From Table VII, Appendix A, $\chi^2_{.05} = 5.99147$. The rejection region is $H >$ 5.99147.

Since the observed value of the test statistic does not fall in the rejection region ($H = 2.258 \not> 5.99147$), H_0 is not rejected. There is insufficient evidence to indicate the probability distributions of zinc measurements for the three notebook locations differ in location at $\alpha = .05$.

e.   The median zinc score for the text line is .374. The median zinc score for the witness line is .296. The median zinc score for the intersection is .319. We found no difference in the distributions of zinc scores among the three notebook locations in part d. We found no difference in the distributions of zinc scores between the text line and the intersection in part b. We also found no difference in the distributions of zinc scores between the witness line and the intersection in part c. Although we did not compare the text line and the witness line zinc scores, we can still conclude these two locations are not different. These two distributions appear to be the furthest apart and there was no difference among all three notebook locations from part d.

14.115 Some preliminary calculations are:

Metal	I	Rank	II	Rank	III	Rank
1	4.6	2	4.2	1	4.9	3
2	7.2	3	6.4	1	7.0	2
3	3.4	1.5	3.5	3	3.4	1.5
4	6.2	3	5.3	1	5.9	2
5	8.4	3	6.8	1	7.8	2
6	5.6	2	4.8	1	5.7	3
7	3.7	1.5	3.7	1.5	4.1	3
8	6.1	1	6.2	2	6.4	3
9	4.9	3	4.1	1	4.2	2
10	5.2	3	5.0	1	5.1	2
		$R_1 = 23$		$R_2 = 13.5$		$R_3 = 23.5$

$$\bar{R}_1 = \frac{R_1}{b} = \frac{23}{10} = 2.3 \qquad \bar{R}_2 = \frac{R_2}{b} = \frac{13.5}{10} = 1.35 \qquad \bar{R}_3 = \frac{R_3}{b} = \frac{23.5}{10} = 2.35 \qquad \bar{R} = \frac{k+1}{2} = \frac{3+1}{2} = 2$$

To determine if there is a difference in the probability distributions of the amounts of corrosion among the three types of sealers, we test:

H_0: The probability distributions of corrosion amounts are identical for the three types of sealers

H_a: At least two of the probability distributions differ in location

The test statistic is

$$F_r = \frac{12b}{k(k+1)}\sum(\bar{R}_j - \bar{R})^2 = \frac{12(10)}{3(3+1)}\left((2.3-2)^2 + (1.35-2)^2 + (2.35-2)^2\right) = 6.35$$

The rejection region requires $\alpha = .05$ in the upper tail of the χ^2 distribution with df = $k - 1$ = 3 − 1 = 2. From Table VII, Appendix A, $\chi^2_{.05} = 5.99147$. The rejection region is $F_r > 5.99147$.

Since the observed value of the test statistic falls in the rejection region ($F_r = 6.35 > 5.99147$), reject H_0. There is sufficient evidence to indicate a difference in the probability distributions among the three types of sealers at $\alpha = .05$.

14.117 Some preliminary calculations are:

List 1	Rank	List 2	Rank	List 3	Rank
48	19	41	14	18	2
43	16	36	10	42	15
39	12	29	5	28	4
57	20	40	13	38	11
21	3	35	9	15	1
47	18	45	17	33	8
58	21	32	7	31	6
	$R_1 = 109$		$R_2 = 75$		$R_3 = 47$

$$\overline{R}_1 = \frac{R_1}{n_1} = \frac{109}{7} = 15.571, \quad \overline{R}_2 = \frac{R_2}{n_2} = \frac{75}{7} = 10.714, \quad \overline{R}_3 = \frac{R_3}{n_3} = \frac{47}{7} = 6.714,$$

$$\overline{R} = \frac{n+1}{2} = \frac{21+1}{2} = 11$$

To determine if there is a difference between at least two of the probability distributions of the numbers of word associates that subjects can name for the three lists, we test:

H_0: The probability distributions of the numbers of word associates named are the same for the three lists

H_a: At least two of the three probability distributions differ in location

The test statistic is

$$H = \frac{12}{n(n+1)} \sum n_j (\overline{R}_j - \overline{R})^2$$

$$= \frac{12}{21(21+1)} \left(7(15.571 - 11)^2 + 7(10.714 - 11)^2 + 7(6.714 - 11)^2 \right) = 7.154$$

The rejection region requires $\alpha = .05$ in the upper tail of the χ^2 distribution with df $= k - 1 = 3 - 1 = 2$. From Table VII, Appendix A, $\chi^2_{.05} = 5.99147$. The rejection region is $H > 5.99147$.

Since the observed value of the test statistic falls in the rejection region ($H = 7.154 > 5.99147$), reject H_0. There is sufficient evidence to indicate a difference in location for at least two of the probability distributions of the numbers of word associates at $\alpha = .05$.

14.119 a. Since $r_s = .643$ is not real close to 1, there is a moderate positive relationship between the public agenda score and the media agenda score.

  b. Since $r_s = .714$ is somewhat close to 1, there is a positive relationship between the public agenda score and the media agenda score, once 'length of war' is removed.

  c. To determine if there is a positive rank correlation between the public agenda score and the media agenda score, we test:

H_0: $\rho_s = 0$

H_a: $\rho_s > 0$

The test statistic is $r_s = .714$.

Reject H_0 if $r_s > r_{s,\alpha}$ where $\alpha = .01$ and $n = 235$. The largest n in Table XIV is $n = 30$. The critical value for $n = 30$ is .432. As n increases, the critical value decreases. Thus, if we reject H_0 with $n = 30$, we would also reject H_o with $n = 235$.

Reject H_0 if $r_s > .432$ (From Table XIV, Appendix A, $n = 30$)

Since the observed value of r_s falls in the rejection region ($r_s = .714 > .432$), H_0 is rejected. There is sufficient evidence of a positive rank correlation between the public agenda score and the media agenda score at $\alpha = .01$.

14.121 a. Answers may vary. Possible answer may include:

The nonparametric test requires no assumptions concerning the population probability distributions. Since each court possesses different requirements and restrictions, the DEF values for the three populations may not be normally distributed with a common variance.

b. The Kruskall-Wallis H-test is appropriate for this analysis because the comparison involves three populations.

H_0: The probability distributions of the DEF values are the same for all three tax courts.
H_a: At least two of the three tax courts have probability distributions of DEF values that differ in location.

c. $\bar{R}_1 = \dfrac{R_1}{n_1} = \dfrac{5,335}{67} = 79.627$, $\bar{R}_2 = \dfrac{R_2}{n_2} = \dfrac{3,937}{57} = 69.070$, $\bar{R}_3 = \dfrac{R_3}{n_3} = \dfrac{3,769}{37} = 101.865$,

$\bar{R} = \dfrac{n+1}{2} = \dfrac{161+1}{2} = 81$

$$H = \frac{12}{n(n+1)}\sum n_j(\bar{R}_j - \bar{R})^2$$
$$= \frac{12}{161(161+1)}\left(67(79.627-81)^2 + 57(69.070-81)^2 + 37(101.865-81)^2\right)$$
$$= 11.20$$

d. The p-value is $p = .0037$. Since the p-value is so small, H_0 would be rejected for any $\alpha > .0037$. There is sufficient evidence to indicate that the mean DEF values for the three tax courts differ for $\alpha > .0037$.

14.123 To determine if more than half of all Al Qaeda attacks against the U.S. involve two or fewer suicide bombings, we test:

H_0: $\eta = 3$
H_a: $\eta < 3$

S = number of measurements less than 3 = 17.

The test statistic is $z = \dfrac{(S-.5)-.5n}{.5\sqrt{n}} = \dfrac{(17-.5)-.5(20)}{.5\sqrt{20}} = 2.91$. (One observation is eliminated because it is equal to 3.)

The *p*-value = $P(z \geq 2.91)$. From Table IV, Appendix A, $p = P(z \geq 2.62) = .5 - .4982$ = .0018.

Since the *p*-value is less than $\alpha = .05$ ($p = .0018 < .05$), H_0 is rejected. There is sufficient evidence to indicate that more than half of all Al Qaeda attacks against the U.S. involve two or fewer suicide bombings at $\alpha = .05$.